Topics in Current Genetics

5

Series Editor: *Stefan Hohmann*

Available online at
SpringerLink.com

Springer

Berlin
Heidelberg
New York
Hong Kong
London
Milan
Paris
Tokyo

Joaquín Ariño · Denis R. Alexander (Eds.)

Protein Phosphatases

With 43 Figures, 16 of Them in Color; and 15 Tables

 Springer

Joaquín Ariño, Ph. D.
Dept. Bioquímica y Biología
Molecular
Facultat de Veterinària
Universitat Autònoma de Barcelona
08193 Bellaterra, Barcelona
Spain

Dr. Denis R. Alexander
Chairman, Molecular Immunology
Programme
The Babraham Institute
Babraham, Cambridge CB2 4AT
United Kingdom

The cover illustration depicts pseudohyphal filaments of the ascomycete *Saccharomyces cerevisiae* that enable this organism to forage for nutrients. Pseudohyphal filaments were induced here in a wild-type haploid MATa Σ1278b strain by an unknown readily diffusible factor provided by growth in confrontation with an isogenic petite yeast strain in a sealed petri dish for two weeks and photographed at 100X magnification (provided by Xuewen Pan and Joseph Heitman).

ISSN 1610-2096
ISBN 3-540-20560-8 Springer-Verlag Berlin Heidelberg New York

Cataloging-in-Publication Data applied for

Bibliographic information published by Die Deutsche Bibliothek
Die Deutsche Bibliothek lists this publication in the Deutsche Nationalbibliografie; detailed bibliographic data is available in the Internet at <http:/dnb.ddb.de>

Springer-Verlag is a part of Springer Science+Business Media
springeronline.com

© Springer-Verlag Berlin Heidelberg 2004
Printed in Germany

The use of general descriptive names, registered names, trademarks, etc. in this publication does not imply, even in the absence of a specific statement, that such names are exempt from the relevant protective laws and regulations and therefore free for general use.

Typesetting: Camera ready by editors
Data-conversion: PTP-Berlin, Stefan Sossna e.K.
Cover Design: Design & Production, Heidelberg
39/3150 - 5 4 3 2 1 0 - Printed on acid-free paper

Table of contents

4 Calcineurin: Roles of the Ca^{2+}/calmodulin-dependent protein phosphatase in diverse eukaryotes.. 73

5 Roles of mammalian protein phosphatase 2C family members in the regulation of cellular functions 91

List of contributors

Alexander, Denis R.
Laboratory of Lymphocyte Signalling and Development, Molecular Immunology Programme, The Babraham Institute, Babraham, Cambridge CB2 4AT, UK
denis.alexander@bbsrc.ac.uk

Alonso, Andres
Program of Signal Transduction, Cancer Research Center, The Burnham Institute, 10901 North Torrey Pines Road, La Jolla, CA 92037, USA

Andersen, Jannik N.
Cold Spring Harbor Laboratory, 1 Bungtown Road, Cold Spring Harbor, NY 11724, USA

Bechmann, Gunther
Institut für Pharmazeutische & Medizinische Chemie, Westfälische Wilhelms-Universität, Hittorfstr. 58-62, 48149 Münster, Germany

Bourdeau , Annie
McGill Cancer Center, McGill University, McIntyre Medical Science Building Room 701, 3655 Promenade Sir William Osler, Montreal, Quebec, Canada, H3G 1Y6

Brunet, Daniel V.
McGill Cancer Center, McGill University, McIntyre Medical Science Building Room 701, 3655 Promenade Sir William Osler, Montreal, Quebec, Canada, H3G 1Y6
Department of Biochemistry, McGill University, Montreal, Quebec, Canada, H3G 1Y6

Chinkers, Michael
Department of Pharmacology, University of South Alabama, Mobile, Alabama 36688, USA
michaelc@jaguar1.usouthal.edu

Cohen, Patricia T.W.
Medical Research Council Protein Phosphorylation Unit, School of Life Sciences, University of Dundee, Dundee DD15EH, Scotland, UK
p.t.w.cohen@dundee.ac.uk

Collas, Philippe
Institute of Medical Biochemistry, University of Oslo, PO Box 1112 Blindern, Oslo 0317, Norway
philippe.collas@basalmed.uio.no

Cunningham, Kyle W.
Department of Biology, Johns Hopkins University, 3400 N. Charles Street, Baltimore, MD 21218, USA
kwc@jhu.edu

den Hertog, Jeroen
Hubrecht Laboratory, Netherlands Institute for Developmental Biology, Uppsalalaan 8, 3584 CT Utrecht, The Netherlands
hertog@niob.knaw.nl

Dombrádi, Viktor
Department of Medical Chemistry, Medical and Health Science Center, Faculty of Medicine, University of Debrecen, Bem tér 18/B, Debrecen, Hungary, H-4026
dombradi@jaguar.dote.hu

Farkas, Ilona
Department of Medical Chemistry, Medical and Health Science Center, Faculty of Medicine, University of Debrecen, Bem tér 18/B, Debrecen, Hungary, H-4026

Feng, Gen-Sheng
Program in Signal Transduction Research, The Burnham Institute, 10901 N. Torrey Pines Road, La Jolla, California 92037, USA
gfeng@burnham.org

Godzik, Adam
Program of Bioinformatics, Cancer Research Center, The Burnham Institute, 10901 North Torrey Pines Road, La Jolla, CA 92037, USA

Haystead, Timothy A. J.
Department of Pharmacology and Cancer Biology, Center for Chemical Biology, Duke University Medical Center, Durham, NC 27710, USA
hayst001@mc.duke.edu

Heinonen, Krista M.
McGill Cancer Center, McGill University, McIntyre Medical Science Building Room 701, 3655 Promenade Sir William Osler, Montreal, Quebec, Canada, H3G 1Y6
Division of Experimental Medicine McGill University, Montreal, Quebec, Canada, H3G 1Y6

Hilioti, Zoe
    Department of Biology, Johns Hopkins University, 3400 N. Charles Street,
    Baltimore, MD 21218, USA

Klumpp, Susanne
    Institut für Pharmazeutische & Medizinische Chemie, Westfälische Wilhelms-
    Universität, Hittorfstr. 58-62, 48149 Münster, Germany
    klumpp@uni-muenster.de

Kobayashi, Takayasu
    Department of Biochemistry, Institute of Development, Aging and Cancer,
    Tohoku University, 4-1 Seiryomachi, Aoba-ku, Sendai 980-8575, Japan

Kókai, Endre
    Department of Medical Chemistry, Medical and Health Science Center, Fac-
    ulty of Medicine, University of Debrecen, Bem tér 18/B, Debrecen, Hungary,
    H-4026

Komaki, Ken-ichiro
    Department of Biochemistry, Institute of Development, Aging and Cancer,
    Tohoku University, 4-1 Seiryomachi, Aoba-ku, Sendai 980-8575, Japan

Krieglstein, Josef
    Institut für Pharmakologie & Toxikologie, Philipps-Universität, Ketzerbach
    63, 35032 Marburg, Germany

Küntziger, Thomas
    Institute of Medical Biochemistry, University of Oslo, PO Box 1112 Blindern,
    Oslo 0317, Norway

Kwiek, Nicole A.
    Department of Pharmacology and Cancer Biology, Center for Chemical Biol-
    ogy, Duke University Medical Center, Durham, NC 27710, USA

Lai, Lisa A.
    Program in Signal Transduction Research, The Burnham Institute, 10901 N.
    Torrey Pines Road, La Jolla, California 92037, USA

Landsverk, Helga B.
    Institute of Medical Biochemistry, University of Oslo, PO Box 1112 Blindern,
    Oslo 0317, Norway

Lapp, Wayne S.
    Department of Physiology, McGill University, Montreal, Quebec, Canada,
    H3G 1Y6

Li, Ming Guang
Department of Biochemistry, Institute of Development, Aging and Cancer, Tohoku University, 4-1 Seiryomachi, Aoba-ku, Sendai 980-8575, Japan

McCain, Daniel F.
Departments of Biochemistry, Albert Einstein College of Medicine, 1300 Morris Park Avenue, Bronx, New York 10461, USA
zyzhang@aecom.yu.edu

Matthews, Reginald James
Section of Infection and Immunity, University of Wales College of medicine, Heath Park, CARDIFF CF14 4XX, Wales, UK.
matthewsrj@cardiff.ac.uk

Mäurer, Anette
Institut für Pharmazeutische & Medizinische Chemie, Westfälische Wilhelms-Universität, Hittorfstr. 58-62, 48149 Münster

Mumby, Marc C.
Department of Pharmacology, University of Texas Southwestern Medical Center, 5323 Harry Hines Boulevard, Dallas, TX 75390-9041, USA
marc.mumby@utsouthwestern.edu

Mustelin, Tomas
Program of Signal Transduction, Cancer Research Center, The Burnham Institute, 10901 North Torrey Pines Road, La Jolla, CA 92037, USA
tmustelin@burnham-inst.org

Rojas, Ana
Program of Bioinformatics, Cancer Research Center, The Burnham Institute, 10901 North Torrey Pines Road, La Jolla, CA 92037, USA

Sasaki, Masato
Department of Biochemistry, Institute of Development, Aging and Cancer, Tohoku University, 4-1 Seiryomachi, Aoba-ku, Sendai 980-8575, Japan

Sathish, Jean Gerard
Section of Infection and Immunity, University of Wales College of medicine, Heath Park, CARDIFF CF14 4XX, Wales, UK.

Tailor, Pankaj
McGill Cancer Center, McGill University, McIntyre Medical Science Building Room 701, 3655 Promenade Sir William Osler, Montreal, Quebec, Canada, H3G 1Y6

Tamura, Shinri
  Department of Biochemistry, Institute of Development, Aging and Cancer, Tohoku University, 4-1 Seiryomachi, Aoba-ku, Sendai 980-8575, Japan
  tamura@idac.tohoku.ac.jp

Tonks, Nicholas K.
  Cold Spring Harbor Laboratory, 1 Bungtown Road, Cold Spring Harbor, NY 11724, USA
  tonks@cshl.edu

Tremblay, Michel L.
  McGill Cancer Center, McGill University, McIntyre Medical Science Building Room 701, 3655 Promenade Sir William Osler, Montreal, Quebec, Canada, H3G 1Y6
  Department of Biochemistry, McGill University, Montreal, Quebec, Canada, H3G 1Y6
  michel.tremblay@mcgill.ca

Zhao, Chunmei
  Program in Signal Transduction Research, The Burnham Institute, 10901 N. Torrey Pines Road, La Jolla, California 92037, USA

Zhang, Eric E.
  Program in Signal Transduction Research, The Burnham Institute, 10901 N. Torrey Pines Road, La Jolla, California 92037, USA

Zhang, Zhong-Yin
  Departments of Biochemistry and Department of Molecular Pharmacology, Albert Einstein College of Medicine, 1300 Morris Park Avenue, Bronx, New York 10461, USA
  zyzhang@aecom.yu.edu

1 Overview of protein serine/threonine phosphatases

Patricia T.W. Cohen

Abstract

Protein phosphatases that cleave phosphate from phosphorylated serine and threonine in proteins are crucial for the regulation of most cellular processes. Three structurally distinct families of protein serine/threonine phosphatases (PPP, PPM, and FCP) are described and the phylogenetic relationships in the PPP and PPM families of *Homo sapiens*, *Drosophila melanogaster*, and *Saccharomyces cerevisiae* are examined. An overview of the functions of the PPP, PPM, and FCP members gained from recent genetic analyses is presented.

1.1 Protein serine/threonine phosphatases and their classification

The past two decades of cDNA and genome sequencing has uncovered nearly 30 bona fide human protein serine/threonine phosphatase catalytic subunit genes. These encode enzymes that catalyse the cleavage of phosphate from serine and threonine residues in proteins that have been phosphorylated by protein kinases. Despite the fact that virtually all the more than 300 serine/threonine kinases are members of the same family, the serine/threonine phosphatases comprise three distinct families, which are structurally distinct from the family of protein tyrosine phosphatases that encompasses the tyrosine and dual specificity (tyrosine and serine/threonine) phosphatases. The current classification of protein serine/threonine protein phosphatases is based upon the amino acid sequences of their catalytic subunits. The PPP family includes the prototypic member protein phosphatase 1 (PP1/Ppp1c), while the PPM family is represented by protein phosphatase 2C (PP2C/Ppm1) and the FCP family was more recently recognised through its founding member (Fcp1), which dephosphorylates the carboxy-terminal domain of RNA polymerase II (Table 1). Members of the three families are widely distributed among eukaryotic phyla (Fig. 1 and 2) and prokaryotes usually contain at least one PPP and/or PPM phosphatase (Kennelly 2002; Cohen 2003).

Topics in Current Genetics, Vol. 5
J. Arino, D.R. Alexander (Eds.) Protein phosphatases
© Springer-Verlag Berlin Heidelberg 2004

Table 1. Three families of protein phosphatases that dephosphorylate phospho-serine and phospho-threonine residues in proteins.

Family	PPP	PPM	FCP
'Signature' motifs	-GDxHG- - GDxVDRG- -GNHE-	-ED- -DGH(A/G)- - GD- -GD- -DG- -DN-	$-\psi\psi\psi$DLDxx$\psi\psi$- -RPxxxF-
Active site	Bimetal (Fe^{3+} +Zn^{2+} in native Ppp3c)	Bimetal (both Mn^{2+} in expressed Ppm1)	Not determined
Catalytic domain	~270 amino acids	~250 amino acids	~150 amino acids
Catalytic mechanism	Metal ion-catalysed dephosphorylation	Metal ion-catalysed dephosphorylation	Not determined
Other characteristics	Some active in absence of metal ions, others activated by Ca^{2+}. Some members inhibited by naturally occurring toxins	Dependent on Mg^{2+} or Mn^{2+} for activity where tested. Some members are also activated by Ca^{2+}.	Dependent on Mg^{2+}, some members can be partially activated by Mn^{2+} or Ca^{2+}.
Human genes	≥ 13	≥ 10	≥ 5

Note: other genes encoding the 'signature' motifs of the PPP and PPM family have been identified, but they do not appear to encode proteins with serine/threonine phosphatase activity. For example, *E. coli* diadenosine tetraphosphatase has a PPP motif and adenyl cyclase has a domain related to the PPM motif. x may be a number of different amino acids; ψ represents one of the hydrophobic amino acids I, L,V,M

1.2 The PPP family

1.2.1 Phylogenetic and structural relationships in the PPP family

Although the PPP catalytic subunits Ppp1c - Ppp7c are present in mammals and *Drosophila melanogaster*, Ppp7c orthologues are absent from yeasts (Fig. 1) and Ppp3c/calcineurin/PP2B orthologues have not been identified in plants. Twelve genes encode PPPs in *Saccharomyces cerevisiae* and 13 in the human species (Stark 1996; Cohen 1997; Andreeva and Kutuzov 1999). Somewhat surprisingly, complete sequencing of the human genome does not appear to have uncovered any more clearly recognisable PPP family members, indicating that the increasing diversity of PPP function from yeast to humans is likely to reside in the regulatory subunits. The amino acid sequences of Ppp1c and Ppp2c have been extremely conserved throughout the evolution of multicellular eukaryotes and these enzymes are among the most slowly evolving proteins known. This may be because they interact with a wide variety of regulatory proteins and also because of their many roles in crucial cellular processes. Interestingly, *D. melanogaster* possesses 19 PPP family members. Two are located on the Y chromosome (Ppp1-Y1 and Ppp1-

Y2) and at least two others are related to male-specific functions (PppY-55A and PppN-58A), all of which appear to lack orthologues in *Homo sapiens* (Fig. 1). Some Ppps in *S. cerevisiae* (Ppq1, Ppz1, Ppz2, Ppg1) also have no orthologues in humans. This suggests that gene duplication within different eukaryotes may be one mechanism used to produce regulation by serine/threonine dephosphorylation of processes specific to particular organisms. In plants, at least 8 Ppp1c isoforms have been produced by gene duplication (Lin et al. 1999).

Many protein serine/threonine phosphatases (Ppp1, 2, 3, 4, 6) in the PPP family are high molecular mass complexes containing one or more regulatory subunits (Table 2). A large number of different regulatory subunits bind in a mutually exclusive manner to the Ppp1c catalytic subunit (Bollen 2001; Ceulemans et al. 2002; Cohen 2002) or to the Ppp2/PP2A 'core' complex of the catalytic subunit and regulatory A subunit (Janssens and Goris 2001). These interactions allow a single PPP catalytic subunit to participate in many different cellular functions. Over fifty different regulatory subunits of Ppp1c are known in humans and at least 20 in *S. cerevisiae*, providing an explanation for the increasing diversity of functions regulated by reversible phosphorylation in higher compared to lower eukaryotes. Many Ppp1c regulatory subunits bind through a conserved '-RVxF-' motif [with the consensus $(R/K)x_1(V/I)x_2(F/W)$ where x_1 may be absent or any residue except large hydrophobic residues and x_2 is any amino acid except large hydrophobic residues, phosphoserine and probably aspartic acid] providing an explanation for mutually exclusive interaction of each regulatory subunit with Ppp1c. A similar mechanism may occur for Ppp2/PP2A, since two conserved domains have been identified in a number of the variable B subunits and shown to interact with the A 'core' regulatory subunit of Ppp2c/PP2Ac (Li and Virshup 2002). Over the last decade there has been a great deal of interest in kinase /phosphatase interactions in the regulation of cellular signalling and their ability to act as signalling modules or molecular switches. Interactions of kinases and phosphatases may mediated by anchoring or adaptor proteins (Pawson and Scott 1998) or may be by direct interaction, as seen in the interaction of Ppp2/PP2A holoenzyme with several different kinases (Millward et al. 1999) and the complex of Ppp1c with Nek2 (NIMA-related kinase) where the kinase could be regarded as the regulatory subunit of the phosphatase catalytic subunit (or vice versa)(Cohen 2002).

PPP family members other than Ppp1, Ppp2/PP2A, Ppp4, Ppp6 may possess fused amino and carboxy terminal domains that, at least in some cases, impart distinct properties to the catalytic domain. Thus, Ppp7c has C-terminal EF hand sequences that confer Ca^{2+} sensitivity and a calmodulin-binding motif that is distinct from that of Ppp3c/calcineurin/PP2B (Andreeva and Kutuzov 1999). Ppp5 has an inhibitory amino terminal domain containing three tetratricopeptide repeats (TPR), which are likely to allow interaction with other proteins (Das et al. 1998). Ppp3c/calcineurin/PP2B found in complex with a regulatory subunit also has a carboxy-terminal domain that allows the enzyme to be activated by Ca^{2+}-calmodulin, as well as an autoinhibitory pseudo-substrate domain at the extreme C-terminus (Klee et al. 1998). Most mammalian PPP members in the

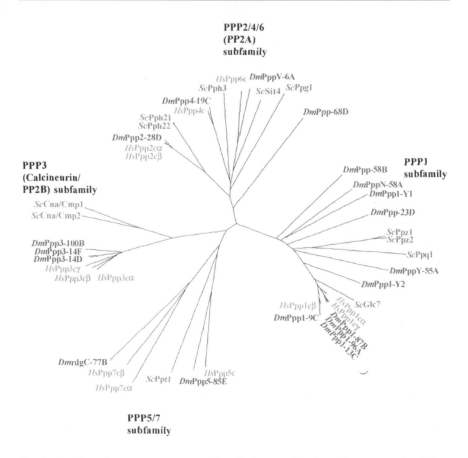

PPP2/4/6 (PP2A) subfamily

*Hs*Ppp6c *Dm*PppV-6A
*Sc*Pph3 *Sc*Sit4 *Sc*Ppg1
*Dm*Ppp4-19C
*Hs*Ppp4c
*Sc*Pph21
*Sc*Pph22
*Dm*Ppp-68D
*Dm*Ppp2-28D
*Hs*Ppp2cα
*Hs*Ppp2cβ

PPP3 (Calcineurin/ PP2B) subfamily

PPP1 subfamily

*Dm*Ppp-58B
*Dm*PppN-58A
*Dm*Ppp1-Y1

*Sc*Cna/Cmp1
*Sc*Cna/Cmp2

*Dm*Ppp-23D

*Dm*Ppp3-100B
*Dm*Ppp3-14F
*Dm*Ppp3-14D
*Hs*Ppp3cγ
*Hs*Ppp3cβ *Hs*Ppp3cα

*Sc*Ppz1
*Sc*Ppz2

*Sc*Ppq1
*Dm*PppY-55A
*Dm*Ppp1-Y2

*Hs*Ppp1cβ
*Dm*Ppp1-9C
*Hs*Ppp1cα
*Hs*Ppp1cγ
*Dm*Ppp1-87B
*Dm*Ppp1-96A
*Dm*Ppp1-13C
*Sc*Glc7

*Dm*rdgC-77B
*Hs*Ppp7cβ
*Sc*Pptl
*Hs*Ppp5c
*Dm*Ppp5-85E
*Hs*Ppp7cα

PPP5/7 subfamily

Ppp1c-Ppp6c subgroups are present in all tissues that have been examined, but Ppp7c is restricted to retina and brain.

The crystal structures of Ppp1c and Ppp3c/calcineurin/PP2B (Barford 1996) reveal the presence of two metal ions at the centre (probably Fe^{2+} and Zn^{2+} in the native enzymes) PPPs are believed to catalyse the dephosphorylation of their substrates in a single step reaction. A water molecule activated by bridging the two metal ions provides a hydroxide for nucleophilic attack on the substrate phosphate. The PPP catalytic domain spans ~270 amino acids and some of the invariant residues play crucial roles in binding divalent metal ions, while others interact with the substrate's phosphate group. Although PPPs mainly act upon phosphoserine and phosphothreonine residues, other phospho amino acids such as phosphohistidine can be dephosphorylated by some PPP family members such as Ppp1 and Ppp2/PP2A (Matthews and MacKintosh 1995). The carboxy-terminal region of the catalytic domain contains a conserved motif -SAxNY- that lies in a flexible loop region and in some Ppps has been implicated in the binding of a number of naturally occurring tumour promoters and toxins, such as okadaic acid found in shellfish and microcystin produced by blue green algae (Table 2) (Holmes et al.

Fig. 1 (overleaf). Phylogenetic tree, depicting the relationships between *Homo sapiens* (*Hs*, red), *Drosophila melanogaster* (*Dm*, blue) and budding yeast *Saccharomyces cerevisiae* (*Sc*, brown) protein serine/threonine phosphatases in the PPP family. The chromosomal location is included for the *Drosophila* PPPs. The unrooted tree is derived from multiple alignment in CLUSTALW (http://www.clustalw.genome.ad.jp/) of the phosphatase catalytic domain [starting ~30 amino acids prior to the invariant GDXHG motif and terminating ~30 amino acids following the conserved SAXNY motif (Barton et al. 1994) and corresponding to residues 29-301 of Ppp1cα$_1$]. Ppp1c and Ppp2c show approximately 40% sequence identity. Isoforms of Ppp1c show >85% sequence identity, while novel *Drosophila* phosphatases in Ppp1 subfamily show <65% sequence identity. Similar values occur in the PPP2A subfamily. Genbank/Swiss-Prot accession numbers for human PPP sequences are Ppp1cα (NP_002699/P08129), Ppp1cβ NP_002700/P37140), PPpp1cγ (NP_002701/P36873), Ppp2cα (NP_002706/P05323), Ppp2cβ (NP_004147/P11082), Ppp4c(NP_002711/P33172), Ppp6c (NP_002712/O00743), Ppp3cα (NP_000935/O08209), Ppp3cβ (NP_066955/P16299), Ppp3cγ (NP_005596), Ppp5c (NP_006238/P53041), Ppp7cα (NP_006231/O14829), Ppp7cβ (NP_006230); *Drosophila* PPP sequences, Ppp1-13C (CAA49594/Q05547), Ppp1-87B (CAA33609/P12982), Ppp1-96A (CAA39820/P48461), Ppp1-9C (CAA39821/P48462), Ppp1Y2 (AF427494), PppY-55A (CAA68808/ P11612), Ppp-23D (AAF51146/Q9VQL9), Ppp-58B (AAF46787/Q9W2A5), Ppp1Y1 (AF427493), PppN-58A (Y17355/ AAF46772/Q9W2C0), Ppp2- 28D (CAA38984/P23696), Ppp4-19C (CAA74606/O76932), PppV-6A (CAA53588/Q27884), Ppp-68D (AY058490/Q9VTP3), Ppp3- 100B(AAA28410/P48456), Ppp3-14F (AAF48623/Q9VXF1) Ppp3-14D (AAC47079/Q27889), Ppp5-85E (AJ271781/ AAF54438), rdgC-77B(M89628/AAB00734); *S. cerevisiae* PPP sequences Glc7 (AAC03231/P32598), Ppz1 (CAA89936/P26570), Ppz2 (AAB64859/P33329), Ppq1 (CAA97886/P32945), Pph21 (CAA98707/P23594), Pph22 (CAA98765/P23595), Pph3 (CAA86797/P32345), Sit4 (CAA98609/P20604), Ppg1 (CAA96312/P32838), Ppt1(CAA97134/P53043) Cna/Cmp1 (AAB67518/P23287), Cna/Cmp2 (CAA86718/P14747).

2002; Honkanen and Golden 2002). In contrast, the complex of Ppp3c/calcineurin/PP2B with its regulatory subunit is the target of the immunosuppressant drugs cyclosporin and FK506, which have been of extreme medical importance in transplantation surgery.

1.2.2 Functional genomic analyses of the PPP family

The challenges of the post genomic era are to identify the function(s) of each PPP complex by gene disruption in both lower and higher eukaryotes and also the physical and genetic interactions that underlie its physiological role(s). Deletion or disruption of genes for analysis of function has been most extensively achieved for the PPP family in the yeast, *S. cerevisiae* (Stark 1996; Janssens and Goris 2001). Mutagenesis in *Drosophila* has shown that isoforms of the catalytic subunit serve different functions; Ppp1-9C 'flapwing' mutants are defective in flight muscle attachments (Raghavan et al. 2000), while mutants of the more abundant Ppp1-87B show defects in chromosome condensation as well as mitosis (Dombrádi et al. 1990; Baksa et al. 1993). Mutagenesis of *PPP4-19C* in *D. melanogaster* and

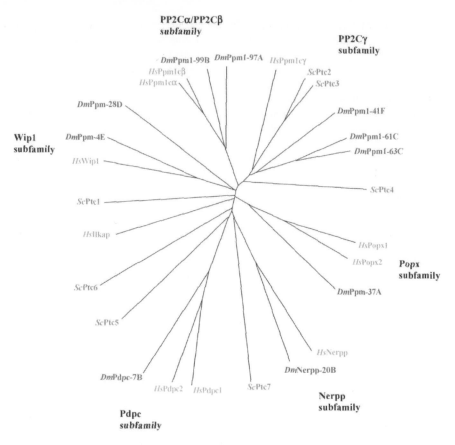

inhibition by RNA-mediated interference (RNAi) of the function of one (*PPH4.1*) of two genes encoding Ppp4c in *Caernorhabditis elegans* identify a role for Ppp4c in the nucleation of microtubules at centrosomes in the developing embryo (Helps et al. 1998; Sumiyoshi et al. 2002). However, inhibition of the function of *PPH4.2* did not enhance the incomplete embryonic lethality caused by loss of function of *PPH4.1*, suggesting that the genes play distinct roles. Targeted gene deletion of Ppp2cα in the mouse showed that Ppp2cβ could not compensate for lack of Ppp2cα during embryonic development despite its 97% identity and a comparable expression levels to that of the total catalytic subunits (Götz et al. 1998).

However, it could be argued that given the wide variety of regulatory subunits with which Ppp1 and Ppp2 catalytic subunits interact, analysis of the function(s) of each PPP complex is likely to be best achieved by deletion or disruption of the regulatory subunits for these PPPs and probably also Ppp4 and Ppp6. Furthermore, it is necessary to target the regulatory gene in more than one organism, since although the catalytic subunits are highly conserved from yeasts to humans, the regulatory subunits show much more variability and many mammalian regulatory subunits do not have readily discernible orthologues in lower organisms. Although

Fig. 2 (overleaf). Phylogenetic tree, depicting the relationships between *Homo sapiens* (*Hs*, red), *Drosophila melanogaster* (*Dm*, blue) and budding yeast *Saccharomyces cerevisiae* (*Sc*, brown) protein serine/threonine phosphatases in the PPM family. The chromosomal location is included for the *Drosophila* PPPs. The unrooted tree is derived from multiple alignment in CLUSTALW (http://www.clustalw.genome.ad.jp/) of the phosphatase catalytic domain [starting 15 amino acids before the conserved motif -ED- and terminating14 amino acids after motif -DN- (Bork et al. 1996) and corresponding to residues 22-297 of Ppm1α.1]. Genbank/EBI Ensembl accession numbers for human PPM sequences are Ppm1α (NP_066283/P35813), Ppm1β (NP_002697/O75688), Ppm1γ (NP_002698/O15355), Popx1 (NP_055721), Popx2 (NP_055449/P49593), Wip1 (NP_003611/O15297), Ilkap (NP_110395/), Nerpp (XP_051093), Pdpc1 (NP_060914), Pdpc2 (ESNG00000172840); *D. melanogaster* PPM sequences, Ppm1-99B (AAF56905), Ppm1-97A(AAF56583), Ppm1-41F (AAF57333), Ppm1-61C (AAF47393), Ppm1-63C (AAF47747), Ppm-37A (AAF53694), Ppm-28D (AAF52565), Ppm-4E (AAC28998), Pdpc-7B (CG12151), Nerpp-20B (CG17598-PA); *S. cerevisiae* PPM sequences, Ptc1p (CAA98562/P35182), Ptc2p (AAB64644/P39966), Ptc3p (CAA84876/P34221), Ptc4p (CAA85082/P38089), Ptc5p (CAA99287), Ptc6p (CAC42991), Ptc7p (AAB68888/P38797).

the evolution of eukaryotic fundamental functions such as the cell cycle, show high conservation across species, there can be considerable divergence in the regulation of less essential processes. In mammals targeted deletions in the Ppp1c regulatory subunit genes for the striated muscle glycogen targeting subunit (G_M/R_{GL}/PPP1R3A) and inhibitor-1 indicate that these complexes with Ppp1c do not mediate the action of insulin on glycogen metabolism (Scrimgeour et al. 1999; Suzuki et al. 2001). In contrast, other studies show that a targeted disruption of G_M/R_{GL}/PPP1R3A leads to insulin resistance and obesity in later life (Delibegovic et al. 2003). Heterozygous deletion of the gene for another glycogen targeting subunit, PTG/R5/PPP1R3C, also leads to insulin resistance in later life (Crosson et al. 2003). Despite these differences, the studies confirm a crucial role for glycogen targeted Ppp1 in the maintenance of normal glycogen levels. Other Ppp1c complexes serve very different functions; for example targeted disruption of dopamine and cAMP regulated phosphoprotein (DARPP-32), an isoform of inhibitor-1 that is particularly abundant in the brain, diminishes responses to the neurotransmitter dopamine, supporting a role for this Ppp1c complex in integration of neuronal signals that modulate responses to dopamine (Fienberg and Greengard 2000).

Decreasing the expression of Ppp2c/PP2A regulatory subunits by RNAi in *Drosophila* S2 cells implicates complexes containing the regulatory B/PR55 subunit in the negative regulation of the classical mitogen-activated protein (MAP) kinase pathway, while complexes containing the B'/B56 subunit negatively regulate the induction of apoptosis (Silverstein et al. 2002; Li et al. 2002). Use of antisense oligodeoxynucleotides to Ppp5 with its fused regulatory domain define a role for this phosphatase in glucocorticoid and p53 mediated cell cycle arrest in certain cell lines (Zuo et al. 1999). While ablation of function in cell lines provides a valuable approach, analysis of many functions requires the use of whole animals. Mutagenesis of rdgC in *Drosophila* and targeted disruption of the Ppp7c with its

Table 2. Structures and properties of the human PPP catalytic subunits

Human gene name	Chromosomal location	Protein names	Protein size (kDa)[1]	Structure	Activators/inhibitors	Functions regulated	Reviews/references
PPP1CA	11q13.3	Ppp1cα/PP1α	α₁ 37; α₂ 39	mainly heterodimeric >50 different regulatory subunits	inhibitor-1 inhibitor-2 okadaic acid microcystin tautomycin	Many, diverse -determined by variable regulatory subunit.	Bollen 2001 Cohen 2002 Cuelemans and Bollen 2002
PPP1CB	2p23.2	Ppp1cβ/PP1β/PP1δ	β 37				
PPP1CC	12q24	Ppp1cγ/PP1γ/PP1cγ	γ₁ 37; γ₂ 38				
PPP2CA	5q31.1	Ppp2cα/PP2Aα/PP2Acα	α 36	mainly heterotrimeric, core regulatory PPPR1 (A) + ≥ 14 different PPP2R2 (B) regulatory subunits dimeric complex with α4	okadaic acid microcystin	Many, diverse -determined by variable regulatory subunit.	Janssens et al. 2001
PPP2CB	8p12	Ppp2cβ/PP2Aβ/PP2Acβ	β 36				
PPP4C	16p11.2	Ppp4c/PP4/PPX	35	heterodimeric+ higher mol. mass structures regulatory subunits: PPP4R1, PPP4R2, α4	okadaic acid microcystin	Organisation of microtubules at centrosomes Signalling pathways?	Helps et al. 1998 Hu et al. 1998 Kloeker et al. 1999 Hastie et al. 2000 Sumiyoshi et al. 2002
PPP6C	9q33.3	Ppp6c/PP6	35	heterodimeric, regulatory subunits: α4 (TAP42, SAP190, SAP185, SAP155, SAP4 in S. cerevisiae)	(okadaic acid in S. cerevisiae)	(G1/S transition, cell shape regulation, translation initiation in S. cerevisiae).	Luke et al. 1996 Di Como et al. 1996 Bastians and Ponstingl 1996 Cohen 1997

Table 2. continued

Human gene name	Chromosomal location	Protein names	Protein (kDa)[1] size	Structure	Activators/ inhibitors	Functions regulated	Reviews/references
PPP3CA	4q24	Ppp3cα/ PP2Bα/ CNAα/ CALNA	α_1 59; α_2 58	heterodimer, with fused calmodulin binding + autoinhibitory domains regulatory subunit: (B) 19 kDa Ca^{2+} binding interacts with calmodulin	Ca^{2+}/calmodulin FKBP12-FK506 Cyclosporin/- cyclophilin	α and β isoforms: T-cell signalling. Neurotransmitter release. Neuroreceptor coupled Ca^{2+} channels γ isoform specific to testis.	Klee et al. 1998 Malleret et al. 2001 Miyakawa et al. 2003
PPP3CB	10q22.2	Ppp3cβ/ PP2Bβ/ CNAβ/ CALNB	β_1 58; β_2 59; β_3 58				
PPP3CC	8p21.3	Ppp3cγ/ PP2Bγ/ CNAγ	57				
PPP5C	19q13.3	Ppp5c/PP5/ PPT	57	monomeric with fused TPR domain	okadaic acid microcystin	Regulation of cell growth & multiple signalling pathways?	Andreeva and Kutuzov 1999 Zuo et al. 1999 Chinkers 2001 Morita et al. 2001
PPP7CA[2]	Xp22	Ppp7cα/ PPEF1	α 76	monomeric?, with fused Ca^{2+}-binding EF hand domains	Ca^{2+}/calmodulin	Specific to retina and certain brain regions. Dephosphorylation of rhodopsin in Drosophila.	Andreeva and Kutuzov 1999 Steele et al. 1992 Ramulu et al. 2001
PPP7CB[2]	4q21.1	Ppp7cβ/ PPEF2	β_1 69; β_2 86	interacts with calmodulin			

[1]Different values for the number of amino acids refers to splice variants; Ppp1cα X70484/S57501; Ppp1cγ, X74008; Ppp3cα, L14778; Ppp3cβ, M29551/AJ488506; Ppp7cβ, AF023456/AF023457.
[2]The names PPP7CA (AF027977) and PPP7CB are used here for genes referred to as PPEF1 and PPEF2 in some databases.

Table 3. Structures and properties of the human PPM (protein phosphatases magnesium or manganese dependent) catalytic subunits.

Human gene name	Chromosomal location	Protein names	Protein size (kDa)	Structure	Functions/enzymes regulated	References
PPM1A	14q23.1	PP2Cα	α_1 42, α_2 36	monomeric	Inactivation of CFTR Negative regulation of stress responsepathways. Dephosphorylation of p38 & MKKs Positive regulation Wnt signalling Dephosphorylation of Cdk2/Cdk6	Travis et al. 1997 Hanada et al. 1998 Takekawa et al. 1998 Strovel et al. 2000 Cheng et al. 2000
PPM1B	2p21	PP2Cβ	β_1 43, β_2 53	monomeric	Negative regulation of stress response pathways Dephosphorylation of p38 and TAK1 Dephosphorylation of Cdk2/Cdk6	Hanada et al. 1998 Hanada et al. 2001 Cheng et al. 2000
PPM1G	2p22.3	PP2Cγ, FIN13	59	monomeric fused collagen homology domain	Negative regulation of cell cycle progression Spliceosome assembly	Guthridge et al. 1997 Murray et al. 1999
ILKAP	2q37.3	ILKAP, PP2Cδ	43	monomeric	Dephosphorylation of ILK1 Negative regulation Wnt signalling	Leung-Hagesteijn et al. 2001
PPM1D	17	Wip1	66	monomeric	Negative regulation of stress response Inactivation of p38 (SAPK) Cell growth and cell cycle	Fiscella et al.1997 Takekawa et al. 2000 Choi et al. 2002
PPM1E	17q23.2	POPX1	83	monomeric fused PIX binding domain	Negative regulation of Cdc42/Rac signalling pathways	Koh et al. 2002

Table 3. continued.

Human gene name	Chromosomal location	Protein names	Protein size (kDa)	Structure	Functions/enzymes regulated	References
PPM1F	22q22	CaMIIKPase/POPX2/hFEM2	49	monomeric fused PIX binding domain	Ca^{2+} signalling and positive regulation of apoptosis Negative regulation of Cdc42/Rac signalling pathways	Tan et al. 2001 Koh et al. 2002
–	–	NERPP-2C	55, 80	monomeric	Negative regulation of neurite extension	Labes et al. 1998
PDPC1	8q22.1	PDP1	61,	heterodimeric	Mitochondrial pyruvate dehydrogenase complex	Lawson et al. 1993 Huang et al. 1998
PDPC2	16q22.1	PDP2	60	heterodimeric	Mitochondrial pyruvate dehydrogenase complex	Huang et al. 1998

Human counterparts of murine PP2Cε and PP2Cζ are not yet clearly identified in the genome databases. CFTR, cystic fibrosis transmembrane conductance regulator; IKKAP, Integrin linked kinase associated phosphatase; CaMIIKPase, Ca^{2+}-calmodulin-dependent protein kinase II phosphatase; POPX, partner of PIX; Wip1, wildtype p53 induced phosphatase; NERPP-2C, neurite extension related protein phosphatase, related to PP2C; PDP, pyruvate dehydrogenase (lipoamide)phosphatase.

fused regulatory domain in mice identifies different regulatory roles for this phosphatase in retinal function in the two organisms (Steele and O'Tousa 1990; Ramulu et al. 2001). Conditional targeted disruption using the Cre-lox system of the regulatory B subunit of Ppp3c/calcineurin/PP2B lead to behavioural defects similar to those described for schizophrenia (Miyakawa et al. 2003). Conditional reversible transient decrease of Ppp3c/calcineurin/PP2B catalytic activity using a tetracycline controlled transcriptional activation system to express a Ppp3c/calcineurin/PP2B inhibitor leads to enhanced learning, memory and long term potentiation (Malleret et al. 2001). The use of such conditional 'knockouts' which allow regional tissue specific decrease of function in adults is advantageous to exclude compensatory mechanisms that may occur during development in higher eukaryotes. Conditional specific 'knock in' mutations that can analyse particular regulatory mechanisms are also advantageous in delineating function.

1.3 The PPM family

1.3.1 Phylogenetic and structural relationships in the PPM family

The PPM (protein phosphatase, magnesium, or manganese dependent) family comprise a group of serine threonine phosphatases that are dependent on Mg^{2+} or Mn^{2+} for activity. The family includes Ppm1/PP2C-like members, Ppm1α /PP2Cα and Ppm1β/PP2Cβ that consist of a catalytic domain with only short amino and carboxy terminal regions (Tamura et al. 1989; Mann et al. 1992; Wenk et al. 1992; Marley et al. 1998). Other members have additional domains, for example Ppm1γ /PP2Cγ has a collagen homology domain (Guthridge et al. 1997) and Popx1 and Popx2 have a region binding to PIX, the Pak [p21(Cdc42/Rac)-activated kinase] interacting guanine nucleotide exchange factor (Koh et al. 2002) (Table 3). Mitochondrial pyruvate dehydrogenase phosphatase catalytic subunit (Pdpc) comprises a PPM catalytic domain preceded by a Ca^{2+} binding domain and is a more divergent member of the PPM family (Lawson et al. 1993) (Fig. 2). Although most PPM activities are usually isolated as monomeric species, PDP is a heterodimer of the catalytic subunit (Pdpc) and a regulatory subunit.

In plants, the PPM catalytic domain is also found fused to domains with signal transduction potential (Rodriguez 1998). For example, the *A. thaliana* kinase associated protein phosphatase KAPP, comprises an amino terminal membrane localisation signal, a kinase receptor interaction domain followed by a PPM catalytic domain. However, the transmembrane domain of KAPP has not been reported among mammalian PPM members. Some plants also differ from mammals in having a large number of PPM genes, with 69 PPM genes reported in *A. thaliana* (Kerk et al. 2002). *S. cerevisiae* adenyl cyclase shares 22% sequence similarity with the PPM catalytic domain in a region amino terminal to the cyclase catalytic domain (Tamura et al. 1989; Das et al. 1996), but it is not included in Fig. 2 since adenyl cyclase is not known to possess phosphatase activity; the PPM-like domain may mediate the ras-GTP activation of adenyl cyclase.

Although the sequences of the protein phosphatases in the PPM family share no similarities with those of the PPP family, their three-dimensional structures and deduced catalytic mechanism are strikingly similar (Das et al. 1996; Barford et al. 1998). Two Mn^{2+} ions at the catalytic site enable an associated water molecule to act as a nucleophile initiating hydrolysis of the substrate phosphomonoester bond. Within the PPM catalytic domain (~250 amino acids), 11 short 'signature' motifs have been recognised (Bork et al. 1996). The six motifs in Table 1 [-ED-, -DGH(A/G)- , -GD-, -GD-, - DG- , -DN-, corresponding to motifs, 1,2 5, 6, 8, and 11] are very highly conserved in eukaryotic PPMs. The invariant aspartic acid residues in motifs 1, 2, 8, and 11 are crucial in binding the divalent metal ions at the catalytic centre.

1.3.2 Functional genomic analyses of the PPM family

Phylogenetic analysis of PPM amino acid sequences suggests that putative *D. melanogaster* homologues can be identified for most human sequences (Fig. 2) but the functions of PPMs do not appear to have been analysed through mutations in *Drosophila*. Although it is less easy to identify precise homologue of each human Ppm in *S. cerevisiae*, members of the PPM family have been implicated in environmental stress responses in yeasts (Maeda et al. 1994; Shiozaki and Russell 1995) and in mammals (Table 3). Functional genomic analyses have shown that Ptc1p, Ptc2p, and Ptc3p in *S. cerevisiae* and their *S. pombe* orthologues participate in the negative regulation of osmosensing MAPK (mitogen-activated protein kinase) cascade (Warmka et al. 2001; Young et al. 2002). Ptc2p negatively regulates the unfolded protein response (Welihinda et al. 1998). Ptc1p and Ptc3p are also involved in the regulation of mitotic growth, dephosphorylating Cdc28 at Thr169, which is phosphorylated by Cdk-activating kinase (Cheng et al. 1999).

By searching for human sequences that could negatively regulate the yeast osmosensing kinase cascade, Takekawa et al (1998) identified human Ppm1α. Overexpressed mammalian Ppm1α /PP2Cα and Ppm1β/PP2Cα negatively regulated the p38/SAPK2A (stress activated protein kinase 2A) and JNK (Jun N-terminal kinase) stress response pathways (Hanada et al. 1998). These Ppm phosphatases were shown to dephosphorylate p38/SAPK2A and upstream components of the stress kinase signalling cascades, some of which are common to both the p38/SAPK2A and JNK pathways (Hanada et al. 1998; Takekawa et al. 1998; Hanada et al. 2001). Ppm1α/PP2Cα and Ppm1β/PP2Cβ have been implicated in control of the cell cycle through dephosphorylation of cyclin dependent kinases (Cheng et al. 2000) and their sequences are most similar to Ptc2p and Ptc3p that perform similar functions. In mammals, only Wip1 in the PPM family has been inactivated by gene targeting. Wip1, induced by ionising radiation in a p53 dependent manner and also by other environmental stresses, mediates negative feedback regulation of p38/SAPK-p53 signalling by dephosphorylation of p38/SAPK2A (Fiscella et al. 1997; Takekawa et al. 2000). Consistent with these functions, fibroblasts from Wip1 null mice embryos have decreased proliferation rates and appear to be compromised entering mitosis. However, they also show a range of other abnormalities in male reproductive organs and immune function (Choi et al. 2002). Phylogenetic analyses also suggest that Ptc5p, Ptc6p and Ptc7p with as yet undefined functions are most closely related to the human Pdpc and Nerpp group. Ptc7p and Pdpc1 exhibit a mitochondrial location (Jiang et al. 2002).

Mammalian Ca^{2+}-calmodulin-dependent protein kinase II (CaMKII) forms a complex with a phosphatase that may act as a signalling module and the interacting phosphatase was shown to be a member of the PPM family (Kitani et al. 1999). The probable orthologue in *C. elegans* Fem2 has also been shown to dephosphorylate CaMKII (Tan et al. 2001). Although *FEM2* was originally identified as a gene involved in sex determination in *C. elegans* (Piekny et al. 2000) this function is probably not conserved to mammals. However, overexpression of both Fem2 and human CaMKII phosphatase in mammalian cells causes caspase dependent apoptosis (Tan et al. 2001) suggesting that the apoptotic function may be evolutionarily conserved.

The p21 (Cdc42/Rac)-activated kinase PAK, which participates in several signalling pathways including the organisation of the actin cytoskeleton, is inactivated by two PPM phosphatases termed Popx1 and Popx2 (Koh et al. 2002). These enzymes can thus inhibit actin stress fibre breakdown. Since Popx2 is identical with CaMKII phosphatase, the data indicate that this phosphatase regulates more than one kinase.

1.4 The FCP family

The TFIIF-stimulated CTD phosphatase 1 (Fcp1) (Archambault et al. 1997; Kobor et al. 1999), the small CTD phosphatases (Scp1, Scp2, Scp3) (Yeo et al. 2003) and its orthologues comprise the FCP family, a distinct group of serine/threonine phosphatases, the only known function of which is to dephosphorylate the carboxy-terminal domain (CTD) of RNA polymerase II (Table 1). Fcp1 comprises an N-terminal catalytic domain in which two aspartic acid residues have been shown to be essential for function and a C-terminal domain that interacts with the transcription factor TFIIF (Archambault et al. 1998). The SCD phosphatases lack the C-terminal domain (Yeo et al. 2003). Fcp1 orthologues are found in organisms from mammals to yeasts and the human SCD phosphatases have a single orthologue in *Drosophila*. Deletion in yeast shows that *FCP1* is an essential gene (Kobor et al. 1999). Fcp1 is dependent on Mg^{2+} for activity, although Ca^{2+} and Mn^{2+} can activate Fcp1 to a lesser extent (Chambers and Kane 1996), while Scp1 is dependent on Mg^{2+}, with Ca^{2+} being unable to substitute for Mg^{2+} (Yeo et al. 2003). The substrate of these phosphatases is the CTD of RNA polymerase II, which comprises 52 repeats of the sequence YSPTSPS in mammals. Fcp1 can dephosphorylate the second and fifth serine residues, while Scp1 preferentially dephosphorylates the fifth serine. CTD dephosphorylation is required to recycle the polymerase at the end of each round of transcription and may also play a role in the regulation of transcription.

Acknowledgements

The author thanks the Medical Research Council, U.K. for financial support.

References

Andreeva AV, Kutuzov MA (1999) RdgC/PP5-related phosphatases: novel components in signal transduction. Cell Signalling 11:555-562

Archambault J, Chambers RS, Kobor MS, Ho Y, Cartier M, Bolotin D, Andrews B, Kane CM, Greenblatt J (1997) An essential component of a C-terminal domain phosphatase that interacts with transcription factor IIF in *Saccharomyces cerevisiae*. Proc Natl Acad Sci USA 94:14300-14305

Archambault J, Pan G, Dahmus GK, Cartier M, Marshall NF, Zhang S, Dahmus ME, Greenblatt J (1998) FCP1, the RAP74-interacting of a human protein phosphatase that dephosphorylates the carboxyterminal domain of RNA polymerase IIO. J Biol Chem 273:27593-27601

Baksa K, Morawietz H, Dombrádi V, Axton JM, Taubert H, Szabó G, Török I, Udvardy A, Gyurlovics H, Szöör B, Glover DM, Reuter G, Gausz J (1993) Mutations in the protein phosphatase 1 gene at 87B can be differentially affect suppression of position-effect variegation and mitosis in *Drosophila melanogaster*. Genetics 135:117-125

Barford D (1996) Molecular mechanisms of the protein serine/threonine phosphatases. Trends Biochem Sci 21:407-412

Barford D, Das AK, Egloff MP (1998) The structure and mechanism of protein phosphatases: insights into catalysis and regulation. Annu Rev Biophys Biomol Struct 27:133-164

Barton GJ, Cohen PTW, Barford D (1994) Conservation analysis and structure prediction of the protein serine/threonine phosphatases: sequence similarity with diadenosine tetraphosphatase from *E. coli* suggests homology to the protein phosphatases. Eur J Biochem 20:225-237

Bastians H, Ponstingl H (1996) The novel human protein serine/threonine phosphatase 6 is a functional homologue of budding yeast Sit4p and fission yeast ppe1, which are involved in cell cycle regulation. J Cell Sci 109:2865-2874

Bollen M (2001) Combinatorial control of protein phosphatase-1. Trends Biochem Sci 26:426-431

Bork P, Brown NP, Hegyi H, Schultz J (1996) The protein phosphatase 2C (PP2C) superfamily: detection of bacterial homologues. Protein Sci 5:1421-5

Ceulemans H, Stalmans W, Bollen M (2002) Regulator-driven functional diversification of protein phosphatase-1 in eukaryotic evolution. BioEssays 24:371-381

Chambers RS, Kane CM (1996) Purification and characterization of an RNA polymerase II phosphatase from yeast. J Biol Chem 271:24498-24504

Cheng A, Kaldis P, Solomon MJ (2000) Dephosphorylation of human cyclin-dependent kinases by protein phosphatase type 2C alpha and beta 2 isoforms. J Biol Chem 275:34744-34749

Cheng A, Ross KE, Kaldis P, Solomon MJ (1999) Dephosphorylation of cyclin-dependent kinases by type 2C protein phosphatases. Genes Dev 13:2946-57

Chinkers M (2001) Protein phosphatase 5 in signal transduction. Trends Endocrinol Metab 12:28-32

Choi J, Nannenga B, Demidov ON, Bulavin DV, Cooney A, Brayton C, Zhang Y, Mbawuike IN, Bradley A, Appella E, Donehower LA (2002) Mice deficient for the wild-type p53-induced phosphatase gene (Wip1) exhibit defects in reproductive organs, immune function, and cell cycle control. Mol Cell Biol 22:1094-1105

Cohen PTW (1997) Novel protein serine/threonine phosphatases: variety is the spice of life. Trends Biochem Sci 22:245-251

Cohen PTW (2002) Protein phosphatase 1-targeted in many directions. J Cell Sci 115:241-256

Cohen PTW (2003) Protein serine/threonine phosphatases and the PPP family. In Bradshaw RA and Dennis EA (eds) Handbook of Cell Signaling Vol 1, pp 593-600, Academic Press, Elsevier Science USA

Crosson SM, Khan A, Printen J, Pessin JE, Saltiel AR (2003) PTG gene deletion causes impaired glycogen synthesis and developmental insulin resistance. J Clin Invest 111:1423-1431

Das AK, Cohen PTW, Barford D (1998) The structure of the tetratricopeptide repeats of protein phosphatase 5: implications for TPR-mediated protein-protein interactions. EMBO J 15:1192-1199

Das AK, Helps NR, Cohen PTW, Barford D (1996) Crystal structure of the protein serine/threonine phosphatase 2C at 2.0 A resolution. EMBO J 15:6798-6809

Delibegovic M, Armstrong CA, Dobbie L, Watt PW, Smith AJH, Cohen PTW (2003) Disruption of the striated muscle glycogen targeting subunit, PPP1R3A, of protein phosphatase 1 leads to increased weight gain, fat deposition and development of insulin resistance. Diabetes 52:596-604

Di Como CJ, Arndt KT (1996) Nutrients, via the Tor proteins, stimulate the association of Tap42 with 2A phosphatases. Genes Devel 10:1904-1916

Dombrádi V, Axton JM, Barker HM, Cohen PTW (1990) Protein phosphatase 1 activity in *Drosophila* mutants with abnormalities in mitosis and chromosome condensation. FEBS Lett 275:39-43

Fienberg AA, Greengard P (2000) The DARP-32 knockout mouse. Brain Research Reviews 31:313-319

Fiscella M, Zhang H, Fan S, Sakaguchi K, Shen S, Mercer WE, Vande Woude GF, O'Connor PM, Appella E (1997) Wip1, a novel human protein phosphatase that is induced in response to ionizing radiation in a p53-dependent manner. Proc Natl Acad Sci USA 94:6048-6053

Götz J, Probst A, Ehler E, Hemmings BA, Kues W (1998) Delayed embryonic lethality in mice lacking protein phosphatase 2A catalytic subunit Cα. Proc Natl Acad Sci USA 95:12370-12375

Guthridge MA, Bellosta P, Tavoloni N, Basilico C (1997) FIN13, a novel growth factor-inducible serine-threonine phosphatase which can inhibit cell cycle progression. Mol Cell Biol 17:5485-5498

Hanada M, Kobayashi T, Ohnishi M, Ikeda S, Wang H, Katsura K, Yanagawa Y, Hiraga A, Kanamaru R, Tamura S (1998) Selective suppression of stress-activated protein kinase pathway by protein phosphatase 2C in mammalian cells. FEBS Lett 437:172-176

Hanada M, Ninomiya-Tsuji J, Komaki K, Ohnishi M, Katsura K, Kanamaru R, Matsumoto K, Tamura S (2001) Regulation of the TAK1 signaling pathway by protein phosphatase 2C. J Biol Chem 276:5753-5759

Hastie CJ, Cohen PTW (1998) Purification of protein phosphatase 4 catalytic subunit: inhibition by the antitumour drug fostriecin and other tumour suppressors and promoters. FEBS Lett 431:357-361

Helps NR, Brewis ND, Lineruth K, Davis T, Kaiser K, Cohen PTW (1998) Protein phosphatase 4 is an essential enzyme required for organisation of microtubules at centrosomes in *Drosophila* embryos. J Cell Sci 111:1331-1340

Holmes CF, Maynes JT, Perreault KR, Dawson JF, James MN (2002) Molecular enzymology underlying regulation of protein phosphatase-1 by natural toxins. Curr Med Chem 9:1981-1989

Honkanen RE, Golden T (2002) Regulators of serine/threonine protein phosphatases at the dawn of a clinical era? Curr Med Chem 9:2055-2075

Hu MC-T, Tang-Oxley Q, Qiu WR, Wang Y-P, Mihindukulasuriya KA, Afshari R, Tan T-H (1998) Protein phosphatase X interacts with c-Rel and stimulates c-Rel/Nuclear Factor κB activity. J Biol Chem 273:33561-33565

Huang B, Gudi R, Wu P, Harris RA, Hamilton J, Popov KM (1998) Isoenzymes of pyruvate dehydrogenase phosphatase. DNA-derived amino acid sequences, expression, and regulation. J Biol Chem 273:17680-17688

Janssens V, Goris J (2001) Protein phosphatase 2A: a highly regulated family of serine/threonine phosphatases implicated in cell growth and signalling. Biochem J 353:417-439

Jiang L, Whiteway M, Ramos C, Rodriguez-Medina JR, Shen SH (2002) The *YHR076w* gene encodes a type 2C protein phosphatase and represents the seventh PP2C gene in budding yeast. FEBS Lett 527:323-325

Kennelly PJ (2002) Protein kinases and protein phosphatases in prokaryotes: a genomic perspective. FEMS Microbiol Lett 206:1-8

Kerk D, Bulgrien J, Smith DW, Barsam B, Veretnik S, Gribskov M (2002) The complement of protein phosphatase catalytic subunits encoded in the genome of *Arabidopsis*. Plant Physiol 129:908-25

Kitani T, Ishida A, Okuno S, Takeuchi M, Kameshita I, Fujisawa H (1999) Molecular cloning of Ca^{2+}/calmodulin-dependent protein kinase phosphatase. J Biochem (Tokyo) 125:1022-1028

Klee CB, Ren H, Wang X (1998) Regulation of the calmodulin-stimulated protein phosphatase, calcineurin. J Biol Chem 273:13367-13370

Kloeker S, Wadzinski BE (1999) Purification and identification of a novel subunit of protein serine/threonine phosphatase 4. J Biol Chem 274:5339-5347

Kobor MS, Archambault J, Lester W, Holstege FCP, Gileadi O, Jansma DB, Jennings EG, Kouyoumdjian F, Davidson AR, Young RA, Greenblatt J (1999) An unusual eukaryotic protein phosphatase required for transcription by RNA polymerase II and CTD dephosphorylation in *S-cerevisiae*. Mol Cell 4:55-62

Koh CG, Tan EJ, Manser E, Lim L (2002) The p21-activated kinase PAK is negatively regulated by POPX1 and POPX2, a pair of serine/threonine phosphatases of the PP2C family. Curr Biol 12:317-321

Labes M, Roder J, Roach A (1998) A novel phosphatase regulating neurite extension on CNS inhibitors. Mol Cell Neurosci 12:29-47

Lawson JE, Niu X-D, Browning KS, Trong HL, Yan J, Reed LJ (1993) Molecular cloning and expression of the catalytic subunit of bovine pyruvate dehydrogenase phosphatase and sequence similarity with protein phosphatase 2C. Biochemistry 32:8987-8993

Leung-Hagesteijn C, Mahendra A, Naruszewicz I, Hannigan GE (2001) Modulation of integrin signal transduction by ILKAP, a protein phosphatase 2C associating with the integrin-linked kinase, ILK1. EMBO J 20:2160-2170

Li X, Scuderi A, Letsou A, Virshup D (2002) B56-associated protein phosphatase 2A is required for survival and protects from apoptosis in *Drosophila melangaster*. Mol Cell Biol 22:3674-3684

Li X, Virshup D (2002) Two conserved domains in the regulatory B subunits mediate binding to the A subunit of protein phosphatase 2A. Eur J Biochem 269:546-552

Lin QL, Buckler ES, Muse SV, Walker JC (1999) Molecular evolution of type 1 serine/threonine protein phosphatases. Mol Phylogenet Evol 12:57-66

Luke MM, Della Seta F, Di Como CJ, Sugimoto H, Kobayashi R, Arndt KT (1996) The SAPs, a new family of proteins, associate and function positively with the SIT4 phosphatase. Mol Cell Biol 16:2744-2755

Maeda T, Wurgler-Murphy SM, Saito H (1994) A two-component system that regulates an osmosensing MAP kinase cascade in yeast. Nature 369:242-245

Malleret G, Haditsch U, Genoux D, Jones MW, Bliss TV, Vanhoose AM, Weitlauf C, Kandel ER, Winder DG, Mansuy IM (2001) Inducible and reversible enhancement of learning, memory, and long-term potentiation by genetic inhibition of calcineurin. Cell 104:675-686

Mann DJ, Campbell DG, McGowan CH, Cohen PTW (1992) Mammalian protein serine/threonine phosphatase 2C: cDNA cloning and comparative analysis of amino acid sequences. Biochim Biophys Acta 1130:100-104

Marley AE, Kline A, Crabtree G, Sullivan JE, Beri RK (1998) The cloning expression and tissue distribution of human PP2Cbeta. FEBS Lett 431:121-124

Matthews HR, MacKintosh C (1995) Protein histidine phosphatase activity in rat liver and spinach leaves. FEBS Lett 364:51-54

Millward T, Zolnierowicz S, Hemmings BA (1999) Regulation of protein kinase cascades by protein phosphatase 2A. Trends Biochem Sci 24:186-191

Miyakawa T, Leiter LM, Gerber DJ, Gainetdinov RR, Sotnikova TD, Zeng H, Caron MG, Tonegawa S (2003) Conditional calcineurin knockout mice exhibit multiple abnormal behaviours related to schizophrenia. Proc Natl Acad Sci USA 100:8987-8992

Morita K, Saitoh M, Tobiume K, Matsuura H, Enomoto S, Nishitoh H, Ichijo H (2001) Negative feedback regulation of ASK1 by protein phosphatase 5 (PP5) in response to oxidative stress. EMBO J 20:6028-6036

Murray MV, Kobayashi R, Krainer AR (1999) The type 2C Ser/Thr phosphatase PP2C□ is a pre-mRNA splicing factor. Genes Develop 13:87-97

Pawson T, Scott JD (1998) Signaling through scaffold, anchoring and adaptor proteins. Science 278:2075-2080

Piekny AJ, Wissmann A, Mains PE (2000) Embryonic morphogenesis in *Caenorhabditis elegans* integrates the activity of LET-502 Rho-binding kinase, MEL-11 myosin phosphatase, DAF-2 insulin receptor and FEM-2 PP2c phosphatase. Genetics 156:1671-1689

Raghavan S, Williams I, Aslam H, Thomas D, Morgan G, Turner J, Fernandes J, Vijayraghavan K, Alphey L (2000) PP1β is required for the maintenance of muscle attachments. Current Biol 10:269-272

Ramulu P, Kennedy M, Xiong W-H, Williams J, Cowan M, Blesh D, Yau K-W, Hurley JB, Nathans J (2001) Normal light response, photoreceptor integrity and rhodopsin dephosphorylation in mice lacking both protein phosphatases with EF hands (PPEF-1 and PPEF-2). Mol Cell Biol 21:8605-8614

Rodriguez PL (1998) Protein phosphatase 2C (PP2C) function in higher plants. Plant Mol Biol 38:919-27

Scrimgeour AG, Allen PB, Feinberg AA, Greengard P, Lawrence JCJ (1999) Inhibitor-1 is not required for the activation of glycogen synthase by insulin in skeletal muscle. J Biol Chem 274:20949-20952

Shiozaki K, Russell P (1995) Counteractive roles of protein phosphatase 2C (PP2C) and a MAP kinase kinase homolog in the osmoregulation of fission yeast. EMBO J 14:492-502

Silverstein AM, Barrow CA, Davis AJ, Mumby MC (2002) Actions of PP2A on the MAP kinase pathway and apoptosis are mediated by distinct regulatory subunits. Proc Natl Acad Sci USA 99:4221-4226

Stark MJR (1996) Yeast protein serine/threonine phosphatases: multiple roles and diverse regulation. Yeast 12:1647-1675

Steele FR, O'Tousa JE (1990) Rhodopsin activation causes retinal degeneration in *Drosophila rdgC* mutants. Neuron 4:883-890

Steele FR, Washburn T, Rieger R, O'Tousa JE (1992) *Drosophila retinal degeneration C (rdgC)* encodes a novel serine/threonine protein phosphatase. Cell 69:669-676

Strovel ET, Wu D, Sussman DJ (2000) Protein phosphatase 2Calpha dephosphorylates axin and activates LEF-1-dependent transcription. J Biol Chem 275:2399-2403

Sumiyoshi E, Sugimoto A, Yamamoto M (2002) Protein phosphatase 4 is required for centrosome maturation in mitosis and sperm meiosis in *C-elegans*. J Cell Sci 115:1403-1410

Suzuki Y, Lanner C, Kim J-H, Vilardo PG, Zhang H, Yang J, Cooper LD, Steele M, Kennedy A, Bock CB, Scrimgeour A, Lawrence JC, Roach AA (2001) Insulin control of glycogen metabolism in knockout mice lacking the muscle-specific protein phosphatase PP1G/R$_{GL}$. Mol Cell Biol 21:2683-2694

Takekawa M, Adachi M, Nakahata A, Nakayama I, Itoh F, Tsukuda H, Taya Y, Imai K (2000) p53-inducible wip1 phosphatase mediates a negative feedback regulation of p38 MAPK-p53 signaling in response to UV radiation. EMBO J 19:6517-6526

Takekawa M, Maeda T, Saito H (1998) Protein phosphatase 2Calpha inhibits the human stress-responsive p38 and JNK MAPK pathways. EMBO J 17:4744-4752

Tamura S, Lynch KR, Larner J, Fox J, Yasui A, Kikuchi K, Suzuki Y, Tsuiki S (1989) Molecular cloning of rat type 2C (1A) protein phosphatase mRNA. Proc Natl Acad Sci 86:1796-1800

Tan KM, Chan SL, Tan KO, Yu VC (2001) The *Caenorhabditis elegans* sex-determining protein FEM-2 and its human homologue, hFEM-2, are Ca^{2+}/calmodulin-dependent protein kinase phosphatases that promote apoptosis. J Biol Chem 276:44193-44202

Travis SM, Berger HA, Welsh MJ (1997) Protein phosphatase 2C dephosphorylates and inactivates cystic fibrosis transmembrane conductance regulator. Proc Natl Acad Sci USA 94:11055-11060

Warmka J, Hanneman J, Lee J, Amin D, Ota I (2001) Ptc1, a type 2C Ser/Thr phosphatase, inactivates the HOG pathway by dephosphorylating the mitogen-activated protein kinase Hog1. Mol Cell Biol 21:51-60

Welihinda AA, Tirasophon W, Green SR, Kaufman RJ (1998) Protein serine/threonine phosphatase Ptc2p negatively regulates the unfolded-protein response by dephosphorylating Ire1p kinase. Mol Cell Biol 18:1967-1977

Wenk J, Trompeter H-I, Pettrich K-G, Cohen PTW, Campbell DG, Mieskes G (1992) Molecular cloning and primary structure of a protein phosphatase 2C isoform. FEBS Lett 297:135-138

Yeo M, Lin PS, Dahmus ME, Gill GN (2003) A novel RNA polymerase II C-terminal domain phosphatase that preferentially dephosphorylates serine 5. J Biol Chem 278:26078-26085

Young C, Mapes J, Hanneman J, Al-Zarban S, Ota I (2002) Role of Ptc2 Type 2C Ser/Thr Phosphatase in Yeast High-Osmolarity Glycerol Pathway Inactivation. Eukaryot Cell 1:1032-1040

Zuo Z, Urban G, Scammell JG, Dean NM, McLean TK, Aragon I, Honkanen RE (1999) Ser/Thr protein phosphatase type 5 (PP5) is a negative regulator of glucocorticoid receptor-mediated growth arrest. Biochemistry 38:8849-8857

2 Protein phosphatase 1

Viktor Dombrádi, Endre Kókai and Ilona Farkas

Abstract

Protein phosphatase 1 (PP1) is one of the first protein phosphatases whose activity was detected and whose catalytic subunit (PP1c) was purified and cloned. It is the representative of the most ancient protein phosphatase family that has a ubiquitous distribution in all eukaryotic organisms. The high level of conservation of the amino acid sequence and protein architecture of the PP1c is remarkable, and the identification of its very similar isoforms was an unexpected result of molecular cloning. The enzyme has a large number of interacting proteins, which tether PP1c to well defined locations and/or regulate its activity. A dynamic exchange of these non-catalytic subunits and the broad substrate specificity of the phosphatase are consistent with a wide range of physiological roles including cell cycle regulation, centrosome separation, interphase chromosome condensation, glycogen metabolism, contractility, morphogenesis, spermatogenesis, learning and memory, as inferred from genetic studies and predicted from biochemical experiments.

2.1 Biochemical characterization of protein phosphatase 1

The name of protein phosphatase 1 (PP1) was coined by Ingebritsen and Cohen (1983) to identify the most important member of the type 1 phosphatase family. PP1 has broad substrate specificity and is expressed in all eukaryotic organisms and cell types studied so far. Based on these properties, it is very likely that PP1 represented a significant component of an enzyme activity detected in early biochemical studies of crude cell extracts that contained an undefined mixture of unidentified phosphatases (for historic accounts see Lee et al. 1980; Fischer and Brautigan 1982). The situation was clarified only after the introduction of suitable methods for the purification of the PP1 catalytic subunit (PP1c). The resistance of PP1c against proteases and organic solvents (Lee et al. 1980), as well as the introduction of heparin-Sepharose (Gergely et al. 1986), poly(L-lysine)-Sepharose (Cohen et al. 1988), and microcystin-Sepharose (Moorhead et al. 1994) affinity chromatography greatly facilitated its isolation. Different PP1c preparations, isolated from a great variety of sources, exhibit a remarkable similarity of physico-chemical and biochemical properties (Table 1), strongly suggesting a high level of structural and functional conservation of PP1c during evolution.

Topics in Current Genetics, Vol. 5
J. Arino, D.R. Alexander (Eds.) Protein phosphatases
© Springer-Verlag Berlin Heidelberg 2004

Table 1. Characteristic properties of PP1c

Purification	resistant to proteases and organic solvents, first phosphatase peak eluted by salt gradient in ion exchange chromatography, retained by heparin-Sepharose affinity column
Molecular mass	~35 kDa
N-terminus	variable, blocked
C-terminus	variable, clipped by proteases
Primary structure	highly conserved 280 amino acid long core
Catalytic site	binuclear metal ion centre
Substrate specificity	broad, but dephosphorylates the β subunit of phosphorylase kinase more efficiently than the α subunit
Metal ion requirement	no external metal ion is required for activity
Inhibitors	inhibitor-1 protein $(IC_{50}: 1.6 \text{ nM})$
	inhibitor-2 protein $(IC_{50}: 3.1 \text{ nM})$
	heparin $(IC_{50}: 0.4 \text{ μM})$
	okadaic acid $(IC_{50}: 2.0 \text{ nM})$
	microcystin-LR $(IC_{50}: 0.1 \text{ nM})$
	calyculin A $(IC_{50}: 0.3 \text{ nM})$
Tertiary structure	globular α/β protein
Quaternary structure	monomer, component of high molecular mass holoenzyme complexes

A micro-heterogeneity of apparently homogeneous PP1c preparations was reported by Silberman et al. (1984). The phenomenon was attributed to limited proteolysis (Villa-Moruzzi and Fischer 1987). Since the N-terminal amino acid of PP1c was blocked (effectively preventing the sequencing of the native protein), it was concluded that a short C-terminal peptide sequence was removed by endogenous proteases. The generation of truncated PP1c can be avoided by the addition of a cocktail of protease inhibitors throughout the purification procedure (Cohen et al. 1988). Amino acid sequencing of internal proteolytic peptides revealed another type of micro-heterogeneity (Johnson et al. 1987). Two overlapping peptides with slightly different amino acid sequences were obtained from the same rabbit skeletal muscle PP1c preparation. Initially this discrepancy was explained by allelic variations. However, in retrospective, we can explain this and other peptide sequence data (Berndt et al. 1987; Alessi et al. 1992; Dent et al. 1992) by the presence of several isoforms, which cannot be resolved by traditional biochemical methods.

2.2 Cloning of the catalytic subunit of protein phosphatase 1

The cloning of PP1c was achieved in two independent ways. The peptide sequences obtained from rabbit muscle PP1c preparations were used for the synthesis of oligonucleotide probes in a structure based screening. The very first cDNA clone of rabbit muscle PP1β (Berndt et al. 1987) unfortunately contained an artefact at its 5'-end. The rest of the sequence was the same as that of the two identical cDNAs (PP1α) cloned independently in two laboratories (Cohen, 1988; Bai et al. 1988). It was concluded that PP1α was generated from PP1β by alternative splicing, while the differences from the original peptide sequences were attributed to allelic variations.

A parallel functional cloning approach of PP1 was initiated by the studies of cell cycle mutants. Ohkura et al. (1988) described a cold and caffeine sensitive *Schizosaccharomyces pombe* mutant that was also defective in the separation of sister chromatids (*dis2*-11) at the restrictive temperature. The recognition of sequence similarity between PP1α and the Dis2 proteins was a great revelation that identified an essential function for protein phosphatase 1(Ohkura et al. 1989). At the same time the overlapping functions of two PP1c isoforms, Dis2 and Sds21 (first suppressor of *dis2*) in fission yeast were established. The general phylogenetic distribution of Dis2 homologues was also described in the same paper. The budding yeast homologue *DIS2S1* and the mouse homologues *dis2 m1* and *dis2 m2* were cloned based on sequence similarity (Ohkura et al. 1989). The same *S. pombe dis2* gene was found by Booher and Beach (1989) in an independent functional screen and was termed *bws1* (bypass of *wee* suppression). The *Saccharomyces cerevisiae* homologue turned out to be identical to *GLC7*, a gene implicated in the control of glycogen metabolism (Clotet et al. 1991; Feng et al. 1991). However, the key to the identification of the encoded protein as PP1c came from studies of the cell cycle regulator gene *BimG* (Doonan and Morris 1989) of *Aspergillus nidulans* (that is called *Emericella nidulans* today). The temperature sensitive conditional mutant, *BimG11* was blocked in mitosis and exhibited overphosphorylation of nuclear proteins at higher temperature. The increased level of phosphorylation hinted the relationship of *BimG* to the protein dephosphorylating enzymes that was convincingly proven by its sequence similarity to PP1α.

Using the mouse *dis2 m1* and *dis2 m2* cDNA probes; their rat homologues termed PP1γ1 and PP2γ2 were also isolated (Sasaki et al. 1990). The two rodent PP1γ forms were generated by alternative splicing of the 3'-end of the primary transcript. Furthermore, the rat PP1α and a fourth PP1c isoform (named PP1δ) were cloned by the same authors.

Meanwhile, a homology based cloning strategy resulted in the identification of four highly similar PP1c *Drosophila melanogaster* isoenzymes (Dombrádi et al. 1989, 1990, 1993), which were called PP1(9C), PP (13C), PP1(87B) and PP1(96A) according to the chromosomal localization of their genes. A sequence comparison revealed that PP1(9C) and PP1δ contained well-conserved sequence motifs that distinguished them from the rest of the animal PP1c forms. It was suggested that PP1(9C) should replace the original PP1β and that PP1δ should be renamed as PP1β (Dombrádi et al. 1990). This suggestion was accepted, however, one can still find the older PP1δ designation in the literature.

One more human PP1α form was reported by Durfee et al. (1993). The cDNA of this so-called PP1α' variant predicts an unusual N-terminal sequence that may be the product of alternative splicing. However, the existence of this splice variant

Table 2. PP1 isoenzymes in eukaryotic model organisms*

Organism	Number of iso- forms	Identification
Homo sapiens	3	PPP1CA /PP1$_\alpha$, PPP1CB/PP1$_\beta$, PPP1CC/PP1$_\gamma$
Mus musculus	3	PP1$_\alpha$, PP1$_\delta$ = PP1$_\beta$, PP1$_\gamma$
Rattus norvegicus	3	PP1$_\alpha$, PP1$_\delta$ = PP1$_\beta$, PP1$_\gamma$
Drosophila melanogaster	4	PP1 (87B), PP1 (96A), PP1 (13C), PP1 (9C)
Anopheles gambiae	2	PP1$_\alpha$, PP1$_\beta$
Caenorhabditis elegans	2	PP1$_\alpha$/CeGLC-7$_\alpha$, PP1$_\beta$/ CeGLC-7$_\beta$
Saccharomyces cerevisiae	1	Glc7
Neurospora crassa	1	PPP1
Schizosaccharomyces pombe	2	PP1-1/Dis2, PP1-2/Sds21
Arabidopsis thaliana	9	TOPP1-8 and PP1
Nicotina tabacum	3	PP1 1, PP1 2, PP1 3
Medicago sativa	5	PP1$_\alpha$, PP1$_\beta$, PP1$_\gamma$, PP1$_\delta$, PP1$_\varepsilon$
Chlamydomas reinhardtii	1	PP1
Plasmodium falciparum	1	PP1/PP$_\alpha$3
Dictyostelium discoideum	1	PP1-like
Paramecium tetraurelia	1	PP1-like

*Based on the NCBI protein database http://www.ncbi.nlm.nih.gov/Entrez

has not been confirmed by any of the available EST sequences, thus it must have been generated by a rather rare splicing event.

As described above, converging approaches set the stage for comparative cloning projects. With the most conserved cDNA sequences as the targets, a large number of PP1c isoforms were identified in a wide range of organisms by the combination of hybridization based screening and PCR methods. EST sequencing and the completion of several genome projects supplied even more comprehensive data for the identification of PP1c genes (Table 2). In accordance with activity assays, at least one gene for PP1c was found in all eukaryotic organisms tested (Dombrádi 1997). All the available sequence data support the thesis that the primary structure of PP1c was extremely well conserved during evolution (Lin et al. 1999). PP1c is one of the most conservative enzymes ever known; they are surpassed only by the core histones and the essential regulatory protein calmodulin in this respect.

2.3 The structure of the protein phosphatase 1 catalytic subunit

The cloning of PP1c was an essential prerequisite of further structural analysis simply because only recombinant proteins were suitable for site directed mutagenesis (Lee et al. 1999) and for crystallization. The PP1c isoforms were expressed in *E. coli* (Zhang et al. 1992), in baculovirus infected S6 insect cells

(Berndt and Cohen 1990) or in *Pichia pastoris* (Szoor et al. 2001). The recombinant PP1c isoforms were indistinguishable from one another according to their biochemical properties and were somewhat different from the freshly isolated "native" enzyme in as much as they requirement metal ions (first of all Mn^{2+}) for activity. In addition, the X-ray diffraction pictures probably represent an inactive conformation of the enzyme, since the crystals were grown in the presence of phosphatase inhibitors. Despite the above natural limitations, the 3D pictures revealed some brand new aspects of the protein architecture. Tungstate (Egloff et al. 1995) takes the place of the substrate's phosphate, while he inhibitors microcystin-LR (Goldberg et al. 1995), okadaic acid (Maynes et al. 2001), and calyculin A (Kita et al. 2002) directly block the entrance to the catalytic site. Quite unexpectedly, the active site embedded two metal ions surrounded by the most conserved amino acids of the protein (Egloff et al. 1995; Goldberg et al. 1995). The identity of the native metal ions remains an open question. In the crystals the metal binding-sites were occupied by manganese ions (Egloff et al. 1995), obviously because the recombinant PP1α and PP1γ1 proteins were refolded in the presence of Mn^{2+}. The structural analogy with the catalytic domain of calcineurin suggests that iron and zinc ions take this place in the native protein (Barford 1999). Nevertheless, the 3D structures clearly indicate the involvement of these metal ions in the catalytic process. They together activate a structural water molecule that can hydrolyze the phosphoester bond of Ser or Thr resides of the substrates. The relatively shallow active site crevice explains how the small phosphorylated side chains of Ser and Thr can reach the activated water. Co-crystallization with a synthetic peptide derived from the glycogen binding subunit (G_M) revealed another well-conserved surface of the PP1c that is found opposite to the catalytic cleft (Egloff et al. 1997). This is the most important recognition site of many regulatory subunits.

As forecasted by the conserved amino acid sequences, the tertiary structures of the PP1α and γ1 isoforms were practically identical. Only slight changes were detected in the presence of different inhibitors. Furthermore, homologues modelling suggested that all of the PP1 isoforms obtained from diverse sources including plants, fungi and animals can be characterized by a nearly uniform tertiary structure. The individual differences are restricted to the N- and C-terminal amino acids (Lin et al. 1999) that are invisible in the X-ray diffraction pictures. These small differences may however have serious consequences, especially at the C-terminal end, where potential phosphorylation sites reside. An inhibitory phosphorylation of Thr320 (in rabbit PP1α) by the cyclin-dependent Cdc2 kinase (Dohadwala et al. 1994; Yamano et al. 1994; Kwon et al. 1997), Nek2 kinase (Helps et al. 2000), and a novel kinase KPI-2 (Wang and Brautigan 2002) was reported. The direct phosphorylation and dephosphorylation of PP1c is a powerful regulatory device that coordinates cell cycle events and renders PP1 inactive at the initial phases of mitosis (Berndt 1999). The dephosphorylation and activation of PP1c by ionizing irradiation has been reported recently (Guo et al. 2002). This regulation is, however, not universal as the phosphorylation site is missing from a number of PP1c isoforms and homologues (Yamano et al. 1994). A more general mechanism for

PP1 regulation is offered by the dynamic binding and modification of associated "non-catalytic" subunits.

2.4 Non-catalytic (regulatory) subunits and interacting proteins of protein phosphatase 1

Even the earliest biochemical studies of PP1 excluded the physiological significance of free PP1c since PP1 activity has always been detected in large molecular mass complexes in fresh tissue extracts (Lee et al. 1980). The classical purification schemes of PP1c unintentionally aimed to destroy the additional non-catalytic subunits of the holoenzyme. The purification of the first PP1c interacting proteins utilized their extreme heat stability (Brandt et al. 1975; Huang and Glinsmann 1975). The thermostable proteins, inhibitor-1 (I-1) and inhibitor-2 (I-2), become the most widely used tools for the identification of PP1 activity (Ingebritsen and Cohen 1983). I-2 has been preferred in the assays since it inhibited PP1 in its dephosphorylated form, while a prior phosphorylation of I-1 by cAMP- dependent protein kinase (PKA) was required for the expression of its inhibitory capacity. I-2 was also found as an integral component of the so-called ATP-Mg-dependent protein phosphatase (Merlevede 1985; Bollen and Stalmans 1992). An activating factor (F_A) and ATP-Mg were required for the (re-)activation of a catalytic factor (F_C) in order to express its phosphatase activity. The latter factor turned out to be the complex of I-2 with PP1c, while F_A was identified as glycogen synthase kinase 3 (GSK3) that affected a transient I-2 phosphorylation. The interaction aids the proper folding of PP1c that is why I-2 was suggested to be a molecular chaperone (Alessi et al. 1993; MacKintosh et al. 1996).

After the discovery of the two heat stable inhibitors, a large number of other PP1c interacting proteins were unravelled by the biochemical purification of the holoenzyme complexes, by affinity chromatography, co-immunoprecipitation, overlay of labelled PP1c in far-Western experiments or in screening expression libraries, by genetic interactions and most efficiently by the yeast two hybrid system (reviewed by Cohen 2002). Even a full fetched proteomic project has been devoted for the identification of PP1c interacting proteins (see the chapter by Tim Haystead in this book). More than 50 PP1c interacting partners were listed in a dedicated homepage (http://pp1signature.pasteur.fr). This list can be further extended by additional recently published proteins. The most important primary and secondary interactions in animal PP1 complexes are depicted in Figure 1 and the interacting proteins of yeast Glc7 are listed in Table 3.

The concept of targeting subunits was formulated (Hubbard and Cohen 1993) and further elaborated (Faux and Scott 1996) to explain the significance of the non-catalytic subunits. The targeting subunits direct PP1c to specific locations, e.g., to the glycogen particles, myosin fibres, nucleus, centrosome, ribosomes, microtubules, mitochondria, membranes, or actin cytoskeleton and control specificity by tethering the enzyme activity towards the substrate proteins located at the

Table 3. Proteins interacting with *S. cerevisiae* PP1c (Glc7/YER133W)*

Protein/Gene	Function
Gac1/YOR178C	Glycogen binding subunit homologue, glycogen metabolism
Pig1/YLR273C	Homologues of Gac1, potential glycogen targeting subunit, interacts with glycogen synthase Gsy2
Pig2/YIL045C	Interacts with glycogen synthase Gsy2
Gip1/YBR045C	Septin formation and meiotic cell division or spore packaging
Gip2/YER054C	Potential glycogen targeting subunit
Reg1/YDR028C	Substrate, binds Snf1 kinase and hexokinase II, negative regulator of glucose repressible genes, necessary for cell growth and/or maintenance
Reg2/YBR050C	Reg1 homologue
Sip5/YMR140W	Reg1 and Snf1 binding protein, facilitates formation of a kinase-phosphatase regulatory complex
Glc8/YMR311C	Homologue of inhibitor-2 that is phosphorylated by Pho85 kinase, molecular chaperone, activates Glc7
----/YFR003C	Homologue of inhibitor-3
Sds22/YKL193C	Essential nuclear targeting subunit
Mhp1/YJL042W	Tubulin homologue, microtubule associated protein
Bni4/YNL233W	Cellular morphology, chitin deposition
Ref2/YDR195W	RNA end formation, maturation of the 3'-end of RNA
Scd5/YOR329C	Actin organization, endocytosis
Bud27/YFL023W	Bud site selection, viability on toxin exposure
Red1/YLR263W	Chromosome condensation, crossing over

*Based on http://genome-www4.stanford.edu/Saccharomyces/SGD

given sub-cellular compartment. In addition, they can also modify specificity by activating the phosphatase against selected substrates.

The evolutionary aspects of the non-catalytic subunits have been considered by Ceulemans et al. (2002). The widespread phylogenetic distribution of inhibitor-3 and Sds22 indicates their most ancient origins. I-2 and the nuclear inhibitor of protein phosphatase 1 (NIPP1) also belong to the oldest representatives of PP1 inhibitors.

The structural requirements for the holoenzyme formation were determined by site directed mutagenesis and X-ray crystallographic studies (see Cohen 2002). The fundamental role of a small peptide region R/K-(X)-V/I-X-F/W, the so-called RVxF motif in the recognition of the regulatory site of PP1c by the non-catalytic subunits was delineated (Egloff et al. 1997). The optimal sequence required for effective interactions was confirmed by phage display panning (Zhao and Lee 1997). Besides this primary binding site, additional sequences like F-X-X-R/K-X-R/K contribute to the cooperative association between PP1c and its regulators. Most but not all PP1c interacting proteins possess the conserved primary binding motif. Notable exceptions to the rule are some of the most ancient regulators, Sds22 and I-2.

The structures of the non-catalytic subunits are quite diverse. Some of them contain separate domains that are associated with other functions. For example,

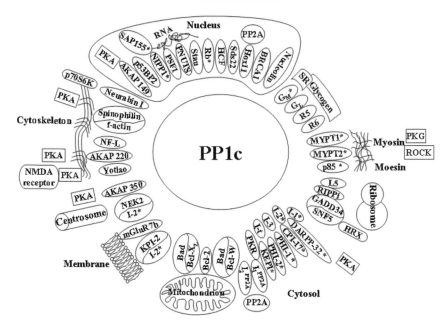

Fig. 1. Protein-protein interactions in animal protein phosphatase 1 holoenzymes. Proteins interacting directly or indirectly with PP1c and the specific cellular compartments where PP1 is localized are depicted. Potential regulatory phosphorylation sites in the proteins are labelled by asterisk.

NIPP1 has an RNA binding domain and a fork-head associated domain that targets it to the spliceosomes (Beullens and Bollen 2002). Other interacting partners (Bad, Bcl2, Bcl-X$_L$, Rb, p53BP2, Hox11, L5, Nek2, Tau, ryanodine receptor, phosphofructokinase, G-substrate, glucose regulated protein 78, etc.) have well defined functions distinct from the regulation of PP1 and probably acquired PP1c binding sites later in their evolution (Ceulemans et al. 2002).

The most interesting associations encompass protein kinase(s) and PP1c (Bauman and Scott 2002; Bennett and Alphey 2002). The interaction of PP1c with kinase(s) brings the two contrasting enzyme activities in one location and allows a quick transient phosphorylation of nearby proteins. The significance of the spatiotemporal coordination of phosphorylation- dephosphorylation is exemplified by these signalling complexes.

As a general trend, the binding of most non-catalytic subunits to PP1c is mutually exclusive. A dynamic exchange of the distinct regulatory proteins is probably the most effective regulatory mechanisms controlling PP1 activity (Bollen 2001). Some of the interacting proteins however, have multiple binding sites and generate ternary or even higher order complexes (Fig. 1). In addition to the displacement of multiple parallel equilibria by the alternating levels of protein synthesis, the process is further modulated by the integration of environmental clues in the structural changes of the PP1c interacting proteins (Aggen et al. 2000). Notably, many PP1 regulators are controlled by phosphorylation. I-1 and DARPP-32 are phosphory-

lated by PKA and are dephosphorylated by calcineurin, thus respond to cAMP and
Ca^{2+}-mediated signals. The phosphorylation of SAP155 splicing factor is required
for its binding to NIPP1 (Boudrez et al. 2002). On the other hand, the phosphorylation of I-2 by GSK3 reduces its potency, and PKA-mediated phosphorylation
NIPP1 inactivates the inhibitor and weakens its interaction with PP1c. Thus,
phosphorylation can either strengthen or weaken the protein-protein interactions.
The potentially important phosphorylation site(s) of the PP1c interacting proteins
are labelled in Figure 1.

According to the most recent findings, some of the interacting proteins are isoform specific. For example neurabins preferentially bind to the PP1α and PP1γ
(Terry-Lorenzo et al. 2002), BRCA1 (Liu et al. 2002), PKP (Tan et al. 2002), and
different Bcl proteins (Ayllon et al. 2002) are associated with PP1α, nucleolin
(Morimoto et al. 2002) binds to the PP1δ=PP1β isoforms, and the glutamate receptor interacts with PP1γ1 (Enz 2002). Other regulators have a more relaxed
specificity; the heat stable proteins I_1^{PP2A} and I_2^{PP2A}, or Hox11 can interact with
both PP1 and PP2A. Most of the interacting partners were shown to be true subunits, regulators or substrates, but others may be artefacts of the assay system. The
physiological significance of many potential interactions has to be tested and substantiated by additional studies.

2.5 Functions of Protein Phosphatase 1

PP1 is an essential enzyme. The disruption of the single PP1c gene *GLC7* in *S.
cerevisiae* is lethal (Clotet et al. 1992); double disruption of the *dis2* and *sds21*
genes in *S. pombe* has the same consequences (Kinoshita et al. 1991). Even the
disruption of the *sds22* gene coding for the nuclear regulatory subunit of PP1 reduces viability (Ohkura and Yanagida 1991). Overexpression of nuclear inhibitor
of PP1 (NIPP1) is also detrimental (Parker et al. 2002). The broad substrate specificity and widespread distribution of the enzyme in different tissues and cellular
compartments suggest that PP1 has multifaceted roles some of which are indispensable for the normal functioning of eukaryotic cells. Different isoforms of PP1
exhibit differential expression in mammalian brain (da Cruz e Silva et al. 1995).
PP1β can be found in skeletal muscles (Dent et al. 1992) and focal adhesions
(Fresu et al. 2001), the PP1α and γ1 forms are concentrated in the dendritic spines
(Ouimet et al. 1995), whereas the expression of PP1γ2 isoform is testis specific
(Kitagawa et al. 1990; Shima et al. 1993). The expression patterns suggest special-
ized functions for these very similar proteins.

2.5.1 Cell cycle regulation

The first documented essential physiological role of PP1 was revealed by the functional cloning of its catalytic subunit (Sect. 2). The cold sensitive *dis2-11* mutant
of *S. pombe* had a drastically reduced PP1 activity at low temperature resulting in

chromatin overcondensation and abnormal disjunction of sister chromatids (Kinoshita et al. 1990). It was assumed that the mutant protein was not only inactive, but somehow inhibited Sds21 the other PP1 isoform in fission yeast as well. In agreement with this suggestion, single disruptants of the *dis2* and *sds21* genes were viable, only the inactivation of both phosphatase genes resulted in the mitotic phenotype (Kinoshita et al. 1991). A similar phenomenon was detected in the temperature sensitive *BimG11* mutant of *E. nidulans* at elevated temperature. Hyperphosphorylation of nuclear proteins was detected in connection with low PP1 activity and block in mitosis (Doonan et al. 1991). The *BimG11* point mutation causes abnormal splicing at the restrictive temperature and produces a truncated PP1 with an unusual C terminal sequence (Hughes et al. 1996). The results obtained in fungal model organisms were strongly supported by the studies of *Drosophila*. A full set of mutants affecting one of the PP1c genes, PP1(87B) had been collected in independent screens before the cloning of the enzyme (reviewed by Dombrádi 1995). The null mutant *e211* (with a large deletion in the PP1 87B gene) and the hypomorphic mutant *hs46* (with a small deletion in the same gene) produced no protein for this isoenzyme and caused a serious loss of PP1 activity in the third larval stage of development (Dombrádi et al. 1990). It was assumed that the maternal deposit of PP1(87B) protein/mRNA (or the other isoforms of PP1c) were sufficient to drive the fruit flies through the earlier stages of morphogenesis, however, the lack of the major PP1 isoform before the preparation for the large metamorphosis in the puparium resulted in lethality. DNA hypercondensation, collapsed microtubule fibres in the metaphase, incomplete anaphase as well as polyploidy and hyperploidy were detected in the actively dividing cells of larval brain and imaginal discs (Axton et al. 1990). Obviously, none of the other three unaffected PP1c isoforms were able to overcome the deficit of PP1(87B) when they were expressed at a normal level. Overexpression of PP1 (96A) alleviated the mutant phenotype suggesting that the defect can be attributed to the decrease in the PP1 dose (Ohkura et al. unpublished data). According to a threshold hypothesis, certain PP1 activity is indispensable for the correct completion of mitosis; low PP1 activity is associated with an anaphase block (Table 4). In accord with this assumption, two other Drosophila PP1(87B) mutants *078* and *Suvar(3)6⁰¹* harbouring point mutations in the codon of the same critical Gly 220 residue have somewhat higher PP1 activity and do not exhibit the cell cycle phenotype described above (Dombrádi and Cohen 1992). Nevertheless, they have another interesting effect on the chromosome condensation in the interphase of the cell cycle (Baksa et al. 1993). The latter mutations caused a strong suppression of position effect variegation that is related to the spreading of the heterochromatic region into the area containing reporter genes. From these studies, we concluded that PP1(87B) was also responsible for the dephosphorylation of non-histone proteins that are important in the determination of the condensation state of interphase chromatin and thus regulate chromosome structure. An additional role of PP1 in the early preparatory stage of mitosis was suggested by the studies of another *dis2* mutant allele *bws1* (Booher and Beach 1989).

The conservation of these essential functions of PP1 in other organisms was demonstrated by the complementation of the *dis2-11* mutant with budding yeast

Table 4. A threshold level of PP1 activity is required for the completion of anaphase during cell cycle

Organism	Mutant	PP1 activity (%)*	Mitosis
Schizosaccharomyces pombe	dis2-11 (20°C)	0	normal
	dis2⁻	20	normal
	sds21⁻	80	normal
	dis2⁻/sds21⁻	0	blocked
Emericella nidulans (*Aspergillus nidulans*)	BimG11(40°C)	15	blocked
Drosophila melanogaster	e211	20	abnormal
	hs46	21	abnormal
	1311	16	abnormal
	1455	17	abnormal
	078	35	normal
	Suvar (3)6^{01}	21	normal

*Specific activity expressed as the percentage of the wild type.

DIS2/GLC7 (Ohkura et al. 1989) and with the *A. thaliana* TOPP1 isoform (Nitschke et al. 1992). TOPP2 (Smith and Walker 1996) and plasmodium PP1 (Bhattacharyya et al. 2002) functionally complemented a *GLC7* mutant of *S. cerevisiae*. The *BimG 11* mutation in *E. nidulans* was partially rescued by *A. thaliana* TOPP8/AtBimG (Arundhati et al. 1995). However, *dis2-11* was not complemented by the *Medicago sativa* MsPP1$_\alpha$ (Páy et al. 1994) while the TOPP2 and TOPP3 plant PP1 isoforms failed to rescue the *GLC7* mutants (Smith and Walker 1996). The missing or partial complementation in some cases indicates functional divergence among the different PP1c isoforms in different organisms.

Under normal conditions, the cell cycle is coupled with the division and movement of centrosomes. According to most recent findings, PP1 is also required in this process. The NimA activated kinase (Nek2) interacts with PP1c and co-localizes to the centrosomes together with the heat stable I-2 protein (Helps et al. 2000; Eto et al. 2002). A cyclic alteration of the I-2 level during the cell cycle was detected earlier (Brautigan et al. 1990). It has been proposed that the ternary complex Nek-2/PP1c/I-2 is the essential coordinator of the centrosome cycle. Transient expression of I-2 supports the hypothesis (Eto et al. 2002).

2.5.2 Glycogen metabolism

The first substrates of PP1 identified in biochemical studies were glycogen phosphorylase and the β subunit of phosphorylase kinase (Lee et al. 1980; Ingebritsen and Cohen 1983). A strong binding of the phosphatase to the glycogen particles was noted and the glycogen binding subunit from skeletal muscle was the first targeting subunit of PP1c to be isolated (Hubbard and Cohen 1993). In addition to these classical observations, several novel glycogen binding subunits of PP1c have been cloned and characterized (Fig. 1).

The ultimate proof for the essential role of PP1 in glycogen metabolism came from genetic experiments. *S. cerevisiae* is exceptionally well suited for these studies since it contains only one essential PP1c gene termed *GLC7* (Clotet et al. 1991; Feng et al. 1991). Several mutations in yeast affected glycogen accumulation (Cannon et al. 1994). One of them, *GLC7* corresponds to PP1c and another one, *GLC8* has sequence similarity to I-2. Interestingly, the latter protein does not inhibit the activity of PP1c (as would have been expected from the sequence comparison) but it is still important for Glc7 stability and may operate as a chaperone of the phosphatase (Nigavekar et al. 2002). A homologue of the animal G subunits was also found in yeast under the name of *GAC1*; this and other potential glycogen binding subunits can regulate glycogen metabolism in *S. cerevisiae* (see Table 3).

Glycogen degradation is catalyzed by glycogen phosphorylase, and glycogen synthase is responsible for the synthetic reaction. Both of them are under the control of kinases (Roach 1990; Cohen 1999) and PP1 localized to the glycogen particle by its targeting subunits (Fig. 2). Biochemical studies indicate that in liver the strong binding of the phosphorylated glycogen phosphorylase $\underline{a}$ (but not the unphosphorylated phosphorylase $\underline{b}$) to the G_L subunit inhibits the dephosphorylation of glycogen synthase (see Cohen 2002). Consequently, the conversion of phosphorylase $\underline{a}$ to phosphorylase $\underline{b}$ by PP1 is a prerequisite for the dephosphorylation and activation of the synthase. This allosteric mechanism prevents the premature activation of glycogen synthesis before the turning off glycogen degradation. The interaction is quite specific for the G_L protein as other glycogen binding subunits lack the short peptide sequence necessary for the binding of phosphorylase *a*.

The function of the muscle specific form of the glycogen binding subunit (G_M) was studied in KO mice. Although the mice lacking G_M had liver glycogen content, no significant disorders in the hormonal regulation of glycogen metabolism have been detected in these animals (Kim and DePaoli-Roach 2002). Therefore, G_M is not involved in the mediation of epinephrine action. On the other hand, G_M is required for the glycogen synthesis during exercise (Aschenbach et al. 2001). It may be important in the regulation of cardiac function, too. It is known that the C-terminus of G_M has a hydrophobic peptide segment that targets the protein to the sarcoplasmic reticulum (Fig. 2). In the heart muscle, phospholambam is dephosphorylated and inactivated by PP1. In correlation with these biochemical properties, overexpression of PP1c or elimination of its inhibitor causes symptoms of heart failure (Carr et al. 2002).

2.5.3 Contractility

A semi-lethal point mutation in the minor PP1c isoform of Drosophila (PP19C) was found in the flapwing mutant (*flw*). The hypomorphic alleles are defective in the locomotion function and exhibit abnormal attachment of the flight muscles (Raghavan et al. 2000). This phosphatase of Drosophila has been the prototype of

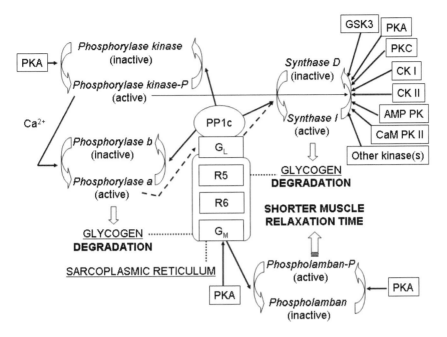

Fig. 2. Coordinated regulation of glycogen metabolism and heart relaxation by protein phosphatase 1. Full arrows indicate activation of a process; broken arrows denote inhibitory actions. Dotted line shows the targeting site of PP1. Brackets connect alternative binding proteins, only one of which is in direct association with PP1c at a given time.

the so-called PP1β isoforms. Albeit it represents less than 20% of PP1 activity in fruit fly, its mutation has profound effect on muscle development. The muscle specific function of PP1β isoforms in vertebrates was suggested by the detection of β specific peptide sequences in the PP1c preparations isolated from rabbit skeletal muscle (Dent et al. 1992).

The LC20 light chain of myosin is phosphorylated in smooth muscle and non-muscle cells by four different kinases in Ca^{2+}-dependent and independent reactions. The dephosphorylation is catalyzed by PP1 targeted to myosin or actin cytoskeleton (Hartshorne et al. 1987). The activity of PP1 is controlled by the phosphorylation of the non-catalytic subunits and inhibitors by a protein kinase network (Fig. 3). The phosphorylation of myosin binding subunits by Rho-dependent kinase (ROCK) integrin-linked protein kinase (ILK) and Raf-1 (Broustas et al. 2002) coordinates Ca^{2+}- independent smooth muscle contraction. The phosphorylation of phosphatase inhibitors potentiates the process even further (Eto et al. 1999; Koyama et al. 2000; Deng et al. 2002; Liu et al. 2002; Takizawa et al. 2002).

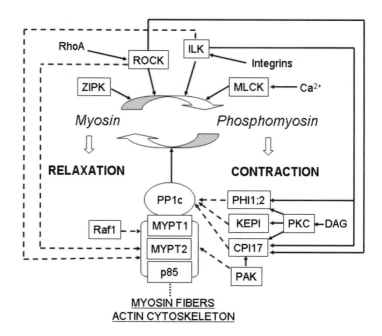

Fig. 3. Regulation of contractility by protein phosphatase 1 The same symbols are used as in Figure 2.

2.5.4 Morphogenesis

The myosin binding subunits of PP1 (MYPTs) were suggested to be crucial in the coordination of Ca^{2+}-independent regulation of smooth muscle contraction. However, the mutation of the *C. elegans* MYPT homologue termed MEL-11 directed attention towards other functions of this subunit. In cooperation with the Rho-dependent protein kinase (LET-502), it can mediate contraction and epidermal cell shape (Wissmann et al. 1999), embryonic morphogenesis (Piekny et al. 2000), as well as cytokinesis (Piekny and Mains 2002). The interaction between PP1c and the *bifocal* gene product is necessary for the proper development of the compound eye in Drosophila (Helps et al. 201). The role of the PP1-neurabin 1 complex in determining cell morphology has been reported recently (Oliver et al. 2002). Obviously, PP1c functions in the remodelling of the cytoskeletal network that is an integral step of many developmental processes.

A side effect of *BimG11* mutation in *E. nidulans* was noted by Borgia (1992). The mutation influenced the development of the filamentous fungus probably via the inhibition of chitin synthesis. In accord with the above observation, we detected a transient increase in PPP1 gene expression in synchronized *N. crassa* cells when the hyphal branching occurred in the conidia and in aerial mycelia (Zeke et al. 2003).

2.5.5 Spermatogenesis

The specific expression of mouse *dis2*m1 (Kitagawa et al. 1990) and its rat homologue PP1γ2 (Shima et al. 1993) in the testes of mature animals was consistent with a possible role of this isoform in the sperm maturation. The question was directly addressed by targeted mutation (Varmuza et al. 1999). Males homozygous for the insertion were sterile and had serious defect in spermatogenesis. Although the testes of these animals contained sufficient PP1α this isoforms was unable to compensate the loss of PP1γ.

2.5.6 Memory and learning

After the discovery of memory mutant Drosophila affected in the genes of cAMP metabolism, the phosphoprotein nature of some memory deposits was appreciated and protein phosphatases were considered as memory erasers. This assumption was investigated in a semi lethal PP1(87B) mutant Survar(3)6[01] in several learning paradigms (Asztalos et al. 1993). The mutant flies with reduced PP1 activity were found deficient in associative learning and visual conditioning tests, but performed better in the habituation of the landing response. A more definite proof of the involvement of PP1 in learning was provided by transgenic mice that expressed a constitutively active I-1 protein in the nerve cells upon steroid treatment (Genoux et al. 2002). The effect of induced I-1 expression was not evident in random learning paradigms but become significant in structured learning-rest-learning cycles. Transgenes learnt more easily and remembered longer than the appropriate controls. Some of the experiments even suggested that PP1 may cause senility of the elderly.

2.6 Conclusions

PP1 is one of the most ancient and most conservative protein phosphatases. Its essential role in the cell cycle progression is in accord with its ubiquitous distribution among all eukaryotic species. The interaction of PP1c with a large number of non-catalytic subunits and regulatory proteins explains why the three dimensional structure and the amino acid sequence of this protein remained nearly unchanged during evolution. Despite this high level of structural conservation, small differences between the very similar isoenzyme forms in the same organism or between the same isoforms in different species can result in distinct functions. The functional diversification of PP1 is partly related to tissue specific expression of the isoforms or regulatory proteins and the development of isoforms specific protein-protein interactions. The specialized roles for the same enzyme in different organisms can be the consequence of the availability of selected substrate and/or interacting molecules, too. In general, the interaction of PP1 with a full array of proteins is a reoccurring theme and represents the most widespread regulatory mechanism for this enzyme. The number of putative interacting proteins is bewildering, and the list still keeps increasing. Future research will focus on these protein complexes, especially in vertebrate models, where the number of PP1c isoforms is limited and other PP1-like novel type 1 phosphatases are not expressed.

PP1 and its homologues have certainly been best characterized in fungi. The situation in the plants is more complex due to a larger number of PP1c isoforms and the scarce information on PP1 interacting proteins. Thus, a full wealth of information is still missing in most of the cases and besides the well-conserved features some surprises may easily be encountered in the future.

Acknowledgements

The work in the authors' laboratory was supported by the grants OTKA 33050 and 38324.

References

Aggen JB, Nairn AC, Chamnerlin R (2000) Regulation of protein phosphatase-1. Chem Biol 7:R13-23

Alessi D, MacDougall LK, Sola MM, Ikebe M, Cohen P (1992) The control of protein phosphatase-1 targeting subunits. Eur J Biochem 210:1023-1035

Alessi DR, Street AJ, Cohen P, Cohen PTW (1993) Inhibitor-2 functions like a chaperone to fold three expressed isoforms of mammalian protein phosphatase-1 into a conformation with the specificity and regulatory properties of the native enzyme. Eur J Biochem 213:1055-1066

Arundhati A, Feiler H, Traas J, Zhang H, Lunnes PA, Doonan J (1995) A novel *Arabidopsis* type 1 protein phosphatase is highly expressed in male and female tissues and functionally complements a conditional cell cycle mutant of *Aspergillus*. Plant J 7:823-834

Aschenbach WG, Suzuki Y, Breeden K, Prats C, Hirshman MF, Dufresne SD, Sakamoto K, Vilardo PG, Steele M, Kim JH, Jing SI, Goodyear LJ, DePaoli-Roach AA (2001) The muscle-specific protein phosphatase PP1G/R_{GL} (G_M) is essential for activation of glycogen synthase by exercise. J Biol Chem 276:39959-39967

Asztalos Z, von Wegerer J, Wustmann G, Dombrádi V, Gausz J, Spatz HC, Friedrich P (1993) Protein phosphatase 1-deficient mutant Drosophila is affected in habituation and associative learning. J Neurosci 13:924-930

Axton JM, Dombrádi V, Cohen PTW, Glover DM (1990) One of the protein phosphatase 1 isoenzymes in Drosophila is essential for mitosis. Cell 63:33-46

Ayllon V, Cayla X, Garcia A, Fleischer A, Rebollo A (2002) The anti-apoptotic molecules Bcl-xL and Bcl-w target protein phosphatase 1alpha to Bad. Eur J Immunol 32:1847-1855

Bai G, Zhang Z, Amin J, Deans-Zirattu SA, Lee EYC (1988) Molecular cloning of a cDNA for the catalytic subunit of rabbit muscle phosphorylase phosphatase. FASEB J 2:3010-3016

Baksa K, Morawietz H, Dombrádi V, Axton M, Taubert H, Szabó G, Török I, Udvardy A, Gyurkovics H, Szöőr B, Glover D, Reuter G, Gausz J (1993) Mutations in the protein phosphatase 1 gene at 87B can differentially affect suppression of position-effect variegation and mitosis in *Drosophila melanogaster*. Genetics 135:117-125

Barford D (1999) Structural studies of reversible protein phosphorylation and protein phosphatases. Biochem Soc Trans 27:751-766

Barton GJ, Cohen PTW, Barford D (1994) Conservaiton analysis and structure prediction of the protein serine/threonine phosphatase. Eur J Biochem 220:225-237

Bauman AL, Scott JD (2002) Kinase- and phosphatase-anchoring proteins: harnessing the dynamic duo. Nat Cell Biol 4:E203-E206

Bennett D, Alphey L (2002) PP1 binds Sara and negatively regulates Dpp signaling in *Drosophila melanogaster*. Nat Genet 4:419-423

Berndt N (1999) Protein dephosphorylation and the intracellular control of the cell number. Front Biosci 4:d22-d42

Berndt N, Cohen PTW (1990) Renaturation of protein phosphatase 1 expressed at high levels in insect cells using a baculovirus vector. Eur J Biochem 190:291-297

Berndt N, Campbell DG, Caudwell FB, Cohen P, da Cruz e Silva EF, da Cruz e Silva OB, Cohen PTW (1987) Isolation and sequence analysis of a cDNA clone encoding a type-1 protein phosphatase catalytic subunit: homology with protein phosphatase 2A. FEBS Lett 223:340-346

Beullens M, Bollen M (2002) The protein phosphatase-1 regulator NIPP1 is also a splicing factor involved in a late step of spliceosome assembly. J Biol Chem 277:19855-19860

Bhattacharyya MK, Hong Z, Kongkasuriyachai D, Kumar N (2002) *Plasmodium falciparum* protein phosphatase type 1 functionally complements a glc7 mutant in *Saccharomyces cerevisiae*. Int J Parasitol 32:739-747

Bollen M (2001) Combinatorial control of protein phosphatase-1. Trends Biochem Sci 26:426-431

Bollen M, Stalmans W (1992) The structure, role, and regulation of type 1 protein phosphatase. Crit Rev Biochem Mol Biol 27:227-281

Booher R, Beach D (1989) Involvement of a type 1 protein phosphatase encoded by bws1[+] in fission yeast mitotic control. Cell 57:1009-1016

Borgia PT (1992) Roles of orAl, tsE and bimG genes of *Aspergillus nidulans* in chitin synthesis. J Bacteriol 174:384-389

Boudrez A, Beullens M, Waelkens E, Stalmans W, Bollen M (2002) Phosphorylation-dependent interaction between the splicing factors SAP155 and NIPP1. J Biol Chem 277:31834-31841

Brandt H, Lee EYC, Killilea SD (1975) A protein inhibitor of rabbit liver phosphorylase phosphatase. Biochem Biophys Res Commun 63:950-956

Brautigan DL, Sunwoo J, Labbe JC, Fernandez A, Lamb NJC (1990) Cell cycle oscillation of phosphatase inhibitor-2 in rat fibroblasts coincident with p34[cdc2] restriction. Nature 344:74-78

Broustas CG, Grammatikakis N, Eto M, Dent P, Brautigan DL, Kasi, U (2002) Phosphorylation of myosin phosphatase by Raf-1 kinase and inhibition of phosphatase activity. J Biol Chem 277:3053-3059

Cannon JF, Pringle JR, Feicher A, Khalil M (1994) Characterization of glycogen-deficient glc mutants of *Saccharomyces cerevisiae*. Genetics 136:485-503

Carr AN, Schmidt AG, Suzuki Y, del Monte F, Sato Y, Lanner C, Breeden K, King SL, Allen PB, Greengard P, Yatani A, Hoit BD, Grupp IL, Hajjar RJ, DePaoli-Roach AA,

Kranias EG (2002) Type 1 phosphatase, a negative regulator of cardiac function. Mol Cell Biol 22:4124-4135

Ceulemans H, Stalmans W, Bollen M (2002) Regulator-driven functional diversification of protein phosphatase-1 in eukaryotic evolution. BioEssays 24:371-381

Clotet J, Posas F, Casamayor A, Schaaff-Gerstenschlager I, Arino J (1991) The gene *DIS2S1* is essential in *Saccharomyces cerevisiae* and is involved in glycogen phosphorylase activation. Curr Genet 19:339-342

Cohen (1999) Identification of a protein kinase cascade of major importance in insulin signal transduction. Philos Trans R Soc Lond B Biol Sci 354:485-495

Cohen P, Alemany S, Hemmings BA, Resink TJ, Stralfors P, Tung HYL (1988) Protein phosphates-1 and protein phosphatase-2A from rabbit skeletal muscle. Meth Enzymol 159:390-408

Cohen PT (1988) Two isoforms of protein phosphatase 1 may be produced from the same gene. FEBS Lett 223:17-23

Cohen PT (2002) Protein phosphatase 1 – targeted in many directions. J Cell Sci 115:241-256

da Cruz e Silva EF, Fox CA, Ouimet CC, Gustafson E, Watson SJ, Greengard P (1995) Differential expression of protein phosphatase 1 isoforms in mammalian brain. J Neurosci 15:3375-3389

Deng JT, Sutherland C, Brautigan DL, Eto M, Walsh MP (2002) Phosphorylation of the myosin phosphatase inhibitors, CPI-17 and PHI-1, by integrin-linked kinase. Biochem J 367:517-524

Dent P, MacDougall LK, MacKintosh C, Campbell DG, Cohen P (1992) A myofibrillar protein phosphatase from rabbit skeletal muscle contains the β isoform of protein phosphatase-1 complexed to a regulatory subunit which greatly enhances the dephosphorylation of myosin. Eur J Biochem 210:1037-1044

Dohadwala M, da Cruz e Silva EF, Hall FL, Williams RT, Carbonaro-Hall DA, Nairn AC, Greengard P, Berndt N (1994) Phosphorylation and inactivation of protein phosphatase 1 by cyclin-dependent kinases. Proc Natl Acad Sci USA 91:6408-6412

Dombrádi V (1995) Biochemistry and molecular biology of protein phosphatase 1 in *Drosophila melanogaster*. Croatica Chimica Acta 68:569-580

Dombrádi V (1997) Comparative analysis of Ser/Thr protein phosphatases. Trends Comp Biochem Physiol 3:23-48

Dombrádi V, Axton JM, Glover DM, Cohen PTW (1989) Cloning and chromosomal localization of Drosophila cDNA encoding the catalytic subunit of protein phosphatase 1α. Eur J Biochem 183:603-610

Dombrádi V, Axton JM, Barker HM, Cohen PTW (1990) Protein phosphatase 1 activity in Drosophila mutants with abnormalities in mitosis and chromosome condensation. FEBS Lett 275:39-43

Dombrádi V, Axton JM, Brewis ND, da Cruz e Silva EF, Alphey L, Cohen PTW (1990) Drosophila contains three genes that encode distinct isoforms of protein phosphatase 1. Eur J Biochem 194:739-745

Dombrádi V, Cohen PTW (1992) Protein phosphorylation is involved in the regulation of chromatin condensation during interphase. FEBS Lett 312:21-26

Dombrádi V, Mann DJ, Saunders RDC, Cohen PTW (1993) Cloning of the fourth functional gene for protein phosphatase 1 in *Drosophila melanogaster* from its chromosomal location. Eur J Biochem 212:177-183

Doonan JH, MacKintosh C, Osmani S, Cohen P, Bai G, Lee EYC, Morris NR (1991) A cDNA encoding rabbit muscle protein phosphatase 1α complements the Aspergillus cell cycle mutation, *bimG11*. J Biol Chem 266:18889-18894

Doonan JH, Morris NR (1989) The bimG gene of *Aspergillus nidulans*, required for completion of anaphase, encodes a homolog of mammalian phosphoprotein phosphatase 1. Cell 57:987-996

Durfee T, Becherer K, Chen PL, Yeh SH, Yang Y, Kilburn AE, Lee WH, Elledge SJ (1993) The retinoblastoma protein associates with the protein phosphatase type 1 catalytic subunit. Genes Dev 7:555-569

Egloff MP, Cohen PTW, Reinemer P, Barford D (1995) Crystal structure of the catalytic subunit of human protein phosphatase 1 and its complex with tungstate. J Mol Biol 254:942-959

Egloff MP, Johnson DF, Moorhead G, Cohen PTW, Cohen P, Barford D (1997) Structural basis for the recognition of regulatory subunits by the catalytic of protein phosphatase 1. EMBO J 16:1876-1887

Enz R (2002) The metabotropic glutamate receptor mGluR7b binds to the catalytic gamma-subunit of protein phosphatase 1. J Neurochem 81:1130-1140

Eto M, Elliot E, Prickett TD, Brautigan DL (2002) Inhibitor-2 regulates protein phosphatase-1 complexed with NimA-related kinase to induce centrosome separation. J Biol Chem 277:44013-44020

Eto M, Karginov A, Brautigan DL (1999) A novel phosphoprotein inhibitor of protein type-1 phosphatase holoenzymes. Biochemistry 38:16952-16957

Faux MC, Scott JD (1996) More on target with protein phosphorylation: conferring specificity by localization. Trends Biochem Sci 21:312-315

Feng Z, Wilson SE, Peng ZY, Schlender KK, Reimann EM, Trumbly RJ (1991) The yeast *GLC7* gene required for glycogen accumulation encodes a type 1 protein phosphatase. J Biol Chem 266:23796-23801

Fisher EH, Brautigan DL (1982) A phosphatase by any other name: from prosthetic group removing enzyme to phosphorylase phosphatase. Trends Biochem Sci 7:3-4

Fresu M, Bianchi M, Parsons JT, Villa-Moruzzi E (2001) Cell-cycle-dependent association of protein phosphatase 1 and focal adhesion kinase. Biochem J 358:407-414

Genoux D, Haditsch U, Knobloch M, Michalon A, Storm D, Mansuy IM (2002) Protein phosphatase 1 is a molecular constraint on learning and memory. Nature 418:970-975

Gergely P, Erdődi F, Bot G (1986) Purification and regulation of protein phosphatases in latent and active forms. Adv Prot Phosphatases III:49-72

Goldberg J, Huang H, Kwon Y, Greengard P, Nairn AC, Kuriyan J (1995) Three-dimensional structure of the catalytic subunit of protein serine/threonine phosphatase-1. Nature 376:745-753

Guo CY, Brautigan DL, Larner JM (2002) Ionizing radiation activates nuclear protein phosphatase-1 by ATM-dependent dephosphorylation. J Biol Chem 277:41756-41761

Hartshorne DJ, Ito M, Erdődi F (1998) Myosin light chain phosphatase: subunit composition, interactions and regulation. J Muscle Res Cell Motility 19:325-341

Helps NR, Luo X, Barker HM, Cohen PTW (2000) NIMA-related kinase 2 (Nek2), a cell-cycle-regulated protein kinase localized to centrosomes, is complexed to protein phosphatase 1. Biochem J 349:509-518

Helps NR, Cohen PTW, Bahri SM, Chia W, Babu K (2001) Interaction with protein phosphatase 1 is essential for bifocal function during the morphogenesis of the Drosophila compound eye. Mol Cell Biol 21:2154-2164

Huang FL, Glinsman WH (1975) Inactivation of rabbit muscle phosphorylase phosphatase by cyclic AMP-dependent kinase. Proc Natl Acad Sci USA 72:3004-3008

Hubbard MJ, Cohen P (1993) On target with a new mechanism for the regulation of protein phosphorylation. Trends Biochem Sci 18:172-177

Hughes M, Arundhati A, Lunnes P, Shaw PJ, Doonan JH (1996) A temperature-sensitive splicing mutation in the bimG gene of Aspergillus produced an N-terminal fragment which interferes with type 1 protein phosphatase function. EMBO J 15:4574-4583

Ingebritsen TS, Cohen P (1983) Protein phosphatases: properties and role in cellular regulation. Science 221:331-338

Johnson GJ, Brautigan DL, Shriner C, Jaspers S, Arino J, Mole JE, Miller TB Jr, Mumby MC (1987) Sequence homologies between type 1 and type 2A protein phosphatases. Mol Endocrinol 1:745-748

Kim JH, DePaoli-Roach AA (2002) Epinephrine control of glycogen metabolism in glycogen-associated protein phosphatase PP1G/R(GL) knockout mice. J Biochem Mol Biol 35:283-290

Kinoshita N, Ohkura H, Yanagida M (1991) Distinct, essential roles of type 1 and 2A protein phosphatases in the control of the fission yeast cell division cycle. Cell 63:405-415

Kita A, Matsunaga S, Takai A, Kataiwa H, Wakimoto T, Fusetani N, Isobe M, Miki K (2002) Crystal structure of the complex between calyculin A and the catalytic subunit of protein phosphatase 1. Structure 10:715-724

Kitagawa Y, Sasaki K, Shima H, Shibuya M, Sugimura T, Nagao M (1990) Protein phosphatases possibly involved in rat spermatogenesis. Biochem Biophys Res Commun 171:230-235

Koyama M, Ito M, Fen J, Seko T, Shiraki K, Takase K, Hartshorne DJ, Nakano T (2000) Phosphorylation of CPI-17, an inhibitory phosphoprotein of smooth muscle myosin phosphatase, by Rho-kinase. FEBS Lett 475:197-200

Kwon YG, Lee SY, Choi Y, Greengard P, Nairn AC (1997) Cell cycle-dependent phosphorylation of mammalian protein phosphatase 1 by cdc2 kinase. Proc Natl Acad Sci USA 94:2168-2173

Lee EYC, Silberman SR, Ganapathi MK, Petrovic S, Paris H (1980) The phosphoprotein phosphatases: properties of the enzymes involved in the regulation of glycogen metabolism. Adv Cyc Nucleot Res 13:95-131

Lee EYC, Zhang L, Zhao S, Wei Q, Zhang J, Qi ZQ, Belmonte ER (1999) Phosphorylase phosphatase: new horizons for an old enzyme. Front Biosci 4:d270-d285

Lin Q, Buckler ES IV, Muse SV, Walker JC (1999) Molecular evolution of type 1 serine/threonine protein phosphatases. Mol Phylogen Evol 12:57-66

Liu QR, Zhang PW, Zhen Q, Walther D, Wang XB, Uhl GR (2002) KEPI, a PKC-dependent protein phosphatase 1 inhibitor regulated by morphine. J Biol Chem 277:13312-13320

Liu Y, Virshup DM, White RL, Hsu LC (2002) Regulation of BRCA1 phosphorylation by interaction with protein phosphatase 1alpha. Cancer Res 62:6357-6361

MacKintosh C, Garton AJ, McDonnell A, Barford D, Cohen PTW, Tonks NK, Cohen P (1996) Further evidence that inhibitor-2 acts like a chaperone to fold PP1 into its native conformation. FEBS Lett 397:235-238

Maynes JT, Bateman KS, Cherney MM, Das AK, Luu HA, Holmes CFB, James MNG (2001) Crystal structure of the tumor-promoter okadaic acid bound to protein phosphatase-1. J Biol Chem 276:44078-44082

Merlevede W (1985) Protein phosphatase and the protein phosphatases. Landmarks in an eventful century. Adv Prot Phosphatases 1:1-18

Moorhead G, MacKintosh RW, Morrice N, Gallagher T, MacKintosh C (1994) Purification of type 1 protein (serine/threonine) phosphatase by microcystin-Sepharose affinity chromatography. FEBS Lett 356:46-50

Morimoto H, Okamura H, Haneji T (2002) Interaction of protein phosphatase 1 delta with nucleolin in human osteoblastic cells. J Histochem Cytochem 50:1187-1193

Nigavekar SS, Tan YS, Cannon JF (2002) Glc8 is a glucose-repressible activator of Glc7 protein phosphatase-1. Arch Biochem Biophys 404:71-79

Nitschke K, Fleig U, Schell J, Palme K (1992) Complementation of the cs *dis2-11* cell cycle mutant of *Schizosaccaromyces pombe* by a protein phosphatase from *Arabidopsis thaliana*. EMBO J 11:1327-1333

Ohkura H, Adachi Y, Kinoshita N, Niwa O, Toda T, Yanagida M (1988) Cold-sensitive and caffeine-supersensitive mutants of the *Schizosaccharomyces pombe* dis gene implicated in sister chromatid separation during mitosis. EMBO J 7:1456-1473

Ohkura H, Kinoshita N, Miyatani S, Toda T, Yanagida M (1989) The fission yeast *dis2+* gene required for chromosome disjoining encodes one of two putative type 1 protein phosphatases. Cell 57:997-1007

Ohkura H, Yanagida M (1991) S. *pombe* gene *sds22+* essential for a midmitotic transition encodes a leucine-rich repeat protein that positively modulates protein phosphatase-1. Cell 64:149-157

Oliver CJ, Terry-Lorenzo RT, Elliot E, Bloomer WA, Li S, Brautigan DL, Colbran RJ, Shenolikar S (2002) Targeting protein phosphatase 1 (PP1) to the actin cytoskeleton: the neurabin I/PP1 complex regulates cell morphology. Mol Cell Biol 22:4690-4701

Ouimet CC, da Cruz e Silva EF, Greengard P (1995) The α and γ1 isoforms of protein phosphatase 1 are highly and specifically concentrated in dendritic spines. Proc Natl Acad Sci USA 92:3396-3400

Parker L, Gross S, Beullens M, Bollen M, Bennett D, Alphey L (2002) Functional interaction between nuclear inhibitor of protein phosphatase type 1 (NIPP1) and protein phosphatase type 1 (PP1) in Drosophila: consequences of over-expression of NIPP1 in flies and suppression by co-expression of PP1. Biochem J 368:789-797

Páy A, Pirck M, Bögre l, Hirt H, Heberle-Bors E (1994) Isolation and characterization of protein phosphatase from alfalfa. Mol Gen Genet 244:176-182

Piekny AJ, Mains PE (2002) Rho-binding kinase (LET-502) and myosin phosphatase (MEL-11) regulate cytokinesis in the early *Caenorhabditis elegans* embryo. J Cell Sci 115:2271-2282

Piekny AJ, Wissmann A, Mains PE (2000) Embryonic morphogenesis in Caenorhabditis elegans integrates the activity of LET-502 Rho-binding kinase, MEL-11 myosin phosphatase, DAF-2 insulin receptor and FEM-2 PP2c phosphatase. Genetics 156:1671-89

Raghavan S, Williams I, Aslam H, Thomas D, Szoor B, Morgan G, Gross S, Turner J, Fernandes J, VijayRaghavan K, Alphey L (2000) Protein phosphatase 1beta is required for the maintenance of muscle attachments. Curr Biol 10:269-272

Roach PJ (1990) Control of glycogen synthase by hierarchal protein phosphorylation. FASEB J 4:2961-2968

Sasaki K, Shima H, Kitagawa Y, Irino S, Sugimura T, Nagao M (1990) Identification of members of the protein phosphatase 1 gene family in the rat and enhanced expression of protein phosphatase 1α gene in rat hepatocellular carcinomas. Jpn J Cancer Res 81:1272-1280

Shima S, Haneji T, Hatano Y, Kasugai I, Sugimura T, Nagao M (1993) Protein phosphatase 1γ2 is associated with nuclei of meiotic cells in rat testis. Biochem Biophys Res Commun 194:930-937

Silberman SR, Speth M, Nemani R, Ganapathi MK, Dombrádi V, Paris H, Lee EYC (1984) Isolation and characterization of rabbit skeletal muscle protein phosphatase C-I and C-II. J Biol Chem 259:2913-2922

Smith RD, Walker JC (1996) Plant protein phosphatases Annu Rev Plant Physiol Plant Mol Biol 47:101-125

Stalmans W, Mvumbi L, Bollen M (1985) Properties and regulation of glycogen synthase phosphatase in the liver. Adv Prot Phosphatases II:333-353

Szoor B, Gross S, Alphey L (2001) Biochemical characterization of recombinant Drosophila type 1 serine/threonine protein phosphatase (PP1c) produced in *Pichia pastoris*. Arch Biochem Biophys 396:213-218

Takizawa N, Koga Y, Ikebe M (2002) Phosphorylation of CPI17 and myosin binding subunit of type 1 protein phosphatase by p21-activated kinase. Biochem Biophys Res Commun 297:773-778

Tan SL, Tareen SU, Melville MW, Blakely CM, Katze MG (2002) The direct binding of the catalytic subunit of protein phosphatase 1 to the PKR protein kinase is necessary but not sufficient for inactivation and disruption of enzyme dimmer formation. J Biol Chem 277:36109-36117

Terry-Lorenzo RT, Carmody LC, Voltz JW, Connor JH, Li S, Smith FD, Milgram SL, Colbran RJ, Shenolikar S (2002) The neuronal actin-binding proteins, neurabin I and neurabin II, recruit specific isoforms of protein phosphatase-1 catalytic subunits. J Biol Chem 277:27716-27724

Varmuza S, Jurisicova A, Okano K, Hudson J, Boekelheide K, Shipp EB (1999) Spermiogenesis is impaired in mice bearing a targeted mutation in the protein phosphatase 1cγ gene. Dev Biol 205:98-110

Villa-Moruzzi E, Fischer E (1987) Phosphorylase phosphatase from skeletal muscles. Eur J Biochem 169:659-667

Wang H, Brautigan DL (2002) A novel transmembrane Ser/Thr kinase complexes with protein phosphatase-1 and inhibitor-2. J Biol Chem 277:49605-49612

Wissmann A, Ingles J, Mains PE (1999) The Caenorhabditis elegans mel-11 myosin phosphatase regulatory subunit affects tissue contraction in the somatic gonad and the embryonic epidermis and genetically interacts with the Rac signaling pathway. Dev Biol 209:111-127

Yamano H, Ishii K, Yanagida M (1994) Phosphorylation of dis2 protein phosphatase at the C-terminal cdc2 consensus and its potential role in cell cycle regulation. EMBO J 13:5310-5318

Zeke T, Kókai E, Szöőr B, Yatzkan E, Yarden O, Szirák K, Fehér, Zs, Bagossi P, Gergely P, Dombrádi V (2003) Expression of protein phosphatase 1 during the asexual development of *Neurospora crassa*. Comp Biochem Physiol Part B 134:161-170

Zhang Z, Bai G, Deans-Zirattu S, Browner MF, Lee EYC (1992) Expression of the catalytic subunit of phosphorylase phosphatase (protein phosphatase-1) in *Escherichia coli*. J Biol Chem 267:1484-1490

Zhao S, Lee EY (1997) A protein phosphatase-1-binding motif identified by the panning of a random peptide display library. J Biol Chem 272:28368-28372

List of abbreviations:

AKAP 149, AKAP 220 and AKAP 350: A-kinase anchoring proteins 149, 220 and 350

AMP PK: AMP-dependent protein kinase

Bad: proapoptotic factor, Bcl-2 antagonist

Bcl–W, Bcl–X_L: Bcl-2-family proteins

Bcl-2: B cell lymphoma type 2 protein, suppressor of apoptosis

BRCA 1: breast cancer protein 1

CaM PK II: calmodulin-dependent protein kinase II

CK I: casein kinase I

CK II: casein kinase II

CPI17: 17 kDa protein kinase C potentiated inhibitor of protein phosphatase 1 (PPP1R14A)

DAG: diacylglycerol

DARPP-32: dopamine and cAMP-regulated phosphoprotein Mr 32000 (PPP1R1B)

GADD34: growth arrest and DNA damage protein 34 (PPP1R15A)

G_L: liver specific glycogen binding subunit (PPP1R3B)

G_M: muscle specific glycogen binding subunit (PPP1R3A)

GSK3: glycogen synthase kinase 3

HCF: human factor C 1, host cell factor

Hox 11: homeodomain transcription factor

HRX: human trithorax-like protein

hSNF5: human RNA polymerase II transcription factor

I-1: inhibitor 1 (PPP1R1A)

$I_1^{PP2A:}$ inhibitor 1 of protein phosphatase 2A (PHAP1)

I-2: inhibitor 2 (PPP1R2)

I_2^{PP2A} : inhibitor 2 of protein phosphatase 2A (SET, PHAP2)

I-3: inhibitor 3 (PPP1R11)

ILK: integrin linked protein kinase

I-t: testis specific inhibitor of protein phosphatase 1

KEPI: kinase enhanced type 1 protein phosphatase inhibitor

KPI–2: kinase / phosphatase / inhibitor 2 transmembrane protein 2

L5: protein 5 of the large ribosomal subunit

mGluR7b: metabotropic glutamate receptor type 7 splice variant b

MLCK: myosin light chain kinase

MYPT1: myosin binding subunits 1 (PPP1R12A)

MYPT2: myosin binding subunits 2 (PPP1R12B)

Nek2: NIMA related protein kinase 2

NF–L: neurofilament-L

NIPP1: nuclear inhibitor of protein phosphatase 1 (PPP1R8)

NMDA: N-methyl-D-aspartate

p53BP2: p53 binding protein 2 (PPP1R13A)

p70S6K: 70 kDa S6 ribosomal protein kinase

p85: myosin binding subunits of protein phosphatase 1 (PPP1R12C)
PAK: p21 activated protein kinase
PHI 1: protein phosphatase 1 holoenzyme inhibitor 1 (PPP1R14B)
PHI 2: protein phosphatase 1 holoenzyme inhibitor 2
PKA: cAMP-dependent protein kinase or A-kinase
PKC: protein kinase C or C-kinase
PKG: cGMP-dependent protein kinase or G-kinase
PKR: single stranded RNA-dependent protein kinase
PNUTS : phosphatase 1 nuclear targeting subunit (PPP1R10)
PP: protein phosphatase
PPP: phosphoprotein phosphatase
PP1c: the catalytic subunit of protein phosphatase 1
PSF1: polypyrimidine tract-binding protein associated splicing factor
R5: protein phosphatase 1 regulatory subunit 5 (PTG/PPP1R3C)
R6: protein phosphatase 1 regulatory subunit 6 (PPP1R3D)
Raf1: Raf-1 kinase
Rb: retinoblastoma protein
RIPP1: ribosomal inhibitor of protein phosphatase 1
ROCK: Rho-activated protein kinase
SAP155: subunit of the U2 snRNP (Sf3b1)
Sds22: second suppressor of dis2 (PPP1R7)
SR: sarcoplasmic reticulum
Stau: staufen, double-stranded RNA-binding protein
ZIPK: zipper interacting protein kinase

3 Protein phosphatase 2A: A multifunctional regulator of cell signaling

Marc C. Mumby

Abstract

Protein phosphatase 2A (PP2A) is a ubiquitously expressed member of the PPP gene family. Genetic studies in yeast, flies, and mice have shown that PP2A is an essential enzyme that functions in the regulation of numerous cellular processes. Despite a wealth of information on its biochemical and structural properties, the precise roles of PP2A in cell signaling have remained obscure. Recent application of genetic methods and the identification of novel interacting proteins have provided new insights into the physiological functions of PP2A. This review provides an overview of the structural organization of PP2A and a discussion of recent progress in understanding how targeting proteins, especially the regulatory subunits, specify PP2A function.

3.1 Introduction

The ability to switch proteins between two functional states by reversible phosphorylation is a fundamental process in cellular signal transduction. The phosphorylation states of proteins are controlled by the coordinated actions of protein kinases and protein phosphatases. Eight distinct classes of serine/threonine phosphatase have been identified in mammals. Protein serine/threonine phosphatases 1, 2A, 2B/calcineurin, 4, 5, 6, and 7 are members of the mammalian PPP gene family that share a conserved serine/threonine phosphatase domain. Members of this gene family differ in regions outside the phosphatase domain and interact with distinct regulatory proteins. Protein phosphatase 2A (PP2A) is a ubiquitously expressed member of the PPP gene family. Genetic studies in yeast, flies, and mice have shown that PP2A is an essential enzyme that functions in the regulation of numerous cellular processes.

Despite a wealth of information on its biochemical and structural properties, the precise roles of PP2A in cell signaling have remained obscure. Recent application of genetic methods and the identification of novel interacting proteins have provided new insights into the physiological functions of PP2A. This review provides an overview of the structural organization of PP2A and a discussion of recent progress in understanding how targeting proteins, especially the regulatory subunits, specify PP2A function.

Topics in Current Genetics, Vol. 5
J. Arino, D.R. Alexander (Eds.) Protein phosphatases
© Springer-Verlag Berlin Heidelberg 2004

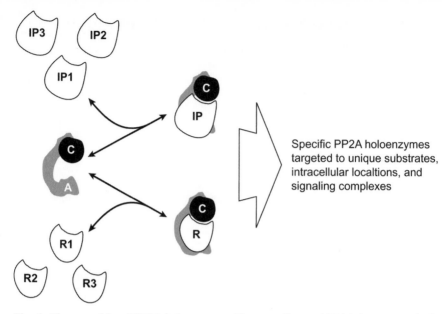

Fig. 1. The assembly of PP2A holoenzymes. The core dimer of PP2A is composed of a complex between the catalytic (C) and scaffold (A) subunits. The core dimer associates with a variety of regulatory subunits (R) and interacting proteins (I) to generate a variety of heterotrimeric holoenzymes. Each individual holoenzyme is targeted to specific substrates, locations, or signaling complexes. This diversity of holoenzymes allows PP2A to function in many aspects of cellular signaling.

3.2 The PP2A core dimer

The term PP2A refers to a collection of phosphatases composed of complexes of conserved catalytic and scaffold subunits with a diversity of regulatory subunits and interacting proteins. The diversity of the regulatory subunits and interacting proteins gives rise to a wide variety of PP2A holoenzymes (Fig. 1). The regulatory subunits and interacting proteins target PP2A to specific substrates and intracellular locations. The generation of many different holoenzymes allows PP2A to participate in a wide variety of signaling pathways. Most, but not all, of the proteins that regulate PP2A interact with a common core enzyme, the core dimmer, composed of a complex between a catalytic subunit (C) and a scaffold subunit (A).

3.2.1 The catalytic subunit

The catalytic subunit of PP2A is composed of 309 amino acids and has a molecular weight of 37 kDa. The amino acid sequence of the phosphatase domain is conserved with other members of the PPP family (Barford et al. 1998; Cohen 1997). The PPP family contains a common metallo-phosphoesterase catalytic domain with two metal ions bound at the active site (Das et al. 1996; Goldberg et al. 1995; Griffith et al. 1995). Regions outside the phosphatase domain of PP2A have little similarity to other members of the PPP family and are likely to be responsible for interactions with PP2A-specific regulatory proteins. The enzymatic core of the PP2A catalytic subunit (residues 1-291) has been modeled on the three-dimensional structure of the protein phosphatase 1 catalytic subunit (Evans et al. 1999). Mutations predicted to disturb the active site of PP1 also impair the catalytic activity of mammalian (Evans et al. 1999; Myles et al. 2001; Ogris et al. 1999) and plant (Lizotte et al. 1999) PP2A catalytic subunits. It is highly likely, that the three-dimensional structure of the enzymatic core of PP2A is very similar to that of protein phosphatase 1.

The catalytic subunit of PP2A undergoes reversible methylation of the carboxy-terminal leucine residue (reviewed in (Mumby 2001). Methylation of Leu-309 is catalyzed by a leucine carboxyl methyltransferase that is highly specific for the catalytic subunits of the PP2A-like phosphatases, including PP2A, PP4, and PP6 (De Baere et al. 1999; Kloeker et al. 1997; Lee and Stock 1993). The methyltransferase that modifies PP2A has been purified and shown to be a unique member of the methyltransferase family (De Baere et al. 1999). The methyl group is removed by a specific methylesterase (Lee et al. 1996; Ogris et al. 1999). Methylation has been reported to cause a modest increase (Favre et al. 1994; Kowluru et al. 1996), a decrease (Zhu et al. 1997), or no change (De Baere et al. 1999; Tolstykh et al. 2000) in the activity of PP2A. The available data are most consistent with the hypothesis that methylation does not directly affect the activity of the catalytic subunit, but rather promotes the assembly of the C subunit into oligomeric complexes.

The role of catalytic subunit methylation in formation of PP2A holoenzymes has been delineated by a combination of biochemical and genetic approaches. Most of the heterotrimeric PP2A in transfected COS cells (Bryant et al. 1999) and brain (Tolstykh et al. 2000) contains methylated C subunit. In contrast, the catalytic subunit in isolated core dimers is not methylated. Mutation of the carboxy-terminal leucine to non-methylatable amino acids results in recovery of the mutant catalytic subunits in core dimers and not heterotrimers (Bryant et al. 1999; Chung et al. 1999). Methylation of the catalytic subunit increases the affinity of the core dimer for the R2 subunit (Tolstykh et al. 2000). However, methylation is not required for holoenzyme formation. AC-R2α trimers still form in the absence of methylation. In addition, methylation effects the interaction of the core dimer with some proteins but not others. For example, methylation has no effect on formation of complexes between the core dimer and polyomavirus middle-tumor antigen (Moreno et al. 2000; Ogris et al. 1997) or the novel PP2A subunits S/G(2) nuclear antigen and striatin (Moreno et al. 2000).

Genetic analysis also supports an important role for methylation in PP2A heterotrimer formation. Inactivation of the PP2A methyltransferase in *S. cerevisiae* causes phenotypes that are similar to those resulting from the loss of heterotrimers due to inactivation of the R2 (*CDC55*) and R5 (*RTS1*) regulatory subunit genes (Wu et al. 2000). These observations are consistent with a decreased ability of non-methylated C subunit to form functional heterotrimers. Interestingly, neither the absence of PP2A methyltransferase activity (Wu et al. 2000) nor mutation of the carboxy-terminal leucine (Evans et al. 1999) is lethal. Thus, at least in yeast, efficient formation of PP2A heterotrimers only appears to be necessary when cells are stressed. While the phenotypes observed in methylation-deficient strains share many features with mutations that inactivate the regulatory subunits, they are not identical. The less severe phenotype of the methylation-deficient strains is consistent with the observation that methylation is not absolutely required for heterotrimer formation.

Regulation of PP2A heterotrimer formation by methylation could have important consequences on dephosphorylation of cellular proteins. Differential effects of methylation on binding of regulatory subunits and interacting proteins could change the distribution of PP2A holoenzymes and alter targeting of PP2A. Changes in C subunit methylation observed during the cell cycle (Turowski et al. 1995) and the ability of cAMP to increase methylation (Floer and Stock 1994) suggest that the PP2A methyltransferase or methylesterase could be subject to regulation. At the present time, however, there are no data showing that PP2A methylation is regulated or that methylation differentially alters PP2A activity toward cellular substrates.

The catalytic subunit of PP2A has also been reported to be phosphorylated. Phosphorylation of a tyrosine residue, tentatively identified as Tyr-307, in the free catalytic subunit of PP2A results in inhibition of phosphatase activity *in vitro* (Chen et al. 1992). Electrophoretic mobility shifts consistent with catalytic subunit phosphorylation are seen in Src kinase transformed cells and following activation of EGF receptors (Chen et al. 1994). It has also been reported that the free catalytic subunit is inhibited by phosphorylation on threonine residues (Guo and Damuni 1993). It is not known whether either of these *in vitro* modifications occurs with the core dimer or PP2A holoenzymes and the physiological roles of these phosphorylation events are unknown.

There are two genes for the catalytic subunit of PP2A (PPP2CA and PPP2CB) that encode the Cα and Cβ isoforms in mammals (Janssens and Goris 2001). Budding yeast contain two genes closely related to mammalian PP2A (*PPH21* and *PPH22*) and a third more distantly related gene (*PPH3*). At least five genes for the PP2A catalytic subunit have been identified in *Arabidopsis* indicating a high degree of complexity of PP2A in plants (Arino et al. 1993). Interestingly, *Drosophila* has a single gene encoding the PP2A catalytic subunit (Mayer-Jaekel et al. 1992). Genetic studies in yeast have shown the two catalytic subunit genes have overlapping functions. While the two mammalian catalytic subunit isoforms are likely to also have some overlapping functions, studies in Cα knockout mice indicate they have distinct functions during embryogenesis. Homozygous Cα null mice exhibit delayed embryonic lethality (Gotz et al. 1998). This lethality occurs even though similar or higher levels of Cβ are expressed during this period. This functional difference between Cα and Cβ may be due to differential localization. During normal embryonic development of the mouse, the Cα isoform is localized on the plasma membrane, while Cβ is localized in the cytoplasm and nucleus

(Gotz et al. 2000). The absence of Cα at the membrane may lead to destabilization of E-cadherin/β-catenin complexes, leading to a failure in mesoderm formation. No differences in catalytic activity or subunit interactions have been detected between the two isoforms (Zhou et al. 2003b). These observations suggest that the differential functions of Cα and Cβ during development are probably due to interactions with distinct targeting proteins.

3.2.2 The scaffold subunit

The scaffold subunit (commonly termed the A subunit) of PP2A functions as an adaptor protein that allows interaction of the regulatory and catalytic subunits. The scaffold subunit has a molecular weight of 65-kDa and is made up of 15 copies of a conserved motif of 38-43 amino acids. These conserved motifs have been termed HEAT repeats based on homology with repeating sequence elements in the Huntingtin, Elongation factor-3, A subunit, and TOR1 proteins. The three-dimensional structure of the alpha isoform of the PP2A scaffold subunit shows that each HEAT-repeat folds into an antiparallel hairpin structure very similar to a model proposed on the basis of secondary structure predictions (Groves et al. 1999; Ruediger et al. 1994). Left-handed rotations at three inter-repeat interfaces generate a novel left-handed superhelical conformation and the A subunit has an overall shape that is reminiscent of a hook. Based on mutagenesis, the intra-repeat turns, especially within repeats 4, 5, 6, 13, 14, and 15, are important in binding regulatory subunits and the catalytic subunit (Ruediger et al. 1994). Residues within these turns form exposed surfaces that are arranged in a continuous ridge along the surface of the A subunit. These hydrophobic surfaces are likely to be sites of interaction with the catalytic and regulatory subunits. The presence of multiple hydrophobic patches along this surface also suggest that protein-protein interactions with the A subunit are likely to involve multiple sites of interaction.

Regulatory subunits bind to common sites on the A subunit. Disruption of the intra-repeat turn within HEAT-repeat 5 abolishes interaction with each of the regulatory subunit families (Ruediger et al. 1994). Site-specific mutagenesis has shown that both common and distinct amino acids within intra-repeat turns 3, 4, 5, and 7 mediate interactions with different regulatory subunits (Ruediger et al. 1999). The different families of regulatory subunits therefore bind to distinct sets of amino acids even though interact with common structures on the A subunit (the intra-repeat turns).

Like the catalytic subunit, the A subunit is present as two isoforms (Aα and Aβ) that are encoded by the PPP2R1A and PPP2R1B genes in humans. While mammals have two A subunit genes both Drosophila (Mayer-Jaekel et al. 1992) and yeast (Stark 1996) have a single gene. The Aα and Aβ subunits are 86% identical in amino acid sequence. The most striking difference between the two proteins is the presence of a unique twelve amino acid sequence at the amino-terminus of Aβ. Aβ mRNA is expressed at lower levels than Aα in most tissues and cell types (Hemmings et al. 1990). Similarly, Aα protein is abundant in all tissues while Aβ levels are at least ten-fold lower (Zhou et al. 2003a). Although the two A subunit isoforms are very similar, the subtle differences in amino acid sequence, especially the N-terminal extension in Aβ, suggest they have distinct functions. The expression of the A subunit switches from Aβ to Aα during early Xenopus development (Bosch et al. 1995). Biochemical differences between the two isoforms have also been identified. The Aβ isoform binds with lower affinity to the catalytic and regulatory subunits than the Aα subunit. Most strikingly, the R2 family of regulatory subunits, which is the major family associated with Aα, does not bind to Aβ (Zhou et al. 2003a). While physiological differences have yet to be identified, these studies suggest that the two A subunit isoforms play distinct roles in directing PP2A function.

An intriguing aspect of the Aβ subunit is its potential role as a tumor suppressor. The human Aβ gene is located in the q22-24 region of chromosome 11, an area where loss of heterozygosity occurs in a wide range of tumors. Analysis of the PPP2R1B gene in cancer-derived cells revealed the presence of mutations in 15% of primary lung tumors, 6% of lung tumor-derived cell lines, and 15% of primary colon tumors. Based on the frequency of mutation, the Aβ gene was postulated to be a tumor suppressor (Wang et al. 1998). Subsequent studies found mutations in the Aβ gene in colorectal (Takagi et al. 2000) and breast cancers (Calin et al. 2000). Loss of functional Aβ due to mutation or deletion would decrease the levels of Aβ-containing PP2A heterotrimers. This would presumably result in increased phosphorylation of proteins dephosphorylated by PP2A forms containing the Aβ scaffold subunit. A number of the mutations found in tumor cells are predicted to alter the structure of the Aβ subunit or block interactions with other subunits (Groves et al. 1999). For example, a large deletion (amino acids 230-518) within the PPP2R1B gene in one small cell lung carcinoma leads to expression of a protein that is defective in interaction with the catalytic subunit (Wang et al. 1998). Most, but not all, of the mutations identified in tumor cells cause a loss of binding of Aβ to the R3 (PR72) regulatory subunit or the catalytic subunit when assayed *in vitro* (Ruediger et al. 2001a). While mutations in Aβ have been detected in several types of tumors, the frequency of Aβ mutations is low. Other studies failed to detect somatic mutations in the PPP2R1B gene in 72 ovarian tumors despite the fact that loss of heterozygosity within 11q22-24 is a common event in these tumors (Campbell and Manolitsas 1999). Thus, the role of the PPP2R1B gene as a tumor suppressor remains uncertain and awaits the demonstration that the wild type protein can restore a normal phenotype in cells containing Aβ mutations.

A low frequency of mutation has also been found in the Aα gene in melanoma, breast cancer, and colon cancer samples (Calin et al. 2000). Each of the four identified Aα mutations leads to loss of interaction between Aα and the regulatory or catalytic subunits *in vitro* (Ruediger et al. 2001b). While the Aα gene is not associated with loss of heterozygosity, these findings raise the possibility that mutations in the alpha isoform of the scaffold subunit can also lead to altered PP2A activity and abnormal cell growth.

Table 1. Nomenclature of mammalian PP2A subunits

Name	Human gene symbol	Common names	Chromosomal location	LocusID/Accession number
Catalytic subunits				
Cα	PPP2CA	Cα, PP2A$_{C\alpha}$	5q23-31	5515
Cβ	PPP2CB	Cβ, PP2A$_{C\beta}$	8p21-12	5516
Scaffold subunits				
Aα	PPP2R1A	Aα, PR65α	19q13	5518
Aβ	PPP2R1B	Aβ, PR65β	11q23	5519
R2 subunits				
R2α	PPP2R2A	Bα, PR55α	8p21	5520
R2β	PPP2R2B	Bβ, PR55β	5q31-5q33	5521
R2γ	PPP2R2C	Bγ	4p16	5522
R2δ		Bδ		AF180350
R3 subunits				
R3α	PPP2R3A	PR72, B"	3q22	5523
R3β	PPP2R3B	PR59	Xp22 and Yp11	192283
R3γ		PR48		28227
R4 subunit	PPP2R4	PTPA	9q34	5524
R5 subunits				
R5α	PPP2R5A	B'α, B56α	1q32	5525
R5β	PPP2R5B	B'β, B56β	11q12-q13	5526
R5γ	PPP2R5C	B'γ, B56γ	14q32	5527
R5δ	PPP2R5D	B'δ, B56δ	6p21	5528
R5ε	PPP2R5E	B'ε, B56ε	14q23	5529

3.3 PP2A regulatory subunits

The ability of the AC core complex to interact with regulatory subunits is critical for PP2A function. For the purposes of this review, proteins that interact with PP2A are separated into "classical" PP2A regulatory subunits and PP2A interacting proteins. The classical subunits interact directly with the AC core complex and many of them were initially identified as components of native PP2A oligomers. PP2A-interacting proteins have generally been identified using other methods. While some PP2A-interacting proteins associate with the AC core complex in a manner analogous to the regulatory subunits, others associate directly with the catalytic subunit or with heterotrimeric forms of the enzyme.

Four families of PP2A regulatory subunits have been identified. A numerical strategy has been adopted for naming the human genes for the subunits of protein serine/threonine phosphatases including PP2A (Cohen 1994). In this nomenclature, the genes encoding the regulatory subunits are designated PPP2R2 to PPP2R5. These names are also the official gene symbols assigned to these proteins. The nomenclature used for the PP2A regulatory subunits in this review is based on their official gene names. The B/PR55 subunit family is R2, the PR72 subunit is R3, the PTPA protein is R4, and the B'/B56 subunit family is R5. Within each gene family, individual isoforms are designated by Greek letters (e.g. R2α). A list of currently known mammalian PP2A subunits is presented in Table 1.

3.3.1 PP2A holoenzyme formation

The major functional forms of protein phosphatase 2A are heterotrimers composed of complexes between the core dimer and one of the regulatory subunits. The mammalian catalytic subunit has also been detected in dimeric complexes with proteins other than the A subunit (discussed below). However, most of the catalytic subunit is present in complexes with the A subunit due to the very high affinity of this interaction (Kamibayashi et al. 1992; Turowski et al. 1997). Formation of heterotrimers involves the interaction of regulatory subunits with the core dimer in a process that can be regulated by catalytic subunit methylation (see above). The molecular mechanisms involved in the binding of regulatory subunits and differential generation of holoenzymes are not known. However, it seems likely that one mechanism involves competition of regulatory subunits for a limiting pool of core dimers.

PP2A regulatory subunits may compete for a limiting amount of core dimer in a simple process of equilibrium dynamics. Increases in the levels of a particular regulatory subunit would change the distribution of PP2A holoenzymes by shifting the equilibrium toward heterotrimers containing that subunit. Whether or not there is a limiting amount of core dimer has been a controversial issue. Although core dimers are generally present in PP2A preparations, PP2A can be isolated under conditions where very little core dimer is detected (Tung et al. 1985b; Zolnierowicz et al. 1994). These data indicate that most of the PP2A in cells and tissues is present as heterotrimeric holoenzymes. If this were the case, the pool of core dimers would likely be small. In contrast, there are data indicating that a substantial pool of core dimers exists. Two thirds of the PP2A in erythrocyte cytosol can be isolated as the core dimer (Umi et al. 1988) and one third of the PP2A in cell lysates is present as core dimers under conditions that minimize proteolysis and dissociation (Kremmer et al. 1997). However, it is difficult to prove that the core dimer detected in these *in vitro* studies is not derived by post-extraction dissociation. Recent data has indicated that it is unlikely that significant amounts of core dimer accumulate *in vivo*. Loss of any one of the PP2A regulatory subunits leads to a corresponding loss of the A and C subunits in *Drosophila* S2 cells where the regulatory subunits have been knocked down using RNA interference (Li et al. 2002a; Silverstein et al. 2002). Ablation of all four regulatory subunits results in a complete loss of PP2A from these cells. Thus, even in the presence of continued expression of A and C subunits, the core dimer is lost from these cells. These experiments argue that the only stable form of PP2A is the heterotrimer. This idea is supported by the observation that the A and C subunits are rapidly degraded when holoenzymes cannot form (Silverstein et al. 2002). Rapid degradation of free subunits is also supported by data showing that mutant forms of the

R2γ subunit that cannot bind to the core dimer are rapidly degraded by ubiquitin-dependent proteolysis in mammalian cells (Strack et al. 2002). While there is some evidence that the levels of the PP2A catalytic subunit can be controlled at the level of translation (Baharians and Schonthal 1998), protein degradation is likely to be a major mechanism controlling the levels of PP2A subunits. It seems likely that the constituency of PP2A holoenzymes at any given time is dictated by the competition of regulatory subunits for a limiting amount of the AC core dimer and that the levels of free subunits and core dimers are kept low by rapid degradation.

The PP2A regulatory subunits have little, or no, sequence similarity despite the fact that they interact with the same core dimer. Accordingly, it has been not been possible to predict regions involved in interaction on the basis of sequence similarity. Interaction of regulatory subunits with the core dimer is likely to be mediated by the assembly of multiple hydrophobic patches or charged amino acids into three-dimensional structures that align with corresponding sites on the surfaces of the A and C subunits. The lack of sequence similarity suggests that these proteins have developed multiple structural organizations to assemble these sites into the appropriate configuration. In fact, the PP2A regulatory subunit families are classified into different structural families by protein fold prediction algorithms (Strack et al. 2002).

The structural organization utilized by one regulatory subunit family may be the β-propeller fold adopted by WD-repeat proteins. WD-repeat proteins comprise a large family of proteins with diverse functions. These proteins contain 4-16 copies of the WD-repeat, a semi-conserved motif of 44-60 amino acids. The R2α subunit was originally identified as a WD-repeat protein containing 6 WD repeats (Smith et al. 1999). The first WD repeat lies within the highly conserved core of the R2 family. Consequently, the other members of the R2 family are also predicted to contain WD-repeats (Strack et al. 2002). The structures of two WD-repeat proteins (the Gβ subunit of heterotrimeric G proteins and the p40 subunit of the ARP2/3 complex) have been determined and many other WD proteins, including R2 subunits, may adopt a similar β-propeller fold. The common functional theme of the β-propeller proteins is the presence of multiple interaction surfaces and formation of reversible complexes with several proteins simultaneously. Consistent with this function of WD-repeat proteins, the R2 subunits interact simultaneously with both the A and C subunits (Kamibayashi et al. 1992; Ruediger et al. 1994). In addition, the R2 subunits mediate additional interactions with substrates and other proteins (discussed below). Additional support for the potential importance of WD-repeats in PP2A-associated proteins has come from the isolation of novel PP2A-interacting proteins that also contain WD-repeats. Striatin and SG2NA are two highly related calcium/calmodulin-binding proteins that contain 7 WD-repeats and form complexes directly with the core dimer (Moreno et al. 2000). The four members of the R2 family, striatin, and SG2NA therefore comprise a group of six WD-repeat proteins that all interact with the PP2A core dimer.

While WD-repeats and β-propeller structures are likely to be important in the interaction of some regulatory subunits with PP2A, it is not the only strategy used to interact with the core dimer. None of the other PP2A regulatory subunits or interacting proteins contain WD-repeats. Therefore, there must be additional structural features that allow binding of proteins to PP2A. Accordingly, Li and Virshup (Li and Virshup 2002) have identified two independent A subunit binding domains that are shared between three families of PP2A regulatory subunits. These two domains correspond to amino acids 188-292 (104 amino acids) and 329-386 (57 amino acids) of the R5α subunit. Common features of these domains include conserved hydrophobic, charged, and polar amino acids. The individual A subunit binding domains from R2α, R3α, and R5α all bind to the A subunit *in vitro*. These first and second A subunit binding domains encompass predicted WD-repeats 3-5 and 5-6 respectively within the R2γ subunit (Strack et al. 2002). The importance of these A subunit binding domains within the context of the intact proteins remains to be established. However, this study provides the first evidence that common structural features are involved in the assembly of PP2A heterotrimers.

3.3.2 The R2 family

The mammalian R2 family currently contains four characterized isoforms (Table 1). These isoforms are 79-87% identical and differ primarily at their N-termini. R2α mRNA is ubiquitously expressed and is the most abundant PP2A regulatory subunit in many cells and tissues. Although expressed in several tissues, the R2β and R2γ isoforms are only expressed at high levels in brain and testis. Three splice variants of R2β have been identified (Schmidt et al. 2002). R2δ is most similar to R2α. Like R2α, R2δ is widely expressed but has a distinct subcellular localization. A detailed analysis of the expression of the different isoforms in brain has shown a largely overlapping pattern of expression for the α and β isoforms (Strack et al. 1998). Although both subunits are expressed in the same populations of neurons, the α and β isoforms are present at different levels in different types of neurons. R2α is distributed mainly in the cell body and is localized in both the cytosol and nucleus. In contrast, the β isoform is excluded from the nucleus and is localized in axons and dendrites in addition to the cell body. A substantial fraction of the γ isoform is associated with an insoluble cytoskeletal fraction, whereas R2α and β are primarily soluble. Another distinction is that R2α and R2β are associated with microtubules while R2γ is not (Price and Mumby 2000). Expression of R2 subunit mRNA is also differentially regulated during development (Strack et al. 1998). An important role of the R2 family of subunits in neural, and other, tissue is to target the holoenzyme to cytoskeletal proteins. The associations of PP2A with the neuronal microtubule binding protein tau (Sontag et al. 1996), neurofilament proteins (Strack et al. 1997), and vimentin (Turowski et al. 1999) are all mediated by R2 subunits. The R2 subunit also mediates the interaction of PP2A with calmodulin-dependent protein kinase IV (Westphal et al. 1998).

Genetic studies have shown that the yeast R2 subunit mediates functions of PP2A in the cell cycle. The single R2 subunit of the budding yeast, *S. cerivisiae*, is encoded by the *CDC55* gene, which is >50% identical to mammalian R2 subunit isoforms. The phenotypes of mutant alleles of R2/*CDC55* indicate that the R2/Cdc55p protein plays multiple roles during mitosis. Inactivating mutations or deletion of R2/*CDC55* causes cold sensitive phenotypes characterized by defects in bud morphogenesis and cell separation during anaphase (Healy et al. 1991). Additional mutant alleles demonstrate an additional requirement for the R2/Cdc55p protein in the mitotic spindle-assembly checkpoint (Minshull et al. 1996; Wang and Burke 1997). The cold sensitivity and budding defects of R2/*cdc55* mutants are suppressed by expression of a mutant form of *S. cerevisiae* cyclin-dependent protein kinase 2 (*cdc28F19*) that cannot be phosphorylated on the inhibitory tyrosine phosphorylation site (Wang and Burke 1997). Under some circumstances *cdc28F19* also restores the spindle-assembly defects in R2/*cdc55* mutants (Minshull et al. 1996), while under others it does not (Wang and Burke 1997). Together, these results indicate that the role of the AC-R2/Cdc55p holoen-zyme form of PP2A in both the budding and spindle-assembly checkpoints in-volves regulation of cdc28p activity.

Dephosphorylation of Tyr-19 by the yeast Cdc25 phosphatase homolog Mih1p normally activates Cdc28p allowing progression through mitosis. The inhibitory tyrosine phosphorylation site on Cdc28p is phosphorylated by the budding yeast homolog of the Wee1 protein kinase, Swe1p. Phosphorylation of Tyr-19 is ele-vated in R2/*cdc55*-null cells (Minshull et al. 1996) indicating that R2/cdc55p is normally involved in promoting dephosphorylation of Tyr-19. Since PP2A does not directly dephosphorylate tyrosine, the most likely targets for R2/Cdc55p are the Cdc25 phosphatase Mih1p or the Wee1 kinase Swe1p. While deletion of R2/*CDC55* causes some increase in phosphorylation of Mih1p that could lead to inhibition of Tyr-19 phosphatase activity, the major effect is to cause accumula-tion of the Swe1p tyrosine kinase (Yang et al. 2000). Although the mechanism is not known, the accumulation of Swe1p may be due to decreased degradation of Swe1p mediated by the Met30-SCF ubiquitin conjugating system responsible for proteasome mediated degradation of Swe1p.

While the *S. cerevisiae* R2/*CDC55* gene is involved in regulation of Cdc25 and the cell cycle, other genetic data indicate it also has additional roles. A tempera-ture sensitive mutation in *cdc20* is partially suppressed by deletion of *cdc55* (Wang and Burke 1997). *CDC20* encodes a protein containing WD-repeats that is involved in mitotic spindle assembly. There are also genetic interactions between *CDC55* and the *YCK1* and *YCK2* genes that encode casein kinase-1-like proteins in *S. cerevisiae*. A *yck1 yck2* double deletion strain exhibits a morphological phe-notype similar to a *cdc55* null mutant. Deletion of *cdc55* is lethal in combination with *yck1 yck2^{ts}* mutations (Robinson et al. 1993). The idea that R2 and casein kinase 1 may function in the same pathway is supported by a report that mammal-ian PP2A can physically associate with casein kinase 1 (Heriche et al. 1997). Mu-tation of *cdc55* also shows a synthetic phenotype with *ELM1*, whose deletion re-sults in constitutive pseudohyphal growth (Blacketer et al. 1993). Deletion of both *elm1* and *cdc55* causes a more severe growth defect and a block in cytokinesis. Finally, a *cdc55* null mutant is suppressed by an extragenic mutation in *BEM2*, a gene necessary for growth polarity during bud emergence (Healy et al. 1991).

Studies in fission yeast, *Schizosaccharomyces pombe*, and mammalian cells also support a role for the R2 subunit in regulating Cdc25 activity. Deletion of the single R2 subunit encoded by the *pab1* gene in *S. pombe* causes multiple defects (Kinoshita et al. 1996). While R2/*pab1* is not essential for viability, its disruption causes alterations in cell shape, disruption of normal of actin and microtubule distribution, and delayed cytokinesis. The Vpr protein of the human immunodeficiency virus type 1 causes G2 arrest in *S. pombe*, which has been used as a model system to study Vpr. The Vpr-mediated G2 arrest is partially suppressed by deletion of either the Wee1 protein kinase or the Cdc25 phosphatase (Elder et al. 2001). Deletion of the R2/*pab1* gene itself also attenuates Vpr-mediated G2 arrest. These data suggest that Vpr acts to upregulate a PP2A activity, mediated by the R2/*pab1* regulatory subunit that activates Wee1 and inhibits Cdc25, resulting in inhibition of Cdc2 protein kinase activity and G2 arrest. The Vpr protein also appears to regulate Cdc25 via the R2 subunit of PP2A in human HEK 293 cells (Hrimech et al. 2000). Vpr binds to the R2 subunit when expressed in these cells and targets the PP2A complex to the nucleus where it enhances the dephosphorylation (and inactivation) of Cdc25. The subsequent enhancement of Cdc2 phosphorylation on the inhibitory tyrosine site would result in an inability to proceed through G2.

An interesting feature of the regulation of Cdc25 by PP2A is the potential involvement of the Pin-1 peptidyl-prolyl *cis/trans* isomerase. Pin-1 specifically catalyzes isomerization of phosphorylated Ser/Thr-Pro bonds from the *cis* to the *trans* configuration. Since Ser/Thr-Pro motifs are targets for the cyclin-dependent kinases, they play important roles in regulating the cell cycle. The PP2A heterotrimer containing the R2α regulatory subunit (AC-R2α) has been shown to be the major phosphatase that dephosphorylates these proline-directed phosphorylation sites (Ferrigno et al. 1993; Zhou et al. 2000). Furthermore, the AC-R2α holoenzyme effectively dephosphorylates only the *trans* pSer/Thr-Pro isomer (Zhou et al. 2000). The critical regulatory phosphorylation sites in Cdc25C (pThr(48)-Pro and pThr(87)-Pro) are in the *cis* configuration. Pin-1 catalyzes the isomerization of these prolyl residues to the *trans* configuration and allows these sites to become substrates for AC-R2α (Zhou et al. 2000). It seems likely that functional interactions between Pin-1 and PP2A play an important role in controlling dephosphorylation of Cdc25 as well as a number of other important cell signaling proteins.

Mutations in the R2 subunit of *Drosophila* PP2A also indicate that this regulatory subunit plays multiple roles including the regulation of the cell cycle. The *aar1* mutation (for abnormal anaphase resolution) contains a P element insertion in the *Drosophila* R2 gene (Mayer-Jaekel et al. 1993). Mutant *aar1* flies die as larvae or early adults with over-condensed chromosomes and abnormal separation of sister chromatids during anaphase. The *aar1* phenotype is reminiscent of the mitotic spindle-assembly checkpoint defects seen in yeast R2 mutants. A second mutant allele of *Drosophila* R2 was identified in a screen for wing imaginal disc abnormalities (Uemura et al. 1993). Also the result of a P element insertion, these mutants termed twins[P], die at an early pupal stage and show pattern duplication of wing imaginal discs, indicative of a cell fate determination defect. Flies harboring a weaker allele, twins[55], survive with a duplication of sensory bristles, the result of a defect in cell fate determination of sensory neurons (Shiomi et al. 1994). The severity of these mutant phenotypes correlates with the level of reduction in R2 protein levels. These data indicate that the *Drosophila* R2 subunit involved in both cell cycle regulation and cell fate determination.

An important role of the R2 subunit is to target PP2A to pathways that regulate MAP kinase activity. Overexpression of the small tumor antigen of SV40 virus disrupts endogenous PP2A complexes containing the R2 subunit. Loss of the R2 subunit leads to enhanced activation of MAP kinase in response to growth factors in some, but not all cell types (Frost et al. 1994; Sontag et al. 1993). The small-t antigen effects may involve protein kinase C and the PI-3 kinase pathway (Sontag et al. 1997). Depletion of the R2 subunit in Drosophila S2 cells using RNA interference also leads to enhanced activation of MAP kinase in response to insulin (Silverstein et al. 2002). These studies indicate that the R2 subunit can play a negative role in regulating MAP kinase activity.

Additional evidence for a role of the R2 subunit in regulating the Ras-Raf-MAP kinase pathway has come from genetic studies in *Caenorhabditis elegans*. The R2 subunit gene, *sur-6*, was isolated as a suppressor of the multivulva phenotype caused by an activated *ras* mutation in *C. elegans* (Sieburth et al. 1999). The *sur-6* protein is the only R2 subunit in *C. elegans* and is most closely related to the mammalian R2α isoform. *Sur-6* mutations by themselves do not cause defects in vulval development but they enhance the vulval induction defects and larval lethality caused by weak mutant alleles of the *C. elegans* Raf protein kinase (*lin-45*). These genetic interactions indicate that *sur-6* (R2) mutations reduce signaling through the Ras pathway at a point downstream of, or in parallel to *ras* and upstream of Raf. Thus, in contrast to the studies described above in *Drosophila* and mammalian cells, the R2 gene appears to be positive regulator of the MAP kinase pathway in the *C. elegans* vulva development pathway. Analysis of double mutants suggests that *sur-6* (R2) and the Kinase suppressor of Raf (KSR) act together in a common pathway to positively regulate signaling through the Ras-Raf-MAPK pathway. Raf and KSR are both phosphoproteins that could be targets of the AC-R2 complex.

In humans, expansion of a novel CAG trinucleotide repeat within the R2β gene has been associated with a form of autosomal dominant spinocerebellar ataxia termed SCA12 (Holmes et al. 1999). SCA12 is caused by neurodegeneration with atrophy of the cortex and cerebellum. The CAG expansion lies near the transcription start site of the R2β gene and could alter expression of this brain-specific isoform. Although mutations in other genes could be important in SCA12, the correlation between the presence of the CAG expansion and disease in a pedigree of SCA12 patients, and the lack of the expansion in non-affected individuals indicates that altered expression of R2β is likely to be causative for this disease. While the mechanism of R2β loss in SCA12 is unknown, this data suggests that R2β is involved in maintenance of neuronal viability.

PP2A holoenzymes containing R2 subunits are targets of the small DNA tumor viruses SV40 and polyoma. The small tumor antigen of SV40, the small tumor antigen of polyoma, and the middle tumor antigen of polyoma interact with the core dimer of PP2A (Pallas et al. 1990; Walter et al. 1990). SV40 small tumor antigen can displace R2 subunits, but not R5 subunits from PP2A holoenzymes, and over-expression of small tumor antigen in cells causes conversion of the AC-R2 complexes to AC-small tumor antigen complexes (Sontag et al. 1993; Yang et al. 1991). Replacement of the R2 subunits with small tumor antigen presumably leads to altered PP2A targeting in SV40 infected cells. Functional consequences of small tumor antigen-mediated changes in PP2A holoenzyme composition include activation of the MAP kinase pathway, activation of the cyclin D promoter, activation of PKCζ, and activation of NF-κB (Sontag et al. 1993; Sontag et al. 1997; Watanabe et al. 1996).

The E4orf4 protein of adenovirus also interacts with PP2A (Kleinberger and Shenk, 1993). A major effect of E4orf4 is the induction of p53-independent apoptosis. The apoptotic activity of E4orf4 is dependent on interaction with PP2A since mutations that disrupt interaction with PP2A also disrupt its apoptotic activity (Shtrichman et al. 1999). Interaction of the E4orf4 protein with PP2A is mediated by regulatory subunits. E4orf4 is capable of interacting with multiple regulatory subunits including R2α and all of the isoforms of the R5 family in cells over-expressing HA-tagged regulatory subunits (Shtrichman et al. 2000). However, only R2α is capable of mediating the apoptotic effect of E4orf4 and only R2 subunits have been identified in E4orf4 complexes (Kleinberger and Shenk 1993). The observation that E4orf4-induced apoptosis requires interaction with PP2A holoenzymes containing the R2 subunits argues that this form of PP2A is involved in pathways that regulate cell survival.

3.3.3 The R3 family

The second family of PP2A regulatory subunits are encoded by the human PPP2R3 genes. Three isoforms of this family have been identified (Table 1.) The gene encoding the R3α subunit produces two alternatively spliced transcripts encoding proteins of 72- and 130-kDa (Hendrix et al. 1993). Both the 72 and 130 kDa isoforms are selectively expressed in skeletal muscle and heart. In vitro, the R3α subunit suppresses the activity of the AC core dimer toward exogenous substrates and increases sensitivity of the enzyme to polycations (Waelkens et al. 1987). A second member of the R3 family (R3β/PR59) was identified on the basis of its interaction with the retinoblastoma-related protein, p107 (Voorhoeve et al. 1999). The third member of the family, R3γ/PR48, was identified through an interaction with the human Cdc6 protein, a component of the DNA replication machinery (Yan et al. 2000). The proteins encoded by the R3 family have a highly conserved carboxy-termini that contain two regions that are nearly perfect matches to the highly conserved EF-hand calcium-binding motif found in many calcium-binding proteins. These motifs were shown to be calcium-binding sites in R3α-72 (Janssens et al. 2003). While the function of calcium binding to R3α is not known, the data indicate that binding of calcium induces a conformational change in R3α that may be important for interaction with the PP2A scaffold subunit and for proper nuclear targeting.

The physiological functions of the R3 family are not as well characterized as those of the R2 family. Yeast does not appear to have genes encoding R3 subunits making it impossible to study the function of R3 in these organisms. The mRNA for the single R3 gene has been knocked out in *Drosophila* S2 cells using RNA interference (Li et al. 2002a; Silverstein et al. 2002). Loss of the R3 protein in these short-term experiments has no obvious effects on cell growth or viability. Most clues about R3 function have come from the identification of proteins that interact with R3 proteins (Silverstein et al. 2003). As mentioned above, R3β/PR59 and R3γ/PR48 interact with p107 and Cdc6 respectively, suggesting these isoforms play roles in targeting PP2A to components of the cell cycle machinery. Forced overexpression of either of these proteins results in cell cycle arrest (Voorhoeve et al. 1999; Yan et al. 2000). The R3 subunit has also been isolated in complexes with protein phosphatase 5 (Lubert et al. 2001) and with the giant scaffolding protein CG-NAP (centrosome and Golgi localized PKN-associated protein)(Takahashi et al. 1999). These observations suggest that R3 might play a role in cross talk between PP2A and PP5, and in a large signaling complex involved in regulating the centrosome and Golgi apparatus.

3.3.4 The R4 family

Like members of the R2 and R3 families, R4 is a protein that appears to enhance the activity of the PP2A core dimer toward a subset of phosphoproteins. The common name for R4 is phosphotyrosyl phosphatase activator protein (PTPA). R4/PTPA is a 37-kDa protein that enhances the nominal tyrosine phosphatase activity of PP2A (Janssens et al. 1998). R4/PTPA is expressed in most tissues (Cayla et al. 1994; Janssens et al. 2000). Homologs of R4/PTPA have been identified in *Drosophila* and *S. cerevisiae* (Van et al. 1998). The tyrosine phosphatase activity of the core dimer of PP2A is stimulated by PTPA, while the heterotrimeric forms are not. This suggests that R4/PTPA competes with other regulatory subunits for interaction with the core dimer. However, physical complexes between R4/PTPA and the core dimer have not been demonstrated. The protein contains two domains that are homologous to ATPases, and ATP hydrolysis is required for the effects on PP2A activity (Cayla et al. 1994). The possible ATPase activity suggests that R4/PTPA might act as a molecular chaperone, altering the active site of PP2A to accommodate phosphotyrosine in addition to phosphoserine and phosphothreonine. R4/PTPA is required in excess over the core dimer to reach maximal phosphotyrosine phosphatase activity indicating that R4/PTPA acts in a stoichiometric rather than a catalytic manner (Cayla et al. 1990).

The physiological functions of the R4/PTPA protein are unknown. The *S. cerevisiae* open reading frame ORF *YIL153w* encodes a protein that is 38% identical to human R4/PTPA. This gene has been termed *YPA1*, *RRD1*, and *NCS1* (Mitchell and Sprague Jr. 2001; Rempola et al. 2000; Van Hoof et al. 2001). A second gene, ORF *YPL152w*, encodes a protein that is 36% identical to vertebrate R4/PTPA but only 25% identical to Ypa1p and has been termed *YPA2*, *RRD2*, and *NOH1*. Disruption of both the *YIL153w/YPA1* and *YPL152w/YPA2* genes is lethal while disruption of either gene alone has no effect on viability at 30°C (Rempola et al. 2000; Van Hoof et al. 2000). Disruption of *YIL153w/YPA1* alone causes a set of pleiotropic phenotypes including: hypersensitivity to calcium, vanadate, ketoconazole, and cycloheximide; resistance to caffeine, nocodazole, and rapamycin; hypersensitivity to DNA damage; enhanced progression through G1, and aberrant bud morphology (Ramotar et al. 1998; Rempola et al. 2000; Van Hoof et al. 2000; Van Hoof et al. 2001). Disruption of *YPL152w/YPA2* causes weaker effects and a more limited set of phenotypes. Evidence for possible interaction of the *YIL153w/YPA1* and *YPL152w/YPA2* genes with PP2A is indirect. Overexpression of one of the yeast PP2A catalytic subunits (*PPH22*) suppresses the lethal phenotype of the *YPA1/YPA2* double mutant (Rempola et al. 2000). Similarities in mutant phenotypes also argue that the yeast R4/PTPA genes interact with PP2A. These studies have suggested that R4/PTPA is required for a subset of functions performed by PP2A (Van Hoof et al. 2000; Van Hoof et al. 2001). No physical or genetic interactions between yeast R4/PTPA and PP2A have been demonstrated. While yeast R4/PTPA may regulate PP2A, possibly through actions as a molecular chaperone, an alternative role for yeast R4/PTPA has been proposed. Genetic and biochemical evidence suggest that the PP2A-like phosphatase Sit4 is the target for the R4/PTPA gene *YIL153w/YPA1*. Yil153w/Ypa1p and YPL152w/Ypa2p both bind to the catalytic subunit of Sit4 *in vitro* and can be immunoprecipitated with Sit4p, but not with PP2A, from cell extracts (Mitchell and Sprague Jr. 2001). Thus, while their roles are still uncertain, the genetic analyses in yeast suggest that the R4/PTPA proteins have targets in addition to PP2A.

3.3.5 The R5 family

The R5 regulatory subunits are components of a PP2A holoenzyme originally termed $PP2A_0$ (Tung et al. 1985a; Zolnierowicz et al. 1994). There are five known isoforms in this family (Table 1). The α and γ isoforms are expressed predominantly in muscle, the β and δ isoforms in brain, and the ϵ isoform in brain and testis. In cardiac muscle, nearly all of the PP2A holoenzyme is composed of the R5α subunit (Zolnierowicz et al. 1994). The R5 family has been subdivided into cytosolic and nuclear types based on localization of transiently expressed proteins (McCright et al. 1996). The R5α, R5β, and R5ϵ isoforms are cytoplasmic while R5γ and R5δ are present in both the cytoplasm and nucleus (Ahmadian-Tehrani et al. 1996; McCright et al. 1996) Transiently expressed R5 subunits are phosphorylated in intact cells. Thus, R5 subunit mediated regulation of PP2A may be modulated by phosphorylation (McCright et al. 1996).

A major role for R5 subunits appears to be targeting of PP2A to substrates and signaling complexes. Members of the R5 subunit family have been shown to mediate interactions of PP2A with a number of proteins including: cyclin G1 (Okamoto et al. 1996) and cyclin G2 (Bennin et al. 2002); components of the Wnt signaling pathway including the adenomatous polyposis coli protein (Seeling et al. 1999) and axin (Li et al. 2001; Yamamoto et al. 2001); the Drosophila homeodomain protein Sex Combs Reduced (Berry and Gehring 2000); L-type calcium channels (Davare et al. 2000); Bcl2 (Ruvolo et al. 2002); paxillin (Ito et al. 2000); and RNA-dependent protein kinase (Xu and Williams 2000). Although the mechanisms are uncertain, the interactions of R5 with cyclin G1 and cyclin G2 are likely to be involved in cell cycle arrest and apoptosis (reviewed in (Silverstein et al. 2003). The R5α subunit mediates the interaction of PP2A with the anti-apoptotic Bcl2 protein (Ruvolo et al. 2002). The interaction of the AC-R5α form of PP2A with Bcl2 appears to play an important role in ceramide-induced apoptosis by dephosphorylating and inactivating Bcl2. Additional evidence for an important role for the R5 subunits in cell survival comes from RNA interference in *Drosophila*. Ablation of both of the R5 subunits results in apoptosis of cultured *Drosophila* cells (Li et al. 2002b; Silverstein et al. 2002). Although the roles of R5-containing isoforms of PP2A in mammalian cells and *Drosophila* appear to be distinct, these observations make a strong case for the involvement of R5 subunits in pathways that modulate cell survival. Interactions of R5 with the adenomatous polyposis coli and axin proteins are likely to underlie the targeting of PP2A to the Wnt pathway where it plays important roles in modulating signaling (reviewed in (Janssens and Goris 2001; Silverstein et al. 2003; Virshup, 2000).

The single R5 gene in *S. cerevisiae*, *RTS1* (for ROX Three Suppressor), was identified as a suppressor of a temperature sensitive allele of *ROX3* and is 50% identical to the mammalian R5 regulatory subunit (Evangelista et al. 1996). Temperature-sensitive alleles of *rox3* block the accumulation of stress-related proteins that normally occurs in response to multiple stress signals including heat shock, osmotic stress, and glucose starvation. *RTS1* (also known as *SCS1*) was also isolated as a high-copy suppressor of a temperature sensitive allele of the mitochondrial chaperonin hsp60 (Shu et al. 1997). *RTS1* is required for the expression of mitochondrial chaperonins, which is defective in hsp60-ts strains. These data suggest that Rts1p functions to regulate expression of mitochondrial proteins in addition to its role in stress responses. Deletion of *RTS1* itself results is a temperature sensitive phenotype (Evangelista et al. 1996) that is rescued by expression of mammalian the R5α and to a lesser degree by the R5γ isoform (Zhao et al. 1997). *RTS1* mutants are also hypersensitive to ethanol, unable to utilize glycerol as a carbon source, and accumulate as large-budded cells with 2N DNA content and a non-divided nucleus (Shu et al. 1997). This G2/M cell cycle block can be suppressed by overexpression of the B-type cyclin *CLB2*.

In contrast to *S. cerevisiae*, *Schizosaccharomyces pombe* has two genes encoding R5 subunits (*par1* and *par2*). While neither gene is essential, loss of both *par1* and *par2* leads to a number of phenotypes including the inability to grow at high or low temperature, the inability to grow under several stress conditions, and defects in cell morphology, septum positioning and cytokinesis (Jiang and Hallberg 2000). Either *par1* or *par2* complement most of the defects associated with mutations of the R5/*RTS1* gene in *S. cerevisiae* indicating they are functional homologs of *RTS1*. While the Par1p protein is present throughout the cytoplasm, localization of the Par2p protein is more localized and depends on the stage of the cell cycle. A detailed analysis of the role of the R5/*par* proteins in septation has suggested that *par1* and *par2* negatively regulate the septation initiation pathway to ensure that multiple rounds of septation do not occur (Jiang and Hallberg 2001). Epistasis experiment indicate that the *par* genes act at or upstream of the Cdc7p protein kinase, which is activated by the small Ras-like GTPase, Spg1p. Based on the similarities between the *S. pombe par* genes and R5 subunits in higher eukaryotes, it is likely that at least some isoforms of mammalian R5 subunits also play a role in cytokinesis.

3.4 PP2A interacting proteins

There are a large group of proteins that interact with PP2A that are not compo-nents of classical PP2A holoenzymes. Some of these interacting proteins are also substrates for PP2A, others act to target PP2A to specific signaling complexes, and some viral proteins alter cellular signaling by disrupting endogenous PP2A complexes. It has become clear in the last ten years that the specificity and physio-logical functions of PP2A are largely mediated by associations with interacting proteins. PP2A-interacting proteins have been identified by a variety of methods, and include viral proteins and a host of cellular proteins that participate is interest-ing aspects of signal transduction. Interactions mediated by specific regulatory subunits are described in the preceding sections. A detailed discussion of all PP2A interacting proteins is beyond the scope of this review. A list of known PP2A in-teracting proteins has recently been compiled (Silverstein et al. 2003). Readers are also referred to recent reviews for additional descriptions of proteins that interact with PP2A (Janssens and Goris 2001; Millward et al. 1999; Virshup 2000).

The preceding sections have emphasized the role of regulatory subunits in me-diating interactions of PP2A with substrates, the cytoskeleton, and various types of signaling complexes. While the regulatory subunits play a prominent role in targeting of PP2A, a number of proteins interact directly with the core dimer or the free catalytic subunit. The direct interaction with these proteins results in the incorporation of PP2A into additional types of signaling complexes that are likely to play important roles in specific signaling pathways. Examples of proteins that have been shown to interact directly with the catalytic subunit of PP2A include: the B cell antigen receptor-associated protein alpha4 and its yeast homolog Tap42p (Chen et al. 1998; DiComo CJ and Arndt 1996; Murata et al. 1997); the eukaryotic translation termination factor eRF1 (Andjelkovic et al. 1996); the HOX11 transcription factor (Kawabe et al. 1997); and the estrogen receptor (Lu et al. 2003).

The direct complex formed between the PP2A catalytic subunit and the Tap42 protein appears to be a required component of the signaling pathway linking nutri-ent availability to protein synthesis in yeast. *S. cerevisiae* Tap42 is phosphorylated by the target of rapamycin (Tor) protein kinase (Jiang and Broach 1999). Rapa-mycin normally inhibits cell growth by inhibiting Tor and shutting off protein syn-thesis. Phosphorylation of Tap42 by Tor results in the association of Tap42 with PP2A catalytic subunits. Nutrient availability stimulates Tor, resulting in phos-phorylation of Tap42 and increased association of Tap42 with catalytic subunits. The increased interaction of Tap42 with PP2A is associated with an increase in the rate of initiation of protein synthesis. The phosphorylated form of Tap42 competes with the yeast A subunit (encoded by the *TPD3* gene) and R2/Cdc55p subunit for binding to the PP2A catalytic subunits (Pph21p/22p). In addition, Tap42 is dephosphorylated by the same R2/Cdc55p-containing PP2A holoenzyme. Inactivation of either the A/Tpd3p or R2/Cdc55p proteins results in rapamycin resistance by blocking Tap42 dephosphorylation and enhancing formation of the Tap42-catalytic subunit complex due to decreased competition for binding.

Another class of interacting proteins act by forming complexes with the PP2A core dimer. In addition to the viral tumor antigens discussed above, proteins that have been reported to form a direct complex with the core dimer include: casein kinase II (Heriche et al. 1997); the EF hand-containing proteins SG2NA and striatin (Moreno et al. 2000); the G protein-coupled chemokine receptor CXCR2 (Fan et al. 2001); and the *C. elegans* SMG-5 protein involved in mRNA surveillance (Anders et al. 2003). While details regarding how the core dimer is involved in regulating these proteins are not known, the formation of complexes with PP2A is likely to play an important role in regulating their activities.

References

Ahmadian-Tehrani M, Mumby MC, Kamibayashi C (1996) Identification of a novel protein phosphatase 2A regulatory subunit highly expressed in muscle. J Biol Chem 271:5164-5170

Anders KR, Grimson A, Anderson P (2003) SMG-5, required for C. elegans nonsense-mediated mRNA decay, associates with SMG-2 and protein phosphatase 2A. EMBO J 22:641-650

Andjelkovic N, Zolnierowicz S, Van Hoof C, Goris J, Hemmings BA (1996) The catalytic subunit of protein phosphatase 2A associates with the translation termination factor eRF1. EMBO J 15:7156-7167

Arino J, Perez-Callejon E, Cunillera N, Camps M, Posas F, Ferrer A (1993) Protein phosphatases in higher plants: multiplicity of type 2A phosphatases in *Arabidopsis thaliana*. Plant Mol Biol 21:475-485

Baharians Z, Schonthal AH (1998) Autoregulation of protein phosphatase type 2A expression. J Biol Chem 273:19019-19024

Barford D, Das AK, Egloff MP (1998) The structure and mechanism of protein phosphatases - insights into catalysis and regulation. Ann Rev Biophys Biomol Struct 27:133-164

Bennin DA, Arachchige Don AS, Brake T, McKenzie JL, Rosenbaum H, Ortiz L, DePaoli-Roach AA, Horne MC (2002) Cyclin G2 associates with protein phosphatase 2A catalytic and regulatory B' subunits in active complexes and induces nuclear aberrations and a G1/S phase cell cycle arrest. J Biol Chem

Berry M, Gehring W (2000) Phosphorylation status of the SCR homeodomain determines its functional activity: essential role for protein phosphatase 2A,B'. EMBO J 19:2946-2957

Blacketer MJ, Koehler CM, Coats SG, Myers AM, Madaule P (1993) Regulation of dimorphism in *Saccharomyces cerevisiae*: involvement of the novel protein kinase homolog Elm1p and protein phosphatase 2A. Mol Cell Biol 13:5567-5581

Bosch M, Cayla X, Van Hoof C, Hemmings BA, Ozon R, Merlevede W, Goris J (1995) The PR55 and PR65 subunits of protein phosphatase 2A from *Xenopus laevis*-- Molecular cloning and developmental regulation of expression. Eur J Biochem 230:1037-1045

Bryant JC, Westphal RS, Wadzinski BE (1999) Methylated C-terminal leucine residue of PP2A catalytic subunit is important for binding of regulatory B alpha subunit. Biochem J 339:241-246

Calin GA, di Iasio MG, Caprini E, Vorechovsky I, Natali PG, Sozzi G, Croce CM, Barbanti-Brodano G, Russo G, Negrini M (2000) Low frequency of alterations of the alpha (PPP2R1A) and beta (PPP2R1B) isoforms of the subunit A of the serine-threonine phosphatase 2A in human neoplasms. Oncogene 19:1191-1195

Campbell IG, Manolitsas T (1999) Absence of PPP2R1B gene alterations in primary ovarian cancers. Oncogene 18:6367-6369

Cayla X, Goris J, Hermann J, Hendrix P, Ozon R, Merlevede W (1990) Isolation and characterization of a tyrosyl phosphatase activator from rabbit skeletal muscle and *Xenopus laevis* oocytes. Biochemistry 29:658-667

Cayla X, Van Hoof C, Bosch M, Waelkens E, Vandekerckhove J, Peeters B, Merlevede W, Goris J (1994) Molecular cloning, expression, and characterization of PTPA, a protein that activates the tyrosyl phosphatase activity of protein phosphatase 2A. J Biol Chem 269:15668-15675

Chen J, Martin BL, Brautigan DL (1992) Regulation of protein serine-threonine phosphatase type 2A by tyrosine phosphorylation. Science 257:1261-1264

Chen J, Parsons S, Brautigan DL (1994) Tyrosine phosphorylation of protein phosphatase 2A in response to growth stimulation and v-*src* transformation of fibroblasts. J Biol Chem 269:7957-7962

Chen J, Peterson RT, Schreiber SL (1998) Alpha 4 associates with protein phosphatases 2A, 4, and 6. Biochem Biophys Res Commun 247:827-832

Chung H, Nairn AC, Murata K, Brautigan DL (1999) Mutation of Tyr307 and Leu309 in the protein phosphatase 2A catalytic subunit favors association with the alpha4 subunit which promotes dephosphorylation of elongation factor-2. Biochemistry 38:10371-10376

Cohen PTW (1994) Nomeclature and chromosomal localization of human protein serine/threonine phosphatase genes. Adv Prot Phosphatases 8:371-374

Cohen PTW (1997) Novel protein serine/threonine phosphatases: variety is the spice of life. Trends Biochem Sci 22:245-251

Das AK, Helps NR, Cohen PT, Barford D (1996) Crystal structure of the protein serine/threonine phosphatase 2C at 2.0 A resolution. EMBO J 15:6798-6809

Davare MA, Horne MC, Hell JW (2000) Protein phosphatase 2A is associated with class C L-type calcium channels (Cav1.2) and antagonizes channel phosphorylation by cAMP-dependent protein kinase. J Biol Chem 275:39710-39717

De Baere I, Derua R, Janssens V, Van Hoof C, Waelkens E, Merlevede W, Goris J (1999) Purification of porcine brain protein phosphatase 2A leucine carboxyl methyltransferase and cloning of the human homologue. Biochemistry 38:16539-16547

DiComo CJ, Arndt KT (1996) Nutrients, via the Tor proteins, stimulate the association of Tap42 with type 2A phosphatases. Genes Dev 10:1904-1916

Elder RT, Yu M, Chen M, Zhu X, Yanagida M, Zhao Y (2001) HIV-1 Vpr induces cell cycle G2 arrest in fission yeast (*Schizosaccharomyces pombe*) through a pathway involving regulatory and catalytic subunits of PP2A and acting on both Wee1 and Cdc25. Virology 287:359-370

Evangelista CC, Rodriguez Torres AM, Limbach MP, Zitomer RS (1996) Rox3 and Rts1 function in the global stress response pathway in baker's yeast. Genetics 142:1083-1093

Evans DR, Myles T, Hofsteenge J, Hemmings BA (1999) Functional expression of human PP2Ac in yeast permits the identification of novel C-terminal and dominant-negative mutant forms. J Biol Chem 274:24038-24046

Fan GH, Yang W, Sai J, Richmond A (2001) Phosphorylation-independent association of CXCR2 with the protein phosphatase 2A core enzyme. J Biol Chem 276:16960-16968

Favre B, Zolnierowicz S, Turowski P, Hemmings BA (1994) The catalytic subunit of protein phosphatase 2A is carboxyl-methylated *in vivo*. J Biol Chem 269:16311-16317

Ferrigno P, Langan TA, Cohen P (1993) Protein phosphatase 2A$_1$ is the major enzyme in vertebrate cell extracts that dephosphorylates several physiological substrates for cyclin-dependent protein kinases. Mol Biol Cell 4:669-677

Floer M, Stock J (1994) Carboxyl methylation of protein phosphatase 2A from *Xenopus* eggs is stimulated by cAMP and inhibited by okadaic acid. Biochem Biophys Res Commun 198:372-379

Frost JA, Alberts AS, Sontag E, Guan K, Mumby MC, Feramisco JR (1994) SV40 small t antigen cooperates with mitogen activated kinases to stimulate AP-1 activity. Mol Cell Biol 14:6244-6252

Goldberg J, Huang H-B, Kwon Y-G, Greengard P, Nairn AC, Kuriyan J (1995) Three-dimensional structure of the catalytic subunit of protein serine/threonine phosphatase-1. Nature 376:745-753

Gotz J, Probst A, Ehler E, Hemmings B, Kues W (1998) Delayed embryonic lethality in mice lacking protein phosphatase 2A catalytic subunit Calpha. Proc Natl Acad Sci USA 95:12370-12375

Gotz J, Probst A, Mistl C, Nitsch RM, Ehler E (2000) Distinct role of protein phosphatase 2A subunit Calpha in the regulation of E-cadherin and beta-catenin during development. Mech Dev 93:83-93

Griffith JP, Kim JL, Kim EE, Sintchak MD, Thomson JA, Fitzgibbon MJ, Fleming MA, Caron PR, Hsiao K, Navia MA (1995) X-ray structure of calcineurin inhibited by the immunophilin-immunosuppressant FKBP12-FK506 complex. Cell 82:507-522

Groves MR, Hanlon N, Turowski P, Hemmings BA, Barford D (1999) The structure of the protein phosphatase 2A PR65/A subunit reveals the conformation of its 15 tandemly repeated HEAT motifs. Cell 96:99-110

Guo H, Damuni Z (1993) Autophosphorylation-activated protein kinase phosphorylates and inactivates protein phosphatase 2A. Proc Natl Acad Sci USA 90:2500-2504

Healy AM, Zolnierowicz S, Stapelton AE, Goebl M, DePaoli-Roach AA, Pringle JR (1991) CDC55, a *Saccharomyces cerevisiae* gene involved in cellular morphogenesis: Identification, characterization, and homology to the B subunit of mammalian type 2A protein phosphatase. Mol Cell Biol 11:5767-5780

Hemmings BA, Adams-Pearson C, Maurer F, Muller P, Goris J, Merlevede W, Hofsteenge J, Stone SR (1990) Alpha and beta-forms of the 65-kDa subunit of protein phosphatase 2A have a similar 39 amino acid repeating structure. Biochemistry 29:3166-3173

Hendrix P, Mayer-Jaekel RE, Cron P, Goris J, Hofsteenge J, Merlevede W, Hemmings BA (1993) Structure and expression of a 72-kDa regulatory subunit of protein phosphatase 2A. Evidence for different size forms produced by alternative splicing. J Biol Chem 268:15267-15276

Heriche JK, Lebrin F, Rabilloud T, Leroy D, Chambaz EM, Goldberg Y (1997) Regulation of protein phosphatase 2A by direct interaction with casein kinase 2a. Science 276:952-955

Holmes SE, O'Hearn EE, McInnis MG, Gorelick-Feldman DA, Kleiderlein, JJ, Callahan C, Kwak NG, Ingersoll-Ashworth RG, Sherr M, Sumner AJ, Sharp AH, Ananth U, Seltzer WK, Boss MA, Vieria-Saecker AM, Epplen JT, Riess O, Ross CA, Margolis RL (1999) Expansion of a novel CAG trinucleotide repeat in the 5' region of PPP2R2B is associated with SCA12. Nature Genet 23:391-392

Hrimech M, Yao XJ, Branton PE, Cohen EA (2000) Human immunodeficiency virus type 1 Vpr-mediated G(2) cell cycle arrest: Vpr interferes with cell cycle signaling cascades by interacting with the B subunit of serine/threonine protein phosphatase 2A. EMBO Journal 19:3956-3967

Ito A, Kataoka TR, Watanabe M, Nishiyama K, Mazaki Y, Sabe H, Kitamura Y, Nojima H (2000) A truncated isoform of the PP2A B56 subunit promotes cell motility through paxillin phosphorylation. EMBO J 19:562-571

Janssens V, Goris J (2001) Protein phosphatase 2A: a highly regulated family of serine/threonine phosphatases implicated in cell growth and signalling. Biochem J 353:417-439

Janssens V, Jordens J, Stevens I, Van Hoof C, Martens E, De Smedt H, Engelborghs Y, Waelkens E, Goris J (2003) Identification and functional analysis of two Ca^{2+}-binding EF-hand motifs in the B"/PR72 subunit of protein phosphatase 2A. J Biol Chem 278:10697-10706

Janssens V, Van Hoof C, Martens E, De Baere I, Merlevede W, Goris J (2000) Identification and characterization of alternative splice products encoded by the human phosphotyrosyl phosphatase activator gene. Eur J Biochem 267:4406-4413

Janssens V, Van Hoof C, Merlevede W, Goris J (1998) PTPA regulating PP2A as a dual specificity phosphatase. Methods Mol Biol 93:103-115

Jiang W, Hallberg RL (2000) Isolation and characterization of par1+ and par2+: Two Schizosaccharomyces pombe genes encoding B' subunits of protein phosphatase 2A. Genetics 154:1025-1038

Jiang W, Hallberg RL (2001) Correct regulation of the septation initiation network in Schizosaccharomyces pombe requires the activities of par1 and par2. Genetics 158:1413-1429

Jiang Y, Broach JR (1999) Tor proteins and protein phosphatase 2A reciprocally regulate Tap42 in controlling cell growth in yeast. EMBO J 18:2782-2792

Kamibayashi C, Lickteig RL, Estes R, Walter G, Mumby MC (1992) Expression of the A subunit of protein phosphatase 2A and characterization of its interactions with the catalytic and regulatory subunits. J Biol Chem 267:21864-21872

Kawabe T, Muslin AJ, Korsmeyer SJ (1997) Hox11 interacts with protein phosphatases PP2A and PP1 and disrupts a G2/M cell-cycle checkpoint. Nature 385:454-458

Kinoshita K, Nemoto T, Nabeshima K, Kondoh H, Niwa H, Yanagida M (1996) The regulatory subunits of fission yeast protein phosphatase 2A (PP2A) affect cell morphogenesis, cell wall synthesis and cytokinesis. Genes Cells 1:29-45

Kleinberger T, Shenk T (1993) Adenovirus E4orf4 protein binds to protein phosphatase 2A, and the complex down regulates E1A-enhanced junB transcription. J Virol 67:756-7560

Kloeker S, Bryant JC, Strack S, Colbran RJ, Wadzinski BE (1997) Carboxymethylation of nuclear protein serine/threonine phosphatase X. Biochem J 327 481-486

Kowluru A, Seavey SE, Rabaglia ME, Nesher R, Metz SA (1996) Carboxylmethylation of the catalytic subunit of protein phosphatase 2A in insulin-secreting cells: evidence for functional consequences on enzyme activity and insulin secretion. Endocrinology 137:2315-2323

Kremmer E, Ohst K, Kiefer J, Brewis N, Walter G (1997) Separation of PP2A core enzyme and holoenzyme with monoclonal antibodies against the regulatory A subunit: abundant expression of both forms in cells. Mol Cell Biol 17:1692-1701

Lee J, Chen Y, Tolstykh T, Stock J (1996) A specific protein carboxyl methylesterase that demethylates phosphoprotein phosphatase 2A in bovine brain. Proc Natl Acad Sci USA 93:6043-6047

Lee J, Stock J (1993) Protein phosphatase 2A catalytic subunit is methyl-esterified at its carboxyl terminus by a novel methyl-transferase. J Biol Chem 26:19192-19195

Li X, Scuderi A, Letsou A, Virshup DM (2002a) B56-associated protein phosphatase 2A is required for survival and protects from apoptosis in *Drosophila melanogaster*. Mol Cell Biol 22:3674-3684

Li X, Virshup DM (2002) Two conserved domains in regulatory B subunits mediate binding to the A subunit of protein phosphatase 2A. Eur J Biochem 269:546-552

Li X, Yost HJ, Virshup DM, Seeling JM (2001) Protein phosphatase 2A and its B56 regulatory subunit inhibit Wnt signaling in *Xenopus*. EMBO J 20:4122-4131

Lizotte DL, McManus DD, Cohen HR, DeLong A (1999) Functional expression of human and arabidopsis protein phosphatase 2A in *Saccharomyces cerevisiae* and isolation of dominant-defective mutants. Gene 234:35-44

Lu Q, Surks HK, Ebling H, Baur WE, Brown D, Pallas DC, Karas RH (2003) Regulation of estrogen receptor alpha-mediated transcription by a direct interaction with protein phosphatase 2A. J Biol Chem 278:4639-4645

Lubert EJ, Hong Y, Sarge KD (2001) Interaction between protein phosphatase 5 and the A subunit of protein phosphatase 2A: evidence for a heterotrimeric form of protein phosphatase 5. J Biol Chem 276:38582-38587

Mayer-Jaekel RE, Baumgartner S, Bilbe G, Ohkura H, Glover DM, Hemmings BA (1992) Molecular cloning and developmental expression of the catalytic and 65- kDa regulatory subunits of protein phosphatase 2A in *Drosophila*. Mol Biol Cell 3:287-298

Mayer-Jaekel RE, Ohkura H, Gomes R, Sunkel CE, Baumgartner S, Hemmings BA, Glover DM (1993) The 55 kd regulatory subunit of *Drosophila* protein phosphatase 2A is required for anaphase. Cell 72:621-633

McCright B, Rivers AM, Audlin S, Virshup DM (1996) The B56 family of protein phosphatase 2A (PP2A) regulatory subunits encodes differentiation-induced phosphoproteins that target PP2A to both nucleus and cytoplasm. J Biol Chem 271:22081-22089

Millward TA, Zolnierowicz S, Hemmings BA (1999) Regulation of protein kinase cascades by protein phosphatase 2A. Trends Biochem Sci 24:186-191

Minshull J, Straight A, Rudner AD, Dernburg AF, Belmont A, Murray AW (1996) Protein phosphatase 2A regulates MPF activity and sister chromatid cohesion in budding yeast. Curr Biol 6:1609-1620

Mitchell DA, Sprague GF Jr (2001) The phosphotyrosyl phosphatase activator, Ncs1p (Rrd1p), functions with Cla4p to regulate the G(2)/M transition in *Saccharomyces cerevisiae*. Mol Cell Biol 21:488-500

Moreno CS, Park S, Nelson K, Ashby D, Hubalek F, Lane WS, Pallas DC (2000) WD40 repeat proteins striatin and S/G(2) nuclear autoantigen are members of a novel family of calmodulin-binding proteins that associate with protein phosphatase 2A. J Biol Chem 275:5257-5263

Mumby M (2001) A new role for protein methylation: switching partners at the phosphatase ball. Science's STKE, http://stke.sciencemag.org/cgi/content/full/OC_sigtrans;2001/79/pe1

Murata K, Wu J, Brautigan DL (1997) B cell receptor-associated protein alpha4 displays rapamycin-sensitive binding directly to the catalytic subunit of protein phosphatase 2A. Proc Natl Acad Sci USA 94:10624-10629

Myles T, Schmidt K, Evans DR, Cron P, Hemmings BA (2001) Active-site mutations impairing the catalytic function of the catalytic subunit of human protein phosphatase 2A permit baculovirus-mediated overexpression in insect cells. Biochem J 357:225-232

Ogris E, Du XX, Nelson KC, Mak EK, Yu XX, Lane WS, Pallas DC (1999) A protein phosphatase methylesterase (PME-1) is one of several novel proteins stably associating with two inactive mutants of protein phosphatase 2A. J Biol Chem 274:14382-14391

Ogris E, Gibson DM, Pallas DC (1997) Protein phosphatase 2A subunit assembly: the catalytic subunit carboxy terminus is important for binding cellular B subunit but not polyomavirus middle tumor antigen. Oncogene 15:911-917

Okamoto K, Kamibayashi C, Serrano M, Prives C, Mumby MC, Beach D (1996) p53-dependent association between cyclin G and the B' subunit of protein phosphatase 2A. Mol Cell Biol 16:6593-6602

Pallas DC, Shahrik LK, Martin BL, Jaspers S, Miller TB, Brautigan DL, Roberts TM (1990) Polyoma small and middle T antigens and SV40 small t antigen form stable complexes with protein phosphatase 2A. Cell 60:167-176

Price NE, Mumby MC (2000) Effects of regulatory subunits on the kinetics of protein phosphatase 2A. Biochemistry 39:11312-11318

Ramotar D, Belanger E, Brodeur I, Masson JY, Drobetsky EA (1998) A yeast homologue of the human phosphotyrosyl phosphatase activator PTPA is implicated in protection against oxidative DNA damage induced by the model carcinogen 4-nitroquinoline 1-oxide. J Biol Chem 273:21489-21496

Rempola B, Kaniak A, Migdalski A, Rytka J, Slonimski PP, di Rago JP (2000) Functional analysis of RRD1 (YIL153w) and RRD2 (YPL152w), which encode two putative activators of the phosphotyrosyl phosphatase activity of PP2A in *Saccharomyces cerevisiae*. Mol Gen Genet 262:1081-1092

Robinson LC, Menold MM, Garrett S, Culbertson MR (1993) Casein kinase I-like protein kinases encoded by YCK1 and YCK2 are required for yeast morphogenesis. Mol Cell Biol 13:2870-2881

Ruediger R, Fields K, Walter G (1999) Binding specificity of protein phosphatase 2A core enzyme for regulatory B subunits and T antigens. J Virol 73:839-842

Ruediger R, Hentz M, Fait J, Mumby M, Walter G (1994) Molecular model of the A subunit of protein phosphatase 2A: interaction with other subunits and tumor antigens. J Virol 68:123-129

Ruediger R, Pham HT, Walter G (2001a) Alterations in protein phosphatase 2A subunit interaction in human carcinomas of the lung and colon with mutations in the A beta subunit gene. Oncogene 20:1892-1899

Ruediger R, Pham HT, Walter G (2001b) Disruption of protein phosphatase 2A subunit interaction in human cancers with mutations in the Aalpha subunit gene. Oncogene 12:10-15

Ruvolo PP, Clark W, Mumby M, Gao F, May WS (2002) A functional role for the B56 alpha-subunit of protein phosphatase 2A in ceramide-mediated regulation of Bcl2 phosphorylation status and function. J Biol Chem 277:22847-22852

Schmidt K, Kins S, Schild A, Nitsch RM, Hemmings BA, Gotz J (2002) Diversity, developmental regulation and distribution of murine PR55/B subunits of protein phosphatase 2A. Eur J Neurosci 16:2039-2048

Seeling JM, Miller JR, Gil R, Moon RT, White R, Virshup DM (1999) Regulation of beta-catenin signaling by the B56 subunit of protein phosphatase 2A. Science 283:2089-2091

Shiomi K, Takeichi M, Nishida Y, Nishi Y, Uemura T (1994) Alternative cell fate choice induced by low-level expression of a regulator of protein phosphatase 2A in the Drosophila peripheral nervous system. Development 120:1591-1599

Shtrichman R, Sharf R, Barr H, Dobner T, Kleinberger T (1999) Induction of apoptosis by adenovirus E4orf4 protein is specific to transformed cells and requires an interaction with protein phosphatase 2A. Proc Natl Acad Sci USA 96:10080-10085

Shtrichman R, Sharf R, Kleinberger T (2000) Adenovirus E4orf4 protein interacts with both B alpha and B ' subunits of protein phosphatase 2A, but E4orf4-induced apoptosis is mediated only by the interaction with B alpha. Oncogene 19:3757-3765

Shu Y, Yang H, Hallberg E, Hallberg R (1997) Molecular genetic analysis of Rts1p, a B' regulatory subunit of *Saccharomyces cerevisiae* protein phosphatase 2A. Mol Cell Biol 17:3242-3253

Sieburth DS, Sundaram M, Howard RM, Han M (1999) A PP2A regulatory subunit positively regulates Ras-mediated signaling during Caenorhabditis elegans vulval induction. Genes & Development 13:2562-2569

Silverstein AM, Barrow CA, Davis AJ, Mumby MC (2002) Actions of PP2A on the MAP kinase pathway and apoptosis are mediated by distinct regulatory subunits. Proc Natl Acad Sci USA 99:4221-4226

Silverstein AM, Davis RJ, Bielinski VA, Esplin ED, Mahmood NA, Mumby MC (2003) Protein Phosphatase 2A. In: Bradshaw R, Dennis E (eds) Handbook of Cell Signaling, Vol 2. Academic Press, San Diego, pp 409-419

Smith TF, Gaitatzes C, Saxena K, Neer EJ (1999) The WD repeat: a common architecture for diverse functions. Trends Biochem Sci 24:181-185

Sontag E, Fedorov S, Kamibayashi C, Robbins D, Cobb M, Mumby M (1993) The interaction of SV40 small tumor antigen with protein phosphatase 2A stimulates the MAP kinase pathway and induces cell proliferation. Cell 75:887-897

Sontag E, Nunbhakdi-Craig V, Lee G, Bloom GS, Mumby MC (1996) Regulation of the phosphorylation state and microtubule-binding activity of tau by protein phosphatase 2A. Neuron 17:1201-1207

Sontag E, Sontag JM, Garcia A (1997) Protein phosphatase 2A is a critical regulator of protein kinase C zeta signaling targeted by SV40 small t to promote cell growth and NF-kappaB activation. EMBO J 16:5662-5671

Stark MJ (1996) Yeast protein serine/threonine phosphatases: multiple roles and diverse regulation. Yeast 12:1647-1675

Strack S, Ruediger R, Walter G, Dagda RK, Barwacz CA, Cribbs JT (2002) Protein phosphatase 2A holoenzyme assembly: identification of contacts between B-family regulatory and scaffolding A subunits. J Biol Chem 277:20750-20755

Strack S, Westphal RS, Colbran RJ, Ebner FF, Wadzinski, BE (1997) Protein serine/threonine phosphatase 1 and 2A associate with and dephosphorylate neurofilaments. Brain Res Mol Brain Res 49:15-28

Strack S, Zaucha JA, Ebner FF, Colbran RJ, Wadzinski BE (1998) Brain protein phosphatase 2A: developmental regulation and distinct cellular and subcellular localization by B subunits. J Comp Neurol 392:515-527

Takagi Y, Futamura M, Yamaguchi K, Aoki S, Takahashi T, Saji S (2000) Alterations of the PPP2R1B gene located at 11q23 in human colorectal cancers. Gut 47:268-271

Takahashi M, Shibata H, Shimakawa M, Miyamoto M, Mukai H, Ono Y (1999) Characterization of a novel giant scaffolding protein, CG-NAP, that anchors multiple signaling enzymes to centrosome and the golgi apparatus. J Biol Chem 274:17267-17274

Tolstykh T, Lee J, Vafai S, Stock JB (2000) Carboxyl methylation regulates phosphoprotein phosphatase 2A by controlling the association of regulatory B subunits. EMBO J 19:5682-5691

Tung HY, Alemany S, Cohen P (1985a) The protein phosphatases involved in cellular regulation. 2. Purification, subunit structure and properties of protein phosphatases- 2A0, 2A1, and 2A2 from rabbit skeletal muscle. Eur J Biochem 148:253-263

Tung HYL, Pelech S, Fisher MJ, Pogson CI, Cohen P (1985b) The protein phosphatases involved in cellular regulation: influence of polyamines on the activities of protein phosphatase-1 and protein phosphatase-2A. Eur J Biochem 149:305-313

Turowski P, Favre B, Campbell KS, Lamb NJ, Hemmings BA (1997) Modulation of the enzymatic properties of protein phosphatase 2A catalytic subunit by the recombinant 65-kDa regulatory subunit PR65alpha. Eur J Biochem 248:200-208

Turowski P, Fernandez A, Favre B, Lamb NJC, Hemmings BA (1995) Differential methylation and altered conformation of cytoplasmic and nuclear forms of protein phosphatase 2A during cell cycle progression. J Cell Biol 129:397-410

Turowski P, Myles T, Hemmings BA, Fernandez A, Lamb NC (1999) Vimentin dephosphorylation by protein phosphatase 2A is modulated by the targeting subunit B55. Mol Biol Cell 10:1997-2015

Uemura T, Shiomi K, Togashi S, Takeichi M (1993) Mutation of *twins* encoding a regulator of protein phosphatase 2A leads to pattern duplication in *Drosophila* imaginal disks. Genes Dev 7:429-440

Umi H, Imazu M, Maeta K, Tsukamoto H, Azuma K, Takeda M (1988) Three distinct forms of type 2A protein phosphatase in human erythrocyte cytosol. J Biol Chem 263:3752-3761

Van Hoof C, Janssens V, De B, I, de Winde JH, Winderickx J, Dumortier F, Thevelein JM, Merlevede W, Goris J (2000) The *Saccharomyces cerevisiae* homologue YPA1 of the mammalian phosphotyrosyl phosphatase activator of protein phosphatase 2A controls progression through the G1 phase of the yeast cell cycle. J Mol Biol 302:103-120

Van Hoof C, Janssens V, De B, I, Stark MJ, de Winde JH, Winderickx J, Thevelein JM, Merlevede W, Goris J (2001) The *Saccharomyces cerevisiae* phosphotyrosyl phosphatase activator proteins are required for a subset of the functions disrupted by protein phosphatase 2A mutations. Exp Cell Res 264:372-387

Van Hoof C, Janssens V, Dinishiotu A, Merlevede W, Goris J (1998) Functional analysis of conserved domains in the phosphotyrosyl phosphatase activator. Molecular cloning of the homologues from *Drosophila melanogaster* and *Saccharomyces cerevisiae*. Biochemistry 37:12899-12908

Virshup DM (2000) Protein phosphatase 2A: a panoply of enzymes. Curr Opin Cell Biol 12:180-185

Voorhoeve PM, Hijmans EM, Bernards R (1999) Functional interaction between a novel protein phosphatase 2A regulatory subunit, PR59, and the retinoblastoma-related p107 protein. Oncogene 18:515-524

Waelkens E, Goris J, Merlevede W (1987) Purification and properties of polycation-stimulated phosphorylase phosphatases from rabbit skeletal muscle. J Biol Chem 262:1049-1059

Walter G, Ruediger R, Slaughter C, Mumby M (1990) Association of protein phosphatase 2A with polyoma virus medium tumor antigen. Proc Natl Acad Sci USA 87:2521-2525

Wang SS, Esplin ED, Li JL, Huang L, Gazdar A, Minna J, Evans GA (1998) Alterations of the PPP2R1B gene in human lung and colon cancer. Science 282:284-287

Wang Y, Burke DJ (1997) Cdc55p, the B-type regulatory subunit of protein phosphatase 2A, has multiple functions in mitosis and is required for the kinetochore/spindle checkpoint in *Saccharomyces cerevisiae*. Mol Cell Biol 17:620-626

Watanabe G, Howe A, Lee RJ, Albanese C, Shu IW, Karnezis AN, Zon L, Kyriakis J, Rundell K, Pestell RG (1996) Induction of cyclin D1 by simian virus 40 small tumor antigen. Proc Natl Acad Sci USA 93:12861-12866

Westphal RS, Anderson KA, Means AR, Wadzinski BE (1998) A signaling complex of Ca2+-calmodulin-dependent protein kinase IV and protein phosphatase 2A. Science 280:1258-1261

Wu J, Tolstykh T, Lee J, Boyd K, Stock JB, Broach JR (2000) Carboxyl methylation of the phosphoprotein phosphatase 2A catalytic subunit promotes its functional association with regulatory subunits in vivo. EMBO J 19:5672-5681

Xu Z, Williams BR (2000) The B56alpha regulatory subunit of protein phosphatase 2A is a target for regulation by double-stranded RNA-dependent protein kinase PKR. Mol Cell Biol 20:5285-5299

Yamamoto H, Hinoi T, Michiue T, Fukui A, Usui H, Janssens V, Van Hoof C, Goris J, Asashima M, Kikuchi A (2001) Inhibition of the Wnt signaling pathway by the PR61 subunit of protein phosphatase 2A. J Biol Chem 276:26875-26882

Yan Z, Fedorov SA, Mumby MC, Williams RS (2000) PR48, a Novel Regulatory Subunit of Protein Phosphatase 2A, Interacts with Cdc6 and Modulates DNA Replication in Human Cells. Mol Cell Biol 20:1021-1029

Yang H, Jiang W, Gentry M, Hallberg RL (2000) Loss of a protein phosphatase 2A regulatory subunit (Cdc55p) elicits improper regulation of Swe1p degradation. Mol Cell Biol 20:8143-8156

Yang S-I, Lickteig RL, Estes RC, Rundell K, Walter G, Mumby MC (1991) Control of protein phosphatase 2A by simian virus 40 small t antigen. Mol Cell Biol 64:1988-1995

Zhao Y, Boguslawski G, Zitomer RS, DePaoli-Roach AA (1997) *Saccharomyces cerevisiae* homologs of mammalian B and B' subunits of protein phosphatase 2A direct the enzyme to distinct cellular functions. J Biol Chem 272:8256-8262

Zhou J, Pham HT, Ruediger R, Walter G (2003a) Characterization of the A-alpha and A-beta subunit isoforms of protein phosphatase 2A: differences in expression, subunit interaction, and evolution. Biochem J 369:387-398

Zhou J, Pham HT, Walter G (2003b) The formation and activity of PP2A holoenzymes do not depend on the isoform of the catalytic subunit. J Biol Chem 278:8617-8622

Zhou XZ, Kops O, Werner A, Lu PJ, Shen M, Stoller G, Kullertz G, Stark M, Fischer G, Lu KP (2000) Pin1-dependent prolyl isomerization regulates dephosphorylation of Cdc25C and tau proteins. Mol Cell 6:873-883

Zhu T, Matsuzawa S, Mizuno Y, Kamibayashi C, Mumby MC, Andjelkovic N, Hemmings BA, Onoe K, Kikuchi K (1997) The interconversion of protein phosphatase 2A between PP2A(1) and PP2A(0) during retinoic acid-induced granulocytic differentiation and a modification on the catalytic subunit in s phase of HL-60 cells. Arch Biochem Biophys 339:210-217

Zolnierowicz S, Csortos C, Bondor J, Verin A, Mumby MC, DePaoli-Roach AA (1994) Diversity in the regulatory B-subunits of protein phosphatase 2A: identification of a novel isoform highly expressed in brain. Biochemistry 33:11858-11867

4 Calcineurin: Roles of the Ca^{2+}/calmodulin-dependent protein phosphatase in diverse eukaryotes

Zoe Hilioti and Kyle W. Cunningham

Abstract

Calcineurin, also known as PP2B and PPP3, is a phospho-serine/threonine specific protein phosphatase that is active only after binding Ca^{2+} and Ca^{2+}/calmodulin in response to physiological Ca^{2+} signals. It is expressed to varying degrees in all tissues and highly conserved among most eukaryotes. Calcineurin has received much attention over the last twelve years since the discovery that the widely used immunosuppressive drugs Cyclosporin A and FK506 specifically bind and inhibit calcineurin after recruiting the cellular chaperones cyclophilin A and FKBP-12. Calcineurin is therefore unique among protein phosphatases both in its dramatic responsiveness to a second messenger and in the clinical utility of its inhibitors in a growing number of therapeutic situations. Calcineurin is widely conserved in nature and recently determined to be a major virulence factor in many pathogenic fungi. This review focuses on the progress towards understanding the broader roles of calcineurin in diverse eukaryotic species, which may offer useful insights into calcineurin function in humans.

4.1 Overview of calcineurin structure and function

The calcineurin holoenzyme is a stable heterodimer composed of a catalytic "A" subunit (58-69 kD) containing a core domain related to the catalytic subunit of PP1 and a regulatory "B" subunit (16-19 kD) related to calmodulin (Rusnak and Mertz 2000). The B subunit binds tightly to the A subunit through interactions with two extensions emerging from the catalytic core, an N-terminal domain forming a cap on the catalytic core and a long C-terminal α-helix termed the B-binding domain (BBD) that makes extensive contacts with the B subunit (Fig. 1, top). Upon a rise in cytosolic free Ca^{2+} concentrations above the resting state of 50 to 100 nM in most cell types, the four Ca^{2+}-binding EF hands of the B subunit become fully occupied concomitant with conformational changes in the A subunit that result in weak activation. Catalytic activity is enhanced at least 100-fold upon the binding of Ca^{2+}/calmodulin to C-terminal calmodulin-binding domain (CBD) and displacement of a short α-helical autoinhibitory peptide (AI) from the active site of the catalytic core. X-ray crystal structures of calcineurin bound to the

Topics in Current Genetics, Vol. 5
J. Arino, D.R. Alexander (Eds.) Protein phosphatases
© Springer-Verlag Berlin Heidelberg 2004

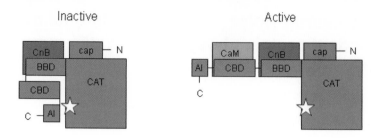

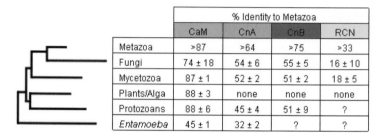

	% Identity to Metazoa			
	CaM	CnA	CnB	RCN
Metazoa	>87	>64	>75	>33
Fungi	74 ± 18	54 ± 6	55 ± 5	16 ± 10
Mycetozoa	87 ± 1	52 ± 2	51 ± 2	18 ± 5
Plants/Alga	88 ± 3	none	none	none
Protozoans	88 ± 6	45 ± 4	51 ± 9	?
Entamoeba	45 ± 1	32 ± 2	?	?

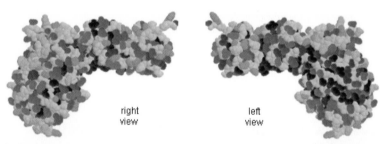

Fig. 1. Structure, Function, and Evolution of Calcineurin. TOP: Domain organization of calcineurin in its inactive (left) and active (right) states. Calcineurin A chain contains the catalytic core domain (CAT) as well as an N-terminal extension and a C-terminal extension that contains the calcineurin B- binding domain (BBD), calmodulin-binding domain (CBD), and auto-inhibitory peptide (AI) bound in the active site (star). Upon a rise in cytosolic Ca2+, calcineurin becomes activated by binding Ca2+ and Ca2+/calmodulin, which removes the auto-inhibitory peptide from the active site. Proteins of the RCN family can interact with both inactive and active forms of calcineurin through the A subunit and exert stimulatory or inhibitory effects. MIDDLE: Distribution and evolution of calcineurin in nature. Sequences homologous to calmodulin, calcineurin A, calcineurin B, and RCN were obtained from BLAST searches of public databases and aligned using ClustalW. The degree of similarity within kingdom Metazoa (animals) and between this kingdom and other eukaryotic kingdoms is shown (mean and range). Phylogenetic tree of the kingdoms were estimated based on recent reports (Baldauf et al. 2000). BOTTOM: Space-filled structure of calcineurin (right and left views) with side chains colored according to phylogenetic conservation. From highest to lowest conservation: dark blue, light blue, green, lime, yellow, orange red.

structurally dissimilar complexes of FK506/FKBP-12 and CsA/Cyclophilin A reveal the same mechanism of inhibition: occupancy of the large cavity that provides access of substrate proteins to the active site nearby (Griffith et al. 1995; Kissinger et al. 1995; Jin and Harrison 2002). The inhibitory complexes only minimally alter the protein conformation surrounding the active site iron and zinc ions, which results in a slight increase of activity on accessible small molecule substrates such as *p*-nitrophenyl phosphate. Owing to the high degree of conservation among all known calcineurins (see below), CsA and FK506 have been used to characterize the functions of calcineurin in diverse species.

Endogenous factors that bind and regulate calcineurin signaling are also being characterized. Genetic studies in fungi recently identified the broadly conserved RCN family of proteins that directly bind and regulate calcineurin from yeast to humans. Initially, the RCNs from yeast and humans were found to inhibit calcineurin activity *in vitro* and *in vivo* when overexpressed (Fuentes et al. 2000; Gorlach et al. 2000; Kingsbury and Cunningham 2000; Rothermel et al. 2000; Vega et al. 2002). Yeast Rcn1 and human Dscr1/Mcip1 (one of three RCN homologs in mammals) are transcriptionally induced upon calcineurin activation. However, the situation is more complex than a simple negative feedback loop because yeast mutants lacking Rcn1 exhibit diminished calcineurin signaling and mouse mutants lacking Dscr1/Mcip1 exhibit decreased calcineurin signaling suggesting these RCNs also stimulate calcineurin (Kingsbury and Cunningham 2000; Vega et al. 2003). Further studies in yeast show that Rcn1 and Dscr1/Mcip1 have dual roles, they stimulate calcineurin signaling when phosphorylated at a conserved serine residue and inhibit calcineurin signaling when dephophorylated, thus explaining the phenotype of the RCN knockout mutants (Hilioti et al. submitted). RCNs therefore appear to behave very much like the Inhibitor-2 molecules, which exert stimulatory or inhibitory effects on PP1 signaling depending on their phosphorylation state (Connor et al. 2000).

4.2 Phylogenetic distribution of calcineurin in nature

The tremendous rise in DNA sequences available in public databanks has heralded a revolution in our understanding of the relationships between species and the structure of conserved elements. Together with the tree of life and searchable DNA sequences from diverse species, it is now possible to estimate the origin of calcineurin and to trace its evolution along multiple paths to present day species. In doing so, one can attempt to distinguish highly conserved functional features of the protein from hypervariable, specialized, or functionless ones. Despite the rapidly changing status of DNA sequence databanks, analysis of current databases leads to a number of observations regarding the origin, regulation, and function of calcineurin.

Sequences homologous to calcineurin have not been found in the completed genomes of any prokaryotes (eubacteria and archaea) but are easily recognizable in the genomes of most types of eukaryotes with notable exceptions. Figure 1 (middle) shows a kingdom-level phylogenetic tree of eukaryotes (Baldauf et al. 2000) and illustrates the distribution of calcineurin A and B orthologs evident to date. All animals retain well-conserved homologs of calcineurin A and B (greater than 64% and 75% identical overall within kingdom Metazoa). All members of kingdom Fungi, the sister kingdom to animals, contain homologs of calcineurin A and B except for *Encephalitizoon cunniculi*, a simplified microsporidian which lacks any true homologs of calcineurin or calmodulin in its completely sequenced genome (Katinka et al. 2001). This species probably dispensed with calcineurin during its massive genome simplification while evolving into an obligate intracellular parasite. As a group, the Fungi are far more diverse than the Metazoa (Baldauf et al. 2000) and this diversity is reflected in the much greater sequence variation for calcineurin A and B (45% and 54% identity overall within Fungi). The kingdom Mycetozoa (amoebas including *Dictyostelium* and *Physarum*) are generally thought to have branched before the Metazoa/Fungi split and clear homologs of calcineurin have been characterized. The complete genomes of two higher plants (*Arabidopsis thaliana* and *Oryza sativa*) and the nearly complete genome of the green alga *Chlamydomonas* do not contain sequences homologous to calcineurin A or B. The many "calcineurin B-like" proteins characterized in higher plants actually are not related to calcineurin B any more than other calcium binding proteins and are now thought to regulate protein kinases rather than protein phosphatases (Luan et al. 2002). Because the branching order of the Plant/Alga kingdom in not firmly established relative to Protozoan groups, it is difficult to determine whether calcineurin was lost during the early evolution of this kingdom or whether this kingdom emerged prior to the origin of calcineurin. Calcineurins are also evident in the genomes of numerous Protozoa, including the ciliophora (*Paramecium tetraurelia* and *Tetrahymena thermophila*), the apicomplexa (*Plasmodium falciparum*, *Toxoplasma gondii*, and *Cryptosporidium parvum*), euglenozoa (*Trypanosoma brucei*, and *Leishmania major*), and heterolobosea (*Naegleria fowleri*, *Naegleria gruberi*). Even in these species, calcineurin A and B subunits exhibit strong similarity to their Metazoan counterparts (41-49% and 51-54% identity respectively). The partial genome of *Entamoeba histolytica*, whose branching position in the eukaryotic tree is extremely early, reveals the most divergent calcineurin A sequences we have been able to find (30-34% identical to Metazoa). This species also contains the most divergent calmodulin homolog known (~45% identical to Metazoan calmodulins) and the most divergent calmodulin-binding domains in its two calcineurin A sequences.

Based on the distribution and conservation of calcineurin among eukaryotes, we conclude that calcineurin evolved extremely early in the eukaryotic domain of life, possibly prior to the common ancestor of all eukaryotes. Since its origin, it has very likely been lost from the *Encephalitozoan* (a microsporidian fungus), and possibly also from the Plant/Alga kingdom. In all species that retain calcineurin, the basic mode of regulation by the binding of Ca^{2+} to the regulatory B subunit, binding of Ca^{2+}/calmodulin to the catalytic A subunit, and the displacement of the C-terminal autoinhibitory peptide from the active site, has been conserved. When sequence variation is mapped onto the three-dimensional structure of calcineurin A/B heterodimers lacking calmodulin and the calmodulin binding domain, one solvent-exposed surface appears to be more highly conserved than all others (Fig. 1, bottom). The β12-β13 strands of the catalytic core domain (left view) are strongly conserved and may represent a binding site for other conserved proteins. These strands in PP1 are thought to bind regulators such as Inhibitor-2. The RCN regulators of calcineurin can be found in all Metazoa, Fungi, and in *Dictyostelium*, but their generally lower degree of sequence conservation has made them difficult to spot in more distant eukaryotes. Perhaps some of the ancient cellular roles of calcineurins are also preserved today in diverse species.

4.3 Calcineurin-sensitive processes

4.3.1 Roles of calcineurin in Metazoa, a brief overview

The role of calcineurin in the immune system of is now legendary. For over thirty years, the natural compound Cyclosporin A has been used to prevent rejection of transplanted organs and tissues due to its remarkable ability to block the activation of T-cells and their precursors. FK506 was later found to have similar properties, though ~100-fold more effectively than CsA. The identification of two distinct receptor proteins that share similar biochemical activities (peptidyl-proline isomerases thought to be involved in protein folding processes) sparked a race to identify their common mechanism of action. In 1991, calcineurin was found to bind both of the drug/immunophilin complexes and to play a pivotal role in T-cell activation (Liu et al. 1991). In T-cells (and many other cell types) calcineurin is now known to dephosphorylate members of the NFAT family of transcription factors in the cytoplasm, rendering these proteins able to traffic into the nucleus and induce expression of target genes (Crabtree and Olson 2002). The relatively low concentration of calcineurin in T-cells makes them particularly sensitive to CsA and FK506, though side effects common to both drugs are observed in numerous other cell types.

The calcineurin-NFAT transcription system is now known to operate in very wide range of cell types in vertebrates, including cardiac and skeletal myocytes. Transgenic mice in which an activated variant of calcineurin (lacking the calmodulin-binding domain and autoinhibitory peptide) was overexpressed in the developing heart results in cardiac hypertrophy, a prelude to cardiomyopathy and heart failure (Molkentin et al. 1998). These findings have stimulated interest in the use of calcineurin inhibitors in therapies to treat heart failure in humans, a leading cause of death in the western world. Calcineurin has also been implicated as a key determinant in the differentiation of skeletal muscle fiber types (Schiaffino and Serrano 2002). In these cells, calcineurin directly regulates the activity of MEF2-family transcription factors in addition to NFATs. NFATs are not present in invertebrate animals such as fruit flies and nematodes. Other transcriptional and non-transcriptional effects of calcineurin may contribute to the effects of calcineurin on the neuromuscular systems of all animal species.

Calcineurin is highly expressed in most types of neurons. Many classes of ion channels have members that appear to be directly or indirectly regulated by calcineurin during rapid neurotransmission (Winder and Sweatt 2001). Another important target of calcineurin is the DARPP32/Inhibitor-1 molecule, which potently inhibits PP1 in its phosphorylated state (Greengard 2001). Calcineurin also dephosphorylates many cellular factors, collectively termed dephosphins, that are involved in the recycling of synaptic vesicles (Cousin and Robinson 2001). The rise of cytosolic Ca^{2+} and the activation of calcineurin during neurotransmission can therefore set off a complex series of effects that collectively lead to the phenomenon of long-term depression, a component of learning and memory mechanisms. Remarkably, genetic studies in mice indicate calcineurin can decrease learning and memory (Mansuy et al. 1998; Malleret et al. 2001).

Side effects shared by patients treated with either CsA or FK506 include nephrotoxicity, neurotoxicity, cancer, and predisposition to opportunistic pathogens (Collier 1990). The former has been attributed to alterations in the Na^+/K^+ ATPase regulation in the proximal tubules of the kidney (Aperia et al. 1992). Although CsA and FK506 remain as spectacularly successful therapeutics already, it is hoped that a better understanding of the functions of calcineurin in different cell types will help to augment their utility and diminish their side effects. As described below, the ability to selectively inhibit calcineurin in non-animal species may prove to be effective as treatment against a wide range of human pathogens.

4.3.2 Roles of calcineurin in Fungi

Most studies of calcineurin function in fungi have been conducted in the unicellular budding and fission yeasts, but filamentous fungi and other pathogenic species are being increasingly studied because of the emerging roles of calcineurin in fungal pathogenesis (Fox and Heitman 2002). The best-studied system is bakers' yeast where a number of calcineurin-sensitive targets and processes have been identified. Homologs of these targets are typically present in the genomes of most other fungi and some, such as the RCN regulators of calcineurin (homologs of yeast Rcn1) and voltage-gated Ca^{2+} channels (homologs of yeast Cch1), are evident even in animals. Yeast appears to be a highly simplified descendent of more complex ancestral species, so there is reason to suspect this yeast may have lost certain ancestral targets of calcineurin that may be preserved in other fungi. The combination of pharmacological and genetic approaches uniquely available in diverse fungi may therefore continue to elucidate new roles for calcineurin for many years to come.

4.3.2.1 Baker's yeast

Calcineurin has been studied most extensively in *Saccharomyces cerevisiae*, a unicellular budding yeast. Yeast expresses two closely related A subunits and a single B subunit (Cyert et al. 1991; Kuno et al. 1991; Cyert and Thorner 1992). In standard laboratory conditions, yeast mutants carrying complete deletions of

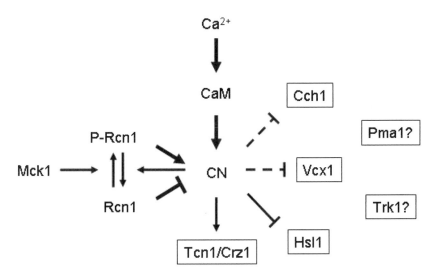

Fig. 2. Calcineurin regulation and functions in bakers' yeast. Calcineurin becomes activated in response to elevated cytosolic Ca^{2+} and binding of calmodulin. The presence of phospho-Rcn1 initially stimulates calcineurin signaling toward many independent targets, including the Tcn1/Crz1 transcription factor, the Hsl1 protein kinase, the Vcx1 H^+/Ca^{2+} exchanger, the Cch1 Ca^{2+} channel, possibly other ion transporters (Pma1 and Trk1), and Rcn1 itself. Dephosphorylation of Rcn1 and up-regulation of Rcn1 expression may secondarily lead to inhibition of calcineurin signaling through accumulation of inhibitory forms.

calcineurin are viable and proliferate at the same rate as wild type strains. Many lines of evidence indicate that Ca^{2+} signals are not generated in standard conditions, so the viability of calcineurin-deficient mutants is not surprisingly. In non-standard growth conditions or in other genetic contexts, mutants lacking calcineurin display a number of phenotypes that can be mimicked in wild type cells by treatment with CsA or FK506. So far, the two A subunits appear to be functionally redundant in all cases and mutants simultaneously lacking both A subunits are phenotypically identical to mutants lacking the B subunit. Thus, the A and B subunits do not have functions independent of one another.

Treatment of yeast cells with high concentrations of $CaCl_2$ has revealed a number of important functions for calcineurin (Fig. 2). Mutants lacking calcineurin A or B subunits proliferate more rapidly than wild type strains in high Ca^{2+} conditions (Cunningham and Fink 1994), and this phenotype has been attributed to the inhibition of Vcx1, a vacuolar H^+/Ca^{2+} exchanger that helps remove excess Ca^{2+} from the cytoplasm (Cunningham and Fink 1996). It remains to be determined whether the regulation of Vcx1 by calcineurin is direct or indirect. In high Ca^{2+} conditions, calcineurin appears to directly dephosphorylate the protein kinase Hsl1, which contributes to its delocalization and destruction (Mizunuma et al. 2001). This protein kinase is not required for the normal regulation of Vcx1 by calcineurin (Bonilla et al. 2002), but Hsl1 plays an important role in regulating mitosis. Hsl1 inactivates the Swe1 protein kinase (a member of the Wee1 family of

protein kinases) which delays cell cycle progression in G2/M phase by phosphorylating the major cyclin-dependent protein kinase Cdc28 (Booher et al. 1993; Lew
and Reed 1995). Calcineurin activation can therefore delay yeast cells in G2/M
phase of the cell cycle in high Ca^{2+} conditions (Mizunuma et al. 1998).

Another well-characterized substrate of calcineurin in yeast is the transcription
factor Tcn1/Crz1 (Matheos et al. 1997; Stathopoulos and Cyert 1997; Mendizabal
et al. 1998). Tcn1 contains three zinc finger domains at its C-terminus that can
specifically bind CDRE sequences in the promoters of target genes. The large N-
terminal domain of Tcn1/Crz1 is dephosphorylated by calcineurin in the cytoplasm, resulting in conformational changes that promote localization to the nucleus and exposure of a transcription activation domain (Stathopoulos-Gerontides
et al. 1999; Polizotto and Cyert 2001; Boustany and Cyert 2002). Global analysis
of gene expression using DNA microarrays revealed at least 111 genes that are
significantly induced in high Ca^{2+} conditions through a Tcn1/Crz1- and calcineurin-dependent mechanism (Yoshimoto et al. 2002). These target genes encode proteins involved in a number of seeming disparate processes such as ion
homeostasis, cell wall maintenance, vesicle trafficking, proteolysis, and miscella-
neous other effects. Many additional phenotypes of calcineurin-deficient mutants
are mimicked by the loss of Tcn1/Crz1 and therefore attributable to the failure to
induce expression of target genes. For example, loss of calcineurin causes increased sensitivity to Ca^{2+}, Mn^{2+}, Na^+, and Li^+ ions due to the failure to induce the
Pmc1 Ca^{2+} pump, the Pmr1 Ca^{2+}/Mn^{2+} pump, and Ena1/Pmr2A Na^+/Li^+ pump.
Failure of calcineurin- and Tcn1/Crz1-deficient mutants to induce transcription of
the Fks2 1,3-β-glucan synthase causes inviability of mutants lacking the functionally redundant Fks1 enzyme (Mazur et al. 1995). More phenotypes are likely to be
found upon deeper analysis of the many other genes regulated by Tcn1/Crz1 and
calcineurin.

One noteworthy target of Tcn1/Crz1 is the yeast RCN regulator of calcineurin
known as Rcn1. Rcn1 enhances calcineurin signaling when expressed at basal levels and inhibits calcineurin signaling when overexpressed (Kingsbury and Cunningham 2000). A model explaining these positive and negative effects of Rcn1
on calcineurin signaling was recently proposed: Rcn1 cycling between phosphorylated and dephophosphorylated states switches its effects on calcineurin between
inhibitory and stimulatory states (Hilioti et al. submitted). Because the phosphorylation site is 100% conserved among animals, fungi, and amoeba, calcineurin
regulation by RCNs may be conserved in all species. This model closely resem-
bles a one for the regulation of PP1 by phospho- and dephospo-Inhibitor-2
(Connor et al. 2000) and may generally serve to fine-tune phosphatase activities.

A variety of physiological stimuli generate Ca^{2+} signals in yeast cells grown in
standard culture medium (lacking Ca^{2+} supplements). In response to mating
pheromones or stress in the endoplasmic reticulum, activation of the Cch1-Mid1
Ca^{2+} channel in the plasma membrane helps to generate Ca^{2+} signals that activate
calcineurin and Tcn1/Crz1 (Locke et al. 2000; Bonilla et al. 2002). The Ca^{2+} influx activity of Cch1-Mid1 appears to be inhibited upon the activation of calcineurin (Matheos et al. 1997; Locke et al. 2000; Muller et al. 2001; Bonilla et al.
2002). Cch1 undergoes mobility shifts on SDS polyacrylamide in a calcineurin-

dependent fashion, consistent with direct inhibition of the Cch1-Mid1 channel by calcineurin (Bonilla and Cunningham in press) although more experiments are necessary to prove this hypothesis. In mammalian cells, however, calcineurin directly dephosphorylates subunits of voltage-gated Ca^{2+} channels subunits homologous to Cch1 and significantly modulates their activities (Yakel 1997).

Finally, calcineurin performs a number of additional functions in yeast that are not as well characterized. Through mechanisms independent of the targets described above, calcineurin strongly inhibits apoptosis-like cell death induced by mating pheromones (Cyert et al. 1991; Cyert and Thorner 1992; Moser et al. 1996; Withee et al. 1997; Severin and Hyman 2002) and stress in the endoplasmic reticulum (Bonilla et al. 2002). In particular, the commonly prescribed antifungal drugs that inhibit sterol biosynthesis in fungi (the azole-class drugs and terbinafine) potently kill calcineurin-deficient yeast cells but not wild type cells (Bonilla et al. 2002; Edlind et al. 2002). Calcineurin also appears to regulate the functional properties of the Trk1 and Trk2 K^+ transporters and potentially the plasma membrane H^+ pump Pma1 (Mendoza et al. 1994; Withee et al. 1998). At present, it is still unclear whether calcineurin directly regulates these enzymes or whether the effects are dependent on other intermediary factors. Much more work is necessary to identify the direct substrates of calcineurin in all these cases.

4.3.2.2 Other Ascomycota (Candida, Schizosaccharomyces, Neurospora, Aspergillus)

Candida albicans is a human pathogen closely related to bakers' yeast. Calcineurin-deficient mutants of *Candida* are viable in culture (Cruz et al. 2002) and die when treated with azole-class antifungal drugs *in vitro* and *in vivo* like wild type cells treated with CsA or FK506 (Marchetti et al. 2000a; Marchetti et al. 2000b; Bonilla et al. 2002; Cruz et al. 2002; Onyewu et al. 2003). The targets of calcineurin involved in azole resistance are not yet known, but this mechanism appears to be conserved in many other fungi. The development of non-immunosuppressive analogs that are specific for fungal calcineurins might therefore improve the efficacy of existing antifungal therapies (Odom et al. 1997a; Onyewu et al. 2003).

The function of calcineurin in *Schizosaccharomyces pombe* (fission yeast) an ascomycete distantly related to the budding yeasts, has been reviewed recently (Sugiura et al. 2002). Mutants lacking the sole A subunit are viable but exhibit several phenotypes such as elongated cell morphology with multiple septa, altered cell polarity, sterility, and abnormal microtubule structures (Yoshida et al. 1994). Calcineurin deficient mutants and wild type cells treated with CsA or FK506 are sensitive to high levels of chloride ions in the environment through unknown mechanisms (Fujita et al. 2002). Mutations in at least 8 genes result in FK506-sensitive growth (Cheng et al. 2002) but it is presently unclear how calcineurin might be important for growth of these mutants and whether these processes are conserved in other species. Recently, the Prz1 transcription factor homologous to yeast Tcn1/Crz1 was found to be directly activated by calcineurin in fission yeast (Hirayama et al. 2003). Prz1 was required for calcineurin-dependent induction of itself and the Pmc1 Ca^{2+} ATPase, as observed earlier in bakers' yeast (Cunningham and Fink 1996; Matheos et al. 1997), but the effects of calcineurin on chloride homeostasis and septation were not attributable to Prz1.

The functions of calcineurin in filamentous fungi such as *Neurospora crassa* (red bread mold), *Aspergillus nidulans*, and *Magnaporthe grisea* (rice blast fungus) have been studied using FK506, CsA, and genetic methods. FK506 and CsA consistently showed a role for calcineurin in the growth and branching of hyphal filaments. Mutants of *Neurospora crassa* deficient in calcineurin B exhibit defects in hyphal growth and branching (Kothe and Free 1998). Additionally, CsA and FK506 altered the Ca^{2+} concentration gradients in the cytoplasm of hyphal tips (Prokisch et al. 1997). Several other mutants with altered hyphal morphology exhibit stronger phenotypes when treated with FK506 or Cyclosporin A and are suppressed by addition of Ca^{2+} (Sone and Griffiths 1999; Bok et al. 2001). The targets of calcineurin are not yet identified, however a promising candidate is a homolog of Rho-dependent protein kinases known as COT-1 which has been genetically implicated in hyphal morphogenesis (Gorovits et al. 1999, 2000). CsA affects growth and formation of infectious structures called appressoria of *Magnaporthe grisea* by a process that depends on a cyclophilin, suggesting a role for calcineurin in these processes (Viaud et al. 2002). Finally, calcineurin-deficient mutants of *Aspergillus nidulans* exhibit strong defects in cell cycle regulation (Rasmussen et al. 1994) and calcineurin has been implicated in the regulation of aflatoxin biosynthesis in *Aspergillus parasiticus* (Jayashree and Subramanyam 2000).

4.3.2.3 Basidiomycota

This phylum of fungi includes the mushrooms, rusts, and smuts. Calcineurin in *Cryptococcus neoformans* (causing fungal meningitis in humans) has been studied extensively by genetic and pharmacological means (Odom et al. 1997a; Odom et al. 1997b; Davidson et al. 2000; Fox et al. 2001; Gorlach et al. 2002). Mutants deficient for either the A or B subunit are viable but fail to grow at physiological temperatures of 37°C, much like the effects of FK506 and CsA (Odom et al. 1997b). They are also defective in mating and in haploid fruiting (Cruz et al. 2001). Few targets of calcineurin are known in this species, except for the Cbp1 protein homologous to RCNs of yeast and mammals (Gorlach et al. 2000). Calcineurin-deficient strains exhibit highly diminished virulence in animal models and augmented sensitivity to azole-class antifungal drugs. Calcineurin may therefore prevent cell death by mechanisms similar to those observed in *Saccharomyces* and *Candida*. Because of the potential for development of new classes of antifungal drugs, identifying the targets of calcineurin involved in fungal pathogenesis remains as an exciting area of ongoing research.

4.3.3 Roles of calcineurin in Mycetozoa

Before the divergence of fungal and animal kingdoms, the Mycetozoa (amoebas) such as the cellular slime mold *Dictyostelium discoideum* branched from the eukaryotic tree (Baldauf et al. 2000). Calcineurin has been characterized biochemically and genetically in *Dictyostelium* (Dammann et al. 1996; Hellstern et al. 1997; Kessen et al. 1999; Aichem and Mutzel 2001). A homolog of Rcn1 is also evident in its nearly complete genome, but homologs of other calcineurin targets from animals and fungi are not yet apparent. FK506 and CsA cause defects in the development of *Dictyostelium* amoebae into multicellular structures (Horn and Gross 1996). The calcineurin inhibitors also prevent growth in high calcium conditions apparently by blocking the ability to transcriptionally up-regulate the expression of PAT1, a member of the plasma membrane Ca^{2+} ATPase family of calcium pumps (Moniakis et al. 1999). *Dictyostelium* therefore appears to retain the use of calcineurin for regulation of ion homeostasis analogous to what has been observed in fungi and animals.

4.3.4 Absence of calcineurin in plants and alga

The completely sequenced genomes of *Arabidopsis thaliana* and *Oryza sativa* (rice) do not contain genes orthologous to calcineurin A or calcineurin B subunits. The incompletely sequenced genome of *Chlamydomonas reinhardtii* (a green algae) also lacks sequences homologous to calcineurin, and no calcineurin-homologous sequences have been reported for any other plants or alga. Quite possibly, calcineurin might not exist in the Plant/Alga kingdom. This tentative conclusion is quite surprising in light of numerous published reports of responses to CsA, FK506, or exogenous calcineurin in higher plants. Until calcineurin has been positively identified in these species, such attributions to calcineurin signaling should be treated with skepticism.

A family of "calcineurin B-like" proteins or CBLs in *Arabidopsis* has been characterized in a number of molecular, genetic, and biochemical experiments (Liu and Zhu 1998; Kudla et al. 1999). These proteins all contain four EF hand motifs, many of which seem capable of binding Ca^{2+}. One such CBL was found to bind rat calcineurin A in the yeast two-hybrid assay. However, the CBL family is not related to calcineurin B subunits any more than they are to other families of four EF-domain proteins such as frequenin and calmodulin (Luan et al. 2002). CBLs are now thought to interact with protein kinases in *Arabidopsis* and function together to promote resistance to high salt conditions (Shi et al. 1999; Halfter et al. 2000; Kim et al. 2003). There is no evidence CBLs interact with protein phosphatases and little evidence for calcineurin A homologs in plants. Nevertheless, expression of a constitutive calcineurin from yeast in tobacco plants induces a salt-resistant phenotype (Pardo et al. 1998). Injection of FKBP-12 and FK506 into guard cells of *Vicia faba* (fava bean) leads to several phenotypic effects (Luan et al. 1993) and CsA can increase phosphorylation of some proteins (Li et al. 1998). These effects might be attributable to phenomena other than calcineurin inhibition. More direct approaches are clearly necessary in order to confirm the possible existence of calcineurin in the Plant/Alga kingdom.

4.3.5 Roles of calcineurin in other eukaryotes

Calcineurins are evident in the incomplete genomes of many other protozoan species, including the free-living Ciolophora (*Paramecium, Tetrahymena*) and human pathogens belonging to the Apicomplexa (*Plasmodium, Cryptosporidium, Toxoplasma, and Eimeria*), the Euglenozoa (e.g. *Trypanosoma, Leishmania*), the Heterolobosea (*Naegleria*) (Fulton et al. 1995; Remillard et al. 1995), and *Entamoeba*. Calcineurin has been purified and biochemically characterized from several of these species, including *Leishmania* (Rascher et al. 1998; Banerjee et al. 1999), and *Paramecium tetraurelia* (Kissmehl et al. 1997; Momayezi et al. 2000) where it may regulate phosphoglucomutases during Ca^{2+}-triggered exocytosis (Kissmehl et al. 1996). FK506 and CsA can modulate the pathogenicity of many of these species in murine models and block the egress of *Toxoplasma gondii* from host cells *in vitro* (Moudy et al. 2001), but in most cases it was not possible to determine whether the calcineurin inhibitors primarily affected the parasites or the hosts. A general limitation of employing these drugs to treat parasitic infections would be their potent immunosuppressive activity. However, a new generation of useful antibiotics is theoretically possible if parasite-specific inhibitors of calcineurin or its targets can be developed.

4.4 Summary and future prospects

Phylogenetic studies indicate a very early origin of calcineurin in the eukaryotic domain of life. The catalytic core domain homologous to PP1 was extended to include an N-terminal cap and C-terminal regulatory domain that binds calmodulin and calcineurin B, both of which bind Ca^{2+} in the physiological range. The overall domain structure and modes of regulation have changed little since its origin. In three related kingdoms: Metazoa, Fungi, and Mycetozoa, calcineurin may regulate expression and function of RCNs, a conserved family of proteins that directly regulate calcineurin. In these same kingdoms, calcineurin activation induces the expression of Ca^{2+} ATPases that help to restore cytosolic Ca^{2+} to low levels. Calcineurin may also have general roles in ion homeostasis, membrane trafficking, and pathways required for cells to survive various types of stress. Other calcineurin-sensitive processes and targets in diverse species will surely be found in the future. Ongoing research may therefore provide deeper insights into the general roles of calcineurin signaling in nature.

The natural small molecule inhibitors of calcineurin, CsA and FK506, have proved remarkably useful as immunosuppressive drugs. New drugs that target calcineurin in selected tissues or interfere with specific targets of calcineurin may be developed to diminish the side effects of the existing drugs and to expand the repertoire of treatable conditions. Non-immunosuppressive analog that inhibit calcineurin (or its targets) specifically in fungi and protozoa may be useful antibiotics against a wide range of human pathogens.

References

Aichem A, Mutzel R (2001) Unconventional mRNA processing in the expression of two calcineurin B isoforms in *Dictyostelium*. J Mol Biol 308:873-882

Aperia A, Ibarra F, Svensson L-B, Klee C, Greengard P (1992) Calcineurin mediates alpha-adrenergic stimulation of Na^+, K^+-ATPase activity in renal tubule cells. Proc Natl Acad Sci USA 89:7394-7397

Baldauf SL, Roger AJ, Wenk-Siefert I, Doolittle WF (2000) A kingdom-level phylogeny of eukaryotes based on combined protein data. Science 290:972-977

Banerjee C, Sarkar D, Bhaduri A (1999) Ca^{2+} and calmodulin-dependent protein phosphatase from *Leishmania donovani*. Parasitology 118:567-573

Bok JW, Sone T, Silverman-Gavrila LB, Lew RR, Bowring FJ, Catcheside DE, Griffiths AJ (2001) Structure and function analysis of the calcium-related gene spray in *Neurospora crassa*. Fungal Genet Biol 32:145-158

Bonilla M, Cunningham KW (2003) Suppression of ER stress-induced cell death by the MAP kinase-dependent stimulation of the Cch1-Mid1 Ca^{2+} channel and calcineurin. Mol Biol Cell 14:4296-4305

Bonilla M, Nastase KK, Cunningham KW (2002) Essential role of calcineurin in response to endoplasmic reticulum stress. EMBO J 21:2343-2353

Booher RN, Deshaies RJ, Kirschner MW (1993) Properties of *Saccharomyces cerevisiae* wee1 and its differential regulation of p34CDC28 in response to G1 and G2 cyclins. EMBO J 12:3417-3426

Boustany LM, Cyert MS (2002) Calcineurin-dependent regulation of Crz1p nuclear export requires Msn5p and a conserved calcineurin docking site. Genes Dev 16:608-619

Cheng H et al. (2002) Role of the Rab GTP-binding protein Ypt3 in the fission yeast exocytic pathway and its connection to calcineurin function. Mol Biol Cell 13:2963-2976

Collier SJ (1990) Immunosuppressive drugs. Curr Opin Immunol 2:854-858

Connor JH et al. (2000) Cellular mechanisms regulating protein phosphatase-1. A key functional interaction between inhibitor-2 and the type 1 protein phosphatase catalytic subunit. J Biol Chem 275:18670-18675

Cousin MA, Robinson PJ (2001) The dephosphins: dephosphorylation by calcineurin triggers synaptic vesicle endocytosis. Trends Neurosci 24:659-665

Crabtree GR, Olson EN (2002) NFAT signaling: choreographing the social lives of cells. Cell 109 Suppl:S67-S79

Cruz MC, Fox DS, Heitman J (2001) Calcineurin is required for hyphal elongation during mating and haploid fruiting in Cryptococcus neoformans. EMBO J 20:1020-1032

:Cruz MC, Goldstein AL, Blankenship JR, Del Poeta M, Davis D, Cardenas ME, Perfect JR, McCusker JH, Heitman J (2002) Calcineurin is essential for survival during membrane stress in *Candida albicans*. EMBO J 21:546-559

Cunningham KW, Fink GR (1994) Calcineurin-dependent growth control in *Saccharomyces cerevisiae* mutants lacking *PMC1*, a homolog of plasma membrane Ca^{2+} ATPases. J Cell Biol 124:351-363

Cunningham KW, Fink GR (1996) Calcineurin inhibits *VCX1*-dependent H^+/Ca^{2+} exchange and induces Ca^{2+} ATPases in yeast. Mol Cell Biol 16:2226-2237

Cyert MS, Kunisawa R, Kaim D, Thorner J (1991) Yeast has homologs (*CNA1* and *CNA2* gene products) of mammalian calcineurin, a calmodulin-regulated phosphoprotein phosphatase. Proc Natl Acad Sci USA 88:7376-7380

Cyert MS, Thorner J (1992) Regulatory subunit (*CNB1* gene product) of yeast Ca^{2+}/calmodulin-dependent phosphoprotein phosphatases is required for adaptation to pheromone. Mol Cell Biol 12:3460-3469

Dammann H, Hellstern S, Husain Q, Mutzel R (1996) Primary structure, expression and developmental regulation of a Dictyostelium calcineurin A homologue. Eur J Biochem 238:391-399

Davidson RC, Cruz MC, Sia RA, Allen B, Alspaugh JA, Heitman J (2000) Gene disruption by biolistic transformation in serotype D strains of *Cryptococcus neoformans*. Fungal Genet Biol 29:38-48

Edlind T, Smith L, Henry K, Katiyar S, Nickels J (2002) Antifungal activity in *Saccharomyces cerevisiae* is modulated by calcium signalling. Mol Microbiol 46:257-268

Fox DS, Cruz MC, Sia RA, Ke H, Cox GM, Cardenas ME, Heitman J (2001) Calcineurin regulatory subunit is essential for virulence and mediates interactions with FKBP12-FK506 in *Cryptococcus neoformans*. Mol Microbiol 39:835-849

Fox DS, Heitman J (2002) Good fungi gone bad: the corruption of calcineurin. Bioessays 24:894-903

Fuentes JJ, Genesca L, Kingsbury TJ, Cunningham KW, Perez-Riba M, Estivill X, de la Luna S (2000) DSCR1, overexpressed in Down syndrome, is an inhibitor of calcineurin-mediated signaling pathways. Hum Mol Gen 9:1681-1690

Fujita M, Sugiura R, Lu Y, Xu L, Xia Y, Shuntoh H, Kuno T (2002) Genetic interaction between calcineurin and type 2 myosin and their involvement in the regulation of cytokinesis and chloride ion homeostasis in fission yeast. Genetics 161:971-981

Fulton C, Lai EY, Remillard SP (1995) A flagellar calmodulin gene of Naegleria, coexpressed during differentiation with flagellar tubulin genes, shares DNA, RNA, and encoded protein sequence elements. J Biol Chem 270:5839-5848

Hilioti Z et al. GSK-3 Kinases enhance calcineurin signaling through phosphorylation of RCNs, submitted

Gorlach J, Fox DS, Cutler NS, Cox GM, Perfect JR, Heitman J (2000) Identification and characterization of a highly conserved calcineurin binding protein, CBP1/calcipressin, in *Cryptococcus neoformans*. EMBO J 19:3618-3629

Gorlach JM, McDade HC, Perfect JR, Cox GM (2002) Antisense repression in *Cryptococcus neoformans* as a laboratory tool and potential antifungal strategy. Microbiology 148:213-219

Gorovits R, Propheta O, Kolot M, Dombradi V, Yarden O (1999) A mutation within the catalytic domain of COT1 kinase confers changes in the presence of two COT1 isoforms and in Ser/Thr protein kinase and phosphatase activities in *Neurospora crassa*. Fungal Genet Biol 27:264-274

Gorovits R, Sjollema KA, Sietsma JH, Yarden O (2000) Cellular distribution of COT1 kinase in *Neurospora crassa*. Fungal Genet Biol 30:63-70

Greengard P (2001) The neurobiology of slow synaptic transmission. Science 294:1024-1030

Griffith JP et al. (1995) X-ray structure of calcineurin inhibited by the immunophilin- immunosuppressant FKBP12-FK506 complex. Cell 82:507-522

Halfter U, Ishitani M, Zhu JK (2000) The *Arabidopsis* SOS2 protein kinase physically interacts with and is activated by the calcium-binding protein SOS3. Proc Natl Acad Sci USA 97:3735-3740

Hellstern S, Dammann H, Husain Q, Mutzel R (1997) Overexpression, purification and characterization of Dictyostelium calcineurin A. Res Microbiol 148:335-343

Hirayama S, Sugiura R, Lu Y, Maeda T, Kawagishi K, Yokoyama M, Tohda H, Giga-Hama Y, Shuntoh H, Kuno T (2003) Zinc finger protein Prz1 regulates Ca^{2+} but not Cl- homeostasis in fission yeast: Identification of distinct branches of calcineurin signaling pathway in fission yeast. J Biol Chem 278:18078-18084

Horn F, Gross J (1996) A role for calcineurin in *Dictyostelium discoideum* development. Differentiation 60:269-275

Jayashree T, Subramanyam C (2000) Oxidative stress as a prerequisite for aflatoxin production by *Aspergillus parasiticus*. Free Radic Biol Med 29:981-985

Jin L, Harrison SC (2002) Crystal structure of human calcineurin complexed with cyclosporin A and human cyclophilin. Proc Natl Acad Sci USA 99:13522-13526

Katinka MD, Duprat S, Cornillot E, Metenier G, Thomarat F, Prensier G, Barbe V, Peyretaillade E, Brottier P, Wincker P, Delbac F, El Alaoui H, Peyret P, Saurin W, Gouy M, Weissenbach J, Vivares CP (2001) Genome sequence and gene compaction of the eukaryote parasite *Encephalitozoon cuniculi*. Nature 414:450-453

Kessen U, Schaloske R, Aichem A, Mutzel R (1999) $Ca^{(2+)}$/calmodulin-independent activation of calcineurin from *Dictyostelium* by unsaturated long chain fatty acids. J Biol Chem 274:37821-37826

Kim KN, Cheong YH, Grant JJ, Pandey GK, Luan S (2003) CIPK3, a calcium sensor-associated protein kinase that regulates abscisic acid and cold signal transduction in *Arabidopsis*. Plant Cell 15:411-423

Kingsbury TJ, Cunningham KW (2000) A conserved family of calcineurin regulators. Genes Dev 13:1595-1604

Kissinger CR et al. (1995) Crystal structures of human calcineurin and the human FKBP12-FK506- calcineurin complex. Nature 378:641-644

Kissmehl R, Treptau T, Hofer HW, Plattner H (1996) Protein phosphatase and kinase activities possibly involved in exocytosis regulation in *Paramecium tetraurelia*. Biochem J 317 (Pt 1):65-76

Kissmehl R, Treptau T, Kottwitz B, Plattner H (1997) Occurrence of a para-nitrophenyl phosphate-phosphatase with calcineurin-like characteristics in *Paramecium tetraurelia*. Arch Biochem Biophys 344:260-270

Kothe GO, Free SJ (1998) Calcineurin subunit B is required for normal vegetative growth in *Neurospora crassa*. Fungal Genet Biol 23:248-258

Kudla J, Xu Q, Harter K, Gruissem W, Luan S (1999) Genes for calcineurin B-like proteins in *Arabidopsis* are differentially regulated by stress signals. Proc Natl Acad Sci USA 96:4718-4723

Kuno T, Tanaka H, Mukai H, Chang CD, Hiraga K, Miyakawa T, Tanaka C. (1991) cDNA cloning of a calcineurin B homolog in *Saccharomyces cerevisiae*. Biochem Biophys Res Commun 180:1159-1163

Lew DJ, Reed SI (1995) A cell cycle checkpoint monitors cell morphogenesis in budding yeast. J Cell Biol 129:739-749

Li J, Lee YR, Assmann SM (1998) Guard cells possess a calcium-dependent protein kinase that phosphorylates the KAT1 potassium channel. Plant Physiol 116:785-795

Liu J, Farmer J Jr, Lane WS, Friedman J, Weissman I, Schreiber SL (1991) Calcineurin is a common target of cyclophilin-cyclosporin A and FKBP-FK506 complexes. Cell 66:807-815

Liu J, Zhu JK (1998) A calcium sensor homolog required for plant salt tolerance. Science 280:1943-1945

Locke EG, Bonilla M, Liang L, Takita Y, Cunningham KW (2000) A homolog of voltage-gated Ca^{2+} channels stimulated by depletion of secretory Ca^{2+} in yeast. Mol Cell Biol 20:6686-6694

Luan S, Kudla J, Rodriguez-Concepcion M, Yalovsky S, Gruissem W (2002) Calmodulins and calcineurin B-like proteins: calcium sensors for specific signal response coupling in plants. Plant Cell 14 Suppl:S389-S400

Luan S, Li W, Rusnak F, Assmann SM, Schreiber SL (1993) Immunosuppressants implicate protein phosphatase regulation of K^+ channels in guard cells. Proc Natl Acad Sci USA 90:2202-2206

Malleret G, Haditsch U, Genoux D, Jones MW, Bliss TV, Vanhoose AM, Weitlauf C, Kandel ER, Winder DG, Mansuy IM (2001) Inducible and reversible enhancement of learning, memory, and long-term potentiation by genetic inhibition of calcineurin. Cell 104:675-686

Mansuy IM, Mayford M, Jacob B, Kandel ER, Bach ME (1998) Restricted and regulated overexpression reveals calcineurin as a key component in the transition from short-term to long-term memory. Cell 92:39-49

Marchetti O, Entenza JM, Sanglard D, Bille J, Glauser MP, Moreillon P (2000a) Fluconazole plus cyclosporine: a fungicidal combination effective against experimental endocarditis due to *Candida albicans*. Antimicrob Agents Chemother 44:2932-2938

Marchetti O, Moreillon P, Glauser MP, Bille J, Sanglard D (2000b) Potent synergism of the combination of fluconazole and cyclosporine in *Candida albicans*. Antimicrob Agents Chemother 44:2373-2381

Matheos DP, Kingsbury TJ, Ahsan US, Cunningham KW (1997) Tcn1p/Crz1p, a calcineurin-dependent transcription factor that differentially regulates gene expression in *Saccharomyces cerevisiae*. Genes Dev 11:3445-3458

Mazur P, Morin N, Baginsky W, el-Sherbeini M, Clemas JA, Nielsen JB, Foor F (1995) Differential expression and function of two homologous subunits of yeast 1,3-beta-D-glucan synthase. Mol Cell Biol 15:5671-5681

Mendizabal I, Rios G, Mulet JM, Serrano R, de Larrinoa IF (1998) Yeast putative transcription factors involved in salt tolerance. FEBS Lett 425:323-328

Mendoza I, Rubio F, Rodriguez-Navarro A, Pardo JM (1994) The protein phosphatase calcineurin is essential for NaCl tolerance of *Saccharomyces cerevisiae*. J Biol Chem 269:8792-8796

Mizunuma M, Hirata D, Miyahara K, Tsuchiya E, Miyakawa T (1998) Role of calcineurin and Mpk1 in regulating the onset of mitosis in budding yeast. Nature 392:303-306

Mizunuma M, Hirata D, Miyaoka R, Miyakawa T (2001) GSK-3 kinase Mck1 and calcineurin coordinately mediate Hsl1 down- regulation by Ca^{2+} in budding yeast. EMBO J 20:1074-1085

Molkentin JD, Lu JR, Antos CL, Markham B, Richardson J, Robbins J, Grant SR, Olson EN (1998) A calcineurin-dependent transcriptional pathway for cardiac hypertrophy. Cell 93:215-228

Momayezi M, Kissmehl R, Plattner H (2000) Quantitative immunogold localization of protein phosphatase 2B (calcineurin) in *Paramecium* cells. J Histochem Cytochem 48:1269-1281

Moniakis J, Coukell MB, Janiec A (1999) Involvement of the Ca^{2+}-ATPase PAT1 and the contractile vacuole in calcium regulation in *Dictyostelium discoideum*. J Cell Sci 112:405-414

Moser MJ, Geiser JR, Davis TN (1996) Ca^{2+}-calmodulin promotes survival of pheromone-induced growth arrest by activation of calcineurin and Ca^{2+}-calmodulin-dependent protein kinase. Mol Cell Biol 16:4824-4831

Moudy R, Manning TJ, Beckers CJ (2001) The loss of cytoplasmic potassium upon host cell breakdown triggers egress of *Toxoplasma gondii*. J Biol Chem 276:41492-41501

Muller EM, Locke EG, Cunningham KW (2001) Differential regulation of two Ca^{2+} influx systems by pheromone signaling in *Saccharomyces cerevisiae*. Genetics 159:1527-1538

Odom A, Del Poeta M, Perfect J, Heitman J (1997a) The immunosuppressant FK506 and its nonimmunosuppressive analog L-685,818 are toxic to *Cryptococcus neoformans* by inhibition of a common target protein. Antimicrob Agents Chemother 41:156-161

Odom A, Muir S, Lim E, Toffaletti DL, Perfect J, Heitman J (1997b) Calcineurin is required for virulence of *Cryptococcus neoformans*. EMBO J 16:2576-2589

Onyewu C, Blankenship JR, Del Poeta M, Heitman J (2003) Ergosterol biosynthesis inhibitors become fungicidal when combined with calcineurin inhibitors against *Candida albicans, Candida glabrata*, and *Candida krusei*. Antimicrob Agents Chemother 47:956-964

Pardo JM, Reddy MP, Yang S, Maggio A, Huh GH, Matsumoto T, Coca MA, Paino-D'Urzo M, Koiwa H, Yun DJ, Watad AA, Bressan RA, Hasegawa PM (1998) Stress signaling through Ca^{2+}/calmodulin-dependent protein phosphatase calcineurin mediates salt adaptation in plants. Proc Natl Acad Sci USA 95:9681-9686

Polizotto RS, Cyert MS (2001) Calcineurin-dependent nuclear import of the transcription factor Crz1p requires Nmd5p. J Cell Biol 154:951-960

Prokisch H, Yarden O, Dieminger M, Tropschug M, Barthelmess IB (1997) Impairment of calcineurin function in *Neurospora crassa* reveals its essential role in hyphal growth, morphology and maintenance of the apical Ca^{2+} gradient. Mol Gen Genet 256:104-114

Rascher C, Pahl A, Pecht A, Brune K, Solbach W, Bang H (1998) Leishmania major parasites express cyclophilin isoforms with an unusual interaction with calcineurin. Biochem J 334 (Pt 3):659-667

Rasmussen C, Garen C, Brining S, Kincaid RL, Means RL, Means AR (1994) The calmodulin-dependent protein phosphatase catalytic subunit (calcineurin A) is an essential gene in *Aspergillus nidulans*. EMBO J 13:2545-2552

Remillard SP, Lai EY, Levy YY, Fulton C (1995) A calcineurin-B-encoding gene expressed during differentiation of the amoeboflagellate *Naegleria gruberi* contains two introns. Gene 154:39-45

Rothermel B, Vega RB, Yang J, Wu H, Bassel-Duby R, Williams RS (2000) A protein encoded within the Down syndrome critical region is enriched in striated muscles and inhibits calcineurin signaling. J Biol Chem 275:8719-8725

Rusnak F, Mertz P (2000) Calcineurin: form and function. Physiol Rev 80:1483-1521

Schiaffino S, Serrano A (2002) Calcineurin signaling and neural control of skeletal muscle fiber type and size. Trends Pharmacol Sci 23:569-575

Severin FF, Hyman AA (2002) Pheromone induces programmed cell death in *S. cerevisiae*. Curr Biol 12:R233-235

Shi J, Kim KN, Ritz O, Albrecht V, Gupta R, Harter K, Luan S, Kudla J (1999) Novel protein kinases associated with calcineurin B-like calcium sensors in *Arabidopsis*. Plant Cell 11:2393-2405

Sone T, Griffiths AJ (1999) The frost gene of *Neurospora crass*a is a homolog of yeast cdc1 and affects hyphal branching via manganese homeostasis. Fungal Genet Biol 28:227-237

Stathopoulos AM, Cyert MS (1997) Calcineurin acts through the *CRZ1/TCN1* encoded transcription factor to regulate gene expression in yeast. Genes Dev 11:3432-3444

Stathopoulos-Gerontides A, Guo JJ, Cyert MS (1999) Yeast calcineurin regulates nuclear localization of the Crz1p transcription factor through dephosphorylation. Genes Dev 13:798-803

Sugiura R, Sio SO, Shuntoh H, Kuno T (2002) Calcineurin phosphatase in signal transduction: lessons from fission yeast. Genes Cells 7:619-627

Vega RB, Rothermel BA, Weinheimer CJ, Kovacs A, Naseem RH, Bassel-Duby R, Williams RS, Olson EN (2003) Dual roles of modulatory calcineurin-interacting protein 1 in cardiac hypertrophy. Proc Natl Acad Sci USA 100:669-674

Vega RB, Yang J, Rothermel BA, Bassel-Duby R, Williams RS (2002) Multiple domains of MCIP1 contribute to inhibition of calcineurin activity. J Biol Chem 277:30401-30407

Viaud MC, Balhadere PV, Talbot NJ (2002) A *Magnaporthe grisea* cyclophilin acts as a virulence determinant during plant infection. Plant Cell 14:917-930

Winder DG, Sweatt JD (2001) Roles of serine/threonine phosphatases in hippocampal synaptic plasticity. Nat Rev Neurosci 2:461-474

Withee JL, Mulholland J, Jeng R, Cyert MS (1997) An essential role of the yeast pheromone-induced Ca^{2+} signal is to activate calcineurin. Mol Biol Cell 8:263-277

Withee JL, Sen R, Cyert MS (1998) Ion tolerance of *Saccharomyces cerevisiae* lacking the Ca^{2+}/CaM- dependent phosphatase (calcineurin) is improved by mutations in *URE2* or *PMA1*. Genetics 149:865-878

Yakel JL (1997) Calcineurin regulation of synaptic function: from ion channels to transmitter release and gene transcription. Trends Pharmacol Sci 18:124-134

Yoshida T, Toda T, Yanagida M (1994) A calcineurin-like gene ppb1[+] in fission yeast: mutant defects in cytokinesis, cell polarity, mating and spindle pole body positioning. J Cell Sci 107 (Pt 7):1725-1735

Yoshimoto H, Saltsman K, Gasch AP, Li HX, Ogawa N, Botstein D, Brown PO, Cyert MS (2002) Genome-wide analysis of gene expression regulated by the calcineurin/Crz1p signaling pathway in *Saccharomyces cerevisiae*. J Biol Chem 277:31079-31088

5 Roles of mammalian protein phosphatase 2C family members in the regulation of cellular functions

Shinri Tamura, Ming Guang Li, Ken-ichiro Komaki, Masato Sasaki and Takayasu Kobayashi

Abstract

The mammalian protein phosphatase 2C (PP2C) family is composed of at least ten different gene products, which share six uniquely conserved motifs. All members have magnesium- and/or manganese-dependent protein phosphatase activities. Biological and biochemical studies have revealed the participation of PP2C family members in the regulation of a variety of cellular functions that include the stress response, the cell cycle, actin cytoskeleton organization, the Wnt signaling pathway, the CFTR chloride ion channel and pre-mRNA splicing. Here, we describe the regulatory mechanisms of these specific cellular functions by PP2C family members.

5.1 Introduction

Protein phosphatase 2C (PP2C) was first identified and purified from rat liver [protein phosphatase IA, (Kikuchi et al. 1977; Hiraga et al. 1981)] and turkey gizzard [SMP-II, (Pato and Adelstein 1980)] as a 43 to 48 kDa magnesium-dependent protein Ser/Thr phosphatase. The presence of two subtypes ($2C_1$ and $2C_2$, later termed 2Cα and 2Cβ, respectively) was demonstrated in skeletal muscle and liver of rabbits and rats (McGowan and Cohen 1987; McGowan et al. 1987). Subsequently, cDNA clones encoding the full-length PP2Cα were isolated from rat kidney (Tamura et al. 1989a), rabbit liver, and human teratocarcinoma (Mann et al. 1992) cDNA libraries and the active enzymatic proteins were successfully expressed in E. coli (Tamura et al. 1989b). A cDNA clone (JW5), encoding PP2Cβ, was first isolated from a rat liver cDNA library by Wenk et al. (1992). Subsequently, several PP2Cβ splicing variants have been reported (Terasawa et al. 1993; Hou et al. 1994; Kato et al. 1995; Marley et al. 1998; Seroussi et al. 2001), with an additional eight novel PP2C genes [2Cγ (FIN13), 2Cδ (ILKAP), 2Cε, 2Cζ, Wip1, CaMKP (hFEM2, POPX2), NERPP-2C, and CaMKP-N (POPX1)] have been identified in mammalian cells (Travis and Welsh 1997; Guthridge et al. 1997; Tong et al. 1998; Leung-Hagesteijin et al. 2001; Li et al. 2003; Kashiwaba

Topics in Current Genetics, Vol. 5
J. Arino, D.R. Alexander (Eds.) Protein phosphatases
© Springer-Verlag Berlin Heidelberg 2004

```
                  I           II            III                      IV
PP2Cα       RVEMED    FFAVYDGHAG    GSTAVGVLISPQHTYFINCGDSRGLL    RVN----GSLAVSRALGDF
PP2Cβ       RVEMED    FFAVYDGHAG    GSTAVGVMVSPTHMYFINCGDSRAVL    RVN----GSLAVSRALGDY
PP2Cγ       RVSMED    MFSVYDGHGG    GTTAVVALIRGKQLIVANAGDSRCVV    RVN----GGLNLSRAIGDH
PP2Cδ       REEMQD    YFAVFDGHGG    GSTATCVLAVDNILYIANLGDSRAIL    RVL----GVLEVSRSIGDG
PP2Cε       RDHMED    IFGIFDGHGG    GTTCLIALLSDKDLTVANVGDSRGVL    RVQ----GILAMSRSLGDY
PP2Cζ       KSRHNED   YWGLFDGHAG    GCCALVVLYLLGKMYVANAGDSRAII    RVM----ATIGVTRGLGDH
Wip1        RKYMED    FFAVCDGHGG    GTTASVVIIRGMKMYVAHVGDSGVVL    GVN----PFLAVARALGDL
CaMKP       RRKMED    YFAVFDGHGG    GTTGVCALITGAALHVAWLGDSQVIL    RVN----GTLAVSRAIGDV
NERPP-2C    KSTHNED   YWSLFDGHAG    GCTALIVVCLLGKLYVANAGDSRAII    RVM----ATIGVTRGLGDH
POPX1       RRKMED    YFAVFDGHGG    GTTGVVTFIRGNMLHVAWVGDSQVML    RVN----GSLSVSRAIGDA

                       V                                  VI
PP2Cα       PEVHDIERSEEDDQFII-LAC-DGIWD        DNMSVILICF
PP2Cβ       PEVYEIVRAEEDE-FVV-LAC-DGIWD        DNMSVVLVCF
PP2Cγ       PDIKVLTLT-DDHEFMV-IAC-DGIWN        DNMTCIIICF
PP2Cδ       PDIRRCQLTPNDR-FIL-LAC-DGLFK        DNVTVMVVRI
PP2Cε       PDILTFDLDKLQPEFMI-LA-SDGLWD        DNITVMVVKF
PP2Cζ       PEVRVYDLTQYEHCPDDVLVLGTDGLWD       DDISVFVIPL
Wip1        PEPDTSVHTLDPRKHKYI-ILGSDGLWN       DNTSAIVICI
CaMKP       ADAASRELTGLED-YLL-LAC-DGFFD        DNITVMVVFL
NERPP-2C    PEVRVYDLSKYEHGADDVLILATDGLWD       DDISVYVIPL
POPX1       ADSASTVLDGTED-YLI-LAC-DGFYD        DNITVIVVFL
```

Fig. 1. Conserved motifs of PP2C family members. The six conserved motifs (I-VI) of mammalian PP2C family members are depicted. The amino acids, which participate in holding manganese ions in the catalytic site are underlined.

et al. 2003; Fiscella et al 1997; Kitani et al. 1999; Tan et al. 2001; Koh et al. 2002; Labes et al. 1998 ; Takeuchi et al. 2001). Properties of these ten PP2C gene products are summarized in Table 1. In this article, we mainly focused on the roles of PP2C family members in cellular function regulation.

5.2 Structure of the PP2C catalytic site

Structurally, all ten PP2C gene products share six conserved motifs (Pilgrim et al. 1995) (Fig. 1). Studies of the PP2Cα crystal structures by Das et al. (1996) revealed that the catalytic domain is composed of a central β–sandwich that binds two manganese ions and is surrounded by α–helices. The location of the six conserved motifs suggests that these motifs, except for motif IV, participate in the formation of the central β–sandwich structure. It was proposed that Glu37, Asp38, Asp60, Gly61, Asp239, and Asp282 play key roles in holding manganese ions in the catalytic site and that the metal-bound water molecule at the catalytic site participates in catalysis as a nucleophile and general acid.

Kusuda et al. (1998) investigated the structure-activity relationship for PP2Cβ−1, which is composed of 390 amino acids, using deletion and point mutations. Analysis of the deletion mutants indicated that the catalytic domain resides in the amino-terminal 300 amino acids. Substitution of each of the 6 specific amino acids caused 98-100% loss of enzymatic activity. Three of these amino acids, Asp38, Asp60, and Asp243, corresponded to Asp38, Asp60, and Asp239 of PP2Cα, respectively. These observations support the idea that the manganese ions play a key role in the catalytic reaction.

Table 1. Properties of mammalian PP2C family members

	Size (kDa)	Isoforms	Tissue distribution	Subcellular localization	References
PP2Cα	42, 36	α-1, α-2	ubiquitous		Tamura et al. 1989a; Mann et al. 1992; Takekawa et al. 1998
PP2Cβ	42-55	β-1(βs)-β-5, βx (βl)	β-1, βx: ubiquitous β-2: brain, heart β-3-5: testis, liver, intestine	β-1, βx: cytoplasm β-1: nucleus	Wenk et al. 1992; Terasawa et al. 1993; Hou et al. 1994; Kato et al. 1995; Marley et al. 1998; Seroussi et al. 2001
PP2Cγ (FIN13)	59		ubiquitous	nucleus	Travis and Welsh 1997; Guthridge et al. 1997
PP2Cδ (ILKAP)	43		ubiquitous		Tong et al. 1998; Leung-Hagesteijin et al. 2001
PP2Cε	34		rich in brain, heart and testis		Li et al. 2003
PP2Cζ	56		testis		Kashiwaba et al. 2003
Wip1	66			nucleus	Fiscella et al. 1997
CaMKP (hFEM2, POPX2)	49		ubiquitous		Kitani et al. 1999; Tan et al. 2001; Koh et al. 2002
NERPP-2C	55 80	NERPP-55 NERPP-80	brain	N-55: cytoplasm N-80: membrane	Labes et al. 1998
CaMKP-N (POPX1)	83		rich in brain, testis	nucleus	Takeuchi et al. 2001; Koh et al. 2002

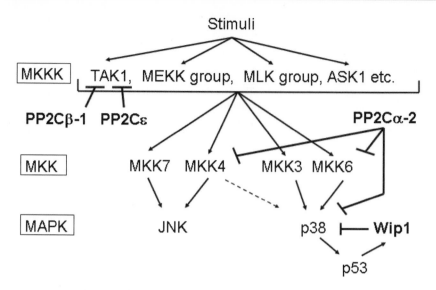

Fig. 2. Regulation of SAPK signaling pathways by PP2C. The protein kinase cascades of SAPK signaling pathways and the points where the phosphatases can interfere with the signals are shown. MKKK, MKK kinase; MKK, MAPK kinase; MAPK, MAP kinase; TAK1, TGF-β-activated kinase 1; MEKK, MEK kinase; MLK, mixed lineage kinase; ASK1, apoptosis signal-regulating kinase 1; JNK, c-Jun N-terminal kinase.

5.3 Regulation of stress-activated protein kinase signaling pathways

5.3.1 PP2Cα

Stress-activated protein kinases (SAPKs), a subfamily of the mitogen-activated protein kinase (MAPK) superfamily, are highly conserved from yeast to mammals. SAPKs relay signals in response to various extracellular stimuli, including environmental stresses and proinflammatory cytokines. In mammalian cells, two distinct classes of SAPKs have been identified, the c-Jun N-terminal kinases (JNK) and the p38 MAPKs (Ip and Davis 1998; Garrington and Johnson 1999) (Fig. 2). SAPK activation requires phosphorylation of conserved tyrosine and threonine residues within the catalytic domain. This phosphorylation is mediated by dual specificity protein kinases, which are members of the MAPK kinase (MKK) family. MKK3 and MKK6 are specific for p38, MKK7 selectively phosphorylates JNK, and MKK4 recognizes either class of the stress-activated kinases (Fig. 2). The MKKs are also activated by the phosphorylation of conserved serine and threonine residues. Several MKK-activating MKK kinases (MKKKs) have been identified, some of which are activated again by phosphorylation (Davis 2000; Kishimoto et al. 2000). In the absence of a signal, the constituents of the

SAPK cascade return to their inactive, dephosphorylated state, suggesting an essential role for phosphatases in SAPK regulation.

PP2Cα participation in the regulation of SAPK signaling pathways was first reported by Takekawa et al. (1998). They isolated a cDNA clone encoding PP2Cα-2 while screening a human cDNA library for genes down-regulating the yeast Hog1 MAPK pathway. They found that PP2Cα-2 overexpression in mammalian cells inhibited stress-induced activation of p38 and JNK, but it had no effect on the PMA-induced activation of ERK, suggesting that PP2Cα-2 selectively inhibits SAPK signaling pathways. Hanada et al. (1998) have found that mouse PP2Cα, corresponding to human PP2Cα-1, exhibited a similar inhibition pattern. PP2Cα-2 dephosphorylated and inactivated MKK4, MKK6, and p38 both in vivo and in vitro (Takekawa et al. 1998). Furthermore, PP2Cα-2 specifically associated with phosphorylated p38.

5.3.2 PP2Cβ

Hanada et al. (1998) have found that in mammalian cells, PP2Cβ-1 selectively suppressed the stress-induced activation of p38 and JNK, but had no effect on the mitogen-induced ERK activation. Investigation of the PP2Cβ-1-mediated suppression of the SAPK pathway revealed that PP2Cβ-1 dephosphorylates and inactivates transforming growth factor-β (TGF-β)-activated kinase (TAK1), a MKKK activated either by stress, TGF-β treatment or interleukin-1 (IL-1) stimulation (Hanada et al. 2001). In addition, PP2Cβ-1 selectively associates with TAK1 in a stable complex. Expression of a PP2Cβ-1 dominant-negative form enhances the IL-1-induced activation of the AP-1 reporter gene, suggesting PP2Cβ-1 negatively regulates TAK1 signaling through TAK1 dephosphorylation in vivo.

5.3.3 Wip1

Fiscella et al. (1997) found that Wip1, a 61-kDa magnesium-dependent protein phosphatase, was induced by ionizing radiation in a p53-dependent manner and was localized to the nucleus. Ectopic expression of Wip1 in a glioblastoma cell line caused growth inhibition, suggesting Wip1 contributes to the p53-mediated signals of DNA damage.

Takekawa et al. (2000) demonstrated that Wip1 expression was also induced by treatment of mammalian cells with methyl methane sulfonate, H_2O_2, or anisomycin. Functional studies of Wip1 revealed its role in the down-regulation of p38-p53-induced signaling during the recovery of damaged cells. Thus, Wip1 induction by stress selectively blocked p38 activation and suppressed subsequent p53 activation. In vitro experiments have shown that Wip1 inactivated p38 by the specific dephosphorylation of a conserved threonine residue; however, it rejected ERK, JNK, MKK4, or MKK6 as a substrate.

5.3.4 PP2Cε

. P2Cε is a novel member of the PP2C family recently identified by Li et al. (2003). They found that ectopic expression of PP2Cε in 293IL-1RI cells inhibited the IL-1- and TAK1-induced activation of the MKK4-JNK or MKK3-p38 signaling pathway. PP2Cε was found to dephosphorylate TAK1 in vitro. Similar to PP2Cβ, PP2Cε expressed in cells was stably associated with TAK1 and attenuated TAK1 binding to MKK4 or MKK6. Ectopic expression of a PP2Cε phosphatase-negative mutant, PP2Cε(D/A), which acted as a dominant negative form, enhanced both the association between TAK1 and MKK4 or MKK6 and the TAK1-induced activation of an AP-1 reporter gene. The association between PP2Cε and TAK1 was transiently suppressed by IL-1 treatment, suggesting that in the absence of IL-1-induced signal, PP2Cε contributes to inactivating the TAK1 signaling pathway by associating with and dephosphorylating TAK1. Functional differences between PP2Cε and PP2Cβ-1 remain to be elucidated.

5.4 Cell cycle regulation

5.4.1 PP2Cα and PP2Cβ

Eukaryotic cell cycle progression is driven by the ordered activation and inactivation of cyclin-dependent protein kinases (CDKs) (Morgan 1996; Morgan 1997; Solomon and Kaldis 1998). CDK full activation, which is necessary for normal cell cycle progression, requires binding of a cyclin, removal of the inhibitory phosphorylation, and the occurrence of activating phosphorylation. Activating phosphorylation, performed by Cdk-activating kinases, is within the "T-loop" on a conserved threonine residue, corresponding to Thr161 in human Cdc2 and Thr160 in human Cdk2.

Cheng et al. (2000) found that two budding yeast PP2C, Ptc2 and Ptc3, are the physiological enzymes that dephosphorylate Thr169 on Cdc28, which is the main CDK involved in regulating the budding yeast cell division cycle. They also observed that PP2C-like activities were responsible for >99% of the phosphatase activity in HeLa cell extracts acting on Cdk2 Thr169. Subsequently, they demonstrated that the Cdk2 and Cdk6 phosphatase activities of the HeLa cell extracts copurified with PP2Cα and PP2Cβ. In addition, both the recombinant PP2Cα and PP2Cβ were found to effectively dephosphorylate monomeric, but not cyclin-bound Cdk2 and Cdk6 in vitro, a situation observed with the budding yeast PP2Cs, supporting the conclusion that PP2Cα and PP2Cβ are the phosphatases that dephosphorylate human CDKs in vivo. Using a similar approach, De Smedt et al. (2002) has concluded that Xenopus PP2Cα is responsible for the dephosphorylation of Thr161 of *Xenopus* monomeric Cdk2.

Another mechanism of cell cycle regulation by PP2Cα has been recently proposed by Ofek et al. (2003). Using the induced expression of exogenous PP2Cα in

293 cells, they demonstrated that PP2Cα expression led to G2/M cell cycle arrest and apoptosis. PP2Cα induced the expression of endogenous p53 and the p53-responsive gene p21, a CDK inhibitor. A PP2Cα mutant lacking phosphatase activity showed no such effects. They concluded that p53 plays an important role in PP2Cα directed cell cycle arrest and apoptosis since perturbation of p53 expression in 293 cells by papilloma virus E6 led to a significant increase in the survival of cells expressing exogenous PP2Cα.

5.4.2 PP2Cδ

Tong et al. (1998) found that transfection with PP2Cδ resulted in inhibition of cell growth, precluding generation of stable 293 or CHO transfectants. Using a modified tetracycline-regulated PP2Cδ–GFP dicistronic expression cassette, they revealed that PP2Cδ overexpression blocked cell cycle progression and arrested cells in early S phase, resulting in inhibition of DNA synthesis, and subsequently leading to cell death. These results suggest that PP2Cδ has a role in the regulation of cell cycle progression, although the mechanistic details are unknown.

5.5 Regulation of actin cytoskeleton organization

5.5.1 PP2Cα

Members of the moesin protein family are localized in membrane structures rich in actin filaments, such as filopodia, membrane ruffles, microvilli, or the cleavage furrow in various cell types (Amieva and Furthmayr 1995; Berryman et al. 1993; Sato et al. 1991). They are proposed to act as linkers between the plasma membrane and the actin cytoskeleton. Moesin is phosphorylated in human platelets by thrombin activation at Thr558, located within or near the F-actin binding domain in the C-terminus. The phosphorylation increase correlates with moesin F-actin binding activity.

Hishiya et al. (1999) demonstrated that calyculin A-sensitive and -insensitive moesin phosphatase activities were present in human platelet lysates. They purified the calyculin A-insensitive phosphatase and immunostained it with anti-PP2Cα antibody. The purified enzyme, as well as the recombinant PP2Cα, efficiently dephosphorylated the phospho-moesin and inactivated F-actin binding, suggesting that PP2Cα influences dynamic interactions between the actin cytoskeleton and the membrane constituents linked to moesin.

5.5.2 POPXs

The p21 (Cdc42/Rac)-activated kinase PAK mediates a number of biological effects downstream of the Rho GTPases, including the organization of the actin cytoskeleton and cell motility (Bagrodia and Cerione 1999). The phosphorylation state of mammalian PAK is highly regulated: upon binding of GTPases, PAK is potently activated by autophosphorylation at multiple sites.

The protein phosphatase (POPX1) responsible for PAK downregulation has been isolated as a binding protein of the PAK interacting guanine nucleotide exchanging factor PIX by Koh et al. (2002). POPX1 was found to be the human ortholog of murine CaMKP-N (Takeuchi et al. 2001). POPX2 was also identified in a database as a putative protein phosphatase, which shares extensive homology with POPX1, and is the human ortholog of murine CaMKP (Kitani et al. 1999). Both POPX1 and POPX2 inactivated the CDC42-activated PAK in the cells and dephosphorylated PAK (at Thr422) in vitro, which is a key regulator of activity in the kinase activation loop. PAK and POPX bind to different regions of PIX and a trimeric complex of POPX-PIX-PAK can be formed in vivo. POPX dephosphorylating activity correlated with an ability to block the in vivo effects of active PAK or active Cdc42, including actin stress fiber breakdown and morphological changes. Since these effects by POPX were further enhanced by the presence of PIX, PIX was suggested to act as a targeting subunit, coupling POPX to PAK.

5.6 Regulation of Wnt signaling pathways

5.6.1 PP2Cα

It is well known that during animal development, the Wnt signaling pathway plays crucial roles in cell adhesion and cell fate determination. Defects in this pathway result in abnormalities of physiological events ranging from early developmental processes to oncogenesis. Wnt proteins constitute a large family of cysteine-rich secreted ligands. The current model of Wnt signal transduction proposes that in the absence of a Wnt signal, β-catenin binds to an APC1-Axin complex, where it is phosphorylated by GSK-3 and subsequently targeted for ubiquitin-mediated proteasomal degradation. When Wnt signaling is activated via ligand binding to the frizzled encoded receptors, GSK-3 activity is inhibited, and cytosolic β-catenin accumulates and interacts with TCF/LEF-1 transcription factors, leading to translocation to the nucleus and transcriptional modulation of target genes (Novak and Dedhar 1999; Kikuchi 1999a; Kikuchi 1999b; Sakanaka et al. 2000). Axin has been identified as a negative regulator of the Wnt signaling pathway. Axin binds to GSK-3, β-catenin and APC, with expression of Axin resulting in the rapid turnover of β-catenin. The disheveled gene family (Dvl) encodes cytoplasmic proteins that are associated with Axin and essential for transmission of the Wnt signal.

Utilizing the yeast two-hybrid system, Strovel et al. (2000) identified PP2Cα as a Dvl-interacting protein. PP2Cα exists in a complex with Dvl, β-catenin, and Axin. In a Wnt-responsive LEF-1 reporter gene assay, expression of PP2Cα activated transcription and also elicited a synergistic response with β-catenin and Wnt-1. In addition, PP2Cα expression relieved Axin-mediated repression of LEF-1-dependent transcription. PP2Cα utilized Axin as a substrate both in vitro and in vivo and decreased its half-life. These results indicate that PP2Cα is a positive regulator of Wnt signal transduction and mediates its effects through Axin dephosphorylation.

5.6.2 ILKAP (PP2Cδ)

Integrin-linked kinase, ILK1, functions to mediate integrin signal transduction (Hannigan et al. 1996). Extracellular matrices and growth factors each stimulate rapid, transient activation of ILK1 activity. ILK1 has been shown to target glycogen synthase kinase 3β (GSK3β), which is an important mediator of developmental signaling of the Wnt/wingless pathway. GSK3β phosphorylation of β-catenin targets the latter for degradation and ILK1 overexpression in epithelial cells induces inhibitory phosphorylation of GSK3β at Ser9, resulting in the stabilization and nuclear translocation of β-catenin, with concomitant activation of Tcf/Lef factors.

Leung-Hagesteijin et al. (2001) isolated an ILK1 binding protein (ILKAP) using a yeast two-hybrid screening, which was identical to PP2Cδ. ILK1 and ILKAP were co-precipitated from HEK 293 lysates, independently of ILK1 or ILKAP catalytic activities. Expression of recombinant, catalytically active ILKAP in HEK 293 cells resulted in the inhibition of protein kinase activity in ILK1 immune complexes, and the catalytically inactive ILKAP mutant did not inhibit ILK1 activity. ILKAP expression strongly inhibited integrin-stimulated phosphorylation of GSK3β at Ser9. ILKAP selectively inhibited activation of Tcf/Lef factors, consistent with the observed inhibition of GSK3β Ser9 phosphorylation, suggesting that ILKAP complexes with ILK1 to selectively inhibit the GSK3β arm of ILK1 signaling.

5.7 Regulation of the cystic fibrosis transmembrane conductance regulator

5.7.1 PP2Cα

Activity of the cystic fibrosis transmembrane conductance regulator (CFTR) chloride ion channel requires ATP and is regulated by cAMP-dependent phosphorylation of multiple serine residues in the R domain (Cheng et al. 1991; Picciotto et al. 1992; Chang et al. 1993). Phosphorylation by protein kinase A (PKA) increases open probability (P0), bursting rate, apparent ATP affinity, and the ATP hydrolysis rate. Channel rundown is caused by dephosphorylation of PKA sites.

Travis et al. (1997) found that in airway and colonic epithelial cells, two major sites of cystic fibrosis, neither okadaic acid nor FK506 prevented inactivation when cAMP was removed, suggesting that a phosphatase distinct from PP1, PP2A, and PP2B was responsible for CFTR inactivation. In their studies, addition of recombinant PP2Cα to membrane patches from HeLa cells expressing CFTR inactivated CFTR chloride ion channels and reactivation required PKA readdition. They further demonstrated that recombinant PP2Cα dephosphorylated CFTR and the R domain peptide in vitro, raising the possibility that PP2Cα is involved in CFTR chloride ion channel regulation.

A specific association between CFTR and PP2C has been demonstrated by Zhu et al. (1999). In their study, a monoclonal anti-CFTR antibody co-precipitated PP2C from baby hamster kidney cells stably expressing CFTR, but did not co-precipitate PP1, PP2A, or PP2B. Conversely, a polyclonal anti-PP2Cα peptide antibody co-precipitated CFTR from baby hamster kidney membrane extracts. Exposing baby hamster kidney cell lysates to dithiobis (sulfosuccinimidyl propionate) caused the cross-linking of the histidine-tagged CFTR (CFTRHis10) and PP2C into high molecular weight complexes that were isolated by chromatography on nickel-nitrilotriacetic acid-agarose. Chemical cross-linking was specific for PP2C because PP1, PP2A, and PP2B did not co-purify with CFTRHis10 after dithiobis (sulfosuccinimidyl propionate) exposure, suggesting that CFTR and PP2C exist in a stable complex facilitating channel regulation.

5.8 Regulation of pre-mRNA splicing

5.8.1 PP2Cγ

Pre-mRNA splicing is a complex reaction that requires five small nuclear ribonucleoprotein particles (snRNPs) and numerous non-snRNP protein factors that assemble into a spliceosome (Kramer 1996; Will and Luhrmann 1997). Although a large number of components required for mammalian splicing have yet to be identified, studies using the complementation assay have revealed the presence of at least four essential factors, SR proteins, SF1, SF3a, and SF3b. SF1, also called mBBP, is involved in branch site recognition, and SF3a and SF3b participate in U2 snRNP binding to the branch site. Multiple cycles of phosphorylation and dephosphorylation may be required for splicing. A number of mammalian kinases have been implicated in splicing, including SRPK1 and SRPK2, Clk/Sty, a CaMK II-like kinase and cyclin E-cdk2. In vitro, most of these kinases phosphorylate the SR carboxy-terminal RS domains, which are extensively phosphorylated in vivo (Fu 1995). RS domain phosphorylation appears to be required for some SR protein function and localization (Xiao and Manley 1997; Misteli and Spector 1998). In addition, experiments with thiophosphorylated U1-70K protein and SF2/ASF suggested that a specific dephosphorylation event(s) is required for splicing (Tazi et al. 1993; Cao et al. 1997).

To identify activities involved in human pre-mRNA splicing, Murray et al. (1999) have developed a procedure to separate HeLa cell nuclear extract into five complementing fractions. An activity called SCF1 was purified from one of these fractions by assaying for splicing reconstitution in the presence of the remaining four fractions. A SCF1 component was shown to be PP2Cγ. PP2Cγ was physically associated with the spliceosome in vitro throughout the splicing reaction, but was first required during the early stages of spliceosome assembly for efficient formation of the A complex, a precursor to the spliceosome. The phosphatase activity was required for the PPCγ splicing function, as an active site mutant did not support spliceosome assembly. The requirement for PP2Cγ was highly specific as the closely related phosphatase PP2Cα was unable to substitute for PP2Cγ. Consistent with a role in splicing, PP2Cγ localized to the nucleus in vivo. These observations indicate that at least one specific PP2Cγ catalyzed dephosphorylation event is required for spliceosome formation.

5.9 Conclusions and Perspectives

In the history of protein Ser/Thr phosphatase research, the membrane permeable inhibitors of PP1 and PP2A (okadaic acid, tautomycin etc.) and PP2B (cyclosporin A and FK506) have played the key roles in elucidating the functions of these phosphatase groups in vivo. In contrast, unsuccessful development of the inhibitors specific for PP2C has been one of the reasons for the slow progress in the functional studies of this protein phosphatase group. However, recent progress in the molecular cell biological studies using the transient or stable expression of the exogenous PP2C together with the biochemical studies has been revealing a variety of unique functions of PP2C family members.

The presence of a variety of genes encoding the PP2C catalytic protein is in contrast with the situation observed in other protein Ser/Thr phosphatase groups, such as PP1 and PP2A. Mammalian cells only have a few genes encoding the catalytic subunit of PP1 or PP2A. However, both PP1 and PP2A have a large number of regulatory proteins, enabling the generation of various holoenzymes. Considering that PP2Cs are essentially monomeric enzymes, it is tempting to speculate that PP2C genomic diversification may have occurred during the course of molecular evolution so that cells have various catalytic proteins with different substrate specificities and regulation, which interact with a large number of protein kinase reactions.

Recent progress in the genome and EST sequencing projects have revealed that Saccharomyces cerevisiae, Drosophila melanogaster, Caenorhabiditis elegans, and Mesembryanthemum crystallinum (ice plant) have eight, twelve, nine, and ten different PP2C genes, respectively (Adams et al. 2000; Miyazaki et al. 1999). In addition, the existence of sixty-nine different PP2C genes has been reported in Arabidopsis thaliana (Kerk et al. 2002). PP2C structural studies from these species have revealed that they contain unique motifs, such as the kinase interaction domain of KAPP (A. thaliana) and the transmembrane domains of KAPP and MPC8 (ice plant), which have not been observed in mammalian PP2C family members (Miyazaki et al. 1999; Stone et al. 2002). These observations raise the possibility that there may be several unidentified PP2C genes in mammalian cells. Molecular cloning and functional clarification of unidentified PP2C family members will be a challenging and exciting topic for the next several years.

References

Adams MD et al.The genome sequence of *Drosophila melanogaster*. Science 287:2185-2195

Amieva MR, Furthmayr H (1995) Subcellular localization of moesin in dynamic filopodia, retraction fibers, and other structures involved in substrate exploration, attachment, and cell-cell contacts. Exp Cell Res 219:180-196

Bagrodia S, Cerione RA (1999) Pak to the future. Trends Cell Biol 9:350-355

Berryman M, Franck Z, Bretscher A (1993) Ezrin is concentrated in the apical microvilli of a wide variety of epithelial cells whereas moesin is found primarily in endothelial cells. J Cell Sci 105:1025-1043

Cao W, Jamison SF, Garcia-Blanco MA (1997) Both phosphorylation and dephosphorylation of ASF/SF2 are required for pre-mRNA splicing in vitro. RNA 3:1456-1467

Chang XB, Tabcharani JA, Hou YX, Jensen TJ, Kartner N, Alon N, Hanrahan JW, Riordan JR (1993) Protein kinase A (PKA) still activates CFTR chloride channel after mutagenesis of all 10 PKA consensus phosphorylation sites. J Biol Chem 268:11304-11311

Cheng A, Kaldis P, Solomon MJ (2000) Dephosphorylation of human cyclin-dependent kinases by protein phosphatase type 2Cα and β 2 isoforms. J Biol Chem275:34744-34749

Cheng SH, Rich DP, Marshall J, Gregory RJ, Welsh MJ, Smith AE (1991) Phosphorylation of the R domain by cAMP-dependent protein kinase regulates the CFTR chloride channel. Cell 66:1027-1036

Das AK, Helps NR, Cohen PT, Barford D (1996) Crystal structure of the protein serine/threonine phosphatase 2C at 2.0 A resolution. EMBO J 15:6798-6809

Davis RJ (2000) Signal transduction by the JNK group of MAP kinases. Cell 103:239-252

De Smedt V, Poulhe R, Cayla X, Dessauge F, Karaiskou A, Jessus C, Ozon R (2002)Thr-161 phosphorylation of monomeric Cdc2. Regulation by protein phosphatase 2C in Xenopus oocytes. J Biol Chem 277:28592-28600

Fiscella M, Zhang H, Fan S, Sakaguchi K, Shen S, Mercer WE, Vande Woude GF, O'Connor PM, Appella E (1997) Wip1, a novel human protein phosphatase that is induced in response to ionizing radiation in a p53-dependent manner. Proc Natl Acad Sci USA 94:6048-6053

Fu XD (1995) The superfamily of arginine/serine-rich splicing factors. RNA 1:663-680

Garrington TP, Johnson GL (1999) Organization and regulation of mitogen-activated protein kinase signaling pathways. Curr Opin Cell Biol 11:211-218

Guthridge MA, Bellosta P, Tavoloni N, Basilico C (1997) FIN13, a novel growth factor-inducible serine-threonine phosphatase which can inhibit cell cycle progression. Mol Cell Biol 17:5485-5498

Hanada M, Kobayashi T, Ohnishi M, Ikeda S, Wang H, Katsura K, Yanagawa Y, Hiraga A, Kanamaru R, Tamura S (1998) Selective suppression of stress-activated protein kinase pathway by protein phosphatase 2C in mammalian cells. FEBS Lett 437:172-176

Hanada M, Ninomiya-Tsuji J, Komaki K, Ohnishi M, Katsura K, Kanamaru R, Matsumoto K, Tamura S (2001) Regulation of the TAK1 signaling pathway by protein phosphatase 2C. J Biol Chem 276:5753-5759

Hannigan GE, Leung-Hagesteijn C, Fitz-Gibbon L, Coppolino MG, Radeva G, Filmus J, Bell JC, Dedhar S. (1996) Regulation of cell adhesion and anchorage-dependent growth by a new β 1-integrin-linked protein kinase. Nature 379:91-96

Hiraga A, Kikuchi K, Tamura S, Tsuiki S (1981) Purification and characterization of Mg^{2+}-dependent glycogen synthase phosphatase (phosphoprotein phosphatase IA) from rat liver. Eur J Biochem 119:503-510

Hishiya A, Ohnishi M, Tamura S, Nakamura F (1999) Protein phosphatase 2C inactivates F-actin binding of human platelet moesin. J Biol Chem 274:26705-26712

Hou EW, Kawai Y, Miyasaka H, Li SS (1994) Molecular cloning and expression of cDNAs encoding two isoforms of protein phosphatase 2Cβ from mouse testis. J Biochem Mol Biol Int 32:773-780

Ip YT, Davis RJ (1998) Signal transduction by the c-Jun N-terminal kinase (JNK)-from inflammation to development. Curr Opin Cell Biol 10:205-219

Kashiwaba M, Katsura K, Ohnishi M, Sasaki M, Tanaka H, Nishimune Y, Kobayashi T, Tamura S (2003) A novel protein phosphatase 2C family member (PP2Cζ is able to associate with ubiquitin conjugating enzyme 9. FEBS Lett 538:197-202

Kato S, Terasawa T, Kobayashi T, Ohnishi M, Sasahara Y, Kusuda K, Yanagawa Y, Hiraga A, Matsui Y, Tamura S (1995) Molecular cloning and expression of mouse Mg^{2+}-dependent protein phosphatase β-4 (type 2Cβ-4). Arch Biochem Biophys 318:387-393

Kerk D, Bulgrien J, Smith DW, Barsam B, Veretnik S, Gribskov M (2002) The complement of protein phosphatase catalytic subunits encoded in the genome of Arabidopsis. Plant Physiol 129:908-925

Kikuchi A (1999a) Modulation of Wnt signaling by Axin and Axil. Cytokine Growth Factor Rev 10:255-265

Kikuchi A (1999b) Roles of Axin in the Wnt signalling pathway. Cell Signal 11:777-788

Kikuchi K, Tamura S, Hiraga A, Tsuiki S (1977) Glycogen synthase phosphatase of rat liver. Its separation from phosphorylase phosphatase on DE-52 columns. Biochem Biophys Res Commun 75:29-32

Kishimoto K, Matsumoto K, Ninomiya-Tsuji J (2000) TAK1 mitogen-activated protein kinase kinase kinase is activated by autophosphorylation within its activation loop. J Biol Chem 275:7359-7364

Kitani T, Ishida A, Okuno S, Takeuchi M, Kameshita I, Fujisawa H (1999) Molecular cloning of Ca^{2+}/calmodulin-dependent protein kinase phosphatase. J Biochem 125:1022-1028

Koh CG, Tan EJ, Manser E, Lim L (2002) The p21-activated kinase PAK is negatively regulated by POPX1 and POPX2, a pair of serine/threonine phosphatases of the PP2C family. Curr Biol 12:317-321

Kramer A (1996) The structure and function of proteins involved in mammalian pre-mRNA splicing. Annu Rev Biochem 65:367-409

Kusuda K, Kobayashi T, Ikeda S, Ohnishi M, Chida N, Yanagawa Y, Shineha R, Nishihira T, Satomi S, Hiraga A, Tamura S (1998) Mutational analysis of the domain structure of mouse protein phosphatase 2Cβ. Biochem J 332:243-250

Labes M, Roder J, Roach A (1998) A novel phosphatase regulating neurite extension on CNS inhibitors. Mol Cell Neurosci 12:29-47

Leung-Hagesteijn C, Mahendra A, Naruszewicz I, Hannigan GE (2001) Modulation of integrin signal transduction by ILKAP, a protein phosphatase 2C associating with the integrin-linked kinase, ILK1. EMBO J 20:2160-2170

Li MG, Katsura K, Nomiyama H, Komaki K, Ninomiya-Tsuji J, Matsumoto K, Kobayashi T, Tamura S (2003) Regulation of the interleukin-1-induced signaling pathways by a novel member of protein phosphatase 2C family (PP2Cε). J Biol Chem 278:12013-12021

Mann DJ, Campbell DG, McGowan CH, Cohen PTW (1992) Mammalian protein serine/threonine phosphatase 2C: cDNA cloning and comparative analysis of amino acid sequences. Biochim Biophys Acta 1130:100-104

Marley AE, Kline A, Crabtree G, Sullivan JE, Beri RK (1998) The cloning expression and tissue distribution of human PP2Cβ. FEBS Lett 431:121-124

McGowan CH, Cohen P (1987) Identification of two isoenzymes of protein phosphatase 2C in both rabbit skeletal muscle and liver. Eur J Biochem 166:713-722

McGowan CH, Campbell DG, Cohen P (1987) Primary structure analysis proves that protein phosphatases $2C_1$ and $2C_2$ are isozymes. Biochim Biophys Acta 930:279-282

Misteli T, Spector DL (1998) The cellular organization of gene expression. Curr Opin Cell Biol 10:323-331

Miyazaki S, Koga R, Bohnert HJ, Fukuhara T (1999) Tissue- and environmental response-specific expression of 10 PP2C transcripts in Mesembryanthemum crystallinum. Mol Gen Genet 261:307-316

Morgan DO (1996) The dynamics of cyclin dependent kinase structure. Curr Opin Cell Biol 8:767-772

Morgan DO (1997) Cyclin-dependent kinases: engines, clocks, and microprocessors. Annu Rev Cell Dev Biol 13:261-291

Murray MV, Kobayashi R, Krainer AR (1999) The type 2C Ser/Thr phosphatase PP2Cγ is a pre-mRNA splicing factor. Genes Dev 13:87-97

Novak A, Dedhar S (1999) Signaling through β-catenin and Lef/Tcf. Cell Mol Life Sci 56:523-537

Ofek P, Ben-Meir D, Kariv-Inbal Z, Oren M, Lavi S (2003) Cell cycle regulation and p53 activation by protein Phosphatase 2Cα. J Biol Chem 278:14299-14305

Pato MD, Adelstein RS (1980) Dephosphorylation of the 20,000-dalton light chain of myosin by two different phosphatases from smooth muscle. J Biol Chem 255:6535-6538

Picciotto MR, Cohn JA, Bertuzzi G, Greengard P, Nairn AC (1992) Phosphorylation of the cystic fibrosis transmembrane conductance regulator. J Biol Chem 267:12742-12752

Pilgrim D, McGregor A, Jackle P, Johnson T, Hansen D (1995) The C. elegans sex-determining gene fem-2 encodes a putative protein phosphatase. Mol Biol Cell 6:1159-1171

Sakanaka C, Sun TQ, Williams LT (2000) New steps in the Wnt/β -catenin signal transduction pathway. Recent Prog Horm Res 55:225-236

Sato N, Yonemura S, Obinata T, Tsukita S, Tsukita S (1991) Radixin, a barbed end-capping actin-modulating protein, is concentrated at the cleavage furrow during cytokinesis. J Cell Biol 113:321-330

Seroussi E, Shani N, Ben-Meir D, Chajut A, Divinski I, Faier S, Gery S, Karby S, Kariv-Inbal Z, Sella O, Smorodinsky NI, Lavi S (2001) Uniquely conserved non-translated regions are involved in generation of the two major transcripts of protein phosphatase 2Cβ. J Mol Biol 312:439-451

Solomon MJ, Kaldis P (1998) in Results and Problems In: Cell Differentiation "Cell Cycle Control" (ed Pagano M) Springer. Heidelberg, Germany, pp79-109

Stone JM, Collinge MA, Smith RD, Horn MA, Walker JC (1994) Interaction of a protein phosphatase with an Arabidopsis serine-threonine receptor kinase. Science 266:793-795

Strovel ET, Wu D, Sussman DJ (2000) Protein phosphatase 2Cα dephosphorylates axin and activates LEF-1-dependent transcription. J Biol Chem 275:2399-2403

Takekawa M, Maeda T, Saito H (1998) Protein phosphatase 2Cα inhibits the human stress-responsive p38 and JNK MAPK pathways. EMBO J 17:4744-4752

Takekawa M, Adachi M, Nakahata A, Nakayama I, Itoh F, Tsukuda H, Taya Y, Imai K (2000) p53-inducible wip1 phosphatase mediates a negative feedback regulation of p38 MAPK-p53 signaling in response to UV radiation. EMBO J 19:6517-6526

Takeuchi M, Ishida A, Kameshita I, Kitani T, Okuno S, Fujisawa H (2001) Identification and characterization of CaMKP-N, nuclear calmodulin-dependent protein kinase phosphatase. J Biochem 130:833-840

Tamura S, Lynch KR, Larner J, Fox J, Yasui A, Kikuchi K, Suzuki Y, Tuiki S (1989a) Molecular cloning of rat type 2C (IA) protein phosphatase mRNA. Proc Natl Acad Sci USA 86:1796-1800

Tamura S, Yasui A, Tuiki S (1989b) Expression of rat protein phosphatase 2C (IA) in Escherichia coli. Biochem Biophys Res Commun 163:131-136

Tan KM, Chan SL, Tan KO and Yu VC (2001) The Caenorhabditis elegans sex-determining protein FEM-2 and its human homologue, hFEM-2, are Ca^{2+}/calmodulin-dependent protein kinase phosphatases that promote apoptosis. J Biol Chem 276:44193-44202

Tazi J, Kornstadt U, Rossi F, Jeanteur P, Cathala G, Brunel C, Luhrmann R (1993) Thiophosphorylation of U1-70K protein inhibits pre-mRNA splicing. Nature 363:283-286

Terasawa T, Kobayashi T, Murakami T, Ohnishi M, Kato S, Tanaka O, Kondo H, Yamamoto H, Takeuchi T, Tamura S (1993) Molecular cloning of a novel isotype of Mg^{2+}-dependent protein phosphtase β (type 2Cβ) enriched in brain and heart. Arch Biochem Biophys 307:342-349

Tong Y, Quirion R, Shen SH (1998) Cloning and characterization of a novel mammalian PP2C isozyme. J Biol Chem 273:35282-35290

Travis SM, Welsh MJ (1997) PP2Cγ: a human protein phosphatase with a unique acidic domain. FEBS Lett 412:415-419

Travis SM, Berger HA, Welsh MJ (1997) Protein phosphatase 2C dephosphorylates and inactivates cystic fibrosis transmembrane conductance regulator. Proc Natl Acad Sci USA 94:11055-11060

Wenk J, Trompeter HI, Pettrich KG, Cohen PTW, Campbell DG, Mieskes G (1992) Molecular cloning and primary structure of a protein phosphatase 2C isoform. FEBS Lett 297:135-138

Will CL, Luhrmann R (1997) snRNP structure and function. In: Eukaryotic mRNA processing (ed Krainer AR) Oxford University Press, New York, NY, pp 130-173

Xiao SH, Manley JL (1997) Phosphorylation of the ASF/SF2 RS domain affects both protein-protein and protein-RNA interactions and is necessary for splicing. Genes & Dev 11:334-344

Zhu T, Dahan D, Evagelidis A, Zheng S, Luo J, Hanrahan JW (1999) Association of cystic fibrosis transmembrane conductance regulator and protein phosphatase 2C. J Biol Chem 274:29102-19107

6 PP5: the TPR phosphatase

Michael Chinkers

Abstract

PP5 is unique among protein serine/threonine phosphatases in containing an amino-terminal TPR domain that functions both to maintain low basal activity and to bind to target proteins. Although PP5 can be activated by partial proteolysis or long-chain fatty acyl compounds *in vitro*, the physiological mechanisms activating this ubiquitous okadaic acid-sensitive protein phosphatase remain unknown. We review the biochemistry and biology of PP5 with an emphasis on putative target proteins. The first genetic studies of PP5 in model organisms are also discussed, as are antisense approaches to determining its function in cultured human cells.

6.1 Introduction

This review will differ from other chapters in this volume on protein serine/threonine phosphatases due to differences in the history of studies on PP5. Unlike PP1, PP2A, PP2B, and PP2C, whose biochemical properties were characterized decades ago, and whose biological functions have been extensively studied, PP5 was discovered relatively recently. Three groups reported its cDNA sequence in 1994, using three different methods to isolate cDNA clones: low-stringency screening using a PP2B probe (Chen et al. 1994), PCR with degenerate oligonucleotides based on conserved protein phosphatase sequences (Becker et al. 1994), and a yeast two-hybrid screen to identify proteins interacting with the ANP receptor (Chinkers 1994). The last approach, which identified a PP5-binding protein without clearly establishing the function of the interaction, foreshadowed the course of much research on PP5 during the ensuing decade. Unlike other phosphatases that were well characterized before their molecular cloning, research on PP5 began with its molecular cloning, and the regulation and function of the enzyme remain poorly understood. Much of the literature on the enzyme describes interactions of unknown function with other proteins. That literature has grown steadily, but our understanding of the biology of PP5 is still in its infancy. The structure of its TPR (tetratricopeptide repeat) regulatory domain has been determined, and putative activators that bind to this domain have been characterized *in vitro*, but we know nothing about the structural basis for modulation of PP5 activity and little about its regulation in the intact cell. In discussing the structure, biochemistry, or genetics of PP5, this article is meant as a critical review of a litera-

Topics in Current Genetics, Vol. 5
J. Arino, D.R. Alexander (Eds.) Protein phosphatases
© Springer-Verlag Berlin Heidelberg 2004

ture that offers us many clues about the structure and function of the enzyme, but that so far has left many key questions about its biology unanswered.

6.2 PP5: a TPR protein

Among protein phosphatases, the distinguishing feature of PP5 is its amino-terminal TPR domain. This domain serves both targeting and regulatory functions, much like the regulatory subunits of related phosphatases. TPR domains, conserved from prokaryotes to mammals in proteins having diverse functions and subcellular localizations, consist of tandem repeats of a 34-amino-acid consensus sequence that mediates protein-protein interactions (Blatch and Lassle 1999). The TPR domain of PP5 contains three such repeats. At the time of this writing, the Interpro web site lists 179 TPR proteins encoded by the human genome, with 132, 84, 63, and 33 TPR proteins encoded by the genomes of *A. thaliana*, *D. melanogaster*, *C. elegans*, and *S. cerevisiae*, respectively. TPR domains constitute one of the most common repeat domains found in the proteomes of most organisms. A recent study using RNA interference to disrupt 86% of the genes of *C. elegans* suggested that TPR proteins as a class tend to be particularly important for viability (Kamath et al. 2003).

Though TPR domains were reviewed elsewhere several years ago (Blatch and Lassle 1999), it may be worth noting their key structural features here. The TPR domain of PP5 was the first to have its crystal structure determined (Das et al. 1998 and Fig. 1). That structure has turned out to be representative of other TPR domain structures determined subsequently (Abe et al. 2000; Gatto et al. 2000; Lapouge et al. 2000; Scheufler et al. 2000; Grizot et al. 2001; Kumar et al. 2001; Taylor et al. 2001; Sinars et al. 2003). In these structures, each TPR repeat forms two antiparallel alpha helices (Fig. 1). When several TPR repeats occur in tandem, they pack together to form a right-handed superhelix. The twist of the superhelix forms a groove that can serve as a binding site for other proteins or for intramolecular interactions (Russell et al. 1999; Abe et al. 2000; Gatto et al. 2000; Lapouge et al. 2000; Scheufler et al. 2000). The similarity of this structure to the structures of HEAT repeats, armadillo repeats, ankyrin repeats, and various leucine-rich repeats has been noted (Kobe and Kajava 2000), as has its similarity to the structure of 14-3-3 proteins (Das et al. 1998). Structural similarity to the Puf repeat domain, which binds RNA (Edwards et al. 2001), is also evident. The PPR domain, a common repeat structure in plants, is similar to TPR domains in its consensus sequence and is predicted to form a similar structure (Small and Peeters 2000).

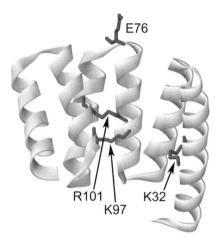

Fig. 1. Ribbon diagram showing the structure of the TPR domain of PP5. Side chains of residues (K32, K97, and R101) required for the binding of hsp90 (Russell et al. 1999) line a groove formed by the α-helices. The side chain of E76 extends away from this groove in a loop connecting helices A and B of TPR2, and is required for autoinhibition of PP5 in its basal state (Kang et al. 2001). This image of the TPR domain of PP5 [Protein Data Bank accession code 1A17] was prepared using DS ViewerPro 5.0.

6.3 Expression and localization

Several studies have described the expression patterns and subcellular localization of PP5. Historically, PP5 was first described as a predominantly nuclear phosphatase in cultured human cells (Chen et al. 1994). Most subsequent work, however, indicated that PP5 in animal cells and yeast is either predominantly cytoplasmic or found at similar levels in the cytoplasm and nucleus (Ollendorff and Donoghue 1997; Russell et al. 1999; Brown et al. 2000; Bahl et al. 2001; Borthwick et al. 2001; Chaudhuri 2001; Morita et al. 2001; Galigniana et al. 2002; Yamaguchi et al. 2002; Jeong et al. 2003). PP5 is predominantly nuclear but detectable in the cytoplasm in *Plasmodium falciparum* (Lindenthal and Klinkert 2002). Signals required for the nuclear import of PP5 have been localized to its C-terminal 80 amino acids (Borthwick et al. 2001). The presence of PP5 in the cytoplasm as well as the nucleus is significant in that many of the proposed targets of the enzyme are present in the cytoplasm or at the plasma membrane. As discussed below, there is evidence that activation of G_{12} family members or copine proteins may induce translocation of PP5 to the plasma membrane (Yamaguchi et al. 2002; Tomsig et al. 2003).

PP5 is expressed ubiquitously in animals, plants, and fungi, but variations in protein and RNA levels as well as in subcellular localization have been noted. In the rat brain, PP5 localization has been studied by both *in situ* hybridization (Becker et al. 1996) and immunocytochemistry (Bahl et al. 2001). PP5 is present at relatively high levels in brain, where its distribution is similar to that of related protein phosphatases. Neuronal PP5 appeared to be exclusively cytoplasmic (Bahl et al. 2001). Recent studies of two parasites, the African trypanosome, *Trypanosoma brucei* and the malarial parasite, *P. falciparum* have indicated that the genomes of both organisms encode unique forms of PP5 (Chaudhuri 2001; Dobson et al. 2001; Lindenthal and Klinkert 2002). The trypanosomal enzyme is found in both the insect and the bloodstream forms of the parasite, and the *P. falciparum* enzyme was found at similar levels in early and late trophozoite, late schizont/segmented, and ring stages. The trypanosomal protein is found throughout the cell, whereas the *P. falciparum* protein is found at higher levels in the nucleus, as noted above. These expression patterns give little basis for inferring the biological function of parasitic PP5, though it is intriguing to speculate that it may play a role in virulence.

An association between PP5 and growth state has been noted in a number of studies that have examined PP5 at the protein or mRNA level. Any correlation between elevated PP5 levels and cell proliferation is imperfect; as noted above, PP5 is expressed at high levels in neurons, which are post-mitotic cells. Further, in *Neurospora crassa*, PPT mRNA levels are higher in conidia than during germination (Yatzkan and Yarden 1997). (For somewhat obscure reasons, fungal PP5 is referred to as PPT, and we will follow that convention here.) In general, though, cells that are actively growing tend to express more PP5. Mammalian cells in culture stained about half as intensely with PP5 antibodies after serum starvation, and cells in confluent cultures stained less intensely than cells in sparse cultures (Chen et al. 1994). In a study examining PP5 mRNA in rat liver, levels were found to be markedly elevated in highly malignant ascites hepatomas, while levels in primary hepatomas were slightly increased, and PP5 levels in regenerating livers were comparable to those in control livers. This did not suggest a correlation between PP5 levels and growth, but a correlation with malignancy was observed (Shirato et al. 2000). In *T. brucei*, PP5 mRNA and protein levels both decrease modestly during serum starvation, and PP5 mRNA levels decrease after several days in stationary phase (Chaudhuri 2001). Similar observations have been made in other organisms. Levels of *D. melanogaster* PP5 protein are higher during embryonic development than in larvae or adults. Parallel measurements of mRNA levels did not always accurately reflect changes in protein levels (Brown et al. 2000), a reminder to use caution when interpreting expression studies based only on mRNA levels. Many such studies have been performed in yeast, but a recent report also examined PPT at the protein level during *S. cerevisiae* growth in culture. Both PPT mRNA and protein were readily detectable by Northern and Western blotting in the early log phase of growth, decreased dramatically by mid-log phase, and became undetectable by stationary phase (Jeong et al. 2003). Genetic studies examining the role of PP5 in growth are discussed below in a separate section.

The expression of PPT mRNA in *S. cerevisiae* has been examined as part of a number of microarray studies. The most dramatic effects on PPT expression were observed in a study of environmental stress responses (Gasch et al. 2000). PPT was found to be part of a cluster of ~600 genes whose levels of expression in response to heat shock, peroxide, menadione, DTT, diamide, hyperosmotic shock, and starvation were repressed in similar patterns. This cluster consists largely of genes involved in growth-related processes, RNA metabolism, nucleotide synthesis, secretion, and other metabolic processes. Two upstream elements conserved in these genes, including the *PPT1* gene, were identified. Interestingly, despite its association with hsp90 (see below) and its relationship to other TPR-containing proteins that bind to hsp90, and whose genes are induced by heat shock (e.g. Cpr6 and Sti1), PPT expression was repressed up to 25-fold by heat shock. This response was distinct from that of two additional hsp90-binding TPR proteins, Cns1 and Cpr7, whose expression was either unaffected or much more modestly repressed by environmental stresses. Consistent with the results described above, PPT expression was repressed 25-fold by growth to stationary phase. Expression of the *PPT1* gene was also repressed 6-8 fold by nitrogen depletion or amino acid starvation (Gasch et al. 2000). In two additional studies of gene expression in *S. cerevisiae*, PPT expression was induced approximately 4-fold by growth in zinc-deficient medium (Lyons et al. 2000) or by treatment with alpha factor (Roberts et al. 2000). In a comprehensive examination of gene expression during *C. elegans* development, PP5 expression was measured but levels were too low for the observed changes to be statistically significant (Hill et al. 2000).

6.4 Regulation of PP5 activity

The TPR domain of mammalian PP5 has been shown to be autoinhibitory (Chen and Cohen 1997; Sinclair et al. 1999). Mammalian PP5 has extremely low phosphatase activity in the basal state, which can be stimulated up to 50-fold upon removal of the TPR domain by partial proteolysis. When we last reviewed the regulation of PP5 (Chinkers 2001), it had been shown that the enzyme could also be activated by removal of the C-terminal 10-13 amino acids (Sinclair et al. 1999) or by the binding of high micromolar concentrations of unsaturated long-chain fatty acids to the TPR domain (Chen and Cohen 1997; Skinner et al. 1997). We have since shown that long-chain fatty acyl-CoA esters can fully activate rat PP5 at low micromolar concentrations that may be physiological. Saturated acyl chains of 16 or more carbons were required for optimal activation by the CoA compounds (Ramsey and Chinkers 2002). It has yet to be determined whether these compounds regulate PP5 in the living cell. Additional mutagenesis studies have suggested that the autoinhibitory face of the TPR domain includes a loop between helices 2a and 2b (Fig. 1), and that the autoinhibitory domain at the C-terminus is limited to the five C-terminal amino acids (Kang et al. 2001). Two reports differ as to whether removal of five additional C-terminal residues has no effect (Sinclair et al. 1999) or interferes with further activation by fatty acids (Kang et al. 2001). The latter report suggested that, in addition to the amino-terminal TPR domain, some C-terminal residues may be important for lipid activation of PP5.

As additional studies of the biochemistry of PP5 from various species have been performed, the roles of lipid activation and of the TPR domain seem less clear. Cauliflower PP5 was activated ten-fold by arachidonic acid, much like its mammalian counterparts (Meek et al. 1999). Since PP5 is extremely susceptible to activating proteolysis during purification, it is not unusual to obtain preparations of the mammalian enzyme that are stimulated only ten-fold by arachidonic acid. Even allowing for difficulties in enzyme purification, however, three reports describing the activity of parasitic PP5 were surprising. Using several different substrates to measure the activity of PP5 from *T. brucei* or *P. falciparum*, only 2-3 fold activation by fatty acids was observed. Similar levels of activation were obtained by deleting the TPR domain of the *P. falciparum* enzyme. It is not clear whether this observation suggests that the parasitic PP5 is constitutively more active than mammalian forms, or whether it has similar basal activity but little ability to be activated by fatty acids. This further calls into question the physiological significance of lipid activation, which might be expected to be conserved if it is physiologically important. A study of *S. cerevisiae* PPT raised similar questions. Purified PPT could be stimulated 20-fold by fatty acids, but this activation was substrate-dependent. In contrast, deletion of the TPR domain or C-terminal sequences resulted in only a two-fold activation of the enzyme. Thus, for PPT, it is not clear that the TPR or C-terminal domains are significantly autoinhibitory, and the low basal activity of the enzyme remains unexplained. The authors of this study also stated that the activation of PPT by long-chain acyl-CoA esters is no more efficient than that by fatty acids, and that the efficient activation of mammalian PP5 by the CoA compounds is substrate-dependent (Jeong et al. 2003). While the physiological importance of lipid activation has always been unclear, these recent studies also call into question the importance of the TPR domain in regulating PP5 activity. In the coming years, it is likely that the question of whether lipids are physiological activators of PP5 will be resolved, and that determination of the crystal structures of the full-length enzyme in its basal and activated forms will reveal the structural bases for low basal activity and enzyme activation.

6.5 Inhibitors of PP5

Although no specific inhibitors of PP5 have been described, it seems worthwhile to briefly review the enzyme's sensitivity to known phosphatase inhibitors. Several reports have described the inhibition of recombinant PP5 from various species by okadaic acid, to which it exhibits sensitivity intermediate between that of PP1 and PP2A. IC_{50} values of 4-5 nM have generally been observed (Becker et al. 1994; Chen et al. 1994; Fukuda et al. 1996; Borthwick et al. 2001; Dean et al. 2001; Dobson et al. 2001; Lindenthal and Klinkert 2002). The inhibition of PP5 by other known phosphatase inhibitors has been less thoroughly characterized. The isolated catalytic domain of PP5 was inhibited at several-fold lower concentrations of fumonisin B_1 than were other phosphatases, but high micromolar to millimolar concentrations of the toxin were required to half-maximally inhibit PP5, PP1, PP2A, PP2B, or PP2C (Fukuda et al. 1996). Two reports described IC_{50} values of 2.5 nM and 15 nM for inhibition by microcystin (Chen et al. 1994; Borthwick et al. 2001). The latter study also examined inhibition of PP5 by four other inhibitors, with IC_{50} values presented for calyculin A (100 nM), tautomycin (120 nM), cantharidin (50 nM), and fostriecin (700 µM). A recent review article lists IC_{50} values approximately one order of magnitude lower for each of these four inhibitors, based on unpublished work (Honkanen and Golden 2002). Thus, unlike okadaic acid, there is no clear consensus regarding what concentrations of these compounds effectively inhibit PP5. An important point noted frequently in the PP5 literature is that studies using okadaic acid and other inhibitors to implicate PP1 or PP2A in biological processes tend to neglect the possibility that these compounds may exert some of their biological effects through inhibition of PP5.

6.6 PP5 binding proteins

Much of the literature on PP5 describes a set of proteins with which it interacts. Some of these interactions have been studied more than others, and the functional implications of some are clearer than those of others. In no case, however, is the function of PP5 binding fully understood. Nonetheless, it seems likely that at least some of the interactions that have been described will prove to be functionally significant, whether due to binding proteins being regulated by PP5 or vice versa. The interaction between PP5 and hsp90 has been confirmed by multiple laboratories and studied extensively; thus, this binding partner is discussed at greatest length below. The other PP5 binding proteins have been described by only one laboratory, often with no further progress reported over the course of several years. In the coming years, the physiological interactions are likely to be sorted from the artifactual ones. Here, we review PP5 binding proteins with an emphasis on proteins for which there is evidence for a functional association.

6.6.1 ANP receptor

We originally identified PP5 by virtue of its TPR domain interacting with the protein kinase-like domain of the ANP receptor in a yeast two-hybrid screen (Chinkers 1994). The ANP receptor is a phosphoprotein, with 6 serine and threonine residues in the juxtamembrane and protein kinase-like domains that must be phosphorylated to sustain the receptor's hormone-stimulated guanylyl cyclase activity (Potter and Hunter 2001). The receptor is desensitized by dephosphorylation of these residues, in a phenomenon thought to be important in pathological conditions in which ANP levels are elevated (Potter and Hunter 2001). Based on its binding to the appropriate region of the ANP receptor, we originally hypothesized that PP5 was the phosphatase responsible for its desensitization (Chinkers 1994). No further evidence supporting this hypothesis has been published, however. In unpublished work, we have not observed co-immunoprecipitation of the two proteins, and while overexpression of PP5 can induce ANP receptor desensitization in cultured cells, similar results can be obtained by overexpressing the catalytic subunits of related phosphatases. A recent report suggested that ANP receptor desensitization may actually be due to decreased activity of a kinase rather than increased activity of a phosphatase (Joubert et al. 2001). It has been difficult to test the biological role of PP5 in this pathway in the absence of a specific inhibitor.

6.6.2 Heat shock protein 90 (hsp90)

Although the ANP receptor did not co-immunoprecipitate with PP5, hsp90 co-immunoprecipitated efficiently with either recombinant or endogenous PP5. Binding to hsp90 was mediated by the enzyme's TPR domain (Chen et al. 1996). We noted in our original report on PP5 that its TPR domain was most closely related to several hsp90-binding TPR proteins found in steroid receptor heterocomplexes (Chinkers 1994). There is now a family of at least a dozen TPR-containing co-chaperones that bind to hsp90, more than half of which are associated with the folding and trafficking of nuclear receptors (Pratt and Toft 2003). We found that PP5 co-immunoprecipitated with the glucocorticoid receptor (GR), and was a major component of hsp90•GR heterocomplexes (Chen et al. 1996; Silverstein et al. 1997). Further, use of the TPR domain as a dominant negative mutant to displace endogenous PP5 from hsp90 blocked glucocorticoid-induced activation of a reporter gene. Based on this observation, we initially suggested that PP5 was required for optimal GR signaling (Chen et al. 1996). It soon became clear, however, that this dominant negative mutant might have unintended effects: the TPR domain of PP5 competed with other TPR-containing co-chaperones for binding to hsp90 (Silverstein et al. 1998). Thus, the dominant negative PP5 mutant might have interfered with GR signaling by displacing other TPR-containing co-chaperones from hsp90. As such, the role of PP5 in GR signaling remained unclear.

Subsequently, antisense studies suggested an inhibitory role for PP5 in GR signaling. Overnight treatment of human A549 cells with a PP5 antisense oligonucleotide designated ISIS 15534 potentiated glucocorticoid activation of a reporter gene, GR binding to DNA, and nuclear translocation of the GR (Zuo et al. 1999; Dean et al. 2001). We regard this work with caution as depletion of PP5 was not monitored at the protein level, and our unpublished data suggest that PP5 in A549 and other mammalian cells does not undergo significant turnover overnight. Nonetheless, modest decreases in PP5 may have dramatic effects on GR signaling, or variations in culture conditions may lead to variations in PP5 half-life. Alternatively, the antisense oligonucleotide may have had unintended effects that stimulated GR signaling through mechanisms unrelated to PP5.

Other experiments have suggested that PP5 has no effect on GR signaling. In *Drosophila* S2 cells, RNA interference can be used to efficiently deplete PP5 within 3 days of incubation with double-stranded RNA (Silverstein et al. 2002). In our hands, this depletion of PP5, confirmed by Western blotting, has no significant effect on nuclear translocation of the GR or on transcriptional activation by glucocorticoids (unpublished results). In additional unpublished work from our laboratory, GR signaling is not altered in yeast lacking the PPT gene. This seems significant, since deletion of genes encoding other hps90-binding TPR proteins from yeast (Sti1, Cpr7) impairs GR signaling (Duina et al. 1996; Chang et al. 1997), and overexpression in yeast of FKBP52, but not of PP5, stimulates GR signaling by an order of magnitude, by increasing the receptor's hormone-binding affinity (Riggs et al. 2003). Overexpression of FKBP51 but not of PP5 blocks this stimulation by FKBP52 (Riggs et al. 2003). Based on the experiments described above, the role of PP5 in GR signaling seems unclear.

Recent studies, however, have suggested a redundant role for PP5 and two other TPR-containing cochaperones, FKBP52 and CyP-40, in GR trafficking. In the model resulting from these studies, depletion of PP5 would not affect GR signaling because FKBP52 and CyP-40 would compensate for its absence. The observation that the intermediate chain of cytoplasmic dynein bound to the peptidyl-prolyl isomerase (PPIase) domain of FKBP52 (Silverstein et al. 1999) inspired a model in which the TPR-containing proteins found in mature GR complexes (FKBP52, PP5, and CyP-40) were important not for the folding of the GR, mediated by hsp70, hsp90, and other TPR proteins, but for nucleocytoplasmic trafficking of the GR (Pratt et al. 1999). FKBP52, PP5, and CyP-40 would anchor the GR to the microtubule-based movement machinery, with hormone binding favoring movement to the nucleus. Data obtained thus far support the model. Competitively disrupting interactions with dynein in living cells by overexpressing the dynein-binding PPIase domain of FKBP52 disrupted hormone-induced nuclear translocation of the GR (Galigniana et al. 2001). PP5, which contains a PPIase-like domain capable of binding to FK506 (Silverstein et al. 1997), and CyP-40, which contains an authentic PPIase domain, have also been shown to interact with the intermediate chain of dynein (Galigniana et al. 2002). Immunofluorescence studies demonstrated that a subpopulation of PP5 co-localized with dynein and microtubules, and that this co-localization was disrupted by the overexpressed PPIase domain of FKBP52 (Galigniana et al. 2002). Thus, PP5 may play a redundant role, along with other TPR-containing hsp90 co-chaperones, in anchoring the GR to the cytoskeleton and in mediating its translocation to the nucleus following hormone binding.

Hsp90 and its co-chaperones are involved in the folding of many client proteins unrelated to steroid receptors, including a number of protein kinases (Pratt and Toft 2003). PP5 has recently been implicated in regulating the maturation of another hsp90 client protein, the heme-regulated eIF2α kinase, also known as the heme-regulated inhibitor of protein synthesis (HRI). HRI is important in coordinating hemoglobin synthesis in reticulocytes and is found in heterocomplexes containing both hsp90 and PP5 (Shao et al. 2002). Under conditions of heme deficiency, phosphorylation and activation of HRI result in phosphorylation of eIF2α, sequestration of eIF2B and inhibition of the initiation of protein synthesis. Phosphorylation and activation of HRI are both hsp90-dependent, and phosphorylation appears to be important in its maturation and activation. Both phosphorylation and activation are enhanced by the phosphatase inhibitors okadaic acid and nodularin at concentrations that block PP5 activity, but are not blocked by the PP2A inhibitor, fostriecin. Fatty acids that activate PP5 inhibited the hsp90-dependent activation of HRI, while related fatty acids that do not activate PP5 were without effect.

Thus, maturation and activation of HRI appear to be inhibited by a protein phosphatase having characteristics consistent with its being PP5, and HRI is physically associated with PP5 in hsp90 heterocomplexes (Shao et al. 2002). It is not clear, whether the point of regulation is the dephosphorylation of HRI, or, since hsp90 and its co-chaperones are phosphoproteins, whether PP5 may regulate the folding of HRI by regulating the phosphorylation state of the chaperone machinery.

We are beginning to understand the structural basis for the interaction between PP5 and hsp90. When the structure of the TPR domain of PP5 was first described, it was speculated that a putative binding groove might be the site of recognition of target proteins such as hsp90 (Das et al. 1998). We tested this hypothesis using a scanning mutagenesis approach in which residues within this groove and elsewhere within the TPR domain were mutated. Three basic residues within the binding groove that are conserved in all hsp90-binding TPR domains were shown to be required for binding to hsp90 (Russell et al. 1999 and Fig. 1). Mutation of residues outside this groove had no effect on hsp90 binding. Subsequent X-ray crystallographic studies of the hsp90-binding TPR domain of Hop in a complex with the C-terminal MEEVD peptide of hsp90 explained the importance of the conserved basic residues (Scheufler et al. 2000). Their side chains interacted with the acidic side chains in the MEEVD peptide and with its C-terminal carboxyl group. The crystal structures of two other hsp90-binding co-chaperones, FKBP51 and CyP-40, have since been solved and show a similar arrangement of the conserved basic residues in their TPR domains (Taylor et al. 2001; Sinars et al. 2003). Site-directed mutagenesis has confirmed the importance of these conserved basic residues in hsp90 binding by Hop, FKBP52, CyP-40, and AIP (Bell and Poland 2000; Ramsey et al. 2000; Ward et al. 2002). Thus, the interactions between the C-terminus of hsp90 and PP5 are similar to those between hsp90 and other TPR-containing co-chaperones. These interactions will be more fully understood when structures of the full-length proteins bound to full-length hsp90 become available.

Preliminary data also suggest that hsp90 might play a role in regulating PP5 activity. When we first described the binding of hsp90 to the TPR domain of PP5, we hypothesized that such binding might disrupt autoinhibitory interactions between the TPR domain and the phosphatase catalytic domain, leading to PP5 activation (Chen et al. 1996). We have not, however, observed any effect of full-length hsp90 on PP5 activity (Ramsey and Chinkers 2002). Indeed, mutagenesis studies suggested that the autoinhibitory face and the hsp90-binding face of the TPR domain were distinct, with hsp90 binding to the groove in the TPR domain and the autoinhibitory region being in a connecting loop between helices (Kang et al. 2001 and Fig. 1). Nonetheless, we observed a ten-fold activation of PP5 by the C-terminal 12-kDa fragment of hsp90 (Ramsey and Chinkers 2002). It remains to be determined whether there might be circumstances under which intact hsp90 can activate PP5 or whether activation by the truncated protein is merely an artifact observed *in vitro*.

6.6.3 ASK1

Apoptosis signal-regulating kinase (ASK1) is a MAP kinase kinase kinase that is phosphorylated and activated in response to oxidative stress (Ichijo et al. 1997; Tobiume et al. 2002). Its activation initiates a phosphorylation cascade leading to activation of JNK and p38 (Ichijo et al. 1997). Studies using knockout mice have demonstrated the importance of ASK1 in sustained oxidant-induced activation of these kinases, confirming studies using cultured cells in which ASK1 mediates oxidant-induced apoptosis (Tobiume et al. 2001). In a two-hybrid screen and in co-immunoprecipitation studies PP5 was identified as an ASK1-interacting protein (Morita et al. 2001). Either endogenous or transfected PP5 co-immunoprecipitated with transfected ASK1, and the interaction was stimulated by reactive oxygen species. Further, PP5 transfected into HeLa cells suppressed ASK1 signaling by dephosphorylating a residue (Thr 845) critical for kinase activity. Control experiments were performed to exclude the possibility that overexpressed PP5 inhibited the ASK1 pathway by nonspecific dephosphorylation of MAP kinases in general. In 293 cells, transfected PP5 inhibited activation of JNK by ASK1, but had no effect on activation of JNK by MEKK1 or on activation of ERK by serum, indicating a specific effect. In addition to inducing dephosphorylation and inactivation of ASK1 in transfected cells, purified PP5 was able to dephosphorylate and inactivate ASK1 *in vitro*. Taken together, these data suggest that PP5 is a physiological regulator of ASK1. PP5 specifically binds to and dephosphorylates the activated form of ASK1 in a negative feedback loop that modulates oxidant-induced signaling through this pathway (Morita et al. 2001). Due to potential artifacts of overexpression, it would be helpful to know the effects of PP5 depletion on this pathway.

6.6.4 Gα_{12} and Gα_{13}

The functions of the heterotrimeric G proteins designated G_{12} and G_{13} are poorly understood. A recent study, however, suggests that the α-subunits of these proteins bind to PP5 when activated, induce PP5 translocation to the plasma membrane, and modestly activate the enzyme (Yamaguchi et al. 2002). The TPR domain of PP5 bound to a constitutively active form of Gα_{12} in a yeast two-hybrid screen, and was then shown to interact with both Gα_{12} and Gα_{13} in pulldown assays. Activation of these G proteins by mutation or by treatment with fluoride stimulated PP5 binding to their α-subunits. PP5 did not bind to Gα_q, showing specificity for the G_{12} family of proteins. Transfection of the activated Gα_{12} into COS-7 cells led to translocation of PP5 to the plasma membrane based on immunofluorescence and subcellular fractionation studies. This translocation required the TPR domain of PP5. Further, both basal and arachidonic acid-stimulated phosphatase activities of purified PP5 were increased 2.5-fold by the activated G_{12} and G_{13} proteins. It is difficult to interpret the role of PP5 in this system; the activators of the G_{12} family of proteins are not well characterized, and if PP5 is an effector, a two-fold stimulation is a modest one. Translocation of the enzyme to the plasma membrane, however, could result in high local concentrations of PP5 and increased dephosphorylation of as-yet-unidentified substrates. Finally, the activated G_{12} and G_{13} proteins have been shown to stimulate ASK1 (Berestetskaya et al. 1998), and one could imagine their recruiting PP5 into the negative feedback loop described above, or, alternatively, sequestering PP5 to prevent its inhibitory dephosphorylation of ASK1.

6.6.5 Copines

The copines are a family of calcium and lipid binding proteins found in plants, animals, and protozoa. They contain C2 domains related to those found in protein kinase C, which mediate binding to phospholipids, and an A domain involved in protein-protein interactions that is related to the A domain that mediates extracellular protein interactions with integrins. There are 7 copine genes in humans, but there are as yet no published data on the function of copines in mammals. The best clue regarding their function comes from an *Arabidopsis* copine mutant that leads to reduced plant size at low temperatures. This phenotype is apparently due to loss of a negative effect of the wild type copine gene on cell death (Tomsig and Creutz 2002). In a recent study, the A domains of human copines I, II, and IV were used as bait in two-hybrid screens for interacting proteins. Twenty-two putative binding partners were identified, one of which was itself a copine. When the A domain of the ubiquitously expressed copine I was used as bait, the TPR domain of PP5 represented 8 of 10 isolates in a two-hybrid screen. All interactions with the copine A domains were confirmed in pulldown assays using GST fusions of the target proteins. PP5 interacted *in vitro* or in the two-hybrid screen with copines I, II, and IV (Tomsig et al. 2003).

The same study identified a consensus sequence for copine binding that corresponded to known or predicted coiled-coil regions of target proteins, including the TPR domain of PP5. Curiously, no other TPR proteins were identified as copine targets. The A domain of copine I mediated calcium-dependent recruitment of PP5 and other target proteins to immobilized phosphatidylserine. This could represent a mechanism for targeting PP5 to plasma membrane substrates. Copine I or isolated copine A domains, but not boiled A domains, stimulated the activity of purified recombinant PP5 by 30-40%. This effect was unimpressive, but was measured in the presence of a large excess of arachidonic acid, using a PP5 preparation having unusually low activity. These observations raised the possibility that copines might regulate PP5 activity as well as its subcellular localization under some circumstances (Tomsig et al. 2003). It will be interesting to see whether PP5 undergoes calcium-induced translocation to the plasma membrane in living cells.

6.6.6 PP2A

Two reports have described an association of PP5, mediated by its TPR domain, with regulatory subunits of PP2A. In the first, the A subunit of PP2A was used as bait in a yeast two-hybrid screen, resulting in isolation of a clone encoding the

TPR domain of PP5. This interaction was confirmed in pulldown assays using bacterially expressed proteins. In co-immunoprecipation experiments, a substantial fraction of the endogenous PP5 in HeLa cells was associated with the A subunit. This was surprising, since a previous study had found most PP5 was associated with hsp90 (Chen et al. 1996). The binding region was mapped to residues 205-426 of the A subunit, and point mutations in residues important for binding to the catalytic subunit of PP2A did not significantly impair PP5 binding. Thus, determinants for binding the catalytic subunit of PP2A and for binding the TPR domain of PP5 may overlap but are different, as might be expected considering their different structures. A B″ subunit of PP2A, PR72, was also shown to co-immunoprecipitate from cell extracts with PP5, leading the authors to postulate the existence of a heterotrimeric A- B″-PP5 complex similar to PP2A complexes. No effect of these regulatory subunits on PP5 activity or subcellular localization were reported (Lubert et al. 2001).

A novel PP2A B″ subunit was identified in a two-hybrid screen for proteins that bound to GANP, a nuclear phosphoprotein thought to be involved in B cell proliferation and differentiation. This B″ subunit, designated G5PR, was found to be expressed ubiquitously and to co-immunoprecipitate with GANP from transfected COS cells. It was therefore speculated that GANP function might be regulated by PP2A. In order to test the specificity of the interaction of G5PR with the catalytic subunit of PP2A, COS cells were co-transfected with G5PR and the FLAG-tagged catalytic subunits of PP1, PP2A, PP2B, PP4, PP5, or PP6. Only PP5 co-immunoprecipitated with G5PR in these experiments, and the association was mediated by its TPR domain. The authors attributed the lack of binding of the C subunit of PP2A to interference by its amino-terminal FLAG tag. In pulldown experiments using a GST-G5PR fusion protein and cell lysates, the A and C subunits of PP2A, as well as PP5, bound to the recombinant G5PR. This was consistent with the homology of G5PR to known PP2A B″ subunits and with the association of PP5 with the A subunit as described above. A direct interaction between PP5 and G5PR was not demonstrated, and it is possible that these experiments simply confirm the ability of PP5 to interact with the A subunit of PP2A. Competition between PP5 and the PP2A catalytic subunit for binding to A subunit-G5PR complexes suggested that such complexes can contain either PP5 or the catalytic subunit of PP2A, but not both. Recombinant G5PR did not affect the catalytic activity of purified PP2A or PP5. Phosphatase activity pulled down from cell lysates with GST-G5PR was inhibited by okadaic acid but not markedly stimulated by arachidonic acid. This suggested that bound PP5 was only a minor component of the recovered phosphatases. The subcellular localizations of overexpressed PP5 and G5PR were similar, consistent with their possible association in living cells. Unfortunately, no such association was detected between the endogenous proteins from cells or tissues (Kono et al. 2002). Thus, it seems clear that PP5 can associate directly with the A subunit of PP2A, and it may associate directly or indirectly with at least two different B″ subunits. Effects of such binding on PP5 activity remain to be demonstrated.

Table 1. Proteins reported to interact with PP5

Protein	2-hybrid	Pulldown	Co-IP	Function
ANPR	+	+	−	−
Hsp90	−	+	+	+
ASK1	+	−	+	+
G_{12}	+	+	+	+
Copine	+	+	−	+
PP2A	+	+	+	−
CDC16, CDC27	+	+	−	−
CRY2	+	+	−	+
ADR1	+	−	−	−
PPG	+	−	−	−

6.6.7 Anaphase-promoting complex

Associations between the TPR domain of PP5 and two components of the ana-phase-promoting complex, CDC16 and CDC27, were observed in a yeast two-hybrid screen. Binding of PP5 to CDC27, but not to CDC16, was confirmed in a pulldown assay. Although overexpressed PP5 was localized to the mitotic spindle, consistent with a hypothesis that it might regulate phosphorylation of the ana-phase-promoting complex at a key point in mitosis, no association between PP5 and the endogenous proteins could be detected, and no functional data were pre-sented (Ollendorff and Donoghue 1997).

6.6.8 CRY2

An interaction between the TPR domain of PP5 and the human blue light receptor designated CRY2, implicated in the regulation of circadian rhythms (Thresher et al. 1998), was detected in a two-hybrid screen and confirmed in a pulldown assay using a PP5 fusion protein and *in vitro*-translated CRY2 (Zhao and Sancar 1997). Inhibition of basal PP5 activity by CRY2 was observed, but basal PP5 activity is extremely low, making the significance of this result difficult to interpret. The same two-hybrid screen identified the hsp70-binding protein, Tpr1, as a CRY2-interacting protein. The TPR domain of Tpr1 is closely related to that of PP5, con-sistent with both proteins binding to CRY2 via similar structural features. It is the EEVD-binding portion of these TPR domains that is most similar, however, and it might reasonably be speculated that the interactions of these proteins with CRY2 is indirect, and mediated by Tpr1 or PP5 binding to the C-terminal EEVD se-quences of hsp70 and hsp90, respectively. This could occur if the hsp70/hsp90 chaperone system is involved in the normal folding of CRY2. The experiments raise the possibility that PP5 may regulate CRY2, or vice versa, but no evidence for any biological function of this interaction was presented.

6.6.9 Other PP5-interacting proteins

In a global two-hybrid analysis designed to identify all interactions between *S. cerevisiae* proteins, PPT was found to interact with the transcription factor, ADR1, and with PPG (Uetz et al. 2000). PPG is a PP1-like phosphatase whose absence leads to impaired glycogen metabolism. PPG also interacted with hsp90 and with Cpr6, another hsp90-binding TPR protein, raising the possibility that its interaction with PPT was indirect and mediated by both proteins binding to hsp90 (Uetz et al. 2000). These interactions have not been further confirmed, and no synthetic phenotype was observed when both the *PPG1* and *PPT1* genes were deleted (Sakumoto et al. 2002). In a more recent analysis of interactions in the *S. cerevisiae* proteome, using pulldown assays and analysis of binding proteins by mass spectrometry, the only PPT-binding proteins identified were the two isoforms of hsp90 (Gavin et al. 2002), consistent with the well characterized binding of PP5 to hsp90 in mammals.

6.7 Genetics of PP5

The use of genetic approaches in model organisms has not yet proceeded far enough to define a biological role for PP5. In yeast and nematodes, no phenotype has yet been associated with the loss of PP5 function, and PP5 mutants in other species have not yet been characterized. The use of a PP5 antisense oligonucleotide to deplete the enzyme in cultured human cells has suggested possible roles for PP5 in regulating growth and hormonal responses, but further experiments are necessary to confirm those roles.

6.7.1 *S. cerevisiae*

The first attempt to define the biological role of PP5 was described by Chen et al., who disrupted the *PPT1* gene in *S. cerevisiae* by standard methods (Chen et al. 1994). Both haploid and diploid cells were viable, and growth rates were indistinguishable from wild type cells in YPD or in synthetic media containing four different carbon sources. Mating and sporulation were normal, and cell size was normal in both log and stationary phases. The mutant cells did not show altered sensitivity to high or low temperatures, various drugs, high concentrations of various salts, or low pH. Glycogen and invertase levels were normal on three different media (Chen et al. 1994). We have independently produced *ppt1Δ* yeast and observed no effects on growth or GR signaling (unpublished results). Knowing that PP5 is physically associated with hsp90, we have collaborated with several laboratories to test whether a synthetic phenotype could be obtained by crossing *ppt1Δ* yeast with hsp90 mutants or mutants lacking various hsp90 co-chaperones. In no case were effects on cell growth or GR signaling observed (unpublished data).

In a study disrupting all 32 of the *S. cerevisiae* protein phosphatase genes, it was confirmed that *ppt1Δ* yeast are viable and grow normally under a variety of conditions, including differing carbon sources, different temperatures, and in the presence of various salts and drugs (Sakumoto et al. 1999). In a second study from the same group, all possible double disruptants of the 30 non-essential protein phosphatase genes were prepared, and phenotypes analyzed as in the first study (Sakumoto et al. 2002). No synthetic phenotype was observed when *ppt1Δ* cells were crossed with any of the other 29 phosphatase mutants (Sakumoto et al. 2002). Thus, *S. cerevisiae* has no obvious requirement for PPT, which presumably plays a redundant role under all growth conditions thus far tested. Although rational approaches to identifying a synthetic phenotype have thus far failed, it remains possible that a synthetic lethal screen will be informative.

6.7.2 *C. elegans*

A recent study described the use of RNA interference in *C. elegans* to systematically inhibit the function of each of the known genes in this organism. As might be expected, a wide variety of defects were observed, leading to the identification of approximately 1000 genes which had not previously been associated with a phenotype. Animals deficient in PP5, however, had a wild type phenotype (Kamath et al. 2003). This result is consistent with the data discussed above demonstrating that PPT is not required for viability or normal growth of *S. cerevisiae*.

6.7.3 *D. melanogaster*

Only two studies have thus far been published regarding *Drosophila* PP5. The *D. melanogaster* PP5 gene was identified by a molecular, not a genetic approach. In order to refine its localization on the third chromosome at locus 85E8-12, determined by in situ hybridization to polytene chromosomes, DNA from three *Drosophila* strains with deletions in the 85D-85F region was analyzed by Southern blotting, using the PP5 cDNA as a probe. Only one of those deletions was found to include the PP5 gene, allowing mapping of the gene to 85E10-12. The DNA containing the deletion encompassing the PP5 gene (85D8-85F1) was obtained from a heterozygous strain, indicating that flies heterozygous for PP5 are viable (Brown et al. 2000). It is not clear, whether animals homozygous for this large deletion are viable or, if not, whether their phenotype would be due to the deletion of PP5 or to that of adjacent genes. Based on the completed physical map of the *Drosophila* genome, 12 genes including PP5 are located on the third chromosome at 85E6 (http://flybase.bio.indiana.edu/).

RNA interference studies in cultured *Drosophila* S2 cells showed no effect of depleting PP5 on cell proliferation. PP5 depletion was confirmed by Western blotting. In the same series of experiments, depletion of PP2A led to apoptosis, and depletion of PP4 led to a reduction in growth rate without apoptosis (Silverstein et al. 2002). These results are consistent with those described above; i.e. there seems to be no requirement for PP5 for the normal growth of *Drosophila* cells, just as has been observed for yeast and nematodes.

6.7.4 Antisense studies

Based on the abilities of okadaic acid to promote transformation and tumor cell growth under some conditions, but to antagonize these processes under other conditions, it was postulated that individual okadaic-acid sensitive phosphatases had opposing effects on cell growth. In this paradigm, a subset of these phosphatases would be promising targets for antineoplastic drugs. PP5 was selected as one such target for antisense oligonucleotides containing modifications allowing them to be effective at nanomolar concentrations (Honkanen and Golden 2002). An oligonucleotide designated ISIS 15534 has been used to test the biological role of PP5 in cultured human cells (Zuo et al. 1998; Zuo et al. 1999; Dean et al. 2001; Urban et al. 2001; Urban et al. 2003). Overnight treatment of human A549 cells with this compound depleted PP5 mRNA in a dose-dependent fashion, and effectively depleted PP5 protein at a concentration of 300 nM. A control oligonucleotide had no effect on PP5 levels (Zuo et al. 1998). For reasons that are unclear, depletion of PP5 at the protein level was not monitored in subsequent experiments. In A549 cells, treatment with 300 nM ISIS 15534 rapidly led to G_1 growth arrest associated with the induction of $p21^{WAF1/Cip1}$. This phenomenon was dependent on the presence of p53, whose phosphorylation was stimulated by ISIS 15534. This suggested that PP5 might ordinarily facilitate cell cycle progression by dephosphorylating p53. In each experiment, depletion of PP5 mRNA was confirmed by Northern blotting (Zuo et al. 1998). The site of ISIS 15534-stimulated p53 phosphorylation has been identified as Ser-15 (Urban et al. 2003). Stimulation of GR signaling by ISIS 15534 was discussed above in the section on hsp90.

The effects of ISIS 15534 on the growth of estrogen-dependent MCF-7 human breast cancer cells has been examined (Urban et al. 2001). These cells proliferate in medium supplemented with fetal calf serum (FCS), but are quiescent in medium containing gelded horse serum (GHS), which lacks estrogen. Medium containing both GHS and estrogen supports MCF-7 cell growth at rates similar to those observed in medium containing FCS, despite the fact that estrogen is relatively ineffective in stimulating PP5 expression: quiescent cells in GHS had low levels of PP5 mRNA that were stimulated >10-fold by FCS, but only 50% by estrogen. These data seemed consistent with studies in yeast, in which PPT levels are 25-fold higher in rapidly growing cells, but deleting its gene does not retard cell growth. Alternatively, PP5 might be required for the growth of mammalian cells, as suggested by the antisense studies with A549 cells, and a 50% increase in PP5 induced by estrogen might suffice to support growth of MCF-7 cells, with the excess PP5 induced by FCS being superfluous. This was the hypothesis pursued by Urban et al. (2001).

Treatment of MCF-7 cells with ISIS 15534 reduced PP5 mRNA to undetectable levels overnight, and arrested their growth within a day. Further, in three stable MCF-7 cell lines, the induction of PP5 with a tet-off system led to estrogen-independent growth. PP5 could be induced 1.2-fold, 1.4-fold, or 2-fold in the three different cell lines tested, as measured by Western blotting. The highest rates of estrogen-independent growth were seen in the lines expressing either 1.2-fold or 2-fold more PP5 than control cells, while the cell line expressing 1.4-fold more PP5 than control cells grew more slowly (Urban et al. 2001). The lack of correlation between PP5 levels and growth rate could be due to clonal variations unrelated to PP5 or to the difficulty in quantitating very small changes in PP5 levels. In all cases, however, PP5 induction led to cell proliferation. Taken together, these experiments suggested that very modest increases in PP5 levels, induced by estrogen treatment in the parental cell line or by withdrawing tetracycline in stable cell lines, stimulated the proliferation of MCF-7 cells. A recent review article from the same group states that *in vivo* overexpression of PP5 did not induce estrogen-independent growth, but did enhance estrogen-dependent tumor growth (Honkanen and Golden 2002). Thus, a role of PP5 in estrogen-independent growth of breast cancer cells may not be universal, but the ISIS 15534 experiments all seem to suggest a growth-stimulatory role for the enzyme in human cells.

At the current time, it is not entirely clear how to reconcile the observation that depleting PP5 in yeast, nematodes, or cultured *Drosophila* cells has no effect on growth, while altering PP5 levels by as little as 20% in human cells seems to have profound effects on growth. A simple explanation might be that the G_1 growth arrest induced by ISIS 15534 is dependent on p53, whose nematode and fly counterparts are functionally different from the human protein. The fly and nematode homologues have pro-apoptotic activity but are unable to induce the G_1 growth arrest seen upon overexpression of mammalian p53 (Ollmann et al. 2000; Derry et al. 2001). Perhaps as p53 acquired a new function during evolution, PP5 also acquired a new function as a p53 regulator. PP5 presumably has other roles that explain its conservation in fungi, plants, and animals, but these roles remain to be uncovered.

6.8 Conclusions

PP5 is a ubiquitous okadaic acid-sensitive phosphatase whose TPR domain targets it to substrates and appears to regulate its enzymatic activity. The enzyme's basal activity is negligible, and while it can be activated *in vitro* by lipid compounds, its normal mechanisms of regulation remain obscure. In the coming years, determination of the structure of the full-length enzyme may aid our understanding of its regulation, and it will be important to identify physiological activators. A number of PP5 binding proteins have been identified, including hsp90, dynein, ASK1, G_{12} proteins, and PP2A subunits. PP5 complexes with these target proteins may play important roles in protein folding, steroid receptor trafficking, and responses to oxidative stress, but their biological functions need to be better defined. Antisense studies suggest a p53-dependent role for PP5 in cell cycle progression in human cells that needs to be confirmed using other methodologies, but which points to a potentially important function for this enzyme. PP5 is highly conserved in fungi, plants, and animals, and its expression is developmentally regulated, but deletion of its gene in *S. cerevisiae* or depletion of its mRNA in *C. elegans* results in no obvious phenotype and its depletion from cultured *Drosophila* cells has no effect on growth. As such, more detailed studies in these organisms as well as genetic data from other model organisms will be important in advancing our understanding of the biology of this protein phosphatase.

References

Abe Y, Shodai T, Muto T, Mihara K, Torii H, Nishikawa S, Endo T, Kohda D (2000) Structural basis of presequence recognition by the mitochondrial protein import receptor Tom20. Cell 100:551-560

Bahl R, Bradley KC, Thompson KJ, Swain RA, Rossie S, Meisel RL (2001) Localization of protein Ser/Thr phosphatase 5 in rat brain. Mol Brain Res 90:101-109

Becker W, Buttini M, Limonta S, Boddeke H, Joost HG (1996) Distribution of the mRNA for protein phosphatase T in rat brain. Mol Brain Res 36:23-28

Becker W, Kentrup H, Klumpp S, Schultz JE, Joost HG (1994) Molecular cloning of a protein serine/threonine phosphatase containing a putative regulatory tetratricopeptide repeat domain. J Biol Chem 269:22586-22592

Bell DR, Poland A (2000) Binding of aryl hydrocarbon receptor (AhR) to AhR-interacting protein. The role of hsp90. J Biol Chem 275:36407-36414

Berestetskaya YV, Faure MP, Ichijo H, Voyno-Yasenetskaya TA (1998) Regulation of apoptosis by α-subunits of G12 and G13 proteins via apoptosis signal-regulating kinase-1. J Biol Chem 273:27816-27823

Blatch GL, Lassle M (1999) The tetratricopeptide repeat: a structural motif mediating protein-protein interactions. Bioessays 21:932-939

Borthwick EB, Zeke T, Prescott AR, Cohen PT (2001) Nuclear localization of protein phosphatase 5 is dependent on the carboxy-terminal region. FEBS Lett 491:279-284

Brown L, Borthwick EB, Cohen PT (2000) *Drosophila* protein phosphatase 5 is encoded by a single gene that is most highly expressed during embryonic development. Biochim Biophys Acta 1492:470-476

Chang HC, Nathan DF, Lindquist S (1997) In vivo analysis of the hsp90 cochaperone Sti1 (p60). Mol Cell Biol 17:318-325

Chaudhuri M (2001) Cloning and characterization of a novel serine/threonine protein phosphatase type 5 from *Trypanosoma brucei*. Gene 266:1-13

Chen MS, Silverstein AM, Pratt WB, Chinkers M (1996) The tetratricopeptide repeat domain of protein phosphatase 5 mediates binding to glucocorticoid receptor heterocomplexes and acts as a dominant negative mutant. J Biol Chem 271:32315-32320

Chen MX, Cohen PT (1997) Activation of protein phosphatase 5 by limited proteolysis or the binding of polyunsaturated fatty acids to the TPR domain. FEBS Lett 400:136-140

Chen MX, McPartlin AE, Brown L, Chen YH, Barker HM, Cohen PT (1994) A novel human protein serine/threonine phosphatase, which possesses four tetratricopeptide repeat motifs and localizes to the nucleus. EMBO J 13:4278-4290

Chinkers M (1994) Targeting of a distinctive protein-serine phosphatase to the protein kinase-like domain of the atrial natriuretic peptide receptor. Proc Natl Acad Sci USA 91:11075-11079

Chinkers M (2001) Protein phosphatase 5 in signal transduction. Trends Endocrinol Metab 12:28-32

Das AK, Cohen PW, Barford D (1998) The structure of the tetratricopeptide repeats of protein phosphatase 5: implications for TPR-mediated protein-protein interactions. EMBO J 17:1192-1199

Dean DA, Urban G, Aragon IV, Swingle M, Miller B, Rusconi S, Bueno M, Dean NM, Honkanen RE (2001) Serine / threonine protein phosphatase 5 (PP5) participates in the regulation of glucocorticoid receptor nucleocytoplasmic shuttling. BMC Cell Biol 2:6

Derry WB, Putzke AP, Rothman JH (2001) *Caenorhabditis elegans* p53: role in apoptosis, meiosis, and stress resistance. Science 294:591-595

Dobson S, Kar B, Kumar R, Adams B, Barik S (2001) A novel tetratricopeptide repeat (TPR) containing PP5 serine/threonine protein phosphatase in the malaria parasite, *Plasmodium falciparum*. BMC Microbiol 1:31

Duina AA, Chang HC, Marsh JA, Lindquist S, Gaber RF (1996) A cyclophilin function in hsp90-dependent signal transduction. Science 274:1713-1715

Edwards TA, Pyle SE, Wharton RP, Aggarwal AK (2001) Structure of Pumilio reveals similarity between RNA and peptide binding motifs. Cell 105:281-289

Fukuda H, Shima H, Vesonder RF, Tokuda H, Nishino H, Katoh S, Tamura S, Sugimura T, Nagao M (1996) Inhibition of protein serine/threonine phosphatases by fumonisin B_1, a mycotoxin. Biochem Biophys Res Commun 220:160-165

Galigniana MD, Harrell JM, Murphy PJ, Chinkers M, Radanyi C, Renoir JM, Zhang M, Pratt WB (2002) Binding of hsp90-associated immunophilins to cytoplasmic dynein: direct binding and in vivo evidence that the peptidylprolyl isomerase domain is a dynein interaction domain. Biochemistry 41:13602-13610

Galigniana MD, Radanyi C, Renoir JM, Housley PR, Pratt WB (2001) Evidence that the peptidylprolyl isomerase domain of the hsp90-binding immunophilin FKBP52 is involved in both dynein interaction and glucocorticoid receptor movement to the nucleus. J Biol Chem 276:14884-14889.

Gasch AP, Spellman PT, Kao CM, Carmel-Harel O, Eisen MB, Storz G, Botstein D, Brown PO (2000) Genomic expression programs in the response of yeast cells to environmental changes. Mol Biol Cell 11:4241-4257

Gatto GJ, Jr., Geisbrecht BV, Gould SJ, Berg JM (2000) Peroxisomal targeting signal-1 recognition by the TPR domains of human PEX5. Nat Struct Biol 7:1091-1095.

Gavin AC, Bosche M, Krause R, Grandi P, Marzioch M, Bauer A, Schultz J, Rick JM, Michon AM, Cruciat CM, Remor M, Hofert C, Schelder M, Brajenovic M, Ruffner H, Merino A, Klein K, Hudak M, Dickson D, Rudi T, Gnau V, Bauch A, Bastuck S, Huhse B, Leutwein C, Heurtier MA, Copley RR, Edelmann A, Querfurth E, Rybin V, Drewes G, Raida M, Bouwmeester T, Bork P, Seraphin B, Kuster B, Neubauer G, Superti-Furga G (2002) Functional organization of the yeast proteome by systematic analysis of protein complexes. Nature 415:141-147.

Grizot S, Fieschi F, Dagher MC, Pebay-Peyroula E (2001) The active N-terminal region of p67^{phox}. Structure at 1.8 Å resolution and biochemical characterizations of the A128V mutant implicated in chronic granulomatous disease. J Biol Chem 276:21627-21631

Hill AA, Hunter CP, Tsung BT, Tucker-Kellogg G, Brown EL (2000) Genomic analysis of gene expression in *C. elegans*. Science 290:809-812

Honkanen RE, Golden T (2002) Regulators of serine/threonine protein phosphatases at the dawn of a clinical era? Curr Med Chem 9:2055-2075

Ichijo H, Nishida E, Irie K, ten Dijke P, Saitoh M, Moriguchi T, Takagi M, Matsumoto K, Miyazono K, Gotoh Y (1997) Induction of apoptosis by ASK1, a mammalian MAPKKK that activates SAPK/JNK and p38 signaling pathways. Science 275:90-94

Jeong JY, Johns J, Sinclair C, Park JM, Rossie S (2003) Characterization of Saccharomyces cerevisiae protein Ser/Thr phosphatase T1 and comparison to its mammalian homolog PP5. BMC Cell Biol 4:3

Joubert S, Labrecque J, De Lean A (2001) Reduced activity of the NPR-A kinase triggers dephosphorylation and homologous desensitization of the receptor. Biochemistry 40:11096-11105

Kamath RS, Fraser AG, Dong Y, Poulin G, Durbin R, Gotta M, Kanapin A, Le Bot N, Moreno S, Sohrmann M, Welchman DP, Zipperlen P, Ahringer J (2003) Systematic functional analysis of the *Caenorhabditis elegans* genome using RNAi. Nature 421:231-237

Kang H, Sayner SL, Gross KL, Russell LC, Chinkers M (2001) Identification of amino acids in the tetratricopeptide repeat and C-terminal domains of protein phosphatase 5 involved in autoinhibition and lipid activation. Biochemistry 40:10485-10490

Kobe B, Kajava AV (2000) When protein folding is simplified to protein coiling: the continuum of solenoid protein structures. Trends Biochem Sci 25:509-515

Kono Y, Maeda K, Kuwahara K, Yamamoto H, Miyamoto E, Yonezawa K, Takagi K, Sakaguchi N (2002) MCM3-binding GANP DNA-primase is associated with a novel phosphatase component G5PR. Genes Cells 7:821-834

Kumar A, Roach C, Hirsh IS, Turley S, deWalque S, Michels PA, Hol WG (2001) An unexpected extended conformation for the third TPR motif of the peroxin PEX5 from *Trypanosoma brucei*. J Mol Biol 307:271-282

Lapouge K, Smith SJ, Walker PA, Gamblin SJ, Smerdon SJ, Rittinger K (2000) Structure of the TPR domain of p67^{phox} in complex with Rac•GTP. Mol Cell 6:899-907

Lindenthal C, Klinkert MQ (2002) Identification and biochemical characterisation of a protein phosphatase 5 homologue from *Plasmodium falciparum*. Mol Biochem Parasitol 120:257-268

Lubert EJ, Hong YL, Sarge KD (2001) Interaction between protein phosphatase 5 and the A subunit of protein phosphatase 2A - Evidence for a heterotrimeric form of protein phosphatase 5. J Biol Chem 276:38582-38587

Lyons TJ, Gasch AP, Gaither LA, Botstein D, Brown PO, Eide DJ (2000) Genome-wide characterization of the Zap1p zinc-responsive regulon in yeast. Proc Natl Acad Sci USA 97:7957-7962

Meek S, Morrice N, MacKintosh C (1999) Microcystin affinity purification of plant protein phosphatases: PP1C, PP5 and a regulatory A-subunit of PP2A. FEBS Lett 457:494-498

Morita K, Saitoh M, Tobiume K, Matsuura H, Enomoto S, Nishitoh H, Ichijo H (2001) Negative feedback regulation of ASK1 by protein phosphatase 5 (PP5) in response to oxidative stress. EMBO J 20:6028-6036

Ollendorff V, Donoghue DJ (1997) The serine/threonine phosphatase PP5 interacts with CDC16 and CDC27, two tetratricopeptide repeat-containing subunits of the anaphase-promoting complex. J Biol Chem 272:32011-32018

Ollmann M, Young LM, Di Como CJ, Karim F, Belvin M, Robertson S, Whittaker K, Demsky M, Fisher WW, Buchman A, Duyk G, Friedman L, Prives C, Kopczynski C (2000) *Drosophila p53* is a structural and functional homolog of the tumor suppressor *p53*. Cell 101:91-101

Potter LR, Hunter T (2001) Guanylyl cyclase-linked natriuretic peptide receptors: structure and regulation. J Biol Chem 276:6057-6060

Pratt WB, Silverstein AM, Galigniana MD (1999) A model for the cytoplasmic trafficking of signalling proteins involving the hsp90-binding immunophilins and p50^{cdc37}. Cell Signal 11:839-851.

Pratt WB, Toft DO (2003) Regulation of signaling protein function and trafficking by the hsp90/hsp70-based chaperone machinery. Exp Biol Med 228:111-133

Ramsey AJ, Chinkers M (2002) Identification of potential physiological activators of protein phosphatase 5. Biochemistry 41:5625-5632.

Ramsey AJ, Russell LC, Whitt SR, Chinkers M (2000) Overlapping sites of tetratricopeptide repeat protein binding and chaperone activity in heat shock protein 90. J Biol Chem 275:17857-17862

Riggs DL, Roberts PJ, Chirillo SC, Cheung-Flynn J, Prapapanich V, Ratajczak T, Gaber R, Picard D, Smith DF (2003) The hsp90-binding peptidylprolyl isomerase FKBP52 potentiates glucocorticoid signaling *in vivo*. EMBO J 22:1158-1167

Roberts CJ, Nelson B, Marton MJ, Stoughton R, Meyer MR, Bennett HA, He YD, Dai H, Walker WL, Hughes TR, Tyers M, Boone C, Friend SH (2000) Signaling and circuitry of multiple MAPK pathways revealed by a matrix of global gene expression profiles. Science 287:873-880

Russell LC, Whitt SR, Chen MS, Chinkers M (1999) Identification of conserved residues required for the binding of a tetratricopeptide repeat domain to heat shock protein 90. J Biol Chem 274:20060-20063

Sakumoto N, Matsuoka I, Mukai Y, Ogawa N, Kaneko Y, Harashima S (2002) A series of double disruptants for protein phosphatase genes in *Saccharomyces cerevisiae* and their phenotypic analysis. Yeast 19:587-599

Sakumoto N, Mukai Y, Uchida K, Kouchi T, Kuwajima J, Nakagawa Y, Sugioka S, Yamamoto E, Furuyama T, Mizubuchi H, Ohsugi N, Sakuno T, Kikuchi K, Matsuoka I, Ogawa N, Kaneko Y, Harashima S (1999) A series of protein phosphatase gene disruptants in *Saccharomyces cerevisiae*. Yeast 15:1669-1679

Scheufler C, Brinker A, Bourenkov G, Pegoraro S, Moroder L, Bartunik H, Hartl FU, Moarefi I (2000) Structure of TPR domain-peptide complexes: critical elements in the assembly of the hsp70-hsp90 multichaperone machine. Cell 101:199-210

Shao J, Hartson SD, Matts RL (2002) Evidence that protein phosphatase 5 functions to negatively modulate the maturation of the hsp90-dependent heme-regulated eIF2α kinase. Biochemistry 41:6770-6779

Shirato H, Shima H, Nakagama H, Fukuda H, Watanabe Y, Ogawa K, Matsuda Y, Kikuchi K (2000) Expression in hepatomas and chromosomal localization of rat protein phosphatase 5 gene. Int J Oncol 17:909-912

Silverstein AM, Barrow CA, Davis AJ, Mumby MC (2002) Actions of PP2A on the MAP kinase pathway and apoptosis are mediated by distinct regulatory subunits. Proc Natl Acad Sci USA 99:4221-4226.

Silverstein AM, Galigniana MD, Chen MS, Owens-Grillo JK, Chinkers M, Pratt WB (1997) Protein phosphatase 5 is a major component of glucocorticoid receptor•hsp90 complexes with properties of an FK506-binding immunophilin. J Biol Chem 272:16224-16230

Silverstein AM, Galigniana MD, Kanelakis KC, Radanyi C, Renoir JM, Pratt WB (1999) Different regions of the immunophilin FKBP52 determine its association with the glucocorticoid receptor, hsp90, and cytoplasmic dynein. J Biol Chem 274:36980-36986.

Silverstein AM, Grammatikakis N, Cochran BH, Chinkers M, Pratt WB (1998) p50^{cdc37} binds directly to the catalytic domain of Raf as well as to a site on hsp90 that is topologically adjacent to the tetratricopeptide repeat binding site. J Biol Chem 273:20090-20095

Sinars CR, Cheung-Flynn J, Rimerman RA, Scammell JG, Smith DF, Clardy J (2003) Structure of the large FK506-binding protein FKBP51, an hsp90-binding protein and a component of steroid receptor complexes. Proc Natl Acad Sci USA 100:868-873

Sinclair C, Borchers C, Parker C, Tomer K, Charbonneau H, Rossie S (1999) The tetratricopeptide repeat domain and a C-terminal region control the activity of Ser/Thr protein phosphatase 5. J Biol Chem 274:23666-23672

Skinner J, Sinclair C, Romeo C, Armstrong D, Charbonneau H, Rossie S (1997) Purification of a fatty acid-stimulated protein-serine/threonine phosphatase from bovine brain and its identification as a homolog of protein phosphatase 5. J Biol Chem 272:22464-22471

Small ID, Peeters N (2000) The PPR motif - a TPR-related motif prevalent in plant organellar proteins. Trends Biochem Sci 25:46-47

Taylor P, Dornan J, Carrello A, Minchin RF, Ratajczak T, Walkinshaw MD (2001) Two structures of cyclophilin 40: folding and fidelity in the TPR domains. Structure 9:431-438

Thresher RJ, Vitaterna MH, Miyamoto Y, Kazantsev A, Hsu DS, Petit C, Selby CP, Dawut L, Smithies O, Takahashi JS, Sancar A (1998) Role of mouse cryptochrome blue-light photoreceptor in circadian photoresponses. Science 282:1490-1494

Tobiume K, Matsuzawa A, Takahashi T, Nishitoh H, Morita K, Takeda K, Minowa O, Miyazono K, Noda T, Ichijo H (2001) ASK1 is required for sustained activations of JNK/p38 MAP kinases and apoptosis. EMBO Rep 2:222-228

Tobiume K, Saitoh M, Ichijo H (2002) Activation of apoptosis signal-regulating kinase 1 by the stress-induced activating phosphorylation of pre-formed oligomer. J Cell Physiol 191:95-104

Tomsig JL, Creutz CE (2002) Copines: a ubiquitous family of Ca^{2+}-dependent phospholipid-binding proteins. Cell Mol Life Sci 59:1467-1477

Tomsig JL, Snyder SL, Creutz CE (2003) Identification of targets for calcium signaling through the copine family of proteins. characterization of a coiled-coil copine-binding motif. J Biol Chem 278:10048-10054

Uetz P, Giot L, Cagney G, Mansfield TA, Judson RS, Knight JR, Lockshon D, Narayan V, Srinivasan M, Pochart P, Qureshi-Emili A, Li Y, Godwin B, Conover D, Kalbfleisch T, Vijayadamodar G, Yang M, Johnston M, Fields S, Rothberg JM (2000) A comprehensive analysis of protein-protein interactions in *Saccharomyces cerevisiae*. Nature 403:623-627

Urban G, Golden T, Aragon IV, Cowsert L, Cooper SR, Dean NM, Honkanen RE (2003) Identification of a functional link for the p53 tumor suppressor protein in dexamethasone-induced growth suppression. J Biol Chem 278:9747-9753

Urban G, Golden T, Aragon IV, Scammell JG, Dean NM, Honkanen RE (2001) Identification of an estrogen-inducible phosphatase (PP5) that converts MCF-7 human breast carcinoma cells into an estrogen-independent phenotype when expressed constitutively. J Biol Chem 276:27638-27646

Ward BK, Allan RK, Mok D, Temple SE, Taylor P, Dornan J, Mark PJ, Shaw DJ, Kumar P, Walkinshaw MD, Ratajczak T (2002) A structure-based mutational analysis of cyclophilin 40 identifies key residues in the core tetratricopeptide repeat domain that mediate binding to hsp90. J Biol Chem 277:40799-40809

Yamaguchi Y, Katoh H, Mori K, Negishi M (2002) $G\alpha_{12}$ and $G\alpha_{13}$ interact with Ser/Thr protein phosphatase type 5 and stimulate its phosphatase activity. Curr Biol 12:1353-1358

Yatzkan E, Yarden O (1997) ppt-1, a Neurospora crassa PPT/PP5 subfamily serine/threonine protein phosphatase. Biochim Biophys Acta 1353:18-22

Zhao S, Sancar A (1997) Human blue-light photoreceptor hCRY2 specifically interacts with protein serine/threonine phosphatase 5 and modulates its activity. Photochem Photobiol 66:727-731

Zuo Z, Dean NM, Honkanen RE (1998) Serine/threonine protein phosphatase type 5 acts upstream of p53 to regulate the induction of p21[WAF1/Cip1] and mediate growth arrest. J Biol Chem 273:12250-12258

Zuo Z, Urban G, Scammell JG, Dean NM, McLean TK, Aragon I, Honkanen RE (1999) Ser/Thr protein phosphatase type 5 (PP5) is a negative regulator of glucocorticoid receptor-mediated growth arrest. Biochemistry 38:8849-8857

7 Protein histidine phosphatases in signal transduction and metabolism

Susanne Klumpp, Gunther Bechmann, Anette Mäurer, Josef Krieglstein

Abstract

Knowledge about the function, structure, and interplay between protein histidine kinases and their response regulators in bacteria is overwhelming. The essential steps are phosphorylation of a histidine residue, phosphotransfer onto aspartate, and dephosphorylation of phosphoaspartate via sensor histidine kinases. Protein histidine phosphatases exist in bacteria as well. Knowledge of N-phosphorylation and dephosphorylation in vertebrates, on the other hand, is still limited. Although there has been unequivocal evidence of the presence of histidine kinases, histidine phosphatases, and N-phosphorylated proteins in mammals for decades, the information is scattered and the few and tiny pieces do not give a complete picture of the puzzle yet. This review on protein histidine phosphatases is intended to give short background information on the bacterial phosphohistidine signal transduction treasures. The major part, however, is directed at the dephosphorylation of phosphohistidine in vertebrate proteins and is culminated in the most recent discovery and cloning of the first protein histidine phosphatase from vertebrates.

7.1 Introduction

Half a century ago when Fischer and Krebs studied glycogen phosphorylase, they recognized the fundamental impact of phosphorylation and dephosphorylation as major cellular regulatory processes (Fischer et al. 1959). In the meantime, it is well known that not only enzymes but also signaling elements (e.g. receptors, channels) and a huge number of regulatory proteins (e.g. transcription factors) are all activated and deactivated by reversible phosphorylation. At the beginning of this era, the interest was mainly focused on protein kinases. Nowadays the importance of protein phosphatases is considered equally significant.

Signal transduction in prokaryotic cells predominantly involves phosphorylation of a histidyl group (N-phosphorylation). On the other hand, literature on protein phosphorylation in eukaryotic cells in the past almost exclusively dealt with serine, threonine, and tyrosine residues (O-phosphorylation). Indeed there were severe doubts as to whether histidine kinases and histidine phosphatases might be present in vertebrates at all (Swanson et al. 1994; Klumpp and Krieglstein 2002). In the meantime, however, we have learned that histidine phosphorylation in

Topics in Current Genetics, Vol. 5
J. Arino, D.R. Alexander (Eds.) Protein phosphatases
© Springer-Verlag Berlin Heidelberg 2004

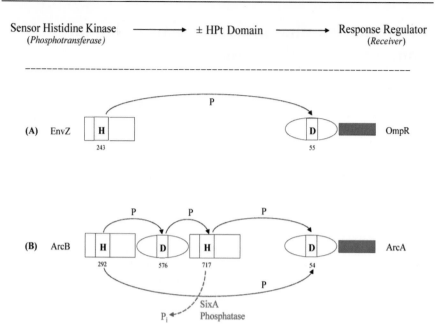

Fig. 1. Systems for histidine and aspartate phosphorylation. Schematic representation of a bacterial two-component system (A) and multicomponent system (B) involving histidine (H) and aspartate (D) phosphorylations on histidine kinases and response regulators, respectively. Both systems operate in *Escherichia coli*. The grey rectangles in OmpR and ArcA represent DNA-binding domains of the receiver proteins acting as transcription factors. The histidine kinase of EnvZ also has aspartate phosphatase activity (see Fig. 2). SixA, shown in red, is a protein histidine phosphatase acting specifically on phosphohistidine-717 of ArcB. Details are given in the text. HPt: Histidine containing phosphotransfer domain.

vertebrates is not limited to the formation of phosphohistidine intermediates in the course of catalysis of metabolic enzymes.

It has been known since the early 1960s that protein histidine phosphates exist in mammals (Boyer et al. 1962), but has taken three decades to discover that proteins involved in signal transduction also undergo reversible phosphorylation on histidine residues, e.g. P-selectin (Crovello et al. 1995), annexin I (Muimo et al. 2000), and G-protein β-subunit (Wieland et al. 1993; Kowluru et al. 1996; Cuello et al. 2003; Hippe et al. 2003). A histidine kinase had been purified from yeast (Huang et al. 1991), and histidine kinases have been described from vertebrates (Besant and Attwood 2000; Kowluru 2002), but for a very long time neither histidine kinases nor histidine phosphatases had been sequenced and cloned from mammals. Information has just recently become available about the first protein histidine phosphatase in vertebrates (Hermesmeier and Klumpp 1999; Accession number CAC16267 (2000); Ek et al. 2002; Klumpp et al. 2002).

7.2 Protein histidine kinases at a glance

The entire genomic sequence of *Escherichia coli* revealed a total of 29 open reading frames to encode putative histidine kinases; *Pseudomonas aeruginosa* has as many as 63. The structure of bacterial histidine kinases is distinct from that of serine/threonine and tyrosine kinases. Instead, it is related to a small class of ATP-binding proteins, referred to as GHKL superfamily, e.g. gyrase B and the chaperone Hsp 90 (Bilwes et al. 1999).

Bacterial histidine kinases act via a "ping-pong" mechanism in which they first phosphorylate themselves on a histidine residue using ATP. Then they transfer the phosphoryl group to a preserved aspartate residue on the response regulator domain, which constitutes the downstream element of this binary module best known as "two-component system" (Fig. 1A). Functional forms of sensor histidine kinases are homodimeric. The response regulator protein often functions as a phosphorylation-dependent DNA-binding transcription factor. A common variant of this two-component system are multiple histidine aspartate relays, which permit incorporation of several distinct signals and increase the specificity of the response (Fig. 1B). Interested readers are advised to consult the specialized book edited by Inouye and Dutta (2003), and a number of recent reviews (Dutta et al. 1999; West and Stock 2001; Foussard et al. 2001).

In eukaryotes homology cloning and genome analyses revealed two-component signal transduction elements in yeasts, ameba, fungi, and plants (for review see Saito 2001). However, animal genome sequences such as those of the fruit fly *Drosophila melanogaster* and the nematode *Caenorhabditis elegans* do not provide any evidence for the existence of two-component systems in these organisms. In vertebrates, neither *bona fide* two-component histidine kinases nor the cognate response regulators were found. However, a highly divergent variance of the bacterial histidine kinase paradigm was observed in mitochondrial protein kinases, such as branched-chain α-ketoacid dehydrogenase kinase and pyruvate dehydrogenase kinase (Popov et al. 1992; Popov 1993). Both lack the autophosphorylated histidine residue preserved in their two-component counterparts, and both catalyze direct transfer of a phosphoryl group from ATP to *serine* residues on an exogenous protein target. Thus, we are left so far with nucleoside diphosphate kinases acting as histidine kinases in vertebrates, transferring a phosphate from their own histidine to histidine residues in other proteins (for review see Roymans et al. 2002). Other than that, mammalian histidine kinases have been described but not yet sequenced (Besant and Attwood 2000; Kowluru 2002).

7.3 Histidine phosphatases in vertebrates

The existence of histidine phosphorylation in vertebrate proteins was first described in the 1960s by Boyer and his colleagues (Boyer et al. 1962; Bieber and Boyer 1966). It took another 30 years until vertebrate histidine phosphatase activity could be detected (Wong et al. 1993; Ohmori et al. 1993). There are several ways to classify the results reported ever since on vertebrate histidine phosphatases. A state of the art overview by means of amino acid alignments or phylogenetic analysis is not feasible. So far, only one histidine phosphatase has been purified and sequenced from vertebrates - independently yet simultaneously by two different research groups, using two different species as enzyme source, with two different substrates, but resulting in identical enzyme proteins (Ek et al. 2002; Klumpp et al. 2002). Therefore, we are left to proceed with what has been discovered about both histidine phosphatases and protein histidine phosphatases according to the *substrates* used.

7.3.1 Proteins as a substrate

Histone H4 was the first protein identified in vertebrates containing phosphorylated histidine residues (Chen et al. 1974). Using histone H4, phosphorylated by a histidine kinase from yeast as a substrate for vertebrate histidine phosphatases, led to serine/threonine phosphatases type 1, 2A, and 2C instead of novel phosphatase proteins specific for phosphohistidine (Kim et al. 1993; Matthews and MacIntosh 1995). Years later, another risky approach was successfully undertaken using one of the easily accessible bacterial proteins phosphorylated on a histidine residue to search again for potential histidine phosphatase proteins in vertebrates. CheA, a bacterial histidine kinase phosphorylating itself on His-48 in the presence of [γ-^{32}P]ATP and Mg^{2+}, was used as a tool to challenge possible vertebrate histidine phosphatases (Klumpp et al. 2002).

A 14 kDa protein dephosphorylating [^{32}P-His]CheA was identified, purified, characterized and cloned (Klumpp et al. 2002). That histidine phosphatase did not hydrolyze *p*-nitrophenyl phosphate, or [^{32}P-serine/threonine]casein, or the [^{32}P-tyrosine]EGF-receptor. Enzyme activity was insensitive to okadaic acid. The amino acid sequence of the 14 kDa protein histidine phosphatase showed no similarity to any of the established phosphatases. A number of database entries of very similar proteins - with unknown function at that time - immediately revealed ubiquitous expression of the novel protein histidine phosphatase. The protein turned out to be identical to the histidine phosphatase acting on succinyl-Ala-His(P)-Pro-Phe-*p*-nitroanilide (Ek et al. 2002). Meanwhile, the first vertebrate substrate protein for this 14 kDa protein histidine phosphatase has been discovered. This enzyme dephosphorylates [^{32}P-His]ATP-citrate lyase (Klumpp et al. 2003).

Going back in history, histone H4 had been used in most cases studying dephosphorylation via potential protein histidine phosphatases from vertebrates (Kim et al. 1993, 1995; Matthews and MacIntosh 1995; Matthews 1995). As a first step towards those dephosphorylation assays histone H4 mostly had to be isolated from commercially available total calf thymus histone. As a second step a 31 kDa protein histidine kinase had to be purified from yeast. The subsequent phosphorylation reaction with [γ-^{32}P]ATP proved specific for histidine residue 75 on histone H4. Histone H4 histidine kinases acting on histidine residues 18 and 75 have been identified in various vertebrate tissues, but not purified and sequenced.

The functional significance of histone H4 histidine phosphorylation is still un-known.

Using such a histone H4 protein substrate containing [^{32}P]phosphohistidine as the sole phosphoamino acid, it has been shown that protein phosphatase types 1, 2A, or 2C resulted in extensive removal of phosphate from the histidine phos-phorylated histone, whereas Ca^{2+}/calmodulin-regulated phosphatase type-2B and protein tyrosine phosphatase type-1B did not dephosphorylate [^{32}P-His]histone H4 (Kim et al. 1993). The influence of okadaic acid, inhibitor protein I$_1$ and Mg^{2+} mirrored the effects of serine/threonine dephosphorylation. Thus, the histidine phosphatase activity of PP1, PP2A, and PP2C is not easily distinguishable from their serine/threonine phosphatase activity.

This phosphohistidine phosphatase activity of purified PP1, PP2A, and PP2C was systematically characterized two years later, allowing some assessment of its significance (Kim et al. 1995). The substrate preference was delineated from the ratio k_{cat}/K_m. Unexpectedly, phosphohistidine in histone H4 was preferred over phosphoserine in phosphorylase a by both PP1 and PP2A. Furthermore, type-2C phosphatase used both histone H4 and the myosin P-light chain equally well as substrates. This suggested that the major cellular protein serine/threonine phos-phatases are likely to act as protein histidine phosphatases in a cell.

Results from studies with whole cell extracts from rat liver - instead of purified phosphatases - were congruent with this hypothesis. Divalent cation-independent protein histidine phosphatase activity was found to be due to serine/threonine phosphatases of the PP1/PP2A family (Matthews and MacKintosh 1995). Essen-tially all the histidine phosphatase activity acting on [^{32}P-His]histone H4 as a sub-strate was inhibited by okadaic acid. This sensitivity to okadaic acid suggested that PP2C plays only a minor role in dephosphorylation of [^{32}P-His]histone H4 in rat liver extracts.

Studying dephosphorylation of yet another phosphohistidine protein substrate, p36, the situation turned out to be just the opposite. PP2C in rat liver extracts was shown to play a major role as protein histidine phosphatase (Motojima and Goto 1994). Mg^{2+} requirement for activity, apparent molecular mass of 45 kDa, and re-sistance to 100 μM okadaic acid suggested that the primary phosphatase acting in vitro on [^{32}P-His]p36 is PP2C. So far, p36 has not been definitely identified.

An overview on vertebrate protein histidine phosphatases and their substrates is presented in Table 1.

7.3.2 Phosphopeptides as a substrate

The phosphohistidine containing peptide succinyl-Ala-His(P)-Pro-Phe-*p*-nitroanilide led to the discovery of a histidine phosphatase in porcine liver cytosol (Ek et al. 2002). A protein with a molecular mass of 14 kDa was purified, charac-terized and cloned. Its phosphohistidine phosphatase activity was insensitive to okadaic acid and related compounds, and insensitive to EDTA. A BLAST search revealed numerous homologues in other species and tissues, yet enzymatic activity

Table 1. Phosphatases in vertebrates hydrolyzing [^{32}P-his]substrate proteins (details given in text)

[^{32}P-his]substrate	Phosphatase / molecular mass	Reference
ATP-citrate lyase	Protein histidine phosphatase (PHP)* 14 kDa	Klumpp et al. 2003
CheA	Protein histidine phosphatase (PHP) 14 kDa	Klumpp et al. 2002
Histone H4	Serine/threonine protein phosphatases types 1, 2A, 2C	Kim et al. 1993; 1995 Matthews and MacIntosh 1995
Nucleoside di-phosphate kinase	Bacteria: 10 kDa Protein histidine phosphatase	Shankar et al. 1995
	Vertebr.: 13 kDa Phosphoamidase/his-phosphatase	Hiraishi et al. 1999
p36	Serine/threonine phosphatase type 2C 45 kDa	Motojima and Goto 1994

* Identical enzymes

had been assigned in just one case. This registered protein histidine phosphatase (CAC16267; Klumpp et al. 2002) has been described above in more detail.

A much longer synthetic peptide corresponding to residues 70-102 of histone H4 had been phosphorylated using protein histidine kinase purified from *Saccharomyces cerevisiae* and was tested for dephosphorylation with the most abundant serine/threonine protein phosphatases (Kim et al. 1993). Phosphate was quantitatively removed from phosphohistidine within this peptide by purified PP1, PP2A and PP2C.

7.3.3 Phosphohistidine as a substrate

There are several reports on histidine phosphatase activity acting on synthetic phosphoamino acid substrates, such as polyphosphohistidine or 3-phosphohistidine (Wong et al. 1993; Ohmori et al. 1993; Ohmori et al. 1994). In both laboratories, the ability to catalyze the dephosphorylation of phosphate *protein* substrates was not tested.

In 1993, Wong and colleagues used high-molecular-mass (> 10 kDa) homopolymers of histidine to search for histidine phosphatases in rat tissue extracts (Wong et al. 1993). Based on the behavior of these activities on ion exchange columns and differences in their susceptibility to heat, thiol modifying reagents, divalent metal ions, and various potential inhibitors they suggested there might be multiple - most likely four - N-phosphatases present. From their data, it seemed unlikely that the classical serine/threonine and tyrosine phosphatases accounted for the novel N-phosphatase activities.

The same year Ohmori and co-workers described a histidine phosphatase from rat brain cytosol acting on 3-phosphohistidine (Ohmori et al. 1993). The enzyme was partially purified and characterized one year later. It is a dimeric enzyme with identical subunits (50 kDa each). According to its pH-optimum and inhibition by tartrate, it was suggested that this hydrolase belongs to the acid phosphatase group (Ohmori et al. 1994).

Phosphoamidases are enzymes classified to hydrolyze phosphocreatine and phosphoarginine. According to that criterion the phosphatases acting on phospho-histidine as a substrate, as described above by the groups of Wong and Ohmori, were declared distinct from phosphoamidases. From time to time, however, scientists still challenge the question as to whether or not phosphoamidases in a broader sense form a group of enzymes distinct from phosphatases. Phosphoamidases may be considered as phosphatases whereas vice versa, not every phosphatase acts as a phosphoamidase. Accordingly, a 13 kDa phosphoamidase from bovine liver was reported capable of acting as a histidine phosphatase and even as a protein his-tidine phosphatase (Hiraishi et al. 1999). The enzyme released P_i from amido-phosphate and from the N-phosphorylated amino acids 3-phosphohistidine and 6-phospholysine. Divalent cations such as Mg^{2+} or Mn^{2+} had no effect on the rate of hydrolysis. Furthermore, the enzyme hydrolyzed autophosphorylated succinic thi-okinase. The question arises as to whether this 13 kDa phosphoamidase/histidine phosphatase (Hiraishi et al. 1999) might be identical to the 14 kDa protein his-tidine phosphatase (Ek et al. 2002; Klumpp et al. 2002). Sequence information is not available from the 13 kDa protein. Substrate specificity, however, suggests that we are dealing with two different proteins of similar size: the 13 kDa phos-phoamidase/histidine phosphatase dephosphorylated nucleoside diphosphate kinase (Hiraishi et al. 1999), whereas the 14 kDa protein histidine phosphatase did not (Klumpp et al. 2003). Alternatively, N- or C-terminal deletions of the 14 kDa protein might affect substrate specificity.

7.4 Histidine and aspartate phosphatases in bacteria

There are a number of different ways to fine tune or stop bacterial signal transduc-tion processes, once initiated by the autophosphorylation of a histidine kinase and continued with the phosphate transfer onto aspartate within a response regulator protein. The mechanisms outlined below are all known to exist in *Escherichia coli.*

a. Most of the bacterial two-component histidine kinases also exhibit aspartate phosphatase activity, e.g. EnvZ (Zhu et al. 2000).
b. In a few cases the response regulator can catalyze its own dephosphorylation, e.g. CheY, and might be helped by other proteins, e.g. CheZ (Alex and Simon 1994; Zhao et al. 2002).
c. Special histidine phosphatases exist as well, e.g. SixA (Hamada et al. 1999; Matsubara and Mizuno 2000).

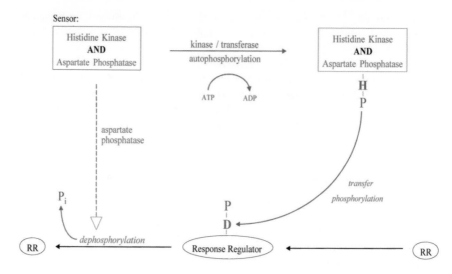

Fig. 2. Histidine kinases acting as aspartate phosphatase. Many of the bacterial two-component sensor histidine kinases act as aspartate phosphatases to catalyze the dephos-phorylation of their cognate response regulator proteins. The activities of bacterial histidine kinases are modulated by environmental stimuli. Sensor histidine kinases first catalyze ATP-dependent autophosphorylation of a specific conserved histidine residue (H). The phosphoryl group (P) is then transferred to a specific aspartate residue (D) located within the conserved regulatory domain of the response regulator (RR). Phosphorylation of the response regulator elicits a specific cellular response. The response regulator is dephosphory-lated by the aspartate phosphatase activity intrinsic to many of the sensor histidine kinases.

7.4.1 Aspartate phosphatases

Most bacterial histidine kinases are bifunctional enzymes having both *histidine kinase* and *aspartate phosphatase* activity (Fig. 2). In the "switch" model the active site shared for both kinase and phosphatase activities is proposed to be in only one of two possible states: either kinase$^+$ phosphatase$^-$ (on) or kinase$^-$ phosphatase$^+$ (off). In an alternative "rheostat" model the kinase/phosphatase domain always possesses both kinase and phosphatase activities and the ratio of these two opposing activities is controlled by an external signal. Mutations that affect kinase but not phosphatase activity, and vice versa, demonstrate that these activities are clearly distinguishable.

The phosphatase activity of bacterial two-component histidine kinases is mostly directed toward their cognate response regulator proteins. Such aspartate phosphatase activity often resides within the histidine containing dimerization domain of the bacterial histidine kinase. The interactions are exemplified here with EnvZ (Fig. 1A, Fig. 2), a transmembrane histidine kinase functioning as an osmosensor in *Escherichia coli* (Zhu et al. 2000). EnvZ phosphorylates OmpR on Asp-55; the phosphorylated product, OmpR-P, acts as a transcription factor. In addition to the

OmpR histidine kinase activity, EnvZ possesses OmpR-P aspartate phosphatase activity as well. In recent years, many of the molecular details have been learned. Domain A (amino acids 223-289) of EnvZ serves as a dimerization and phosphotransfer domain and has Mg^{2+}-dependent phosphatase activity by itself. Domain B (amino acids 290-450) functions as an ATP-binding and catalytic domain. It is assumed that the EnvZ phosphatase activity is regulated by the relative positioning of domains A and B, which in turn is controlled by external signals. His-243 of EnvZ participates in both kinase and phosphatase reactions (Dutta et al. 2000).

In a few cases, the response regulator can also catalyze its own dephosphorylation. In chemotaxis, chemical attractants or repellants bind to transmembrane receptors and regulate the autophosphorylation of the *Escherichia coli* sensor kinase CheA. The phosphoryl group is transferred from a histidyl residue on CheA to an aspartyl residue on the response regulator CheY. Interaction of phospho-CheY with the flagellar motor dictates cellular swimming behavior. Although CheY can catalyze its own dephosphorylation, the protein CheZ is critical for the rapid response of *Escherichia coli* to stimuli. Whether CheZ acts as an allosteric activator of the intrinsic autophosphorylation activity of CheY or as a true phosphatase in its own right has long been debated. Just recently the co-crystal structure of CheZ with CheY has been reported (Zhao et al. 2002). It has revealed an essential catalytic role for Gln-147 of CheZ which inserts directly into the CheY active site. Thus, CheZ seems to use the existing mechanism of CheY autodephosphorylation and renders it more efficient.

7.4.2 Histidine phosphatases

In addition to the aspartate phosphatase activity intrinsic to bacterial histidine kinases and/or response regulators, histidine phosphatase proteins are present in bacteria as well. SixA has been isolated from *Escherichia coli* as the first protein to exhibit such histidine phosphatase activity (Matsubara and Mizuno 2000). The protein consists of 161 amino acid residues (17.2 kDa) and has an arginine-histidine-glycine (RHG) signature at the N-terminus. This RHG motive was suggested to function as a nucleophilic phosphoacceptor and was reported for several phosphatases including eukaryotic fructose-2,6-bisphosphatase and acid phosphatases. SixA phosphatase is a crucial regulatory factor that is involved in the ArcB signaling, particularly under certain anaerobic respiratory growth conditions. ArcB is a histidine kinase, which has multiple phosphorylation domains (Fig. 1B). First, His-292 in the ArcB histidine kinase acquires the γ-phosphoryl group from ATP through its own catalytic activity. Second, the phosphoryl group on His-292 moves under its intrinsic phosphor-accepting Asp-576 in the receiver part of ArcB. Third, His-717 in the phosphotransmitter domain is modified by phosphorylation through His-292 and Asp-576. The final destination of the phosphoryl group on His-717 is Asp-54 in the next protein, the ArcA receiver. Alternatively, ArcA can also receive the phosphoryl group directly from His-292. SixA can drain the phosphoryl group from the phosphohistidine of the HPt phosphotransmitter domain (His-717) resulting in down-regulation of the ArcB→ArcA phosphorelay at an intermediate step (Fig. 1B). This additional regulatory mechanism is brilliant because the ArcB histidine kinase, in conjunction with the SixA histidine phosphatase protein, can propagate certain anaerobic respiratory signals in a very sophisticated manner.

7.5 Nucleoside diphosphate kinases

Nucleoside diphosphate kinases (NDPKs, nm23 proteins) are low molecular weight proteins (~ 20 kDa) that catalyze the phosphorylation of nucleoside 5′-diphosphates to triphosphates (for reviews see Postel 1998; Roymans et al. 2002; Steeg et al. 2002). NDPKs are emphasized here for several reasons: (i) The enzymes found in bacteria have also been preserved in humans with a 43% identity. (ii) NDPKs can bind DNA and regulate gene expression. (iii) NDPKs autophosphorylate on a histidine residue at their catalytic site with either $[\gamma-^{32}P]ATP$ or $[\gamma-^{32}P]GTP$. (iv) The phosphate from histidine-phosphorylated NDPKs can be used to phosphorylate histidine and aspartate residues in other proteins. Even O-phosphorylation on serine has been reported. Thus, phosphate for phosphorylation reactions can be derived from [P-His]NDPK in some cases, instead of ATP or GTP. NDPKs, therefore, were reported to have "kinase" and/or "transferase" activity. (v) In analogy to the well-known signal transduction pathways in bacteria called the "two-component-system," this term might be applicable for NDPK catalyzed phosphorylation reactions as well. A component I (transmitter) NDPK transfers phosphate from its own histidine residue to an amino acid in another protein, component II (receiver). However, one must emphasize that NDPKs have no sequence similarity to the two-component histidine kinases from bacteria, and the proteins phosphorylated by NDPKs do not resemble the bacterial two-component response regulators.

In addition to the important role in nucleotide metabolism, NDPKs are involved in a number of cellular regulatory functions. In *Escherichia coli*, NDPKs can act as a kinase to phosphorylate histidine protein kinases, such as EnvZ or CheA, suggesting that NDPKs are major physiological players in bacterial signal transduction (Lu et al. 1996). Vertebrate NDPKs have been shown to transfer the phosphate from its own histidine to histidines of the catalytic sites of ATP-citrate lyase (Wagner and Vu 1995) and succinic thiokinase (Freije et al. 1997). NDPKs are also capable of transferring their phosphate from histidine to aspartate, e.g. in aldolase C (Wagner and Vu 2000). Activation of heterotrimeric G proteins by NDPK with concomitant phosphorylation of histidine residue 266 in G_β has even been reported more recently (Hippe et al. 2003; Cuello et al. 2003).

As described above, the phosphorylated NDPKs can continue to use their phosphate on histidine for phosphorylation of amino acids within other proteins. Alternatively, phosphohistidine of NDPKs may be hydrolyzed. For instance, a special phosphoprotein phosphatase has been reported from *Pseudomonas aeruginosa* to dephosphorylate NDPKs (Shankar et al. 1995). The result was a loss of NDPK enzyme activity. This 10 kDa histidine phosphatase showed 66% homology with the human Bax protein identified as an effecter in programmed cell death. The enzyme required Mg^{2+} for activity. Its activity towards other histidine phosphoproteins, e.g. CheA was limited. A 13 kDa phosphoamidase as well as a 56 kDa inorganic pyrophosphatase were also reported to dephosphorylate NDPKs in vertebrates (Hiraishi et al. 1999).

7.6 Concluding remarks

Knowledge of histidine phosphatases in vertebrates is also in its beginnings. The substrates used to study those enzymes were quite heterogeneous, ranging from mere 3-phosphohistidine and peptides to a few selected proteins such as histone H4, p36 of unknown identity, or the histidine autokinase CheA from bacteria. Two of the metabolic enzymes undergoing histidine autophosphorylation have been examined as well - nucleoside diphosphate kinase and ATP-citrate lyase. However, molecules involved in signal transduction processes have not yet been studied. Nothing is still known about dephosphorylation of phosphohistidine in P-selectin, Annexin I or G_β.

The precise function of protein histidine phosphorylation in vertebrate signal transduction has also not been established yet. However, the field seems ripe for rapidly expanding studies.

Acknowledgements

Work in our laboratories (S.K. and J.K.) was supported by grants from the Deutsche Forschungsgemeinschaft and from the Fonds der Chemischen Industrie.

References

Alex LA, Simon MI (1994) Protein histidine kinases and signal transduction in prokaryotes and eukaryotes. Trends Genet 10:133-138

Besant PG, Attwood PV (2000) Detection of a mammalian histone H4 kinase that has yeast histidine kinase-like enzymic activity. Int J Biochem Cell Biol 32:243-253

Bieber LL, Boyer PD (1966) ^{32}P-labeling of mitochondrial protein and lipid fractions and their relation to oxidative phosphorylation. J Biol Chem 241:5375-5383

Bilwes AM, Alex LA, Crane BR, Simon MI (1999) Structure of CheA, a signal-transducing histidine kinase. Cell 96:131-141

Boyer PD, DeLuca M, Ebner KE, Hultquist DE, Peter JB (1962) Identification of phosphohistidine in digests from a probable intermediate of oxidative phosphorylation. J Biol Chem 237:3306-3308

Chen C-C, Smith DL, Bruegger BB, Halpern RM, Smith RA (1974) Occurrence and distribution of acid-labile histone phosphates in regenerating rat liver. Biochemistry 13:3785-3789

Crovello CS, Furie BC, Furie B (1995) Histidine phosphorylation of P-selectin upon stimulation of human platelets: A novel pathway for activation-dependent signal transduction. Cell 82:279-286

Cuello F, Schulze RA, Heemeyer F, Meyer HE, Lutz S, Jakobs KH, Niroomand F, Wieland T (2003) Activation of heterotrimeric G proteins by a high energy phosphate transfer via nucleoside diphosphate kinase (NDPK) B and Gß subunits. J Biol Chem 278:7220-7226

Dutta R, Qin L, Inouye M (1999) Histidine kinases: Diversity of domain organization. Mol Microbiol 34:633-640

Dutta R, Yoshida T, Inouye M (2000) The critical role of the conserved Thr^{247} residue in the functioning of the osmosensor EnvZ, a histidine kinase/phosphatase, in *Escherichia coli*. J Biol Chem 275:38645-38653

Ek P, Pettersson G, Ek B, Gong F, Li J-P, Zetterqvist Ö (2002) Identification and characterization of a mammalian 14-kDa phosphohistidine phosphatase. Eur J Biochem 269:5016-5023

Fischer EH, Graves DJ, Crittenden ERS, Krebs EG (1959) Structure of the site phosphorylated in the phosphorylase *b* to *a* reaction. J Biol Chem 234:1698-1704

Foussard M, Cabantous S, Pédelacq J-D, Guillet V, Tranier S, Mourey L, Birck C, Samama J-P (2001) The molecular puzzle of two-component signaling cascades. Microb Infect 3:417-424

Freije JMP, Blay P, MacDonald NJ, Manrow RE, Steeg PS (1997) Site-directed mutation of Nm23-H1. J Biol Chem 272:5525-5532

Hamada K, Kato M, Mizuno T, Hakoshima T (1999) Crystallographic characterization of a novel protein SixA which exhibits phospho-histidine phosphatase activity in the multistep His-Asp phosphorelay. Acta Crystallogr D-Biol Crys 55:269-271

Hermesmeier J, Klumpp S (1999) Histidine phosphatase activity in vertebrates. Arch Pharm Pharm Med Chem 331:24

Hippe H-J, Lutz S, Cuello F, Knorr K, Vogt A, Jakobs KH, Wieland T, Niroomand F (2003) Activation of heterotrimeric G proteins by a high energy phosphate transfer via nucleoside diphosphate kinase (NDPK) B and Gß subunits. J Biol Chem 278:7227-7233

Hiraishi H, Yokoi F, Kumon A (1999) Bovine liver phosphoamidase as a protein histidine/lysine phosphatase. J Biochem 126:386-374

Huang J, Wei Y, Kim Y, Osterberg L, Matthews HR (1991) Purification of a protein histidine kinase from the yeast *Saccharomyces cerevisiae*. J Biol Chem 266:9023-9031

Inouye M, Dutta R (2003) Histidine kinases in signal transduction. Elsevier Science, San Diego

Kim Y, Huang J, Cohen P, Matthews HR (1993) Protein phosphatases 1, 2A, and 2C are protein histidine phosphatases. J Biol Chem 268:18513-18518

Kim Y, Pesis KH, Matthews HR (1995) Removal of phosphate from phosphohistidine in proteins. Biochim Biophys Acta 1268:221-228

Klumpp S, Bechmann G, Mäurer A, Selke D, Krieglstein J (2003) ATP-citrate lyase as a substrate of protein histidine phophatase in vertebrates. Biochem Biophys Res Commun 306:110-115

Klumpp S, Hermesmeier J, Selke D, Baumeister R, Kellner R, Krieglstein J (2002) Protein histidine phosphatase: A novel enzyme with potency for neuronal signaling. J Cereb Blood Flow Metabol 22:1420-1424

Klumpp S, Krieglstein J (2002) Phosphorylation and dephosphorylation of histidine residues in proteins. Eur J Biochem 269:1067-1071

Kowluru A (2002) Identification and characterization of a novel protein histidine kinase in the islet ß cell: evidence for its regulation by mastoparan, an activator of G-proteins and insulin secretion. Biochem Pharmacol 63:2091-2100

Kowluru A, Seavey SE, Rhodes CJ, Metz SA (1996) A novel mechanism for trimeric GTP-binding proteins in the membrane and secretory granule fractions of human and rodent ß cells. Biochem J 313:97-107

Lu Q, Park H, Egger LA, Inouye M (1996) Nucleoside-diphosphate kinase-mediated signal transduction via histidyl-aspartyl phosphorelay systems in *Escherichia coli*. J Biol Chem 271:32886-32893

Matsubara M, Mizuno T (2000) The SixA phospho-histidine phosphatase modulates the ArcB phosphorelay signal transduction in *Escherichia coli*. FEBS Letters 470:118-124

Matthews HR (1995) Protein kinases and phosphatases that act on histidine, lysine, or arginine residues in eukaryotic proteins: A possible regulator of the mitogen-activated protein kinase cascade. Pharmacol Ther 67:323-350

Matthews HR, MacKintosh C (1995) Protein histidine phosphatase activity in rat liver and spinach leaves. FEBS Lett 364:51-54

Motojima K, Goto S (1994) Histidyl phosphorylation and dephosphorylation of p36 in rat liver extract. J Biol Chem 269:9030-9037

Muimo R, Hornickova Z, Riemen CE, Gerke V, Matthews H, Mehta A (2000) Histidine phosphorylation of annexin I in airway epithelia. J Biol Chem 275:36632-36636

Ohmori H, Kuba M, Kumon A (1993) Two phosphatases for 6-phospholysine and 3-phosphohistidine from rat brain. J Biol Chem 268:7625-7627

Ohmori H, Kuba M, Kumon A (1994) 3-phosphohistidine/6-phospholysine phosphatase from rat brain as acid phosphatase. J Biochem 116:380-385

Popov KM, Kedishvili NY, Zhao Y, Shimomura Y, Crabb DW, Harris RA (1993) Primary structure of pyruvate dehydrogenase kinase establishes a new family of eukaryotic protein kinases. J Biol Chem 268:26602-26606

Popov KM, Zhao Y, Shimomura Y, Kuntz MJ, Harris RA (1992) Branched-chain α-ketoacid dehydrogenase kinase. J Biol Chem 267:13127-13130

Postel EH (1998) NM23-NDP kinase. Int J Biochem Cell Biol 30:1291-1295

Roymans D, Willems R, Van Blockstaele DR, Slegers H (2002) Nucleoside diphosphate kinase (NDPK/NM23) and the waltz with multiple partners: Possible consequences in tumor metastasis. Clin Exp Metastasis 19:465-476

Saito H (2001) Histidine phosphorylation and two-component signaling in eukaryotic cells. Chem Rev 101:2497-2509

Shankar S, Kavanaugh-Black A, Kamath S, Chakrabarty AM (1995) Characterization of a phosphoprotein phosphatase for the phosphorylated form of nucleoside-diphosphate kinase from *Pseudomonas aeruginosa*. J Biol Chem 270:28246-28250

Steeg PS, Palmieri D, Ouatas T, Salerno M (2003) Histidine kinases and histidine phosphorylated proteins in mammalian cell biology, signal transduction and cancer. Cancer Lett 190:1-12

Swanson RV, Alex LA, Simon MI (1994) Histidine and aspartate phosphorylation: Two-component systems and the limits of homology. Trends Biochem Sci 19:485-490

Wagner PD, Vu N-D (1995) Phosphorylation of ATP-citrate lyase by nucleoside diphosphate kinase. J Biol Chem 270:21758-21764

Wagner PD, Vu N-D (2000) Histidine to aspartate phosphotransferase activity of nm23 proteins: Phosphorylation of aldolase C on Asp-319. Biochem J 346:623-630

West AH, Stock AM (2001) Histidine kinases and response regulator proteins in two-component signaling systems. Trends Biochem Sci 26:369-376

Wieland T, Nürnberg B, Ulibarri I, Kaldenberg-Stasch S, Schultz G, Jakobs KH (1993) Guanine nucleotide-specific phosphate transfer by guanine nucleotide-binding regulatory protein ß-subunits. J Biol Chem 268:18111-18118

Wong C, Faiola B, Wu W, Kennelly PJ (1993) Phosphohistidine and phospholysine phosphatase activities in the rat: Potential protein-lysine and protein-histidine phosphatases? Biochem J 296:293-296

Zhao R, Collins EJ, Bourret RB, Silversmith RE (2002) Structure and catalytic mechanism of the E. coli chemotaxis phosphatase CheZ. Nature Struct Biol 9:563-564

Zhu Y, Qin L, Yoshida T, Inouye M (2000) Phosphatase activity of histidine kinase EnvZ without kinase catalytic domain. Proc Natl Acad Sci USA 97:7808-7813

8 Anchoring of protein kinase and phosphatase signaling units

Philippe Collas, Thomas Küntziger, and Helga B. Landsverk

Abstract

The specificity of action of protein kinases and phosphatases can be achieved by their recruitment into multiprotein complexes at discrete subcellular loci. One class of molecules targeting kinases and phosphatases within single complexes includes A-kinase (PKA) anchoring proteins or AKAPs. Interestingly, AKAPs not only anchor enzymes but may also regulate their activity. Intermolecular interactions within the AKAP complex, or interaction with a substrate, also serve to modulate activity of an associated phosphatase. This communication highlights the role of cytoplasmic and nuclear AKAPs as protein kinase/phosphatase adaptors. Non-AKAP kinase/phosphatase adaptor proteins of the centrosomal complex are also presented.

8.1 Introduction

Multiple hormones signal through common second messengers to activate the same protein kinase and phosphatase cascades, but elicit distinct intracellular responses. Many protein kinases and phosphatases have broad substrate specificities and may be used in various combinations to achieve distinct biological responses. Mechanisms must exist to properly guide the repertoire of enzymes into specific signaling pathways. One form of regulation of these various effects is the intracellular compartmentalization of the kinases and phosphatases involved to specific sites of action. This function can be achieved by the recruitment of signaling molecules into multiprotein signaling networks or the activation of silent enzymes already positioned in the vicinity of their substrate (Pawson and Scott 1997).

Paradigms of compartmentalized protein kinases and phosphatases are the cAMP-dependent protein kinase (PKA) and protein phosphatase 1 (PP1). Both are broad specificity enzymes with ubiquitous patterns of expression. These enzymes have been extensively described elsewhere (Bollen 2001; Bollen and Beullens 2002; Cohen 2002; Tasken et al. 1997; see also review 2 in this volume). There is a large body of evidence to indicate that substrate specificity of both enzyme classes is mediated by subcellular localization. Families of targeting and anchoring proteins that simultaneously tether PKA and PP1, as well as other protein kinases and phosphatases, have been identified. This communication focuses on (i) a specific class of protein kinase and phosphatase adaptors, the so-called A-

Topics in Current Genetics, Vol. 5
J. Arino, D.R. Alexander (Eds.) Protein phosphatases
© Springer-Verlag Berlin Heidelberg 2004

kinase anchoring proteins or AKAPs, and (ii) kinase /phosphatase adaptors of the centrosomal complex.

8.2 Intracellular targeting of cAMP signaling by A-kinase anchoring proteins

The biological effects of cAMP are mediated by PKA. Type-I and II PKA holoenzymes consist of two catalytic and two regulatory (RI or RII, respectively) subunits (Tasken et al. 1997). Specificity of PKA is largely determined by the structure and properties of the R subunits, whereas the catalytic subunits exhibit similar kinetics features and substrate specificities. PKA is activated by binding of two cAMP molecules to each R subunit followed by release of the two catalytic subunits from the R-cAMP complex. Activated catalytic subunits phosphorylate specific substrates, an event that enhances hormonal responses by altering the activity of key enzymes and structural proteins of the cytoplasm and the nucleus. Free catalytic subunits also translocate to the nucleus where they play a role in the activation of cAMP-responsive genes (Riabowol et al. 1988).

The specificity of cellular and nuclear responses to cAMP results from interactions of the RII subunits of PKA with AKAPs in various cellular compartments. A variety of techniques have identified over 30 AKAPs, some of which are members of gene families with several splice variants. AKAPs are functionally defined by their binding to the RI and/or RII subunit dimers of the PKA holoenzymes. Classically, RI or RII binding occurs via a well-characterized amphipathic helix in which the hydrophobic side chains constitute the main binding determinants (Edwards and Scott 2000; Vijayaraghavan et al. 1999); however some exceptions have been identified (Diviani et al. 2000).

The targeting domain of AKAPs determines localization of the PKA-mediated cAMP signal to discrete subcellular compartments (Fig. 1A). Targeting is mediated by protein-protein interactions to structural components or protein-lipid interactions with membranes. Consequently, AKAPs have been identified in many cellular compartments, including plasma membrane, Golgi, endoplasmic or sarcoplasmic reticulum, vesicles, mitochondria, microtubules, centrosome, nuclear envelope, and nuclear matrix/chromatin (Edwards and Scott 2000; Smith and Scott 2002). Targeting may also be dictated by alternative splicing of AKAP genes (Edwards and Scott 2000).

Perhaps the most exciting property of AKAPs is in their ability to simultaneously anchor other protein kinases such as protein kinase C (PKC) or glycogen synthase kinase-3 (GSK-3), protein phosphatases (including PP1, PP2A, or PP2B) and phosphodiesterases (Tasken et al. 2001) (Fig. 1B). Consequently, AKAPs have emerged as scaffolding proteins that can integrate, and perhaps orchestrate, multiple signaling pathways (Colledge and Scott 1999; Edwards and Scott 2000; Smith and Scott 2002).

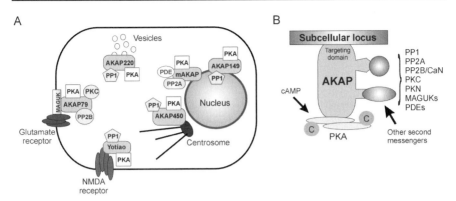

Fig. 1. AKAPs as kinase/phosphatase adaptor proteins. (A) Subcellular localization of AKAPs discussed in this paper and associated signaling molecules. Note that AKAP149 and mAKAP are associated with the nuclear envelope. AKAP149, however, is also found in the endoplasmic reticulum network and in mitochondria. (B) Schematic representation of a generic AKAP.

8.3 A signaling complex formed by AKAP79, PP2B/calcineurin, PKA, PKC, and MAGUK

AKAP79 has been shown to be enriched in neurons but also exists in T cells. AKAP79 constitutes a scaffold protein anchoring PKA, PKC, and protein phosphatase-2B(PP2B)/calcineurin (CaN). AKAP79 is localized to excitatory neuronal synapses where it is recruited to glutamate receptors by interactions with membrane-associated guanylate kinase (MAGUK) scaffold proteins (Fig. 1A). Anchored PKA and CaN in these complexes have been suggested to have important functions in regulating glutamate receptors in synaptic plasticity (Colledge et al. 2000). CaN is a phosphatase that plays essential roles in many signaling pathways. CaN consist of a catalytic A subunit (CnA) and a Ca^{2+}-binding regulatory B subunit (CnB). In activated T cells, CaN dephosphorylates the transcription factor NFAT, enabling its translocation into the nucleus where it contributes to activating the *IL2* gene (Northrop et al. 1994). Immunosuporessants can inhibit CaN phosphatase activity, preventing NFAT activation and cytokine gene induction (Rao et al. 1997). Thus, correct targeting of CaN is important to elicit the appropriate activating or inhibitory function. Interaction between AKAP79 and CaN has been biochemically characterized (Kashishian et al. 1998). Two regions of CnA are involved in binding to AKAP79 and binding to CaN does not require CnB. Furthermore, overexpression of AKAP79 inhibits NFAT dephosphorylation, which results in a decrease in NFAT activation (Kashishian et al. 1998). Therefore, AKAP79 not only binds CaN but also inhibits its phosphatase activity in vitro.

AKAP79 also binds PKA (Carr et al. 1992) and PKC (Faux and Scott 1997; Faux et al. 1999). Interestingly, unlike other PKC anchors, lipid activators or PKC activation do not seem to be required for association with AKAP79 (Faux et al.

1999). The AKAP binds and inhibits the conserved catalytic core of PKCβII. AKAP79 also associates with conventional, novel, and atypical isoforms of PKC in vitro and in vivo. Moreover, immunolabeling of rat hippocampal neurons showed that the murine AKAP79 homologue, AKAP150, is co-localized with PKCαβ, PKCε or PKCι (Faux et al. 1999). Therefore, AKAP79 may associate with multiple PKC isoforms and possibly modulate their activities through a mechanism involving protein-protein interactions at the catalytic core of the kinase.

Individual enzymes in the AKAP79 signaling unit are regulated by distinct second messengers. Nevertheless, both PKC and CaN are inhibited when associated with AKAP79, suggesting that additional regulatory signals must be required to release active enzymes. One such regulatory molecule was shown to be calmodulin. Calmodulin regulates AKAP79-PKC interaction (Faux and Scott 1997). AKAP79 binds calmodulin with nanomolar affinity in a Ca^{2+}-dependent manner. Both AKAP79 and PKC display overlapping immunolabeling patterns in cultured hippocampal neurons. Furthermore, calmodulin reverses the inhibition of PKCβII by an AKAP79(31-52) peptide and reduces inhibition by full-length AKAP79 (Faux and Scott 1997). The effect of calmodulin on inhibition of a constitutively active PKC peptide by AKAP79(31-52) is partially dependent on Ca^{2+}. Additionally, Ca^{2+}/calmodulin reduces the amount of PKC that co-precipitates with AKAP79 and promotes an increase in PKC activity in a preparation of postsynaptic vesicles. These observations suggest that Ca^{2+}/calmodulin competes with PKC for binding to AKAP79, releasing the inhibited kinase from its association with the anchoring protein (Faux and Scott 1997).

Direct evidence for the formation of a complex containing PKA, CaN, AKAP79, and MAGUKs in living cells was recently provided (Oliveria et al. 2003). Immunofluorescence and fluorescence resonance energy transfer analyses have demonstrated the assembly of such a complex at the level of the plasma membrane/cytoskeleton (Oliveria et al. 2003). Notably, binding of CnA and PKA-RII to membrane-targeted AKAP79 was detected. Fluorescence resonance energy transfer between CnA and PKA-RII simultaneously bound to AKAP79 within 50 Å of each other was also reported, illustrating the existence of a ternary PKA-CaN-AKAP79 complex in living cells. Moreover, glutamate receptors and PKA are recruited into a macromolecular signaling complex through direct interaction between the MAGUK proteins (molecules involved in clustering glutamate receptors) PSD-95 and SAP97, and AKAP79/150 (Colledge et al. 2000). The complex modulates phosphorylation of AMPA receptors by PKA, a process believed to be implicated in regulating synaptic plasticity (Colledge et al. 2000).

8.4 AKAP220 anchors PKA, PP1, and GSK-3β in a quaternary complex

A 220-kDa AKAP, AKAP220, contributes to coordinating the intracellular localization of PKA and PP1 (Schillace and Scott 1999). The localization of AKAP220 remains elusive but was proposed to be restricted to peroxisomal vesicles in the testis (Lester et al. 1996). AKAP220 has also been identified in the midpiece of elongated spermatids and mature human spermatozoa (Reinton et al. 2000).

AKAP220 harbors a canonical PP1-binding RVXF motif (Egloff et al. 1997; Zhao and Lee 1997; Bollen 2001), where X is any amino acid, which is involved in PP1 anchoring (Schillace and Scott 1999). AKAP220 binds PP1α with nanomolar affinity in vitro and is able to interact with PP1β and PP1γ isoforms in cells, suggesting that AKAP220 can form a complex with multiple PP1 variants (Schillace and Scott 1999). A recombinant AKAP220 fragment acts as a competitive inhibitor of PP1 activity (Schillace et al. 2001). Mapping PP1 binding sites and enzyme assays indicate that several interaction sites act synergistically to inhibit PP1. Interestingly, the RVXF motif of AKAP220 binds PP1 but does not affect enzyme activity whereas additional COOH-terminal determinants inhibit the phosphatase (Schillace et al. 2001). In this respect, AKAP220 seems to inhibit PP1 activity differently from other PP1 regulatory subunits. The latter classically control phosphatase activity via the RVXF motif although additional motives can be involved in anchoring PP1 to their regulatory subunit (Bollen 2001).

Association of PKA and PP1 with AKAP220 is not exclusive. Addition of PKA-RII to the AKAP220 complex enhances the inhibitory effect of AKAP220 towards associated PP1 by ~4-fold (Schillace et al. 2001). Biochemical studies have suggested that AKAP220 and PKA-RII are likely to function additively to inhibit PP1 by binding to two separate sites on PP1. This finding illustrates an emerging concept of AKAP-mediated signaling in which intermolecular interactions within the complex affect the activity of the other anchored enzyme (Schillace et al. 2001).

AKAP220 also anchors glycogen synthase kinase-3 (GSK-3) (Tanji et al. 2002). GSK-3 regulates several physiological responses including gene expression, protein synthesis, protein localization, and protein degradation in mammalian cells, by phosphorylating many substrates. GSK-3 is regulated by various extracellular ligands and its substrates include neuronal cell adhesion molecules, neurofilaments, synapsin I, Tau, β-catenin, cyclin D1, and various transcription factors. The GSK-3β isoform was found to bind AKAP220 in a yeast two-hybrid screen. In intact cells, GSK-3β forms a quaternary complex with AKAP220, PKA and PP1. PKA phosphorylates GSK-3β directly and thereby reduces GSK-3β activity. This also takes place within with AKAP220 complex (Tanji et al. 2002). Conversely, phosphatase inhibitor studies have shown that PP1 enhances GSK-3β activity even when bound to AKAP220. These studies suggest that both PKA and PP1 can regulate GSK-3β activity by forming a complex with AKAP220. The effects of stimulating PKA activity by enhancing cAMP levels and of inhibiting AKAP220-protein phosphates activity are not additive (Tanji et al. 2002), suggesting that PKA and PP1 regulate AKAP220-bound GSK-3 kinase through pathways involving common components. Together with the observation that binding of PKA-RII to AKAP220 enhances the inhibitory activity of the AKAP towards PP1 (Schillace et al. 2001), this suggests two alternatives for the regulation of GSK-3β activity by AKAP220. PKA may directly phosphorylate and inhibit GSK-3 in the

AKAP220 complex. Another possibility is that AKAP220 and PKA inhibit PP1 cooperatively, thus enhancing phosphorylation of GSK-3β and inhibiting GSK-3β activity (Tanji et al. 2002). The physiological significance of targeting PKA, PP1, and GSK-3β remains to the explored.

8.5 Yotiao anchors PKA and PP1 at the NMDA receptor

Modulation of N-methyl-D-aspartate (NMDA) receptor activity in the brain by protein kinases and phosphatases contributes to the regulation of synaptic transmission. Targeting of these enzymes near the substrate has been suggested to enhance phosphorylation-dependent modulation of activity. Yotiao, an NMDA receptor-associated protein, binds PKA (Lin et al. 1998) as well as PP1 (Westphal et al. 1999). Yotiao has been proposed to be a splice-variant of the large AKAP450 (see below) containing only one RII-binding motif (Feliciello et al. 1999); however, northern blotting analysis suggests that yotiao may also represent a partial clone (Lin et al. 1998). PP1 associated with yotiao is active, thereby downregulating the activity of the channel, whereas PKA activation overcomes constitutive PP1 activity and confers rapid enhancement of NMDA receptor currents (Westphal et al. 1999). Therefore, yotiao appears to be a scaffold protein that anchors PKA and PP1 to NMDA receptors to regulate channel activity.

8.6 Kinase/phosphatase adaptors at the centrosome/microtubule system

In animal cells, the centrosome is a nucleus-associated organelle responsible for the nucleation of microtubules and their organization in a dynamic network (Kuntziger and Bornens 2000; Bornens 2002). Microtubules (MTs) organize the cytoplasm, anchor and position the cell organelles and form the tracks for vesicular transport between the different cell compartments. In mitosis, the centrosome/microtubule system undergoes a dramatic reorganization in terms of structure and activities and is directly involved in the processes that allow cell division to take place timely and accurately.

8.6.1 Microtubule association

The microtubule network spans the whole cytoplasm and shows in interphase a total surface of proteins estimated to be around 1,000 μm^2 for a fibroblast, in the same range as the plasma membrane (Gundersen and Cook 1999). These features make MTs particularly appropriate for the spatial organization of signal transduction processes. A large number of protein kinases and phosphatases have been found to physically interact with MTs, mostly through interaction with

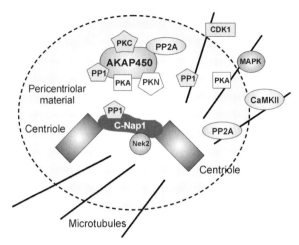

Fig. 2. Protein kinase/phosphatase adaptors of the centrosome and microtubules. Schematic representation of centrioles, pericentriolar material and microtubules with associated protein kinase and phosphatase anchor proteins (AKAP450, C-Nap1) and signaling molecules. Note that microtubules themselves anchor a variety of components, often via MAPs.

microtubule associated proteins (MAPs) (Fig. 2). MAPs bind MTs with a high affinity and in a reversible manner. In general, they enhance nucleation and stabilize MTs, but when phosphorylated (for example upon entry into mitosis), the MT-stabilizing effect of MAPs is often diminished (Desai and Mitchison 1997). Some of the most studied MAPs, such as MAP1, MAP2, and Tau, are only expressed in neuronal cells where they are involved in MT organization during formation of dendrites and axons (Drewes et al. 1998).

MAP2 binds a small (~4%) but constitutively active pool of MAP kinase (MAPK) in primary hippocampal neurons (Morishima-Kawashima and Kosik 1996). In NIH 3T3 cells, one third of the total MAPK pool is associated with MTs (Reszka et al. 1995). MAP2 is also responsible for PKA co-purification with MTs in calf brain extracts (Vallee et al. 1981). In vitro experiments have shown that MAP2 is a substrate of PKA and that the resulting phosphorylation reduces the MT-nucleating activity of MAP2 (Itoh et al. 1997). Hippocampal neurons of *Map2*[-/-] mice present a significant reduction of the MT-associated and total amounts of PKA, and a reduced activation of the PKA-activated cAMP-responsive element binding protein after forskolin stimulation (Harada et al. 2002).

The ubiquitously expressed MAP4 allows in vitro and in vivo association of the mitotic kinase CDK1 to the spindle MTs at mitosis, through binding of associated cyclin B (Bailly et al. 1989; Ookata et al. 1995). Furthermore, MAP4 is phosphorylated at multiple sites by CDK1 in HeLa cells, probably resulting in the abolishment of the MT-stabilizing activity of MAP4 and thus allowing a direct regulation of MT dynamics by CDK1 at mitosis (Ookata et al. 1997). An association to the mitotic spindle has also been shown for Ca^{2+}/calmodulin-dependent kinase II (CaMKII), but whether it is mediated by a MAP is unknown (Ohta et al. 1990).

An important fraction of PP2A is associated to MTs in several cell types. In neurons, this interaction is mediated by Tau (Sontag et al. 1996). MT-associated PP2A shows cell cycle-dependent variations in activity (but not quantity), with maximal and minimal values in S and G2/M respectively (Sontag et al. 1995). Tau has also been reported to mediate interaction between MTs and a pool of PP1 (Liao et al. 1998). Several other important kinases, such as casein kinase II, cMos, and PI-3 kinase are also physically associated to microtubules (Gundersen and Cook 1999).

An interesting situation is when the activity of the kinase/phosphatase associated to MTs is influenced by a modification of the assembly or dynamics of the MT network. For example, MT destabilization by two MT-disrupting drugs (colchicine and vinblastine) induces activation of MAPK, which is able to phosphorylate MAP2 in vitro (Shinohara-Gotoh et al. 1991). This suggests a mechanism where MT depolymerization stimulates a MAP-phosphorylating activity. This would decrease the MT-stabilizing activity of MAPs and thus amplify MT depolymerization. The complementary situation is provided by a report showing that MT assembly results in the hyperphosphorylation of the MT-destabilizing factor stathmin/Op18 (Kuntziger et al. 2001). This results in the inactivation of stathmin (Cassimeris 2002), thus amplifying MT polymerization. Although the kinase responsible for the MT-dependent phosphorylation of stathmin was not identified, it was shown to be MT-associated (Kuntziger et al. 2001). In another example, microtubule damage provoked by different MT-disrupting drugs (paclitaxel and vincristine) was shown to induce apoptosis through activation of PKA (Srivastava et al. 1998). It thus appears that one role of the MTs is to organize a network that anchors key signal transduction complexes, but that in some circumstances (cell cycle transitions or local regulations) the signals mediated by these complexes originate from the MT network itself.

8.6.2 Centrosome association

Anchoring proteins are found in the protein matrix (often referred as pericentriolar material or PCM) that surrounds the two centrioles of the centrosome (Fig. 2). The PCM is organized by large coiled-coil proteins involved in nucleation and organization of MTs, as well as in anchoring and clustering components of signaling pathways (Bornens 2002; Doxsey 2001).

The centrosomal Nek2-associated protein (C-Nap1) has been found in a ternary complex with the cell cycle-regulated NIMA–related protein kinase Nek2 and with PP1 (Helps et al. 2000). All three proteins are found in centrosomes (Andreassen et al. 1998; Fry et al. 1998a; Fry et al. 1998b) (Fig. 2). This suggests that this ternary complex could be present in the PCM and regulate centrosome organization, possibly by controlling in a phosphorylation-dependent manner the distance between the two centrioles (Fry et al. 1998b). To support this idea, it was shown that C-Nap1 is a substrate for both Nek2 and PP1, and that PP1 is a substrate for Nek2 (Helps et al. 2000).

AKAP450 is a 453-kDa AKAP located in human centrosomes (Keryer et al. 1993; Witczak et al. 1999). AKAP450 has been characterized independently under different names, AKAP350 (Schmidt et al. 1999), CG-NAP (Takahashi et al. 1999), and hyperion (W. Kemmner and U. Schwarz, personal communication). AKAP450 anchors PKA-type II at the centrosome and Golgi area (Keryer et al. 1993; Witczak et al. 1999) (Fig. 2). It also harbors a binding domain for PKN, a protein kinase with a catalytic domain homologous to the PKC family, and activated by the small GTPase Rho and by unsaturated fatty acids (Takahashi et al. 1999). AKAP450 binds also PP1 and PP2A (Takahashi et al. 1999). It was recently shown that AKAP450 anchors hypophosphorylated PKCε at the Golgi/centrosome area and acts as a scaffold for the maturation by phosphorylation of PKCε (Takahashi et al. 2000). In addition to binding protein kinases and phosphatases, AKAP450 binds to calmodulin in vitro and the calmodulin-interacting domain found in the C-terminal part of the protein may be involved in targeting AKAP450 to the centrosome (Gillingham and Munro 2000; Takahashi et al. 2002; Keryer et al. 2003). If calmodulin actually interacts with AKAP450 at centrosomes, it could affect PKC binding to AKAP450 in a Ca^{2+}-dependent manner, as is the case for PKC binding to AKAP79 (Faux and Scott 1997). AKAP450 also anchors PDE4D3 together with PKA at centrosomes of testicular Sertoli cells (Tasken et al. 2001). Therefore, AKAP450 forms a signaling complex harboring PKA and one isoform of the enzyme that catabolizes cAMP into AMP and thus regulates, together with adenylate cyclases, intracellular local cAMP concentration.

With so many binding partners, AKAP450 potentially plays an important regulatory role in the various activities at the centrosome. For example, AKAP450 has been reported to anchor the γ–tubulin ring complexes responsible for microtubule-nucleation to the centrosomal matrix (Takahashi et al. 2002). Interestingly, AKAP450 shares this property with pericentrin/kendrin (Dictenberg et al. 1998; Takahashi et al. 2002), another protein that binds PKA at the PCM and shows homology with AKAP450, but that does not contain the conventional RII-binding sites found in AKAPs (Diviani et al. 2000). To evaluate possible role of AKAP450 in centrosome functions, displacement of endogenous centrosomal AKAP450 was achieved in HeLa cells through expression of a C-terminal construct. As a result, cells showed delocalization of centrosomal PKA-type II, impaired cytokinesis – the completion of which has recently been found to be controlled by the centrosome in somatic cells (Piel and Bornens 2001) – and centrosome reproduction (Keryer et al. 2003). Thus, the centrosomal scaffolding of many regulatory proteins by AKAP450 could be important not only for the activities of the centrosome as a microtubule organizing center and as a regulator of cell cycle progression, but also for its stability and biogenesis.

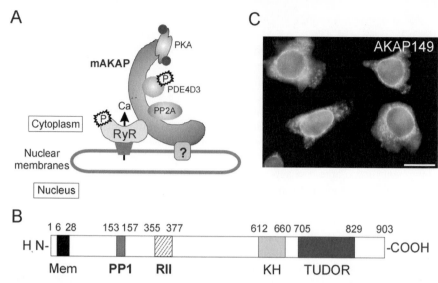

Fig. 3. mAKAP and AKAP149 at the nuclear envelope: models. (A) A model of the mAKAP complex. mAKAP associates with PKA, PDE4D3, PP2A, and the ryanodine receptor (RyR) at the nuclear envelope of cardiomyocytes. PKA can phosphorylate RyR and PDE4D3, promoting an increase in ion channel activity. PP2A is proposed to reverse RyR and PDE4D3 phosphorylation. Redrawn from Kapiloff et al. (2002) with permission. (B) Identified functional and binding domains of human AKAP149. Mem, membrane-association domain. PP1 and RII refer to motifs binding to PP1 and RII, respectively. Numbers indicate amino acid positions. (C) Immunofluorescence localization of AKAP149 in HeLa cells using a monoclonal antibody directed against residues 66-212 of human AKAP149. Bar, 10 μm. Taken from Steen et al. (2000).

8.7 Integration of cAMP and Ca^{2+} signaling at the cardiomyocyte nuclear envelope by mAKAP

The cell nucleus also harbors AKAPs, few of which however have been shown to bind both protein kinases and phosphatases. mAKAP is found predominantly in heart and skeletal muscle where it targets PKA to the nuclear envelope of differentiated myocytes. Two independent mAKAP regions containing spectrin-like repeat motifs are sufficient for targeting to the nuclear envelope (Kapiloff et al. 1999). Displacement studies have shown that targeting of mAKAP can be saturated (Kapiloff et al. 1999). These findings are consistent with the idea that targeting involves interaction with another protein expressed at limiting concentrations at the nuclear envelope (Kapiloff et al. 1999).

mAKAP is a part of a multicomponent signaling complex at the nuclear envelope, which includes the ryanodine receptor (RyR), a phosphodiesterase (PDE4D3 isoform) and a protein phosphatase (PP2A) (Kapiloff et al. 2001; Dodge et al. 2001) (Fig. 3A). The RyR is a Ca^{2+} channel that is also found in the sarcoplasmic reticulum where it is involved in striated muscle excitation-contraction (Fill and Copello 2002). At the nuclear envelope, the RyR may be involved in controlling nucleoplasmic Ca^{2+} concentrations (Kapiloff et al. 2001). The RyR is a substrate for PKA and phosphorylation of the RyR by PKA has been reported to increase Ca^{2+} conductance of the RyR (Fill and Copello 2002) (Fig. 3A). It is possible that mAKAP-associated PP2A turns off the cAMP signal by dephosphorylating the RyR (Kapiloff et al. 2001), in a manner similar to the antagonistic actions of PKA and PP2A on L-type calcium channels (Davare et al. 2000). PP2A has also been shown to be involved in cAMP-stimulated dephosphorylation (Feschenko et al. 2002) in a PKA-independent manner. The mAKAP-associated PDE is involved in attenuating the cAMP signal by hydrolyzing cAMP, thus bringing cAMP back to resting levels (Houslay and Adams 2003). In addition, PDE4D3 is a substrate for PKA and as PKA in the mAKAP complex becomes activated by an increase in cAMP concentrations, PDE4D3 activity increases, resulting in a negative feed-back (Dodge et al. 2001). It is thus possible that cAMP stimulates both activation – through PKA (Kapiloff et al. 2001) – and inactivation (through PDE4D3 and PP2A) (Kapiloff et al. 2001; Dodge et al. 2001) of the RyR. The mAKAP Ca^{2+} complex may function in the integration of these responses. Finally, the RyR is regulated by Ca^{2+} (Fill and Copello 2002), therefore the PKA-mAKAP complex may also permit integration of cAMP and Ca^{2+} signals at the cardiomyocyte nuclear envelope (Kapiloff et al. 2001).

8.8 Targeting of PP1 to the nuclear envelope by AKAP149: implications on cell cycle progression

Human AKAP149 (Trendelenburg et al. 1996) is an integral membrane protein of the endoplasmic reticulum-nuclear envelope membrane network (Fig. 3B). AKAP149 binds PKA-RII in vitro (Steen et al. 2000b) and co-immunoprecipitates PKA and PKA activity from HeLa nuclear envelope preparations (our unpublished data). AKAP149 contains an RVXF motif and binds PP1 in vitro and in vivo (Steen et al. 2000b). Finally, AKAP149 contains a single KH and a TUDOR motif (Fig. 3C), suggestive of an RNA-binding property. The mouse homologue of AKAP149, AKAP121, tethers PKA to the outer mitochondrial membrane (Feliciello et al. 1998). AKAP121 has been shown to bind specific 3'UTR sequences of the mRNA encoding Mn superoxide dismutase (MnSOD), a mitochondrial protein (Ginsberg et al. 2003). RNA binding requires a structural motif in the 3'UTR and is stimulated by cAMP (Ginsberg et al. 2003). AKAP121 overexpressed in HeLa cells promotes translocation of MnSOD mRNA from the cytosol to mitochondria and leads to an increase in mitochondrial MnSOD (Ginsberg et al. 2003). It is thus possible that the KH motif of human AKAP149 has a similar role; alternatively, it may be involved in other RNA-binding functions.

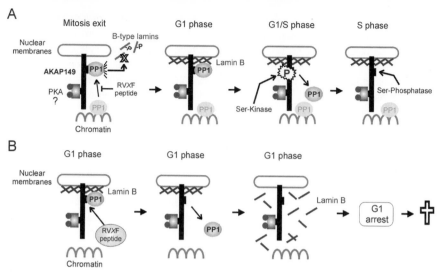

Fig. 4. A model for the role of AKAP149 in nuclear lamina assembly. (A) At the end of mitosis, membranes harboring AKAP149 are targeted to chromatin. Whether AKAP149 binds PKA at this stage is uncertain. AKAP149 recruits a fraction of chromatin-bound PP1, a process inhibited by peptides containing the PP1-binding RV*X*F motif of AKAP149. AKAP149 acts as a B-type lamin specifier and thereby stimulates the phosphatase activity of PP1 towards the phosphorylated, depolymerized, B-type lamins. PP1-mediated lamin dephosphorylation induces the lamins to polymerize and assemble into a nuclear lamina. Only B-type lamins are represented. A-type lamins assemble through a different pathway and whether PP1 also elicits A-type lamin dephosphorylation is unknown. PP1 remains associated with AKAP149 throughout G1. At the G1/S transition, serine-phosphorylation of AKAP149 coincides with release of PP1 from the AKAP. To where released PP1 is retargeted is not known. In the course of S phase, a serine-phosphatase dephosphorylates AKAP149. This, however, is not sufficient to promote PP1 retargeting to the AKAP, which only takes place upon exit from the next mitosis. (B) Association of PP1 with nuclear membrane-bound AKAP149 during G1 is essential for maintenance of nuclear integrity. Introduction of AKAP149-derived RV*X*F peptides into G1 nuclei in vitro or in vivo causes dissociation of PP1 from the AKAP. As a result, both A- and B-type lamins become phosphorylated and solubilize into the nuclear interior. Only B-type lamins are shown. The cells arrest with a G1 morphology for several hours and eventually undergo apoptosis.

8.8.1 Targeting of PP1 to the nuclear envelope by AKAP149 is essential for reentry into interphase

AKAP149 and PP1 are associated at the nuclear envelope in a cell cycle-dependent manner (Fig. 4A). At the end of mitosis, AKAP149 recruits a fraction of chromatin-bound PP1 to the reforming nuclear membranes (Steen et al. 2000b; Steen and Collas 2001). This process is inhibited by competitor peptides harboring the PP1-binding RV*X*F motif of AKAP149. Mistargeting of PP1 to the nuclear membranes inhibits the assembly of B-type lamins into a nuclear lamina (Steen and Collas 2001), presumably as a result of absence of dephosphorylation of the lamins by PP1 (Thompson et al. 1997). Moreover, addition of the specific PP1 inhibitor, Inhibitor-2, during nuclear reassembly in vitro does not alter the nuclear envelope targeting of PP1, but it inhibits nuclear lamina assembly indicating that, in addition to proper targeting, PP1 activity is required (Steen et al. 2003). Failure to assemble B-type lamins into the nuclear envelope is followed by rapid apoptosis and extensive proteolytic degradation of nuclear envelope markers (Steen and Collas 2001). As lamin B knockouts in *Caenorhabditis elegans* are embryonic lethal (Liu et al. 2000), this supports the view that proper assembly of B-type lamins is essential for cell survival. Nevertheless, at present, it cannot be excluded that RV*X*F motif-containing peptides sequester PP1 and prevent it from dephosphorylating additional targets that are directly responsible for inducing apoptosis (Steen and Collas 2001).

PP1 remains associated with AKAP149 throughout G1 phase and is released from the AKAP (and the nuclear envelope) upon S phase entry (Steen et al. 2000; Steen and Collas 2001) (Fig. 4A). Association of PP1 with its regulatory subunits has been shown to involve, in some instances, phosphorylation of serine residues near the RV*X*F motif (Beullens et al. 1999; Liu and Brautigan 2000; McAvoy et al. 1999). Regulation of the interaction between AKAP149 and PP1 at the nuclear envelope correlates with phosphorylation of AKAP149 (Steen et al. 2003).

8.8.2 AKAP149 is a B-type lamin specifier

We recently found that AKAP149 is not only a novel anchoring protein for PP1, but also a substrate-dependent regulator of associated PP1 activity (Steen et al. 2003). Indeed, AKAP149 stimulates the phosphatase activity of associated PP1 towards immunoprecipitated B-type lamins, whereas it inhibits PP1 activity towards an irrelevant substrate such as phosphorylase *a* (Steen et al. 2003). This indicates that AKAP149 acts as a B-type lamin specifying subunit of PP1.

8.8.3 AKAP149-PP1 association in G1: control of nuclear integrity

PP1 remains associated with AKAP149 at the nuclear envelope throughout G1 phase in HeLa cells (Steen et al 2003). Selective displacement of PP1 from AKAP149 by intranuclear microinjection of RV*X*F-containing AKAP149 peptides during G1 in living cells leads to a G1 arrest (Fig. 4B). This argues that the association of PP1 with AKAP149 is required for cell cycle progression into S phase. The G1 arrest phenotype is maintained for up to ~8 h before the cells enter apoptosis (Steen et al. 2003) (Fig. 4B). Furthermore, the lamin distribution is dramatically affected in the G1-arrested cells. Both A- and B-type lamins are displaced from their perinuclear localization, and are phosphorylated and solubilized in the nucleoplasm (Steen et al. 2003). These observations suggest that AKAP149-anchored PP1 continuously contributes to dephosphorylating B-type lamins during

G1 and that this is necessary to maintain the lamins in a polymerized form. Anchoring of PP1 to AKAP149 seems to be essential for this function, as overloading G1 cells containing the inhibitor peptide with excess active PP1 cannot rescue the phenotype. It is therefore likely that accurate anchoring of PP1 to specific subnuclear loci is important for the proper modulation of PP1 activity towards its nuclear G1-phase substrates.

8.9 Anchoring of AMY-1 to S-AKAP84/AKAP149

AMY-1 is a c-Myc binding protein that is ubiquitously expressed in human tissues, is mostly located in the cytoplasm, but translocates to the nucleus during S phase (Taira et al., 1998). It has been found to bind in vitro and in vivo to either the RII binding region of S-AKAP84 (a splice variant of AKAP149), AKAP149 at the mitochondria or AKAP95 in the nucleus of HeLa cells in a competitive manner (Furusawa et al., 2001;Furusawa et al., 2002). Presence of AMY-1 in a ternary complex with S-AKAP84 and RII inhibits PKA activity in the complex by preventing binding of its catalytic subunit. In this way, AMY-1 may serve to modulate PKA activity (Furusawa et al., 2002). AAT-1 is an AMY-1 binding protein that is specifically expressed in testis. It binds to the N-terminal region of S-AKAP84/AKAP149, within which the PP1 binding RVXF motif is also located, possibly indicating that AAT-1 and PP1 may bind S-AKAP84/AKAP149 competitively. AAT-1 is a PKA substrate and weakly stimulates PKA activity when it is present in a quaternary complex with AMY-1, S-AKAP84/AKAP149 and PKA (Yukitake et al., 2002). It would be interesting to determine whether nuclear envelope-associated AKAP149 also binds AMY-1 and/or AAT-1. One may speculate that AMY-1 preferentially binds AKAP149 or AKAP95 during various stages of S-phase in the nucleus.

8.10 Perspectives

AKAPs constitute an emerging class of adaptor molecules for protein kinases and phosphatases. This adaptor function appears to be restricted spatially and temporally to mediate accurate cellular responses. AKAPs can also function as regulators of associated phosphatase (PP1) activity, acting as inhibitors or as substrate-specifiers. Regulation of the activity may also be modulated by intermolecular interactions within the AKAP complex. It should also be noted that AKAPs may not only integrate signaling networks but also perform important tasks during the cell cycle, possibly independently of their role in intracellular signaling (Collas et al. 1999; Eide et al. 2002; Steen et al. 2000a).

Recent progress in our understanding of cell signaling has underlined the complexity of the intracellular signal networks. Many components of these pathways are continuously being identified, but it remains uncertain how many such pathways exit, how they interact with each other and how each is controlled (Smith and Scott 2002). Several pathways may share common molecules, and specificity of signaling events is at least in part achieved by tethering components to discrete sites or organelles, preferably in the vicinity of substrates. This tethering may be cell cycle dependent (Carlson et al. 2001; Landsverk et al. 2001). The growing use of techniques such as mass spectrometry, peptide arrays, genome two-hybrid screens, and 4-dimensional live imaging analyses are expected to allow the discovery of many more anchors for signaling networks.

Acknowledgments

Our work is supported by the Norwegian Cancer Society, the Research Council of Norway, European Union Grants MCFI-2001-01266 and QLK3-CT-2002-02149, and the Human Frontiers Science Program.

References

Andreassen PR, Lacroix FB, Villa-Moruzzi E, Margolis RL (1998) Differential subcellular localization of protein phosphatase-1 alpha, gamma1, and delta isoforms during both interphase and mitosis in mammalian cells. J Cell Biol 141:1207-1215

Bailly E, Doree M, Nurse P, Bornens M (1989) p34cdc2 is located in both nucleus and cytoplasm; part is centrosomally associated at G2/M and enters vesicles at anaphase. EMBO J 8:3985-3995

Beullens M, Van Eynde A, Vulsteke V, Connor J, Shenolikar S, Stalmans W, Bollen M (1999) Molecular determinants of nuclear protein phosphatase-1 regulation by NIPP-1. J Biol Chem 274:14053-14061

Bollen M (2001) Combinatorial control of protein phosphatase-1. Trends Biochem Sci 26:426-431

Bollen M, Beullens M (2002) Signaling by protein phosphatases in the nucleus. Trends Cell Biol 12:138-145

Bornens M (2002) Centrosome composition and microtubule anchoring mechanisms. Curr Opin Cell Biol 14:25-34

Carlson CR, Witczak O, Vossebein L, Labbe JC, Skålhegg BS, Keryer G, Herberg FW, Collas P, Tasken K (2001) CDK1-mediated phosphorylation of the RIIalpha regulatory subunit of PKA works as a molecular switch that promotes dissociation of RIIalpha from centrosomes at mitosis. J Cell Sci 114:3243-3254

Carr DW, Stofko-Hahn RE, Fraser ID, Cone RD, Scott JD (1992) Localization of the cAMP-dependent protein kinase to the postsynaptic densities by A-kinase anchoring proteins. Characterization of AKAP 79. J Biol Chem 267:16816-16823

Cassimeris L (2002) The oncoprotein 18/stathmin family of microtubule destabilizers. Curr Opin Cell Biol 14:18-24

Cohen PT (2002) Protein phosphatase 1- targeted in many directions. J Cell Sci 115:241-256

Collas P, Le Guellec K, Tasken K (1999) The A-kinase anchoring protein, AKAP95, is a multivalent protein with a key role in chromatin condensation at mitosis. J Cell Biol 147:1167-1180

Colledge M, Dean RA, Scott GK, Langeberg LK, Huganir RL, Scott JD (2000) Targeting of PKA to glutamate receptors through a MAGUK-AKAP complex. Neuron 27:107-119

Colledge M, Scott JD (1999) AKAPs: from structure to function. Trends Cell Biol 9:216-221

Davare MA, Horne MC, Hell JW (2000) Protein phosphatase 2A is associated with class C L-type calcium channels (Cav1.2) and antagonizes channel phosphorylation by cAMP-dependent protein kinase. J Biol Chem 275:39710-39717

Desai A, Mitchison TJ (1997) Microtubule polymerization dynamics. Annu Rev Cell Dev Biol 13:83-117

Dictenberg JB, Zimmerman W, Sparks CA, Young A, Vidair C, Zheng Y, Carrington W, Fay FS, Doxsey SJ (1998) Pericentrin and gamma-tubulin form a protein complex and are organized into a novel lattice at the centrosome. J Cell Biol 141:163-174

Diviani D, Langeberg LK, Doxsey SJ, Scott JD (2000) Pericentrin anchors protein kinase A at the centrosome through a newly identified RII-binding domain. Curr Biol 10:417-420

Dodge KL, Khouangsathiene S, Kapiloff MS, Mouton R, Hill EV, Houslay MD, Langeberg LK, Scott JD (2001) mAKAP assembles a protein kinase A/PDE4 phosphodiesterase cAMP signaling module. EMBO J 20:1921-1930

Doxsey SJ (2001) Re-evaluating centrosome function. Nat Rev Mol Cell Biol 2:688-698

Drewes G, Ebneth A, Mandelkow EM (1998) MAPs, MARKs and microtubule dynamics. Trends Biochem Sci 23:307-311

Edwards AS, Scott JD (2000) A-kinase anchoring proteins: protein kinase A and beyond. Curr Opin Cell Biol 12:217-221

Egloff MP, Johnson DF, Moorhead G, Cohen PT, Cohen P, Barford D (1997) Structural basis for the recognition of regulatory subunits by the catalytic subunit of protein phosphatase 1. EMBO J 16:1876-1887

Eide T, Carlson C, Tasken KA, Hirano T, Tasken K, Collas P (2002) Distinct but overlapping domains of AKAP95 are implicated in chromosome condensation and condensin targeting. EMBO Rep 3:426-432

Faux MC, Rollins EN, Edwards AS, Langeberg LK, Newton AC, Scott JD (1999) Mechanism of A-kinase-anchoring protein 79 (AKAP79) and protein kinase C interaction. Biochem J 343:443-452

Faux MC, Scott JD (1997) Regulation of the AKAP79-protein kinase C interaction by Ca2+/Calmodulin. J Biol Chem 272:17038-17044

Feliciello A, Cardone L, Garbi C, Ginsberg MD, Varrone S, Rubin CS, Avvedimento EV, Gottesman ME (1999) Yotiao protein, a ligand for the NMDA receptor, binds and targets cAMP-dependent protein kinase II(1). FEBS Lett 464:174-178

Feliciello A, Rubin CS, Avvedimento EV, Gottesman ME (1998) Expression of a kinase anchor protein 121 is regulated by hormones in thyroid and testicular germ cells. J Biol Chem 273:23361-23366

Feschenko MS, Stevenson E, Nairn AC, Sweadner KJ (2002) A novel cAMP-stimulated pathway in protein phosphatase 2A activation. J Pharmacol Exp Ther 302:111-118

Fill M, Copello JA (2002) Ryanodine receptor calcium release channels. Physiol Rev 82:893-922

Fry AM, Mayor T, Meraldi P, Stierhof YD, Tanaka K, Nigg EA (1998a) C-Nap1, a novel centrosomal coiled-coil protein and candidate substrate of the cell cycle-regulated protein kinase Nek2. J Cell Biol 141:1563-1574

Fry AM, Meraldi P, Nigg EA (1998b) A centrosomal function for the human Nek2 protein kinase, a member of the NIMA family of cell cycle regulators. EMBO J 17:470-481

Furusawa M, Ohnishi T, Taira T, Iguchi-Ariga SM, Ariga H (2001) AMY-1, a c-Myc-binding protein, is localized in the mitochondria of sperm by association with S-AKAP84, an anchor protein of cAMP-dependent protein kinase. J Biol Chem 276:36647-36651

Furusawa M, Taira T, Iguchi-Ariga SM, Ariga H (2002) AMY-1 interacts with S-AKAP84 and AKAP95 in the cytoplasm and the nucleus, respectively, and inhibits cAMP-dependent protein kinase activity by preventing binding of its catalytic subunit to A-kinase anchoring protein (AKAP) complex. J Biol Chem 277:50885-50892

Gillingham AK, Munro S (2000) The PACT domain, a conserved centrosomal targeting motif in the coiled-coil proteins AKAP450 and pericentrin. EMBO Rep 1:524-529

Ginsberg MD, Feliciello A, Jones JK, Avvedimento EV, Gottesman ME (2003) PKA-dependent Binding of mRNA to the Mitochondrial AKAP121 Protein. J Mol Biol 327:885-897

Gundersen GG, Cook TA (1999) Microtubules and signal transduction. Curr Opin Cell Biol 11:81-94

Harada A, Teng J, Takei Y, Oguchi K, Hirokawa N (2002) MAP2 is required for dendrite elongation, PKA anchoring in dendrites, and proper PKA signal transduction. J Cell Biol 158:541-549

Helps NR, Luo X, Barker HM, Cohen PT (2000) NIMA-related kinase 2 (Nek2), a cell-cycle-regulated protein kinase localized to centrosomes, is complexed to protein phosphatase 1. Biochem J 349:509-518

Houslay MD, Adams DR (2003) PDE4 cAMP phosphodiesterases: modular enzymes that orchestrate signalling cross-talk, desensitization and compartmentalization. Biochem J 370:1-18

Itoh TJ, Hisanaga S, Hosoi T, Kishimoto T, Hotani H (1997) Phosphorylation states of microtubule-associated protein 2 (MAP2) determine the regulatory role of MAP2 in microtubule dynamics. Biochemistry 36:12574-12582

Kapiloff MS, Jackson N, Airhart N (2001) mAKAP and the ryanodine receptor are part of a multi-component signaling complex on the cardiomyocyte nuclear envelope. J Cell Sci 114:3167-3176

Kapiloff MS, Schillace RV, Westphal AM, Scott JD (1999) mAKAP: an A-kinase anchoring protein targeted to the nuclear membrane of differentiated myocytes. J Cell Sci 112:2725-2736

Kashishian A, Howard M, Loh C, Gallatin WM, Hoekstra MF, Lai Y (1998) AKAP79 inhibits calcineurin through a site distinct from the immunophilin-binding region. J Biol Chem 273:27412-27419

Keryer G, Rios RM, Landmark BF, Skalhegg B, Lohmann SM, Bornens M (1993) A high-affinity binding protein for the regulatory subunit of cAMP- dependent protein kinase II in the centrosome of human cells. Exp Cell Res 204:230-240

Keryer G, Witczak O, Delouvee A, Kemmner WA, Rouillard D, Tasken K, Bornens M (2003) Dissociating the centrosomal matrix protein AKAP450 from centrioles impairs centriole duplication and cell cycle progression. Mol Biol Cell 14:2436-2446

Küntziger T, Bornens M (2000) The centrosome and parthenogenesis. Curr Top Dev Biol 49:1-25

Küntziger T, Gavet O, Manceau V, Sobel A, Bornens M (2001) Stathmin/Op18 phosphorylation is regulated by microtubule assembly. Mol Biol Cell 12:437-448

Landsverk HB, Carlson CR, Steen RL, Vossebein L, Herberg FW, Tasken K, Collas P (2001) Regulation of anchoring of the RIIalpha regulatory subunit of PKA to AKAP95 by threonine phosphorylation of RIIalpha: implications for chromosome dynamics at mitosis. J Cell Sci 114:3255-3264

Lester LB, Coghlan VM, Nauert B, Scott JD (1996) Cloning and characterization of a novel A-kinase anchoring protein, AKAP 220, association with testicular peroxisomes. J Biol Chem 271:9460-9465

Liao H, Li Y, Brautigan DL, Gundersen GG (1998) Protein phosphatase 1 is targeted to microtubules by the microtubule-associated protein Tau. J Biol Chem 273:21901-21908

Lin JW, Wyszynski M, Madhavan R, Sealock R, Kim JU, Sheng M (1998) Yotiao, a novel protein of neuromuscular junction and brain that interacts with specific splice variants of NMDA receptor subunit NR1. J Neurosci 18:2017-2027

Liu J, Ben-Shahar TR, Riemer D, Treinin M, Spann P, Weber K, Fire A, Gruenbaum Y (2000) Essential roles for caenorhabditis elegans lamin gene in nuclear organization, cell cycle progression, and spatial organization of nuclear pore complexes. Mol Biol Cell 11:3937-3947

Liu J, Brautigan DL (2000) Glycogen synthase association with the striated muscle glycogen-targeting subunit of protein phosphatase-1. Synthase activation involves scaffolding regulated by beta-adrenergic signaling. J Biol Chem 275:26074-26081

McAvoy T, Allen PB, Obaishi H, Nakanishi H, Takai Y, Greengard P, Nairn AC, Hemmings HC Jr (1999) Regulation of neurabin I interaction with protein phosphatase 1 by phosphorylation. Biochemistry 38:12943-12949

Morishima-Kawashima M, Kosik KS (1996) The pool of map kinase associated with microtubules is small but constitutively active. Mol Biol Cell 7:893-905

Northrop JP, Ho SN, Chen L, Thomas DJ, Timmerman LA, Nolan GP, Admon A, Crabtree GR (1994) NF-AT components define a family of transcription factors targeted in T-cell activation. Nature 369:497-502

Ohta Y, Ohba T, Miyamoto E (1990) Ca2+/calmodulin-dependent protein kinase II: localization in the interphase nucleus and the mitotic apparatus of mammalian cells. Proc Natl Acad Sci USA 87:5341-5345

Oliveria SF, Gomez LL, Dell'Acqua ML (2003) Imaging kinase--AKAP79--phosphatase scaffold complexes at the plasma membrane in living cells using FRET microscopy. J Cell Biol 160:101-112

Ookata K, Hisanaga S, Bulinski JC, Murofushi H, Aizawa H, Itoh TJ, Hotani H, Okumura E, Tachibana K, Kishimoto T (1995) Cyclin B interaction with microtubule-associated protein 4 (MAP4) targets p34cdc2 kinase to microtubules and is a potential regulator of M-phase microtubule dynamics. J Cell Biol 128:849-862

Ookata K, Hisanaga S, Sugita M, Okuyama A, Murofushi H, Kitazawa H, Chari S, Bulinski JC, Kishimoto T (1997) MAP4 is the in vivo substrate for CDC2 kinase in HeLa cells: identification of an M-phase specific and a cell cycle-independent phosphorylation site in MAP4. Biochemistry 36:15873-15883

Pawson T, Scott JD (1997) Signaling through scaffold, anchoring, and adaptor proteins. Science 278:2075-2080

Piel M, Nordberg J, Euteneuer U, Bornens M (2001) Centrosome-dependent exit of cytokinesis in animal cells. Science 291:1550-1553

Rao A, Luo C, Hogan PG (1997) Transcription factors of the NFAT family: regulation and function. Annu Rev Immunol 15:707-47

Reinton N, Collas P, Haugen TB, Skalhegg BS, Hansson V, Jahnsen T, Tasken K (2000) Localization of a novel human A-kinase-anchoring protein, hAKAP220, during spermatogenesis. Dev Biol 223:194-204

Reszka AA, Seger R, Diltz CD, Krebs EG, Fischer EH (1995) Association of mitogen-activated protein kinase with the microtubule cytoskeleton. Proc Natl Acad Sci USA 92:8881-8885

Riabowol KT, Fink JS, Gilman MZ, Walsh DA, Goodman RH, Feramisco JR (1988) The catalytic subunit of cAMP-dependent protein kinase induces expression of genes containing cAMP-responsive enhancer elements. Nature 336:83-86

Schillace RV, Scott JD (1999) Association of the type 1 protein phosphatase PP1 with the A-kinase anchoring protein AKAP220. Curr Biol 9:321-324

Schillace RV, Voltz JW, Sim AT, Shenolikar S, Scott JD (2001) Multiple interactions within the AKAP220 signaling complex contribute to protein phosphatase 1 regulation. J Biol Chem 276:12128-12134

Schmidt PH, Dransfield DT, Claudio JO, Hawley RG, Trotter KW, Milgram SL, Goldenring JR (1999) AKAP350, a multiply spliced protein kinase A-anchoring protein associated with centrosomes. J Biol Chem 274:3055-3066

Shinohara-Gotoh Y, Nishida E, Hoshi M, Sakai H (1991) Activation of microtubule-associated protein kinase by microtubule disruption in quiescent rat 3Y1 cells. Exp Cell Res 193:161-166

Smith FD, Scott JD (2002) Signaling complexes: junctions on the intracellular information super highway. Curr Biol 12:R32-R40

Sontag E, Nunbhakdi-Craig V, Bloom GS, Mumby MC (1995) A novel pool of protein phosphatase 2A is associated with microtubules and is regulated during the cell cycle. J Cell Biol 128:1131-1144

Sontag E, Nunbhakdi-Craig V, Lee G, Bloom GS, Mumby MC (1996) Regulation of the phosphorylation state and microtubule-binding activity of Tau by protein phosphatase 2A. Neuron 17:1201-1207

Srivastava RK, Srivastava AR, Korsmeyer SJ, Nesterova M, Cho-Chung YS, Longo DL (1998) Involvement of microtubules in the regulation of Bcl2 phosphorylation and apoptosis through cyclic AMP-dependent protein kinase. Mol Cell Biol 18:3509-3517

Steen RL, Beullens M, Landsverk HB, Bollen M, Collas P (2003) AKAP149 is a novel PP1 specifier required to maintain nuclear envelope integrity in G1 phase. J Cell Sci 116:in press

Steen RL, Collas P (2001) Mistargeting of B-type lamins at the end of mitosis: implications on cell survival and regulation of lamins A/C expression. J Cell Biol 153:621-626

Steen RL, Martins SB, Tasken K, Collas P (2000) Recruitment of Protein Phosphatase 1 to the Nuclear Envelope by A-Kinase Anchoring Protein AKAP149 Is a Prerequisite for Nuclear Lamina Assembly. J Cell Biol 150:1251-1262

Taira T, Maeda J, Onishi T, Kitaura H, Yoshida S, Kato H, Ikeda M, Tamai K, Iguchi-Ariga SM, Ariga H (1998) AMY-1, a novel C-MYC binding protein that stimulates transcription activity of C-MYC. Genes Cells 3:549-565

Takahashi M, Shibata H, Shimakawa M, Miyamoto M, Mukai H, Ono Y (1999) Characterization of a novel giant scaffolding protein, CG-NAP, that anchors multiple signaling enzymes to centrosome and the golgi apparatus. J Biol Chem 274:17267-17274

Takahashi M, Mukai H, Oishi K, Isagawa T, Ono Y (2000) Association of immature hypophosphorylated protein kinase cepsilon with an anchoring protein CG-NAP. J Biol Chem 275:34592-34596

Takahashi M, Yamagiwa A, Nishimura T, Mukai H, Ono Y (2002) Centrosomal proteins CG-NAP and kendrin provide microtubule nucleation sites by anchoring gamma-tubulin ring complex. Mol Biol Cell 13:3235-3245

Tanji C, Yamamoto H, Yorioka N, Kohno N, Kikuchi K, Kikuchi A (2002) A-kinase anchoring protein AKAP220 binds to glycogen synthase kinase-3beta (GSK-3beta) and mediates protein kinase A-dependent inhibition of GSK-3beta. J Biol Chem 277:36955-36961

Tasken K, Skålhegg BS, Tasken KA, Solberg R, Knutsen HK, Levy FO, Sandberg M, Ørstavik S, Larsen T, Johansen AK, Vang T, Schrader HP, Reinton NT, Torgersen KM, Hansson V, Jahnsen T (1997) Structure, function, and regulation of human cAMP-dependent protein kinases. Adv Second Messenger Phosphoprotein Res 31:191-204

Tasken KA, Collas P, Kemmner WA, Witczak O, Conti M, Tasken K (2001) Phosphodiesterase 4D and protein kinase a type II constitute a signaling unit in the centrosomal area. J Biol Chem 276:21999-22002

Thompson LJ, Bollen M, Fields AP (1997) Identification of protein phosphatase 1 as a mitotic lamin phosphatase. J Biol Chem 272:29693-29697

Trendelenburg G, Hummel M, Riecken EO, Hanski C (1996) Molecular characterization of AKAP149, a novel A kinase anchor protein with a KH domain. Biochem Biophys Res Commun 225:313-319

Vallee RB, DiBartolomeis MJ, Theurkauf WE (1981) A protein kinase bound to the projection portion of MAP 2 (microtubule-associated protein 2). J Cell Biol 90:568-576

Vijayaraghavan S, Liberty GA, Mohan J, Winfrey VP, Olson GE, Carr DW (1999) Isolation and molecular characterization of AKAP110, a novel, sperm-specific protein kinase A-anchoring protein. Mol Endocrinol 13:705-717

Westphal RS, Tavalin SJ, Lin JW, Alto NM, Fraser ID, Langeberg LK, Sheng M, Scott JD (1999) Regulation of NMDA receptors by an associated phosphatase-kinase signaling complex. Science 285:93-96

Witczak O, Skålhegg BS, Keryer G, Bornens M, Taskén K, Jahnsen T, Ørstavik S (1999) Cloning and characterization of a cDNA encoding an A-kinase anchoring protein located in the centrosome, AKAP450. EMBO J 18:1858-1868

Yukitake H, Furusawa M, Taira T, Iguchi-Ariga SM, Ariga H (2002) AAT-1, a novel testis-specific AMY-1-binding protein, forms a quaternary complex with AMY-1, A-kinase anchor protein 84, and a regulatory subunit of cAMP-dependent protein kinase and is phosphorylated by its kinase. J Biol Chem 277:45480-45492

Zhao S, Lee EY (1997) A protein phosphatase-1-binding motif identified by the panning of a random peptide display library. J Biol Chem 272:28368-28372

List of abbreviations

AKAP: A-kinase anchoring protein

CaMKII: calcium/calmodulin-dependent kinase II
CaN: calcineurin
CDK1: cyclin-dependent kinase 1
GSK-3: glycogen synthase kinase-3
MAGUK: membrane-associated guanylate kinase
MAP: microtubule-associated protein
MAPK: mitogen-activated protein kinase
MT: microtubule
Nek2: NIMA-related protein kinase 2
NMDA: N-methyl-D-aspartate
PCM: pericentriolar material
PKA: cAMP-dependent protein kinase
PKC: protein kinase C
PKN: protein kinase N
PP1: protein phosphatase 1
PP2A: protein phosphatase 2A
PP2B: protein phosphatase 2B
R: regulatory subunit

9 Functional proteomics in phosphatase research

Nicole C. Kwiek, Timothy A. J. Haystead

Abstract

With the technological development of mass spectrometry as a powerful tool in biology and the completion of the human genome sequencing project, proteomics has emerged as a field that promises to revolutionize the face of biology. In particular, we may now rapidly and thoroughly analyze signal transduction networks in a variety of systems in order to characterize individual components. Information garnered from such studies will undoubtedly aid in our understanding of the biology of a system as well as promote drug development. Many aspects of protein phosphatase action remain unclear due to paucity in our understanding of in vivo activity and regulation. Emerging proteomics studies and technology is poised to fill such gaps.

9.1 Introduction to functional proteomics

The scientific community now benefits from the remarkable success of entire genome sequencing projects including, among others, model organisms such as yeast and fruit flies to higher eukaryotes such as mice and humans. This primary data serves as a platform for the progress of an adjunct field, proteomics, that involves the large-scale characterization of the protein complement, or proteome, within a cell, tissue, or organism (Wasinger et al. 1995; Wilkins et al. 1996). Initiated in the 1970s, early proteomic studies used two-dimensional gel electrophoresis (2DGE) to develop maps of all proteins within a given system (Klose 1975; O'Farrell 1975; Scheele 1975). Although progressive, these studies lacked identification of the resolved proteins because of comparatively crude resources in sequencing technology. However, with the use of state-of-the-art mass spectrometry in biological applications and the wealth of data from genome projects, proteomics emerges as a powerful enterprise that promises to advance current biology through an increased understanding of global protein content, activity, and changes during normal development or disease pathogenesis.

Interest in the proteomics field is further precipitated by our realization that the biology of an organism simply cannot be elucidated from DNA sequence alone. Despite advances in bioinformatics, our ability to predict genes from primary genomic sequence remains poor. Furthermore, the mere existence of an open reading frame does not ensure the subsequent formation of the cognate protein. Several studies demonstrated that even mRNA levels do not necessarily correlate with

Topics in Current Genetics, Vol. 5
J. Arino, D.R. Alexander (Eds.) Protein phosphatases
© Springer-Verlag Berlin Heidelberg 2004

protein formation (Futcher et al. 1999; Gygi et al. 1999). Most importantly, proteins, and not nucleic acids, remain the functional unit of the cell. In contrast to the genome of a particular organism, the proteome is dynamic and subject to a variety of modifications in response to distinct environmental cues. Such alterations (e.g. phosphorylation, acetylation, proteolytic processing, etc.) can then dramatically change the protein "phenotype" of a cell without a corresponding change of the genome (Godovac-Zimmermann and Brown 2001). Thus, although only 30,000 genes are predicted within the human genome, nearly 2 million functional proteins may exist once splice variants and post-transcriptional modifications are accounted.

The proteomics field is divided into several subtypes including expression, structural, and functional studies. Unlike other disciplines, the goal of functional proteomics in its broadest terms is to fully characterize the proteome of a specimen. This includes unambiguously identifying each protein, determining its abundance, detecting post-translational modifications, and classifying the interactions and complexes that are formed among individual protein members. In achieving this, multidiscipline approaches including classical biochemistry, elegant genetics, and high-throughput arrays are commonly used. In this chapter, we attempt to address complex problems within the phosphatase field through the use of functional proteomics applications, including examples from previously published work in addition to current undertakings. Although much attention will be focused on the regulation of serine/threonine phosphatases, some technologies also apply to tyrosine phosphatase research.

9.1.1 Protein phosphatase-1 research: Identification of novel regulatory subunits

The critical function of reversible phosphorylation in cellular signaling is exemplified by its perturbation in virtually all disease processes including neurological, metabolic, respiratory, and proliferative disorders. In a cell, the phosphorylation state of a particular protein is governed by the opposing actions of kinases and phosphatases. Interestingly, the human genome encodes for more than 300 serine/threonine kinases but only 20 respective serine/threonine-specific phosphatases (Ceulemans et al. 2002; Manning et al. 2002). In the wake of such an imbalanced ratio of kinases to phosphatases, how then is the specificity of regulation maintained?

Studies of one major eukaryotic phosphatase, protein phosphatase-1 (PP1) provided some insight into this conundrum. PP1 dephosphorylates serine and threonine residues of proteins implicated in diverse processes such as glycogen metabolism, cell cycle progression, muscle contraction, and neuronal activities. The ability of this enzyme to be regulated independently within the cell results from the mutually exclusive interaction of its catalytic subunit (PP1c) with an assortment of regulatory subunits. This complex formation then dictates PP1 function through alteration of substrate specificity, subcellular localization, and/or subsequent regulation (reviewed in this volume by Dombradi et. al). Therefore, PP1 maintains its ability to be regulated in a specific manner despite its role in a plethora of distinct processes. The collection of putative or established PP1 regulatory subunits now exceeds 45 in mammals, yet very few of these proteins share significant sequence similarity (Cohen 2002). However, a conserved peptide sequence, the (R/K)(V/I)XF motif within a putative regulatory subunit, appears critical for PP1c interaction (Egloff et al. 1997). Both in vitro and in vivo studies demonstrated that the mutation of valine or phenylalanine residues within this sequence weakens or ablates PP1c interaction with several regulatory subunits including G_m, M_{110} (Egloff et al. 1997), AKAP 149 (Steen et al. 2000), spinophilin (Hsieh-Wilson et al. 1999; Yan et al. 1999) and others (Alms et al. 1999; Rudenko et al. 2003). Although the (R/K)(V/I)XF motif appears to specify PP1c interaction, database searching reveals that this sequence occurs in approximately 10% of all proteins. Many of these candidate proteins will unlikely bind PP1c because the motif is inaccessible to interaction based on physical conformation. In addition, the X position in this motif may not tolerate some amino acids including large hydrophobic residues. Furthermore, as the identification and characterization of bona fide regulatory subunits advance, the (R/K)(V/I)XF motif may necessarily be expanded to accommodate unique interacting proteins. For example, the consensus sequence for PP1 binding now exists as $(K/R/H/N/S)X_1(V/I/L)X_2(F/W/Y)$ (Bollen 2001; Cohen 2002).

Thus, the assignment of PP1-binding proteins through database searching alone is difficult. Rather, in order to more fully understand PP1 action and regulation, a functional proteomics approach may be utilized to identify novel regulatory subunits. Here, we discuss two distinct applications including the yeast two-hybrid system and microcystin affinity chromatography with an additional description of prevailing techniques in protein identification.

9.1.2 Yeast two-hybrid systems and arrays

Originally developed by Fields and Song over a decade ago, the yeast two-hybrid system has emerged as a powerful tool by which to detect protein-protein interactions in vivo (Fields and Song 1989). Simple and inexpensive, this method exploits the modular nature of eukaryotic transcriptional activators. In the classical approach, the protein of interest ("bait") is fused to the DNA-binding domain (DBD) of a transcription factor whereas a protein partner ("prey") is fused to the activation domain (AD) of the same transcription factor. Following co-expression of the hybrid proteins in yeast, the interaction of the two fusion proteins results in the reconstitution of a functional transcriptional activator, thereby facilitating the expression of one or more reporter genes. The identification of the prey is then determined by DNA sequencing. Often, the bait protein is screened against a total genomic or cDNA library, thereby enabling the unbiased identification of potential binding partners from a large pool of candidate genes. In several insightful studies, the yeast two-hybrid approach was implemented as a means to identify novel regulatory subunits of PP1 (Allen et al. 1998; Bennett and Alphey 2002; Dunaief et al. 2002; Monshausen et al. 2002; Rudenko et al. 2003). In this work,

PP1c was used as bait in order to probe cDNA libraries from yeast, fruit fly, or mammalian cells. Subsequent biochemical and genetic experiments verified the interaction, thereby validating this approach.

However, despite its successes, the classical yeast two-hybrid system possesses several limitations. In principle, it cannot detect interactions involving three or more proteins or those proteins that bind DNA directly. Furthermore, membrane-anchored proteins are typically underrepresented in yeast two-hybrid results. In addition, this system is unable to detect those interactions that depend on post-translational modifications of the participating proteins. This weakness is particularly relevant to PP1-binding proteins as regulatory subunits are often regulated themselves by phosphorylation. Most importantly, this methodology still suffers from false positives (those interactions of no physiological relevance) and negatives (the lack of recovery of expected interactions). In order to address these limitations, several variations of the classical assay emerged including the yeast one- and three-hybrid as well as reverse and small molecule-hybrid screens (Topcu and Borden 2000). Furthermore, in post-genome fashion, recent efforts by several groups have focused on the generation of protein interaction maps through the use of high-throughput yeast two-hybrid arrays. In an effort to define an organism's "interactome", this format examines all of the possible binary combinations between proteins encoded by any single genome (Ito et al. 2002). Still in its infancy, the results of such studies remain to be validated.

9.1.3 Microcystin affinity chromatography

Although powerful, the inherent limitations of the yeast two-hybrid system necessitate the use of alternative approaches to identify novel PP1 regulatory subunits. In our laboratory, we implemented the use of the toxin microcystin-LR (MC-LR) conjugated to various matrices in order to directly examine phosphatase-protein interactions in a biochemical context. Originally isolated from cyanobacteria, MC-LR belongs to the closely related family of cyclic peptide microcystin toxins. Potent and highly specific, the toxicity of these molecules derives from their ability to equally inhibit both PP1 and PP2A protein phosphatases (IC50 = 0.2nM) (Eriksson et al. 1990; Toivola et al. 1994; Dawson and Holmes 1999). In its two-step inhibition, MC binds to the active center of the phosphatase, effecting loss of enzymatic activity. Then, the phosphatase covalently binds to the methyl-dehydro-alanine residue of MC through a cysteine residue (Dawson and Holmes 1999; Holmes et al. 2002; Mikhailov et al. 2003). Immobilization of MC to matrices such as Sepharose or biotin does not interfere with inhibition. In addition, because MC is reduced prior to attachment, a covalent bond between toxin and phosphatase is not formed, facilitating the recovery of proteins through the use of biotin or chaotropic agents. Furthermore, the (R/K)(V/I)XF motif is situated distinctly from the catalytic site of PP1, thereby allowing the simultaneous capture of both phosphatase and regulatory subunit. Indeed, initial studies using MC-Sepharose demonstrated a successful single-step purification of both PP1c and glycogen-targeting subunits from skeletal muscle myofibrils (Moorhead et al. 1994). For the most recent protocols for MC-resin synthesis, refer to Campos et al. (1996) and Damer et al. (1998).

In our laboratory, we apply both MC-Sepharose and MC-biotin techniques in the quest for novel PP1 regulatory subunits (Campos et al. 1996; Damer et al. 1998). Although similar in principle, MC-biotin allows for elution by excess biotin rather than a chaotropic agent, thereby creating milder conditions for bound proteins and enabling holoenzymes to remain intact. Following the adsorption of a tissue extract onto resin beads, stringent column washing (increased salt and detergent concentrations) is employed to eliminate nonspecific ionic and hydrophobic interactions. Eluted proteins are subsequently resolved by either one- or two-dimensional gel electrophoresis and submitted for identification by either Edman sequencing or mass spectrometry (Fig. 1).

Edman degradation chemistry involves the stepwise chemical degradation of a protein or peptide at its N-terminus and the subsequent identification of the released amino acids from each cycle to obtain de novo sequence (Edman 1949). One of the earliest techniques for amino acid microsequencing, the methodology remains largely unchanged since its inception in 1949. However, we developed a novel Edman approach termed mixed peptide sequencing to identify proteins (Damer et al. 1998). Briefly, this technique involves the simultaneous sequencing of all peptides derived from a given protein following chemical cleavage. Resultant peptide fragments are then submitted directly into an automated Edman sequencer without further manipulation. Following multiple cycles of Edman degradation, unordered sequence data are then deconvoluted and matched through the FASTF or TFASTF algorithms (described in detail later) to relevant databases, resulting in the unambiguous identification of a protein. This technique proves to be a highly sensitive means of analysis, rapidly identifying proteins amounting less than 1 pmol. Using mixed peptide sequencing, our laboratory was able to identify 36 proteins obtained by microcystin-affinity chromatography including the catalytic subunits of PP1 and PP2A as well as the characterized myosin phosphatase targeting subunit (Damer et al. 1998). In addition, peptides were matched to seven ESTs in the database, thereby representing novel proteins whose functions are entirely unknown. The characterization of these putative PP1 regulatory subunits is ongoing.

Today, mass spectrometry (MS) supersedes the use of Edman chemistry in proteomics endeavors. Like mixed peptide sequencing, MS provides amino acid sequences that can then be searched against a given database in order to identify a particular protein. However, MS possesses distinct advantages including increased sensitivity (low femtomolar) and higher throughput (Pandey and Mann 2000). Furthermore, because MS measures the inherent mass of a molecule, potentially any posttranslational modification (e.g. phosphorylation, glycosylation, ubiquitination, etc.) can be detected and characterized. Rather than engage in a lengthy description of the current advances in mass spectrometry, in this section we highlight significant features that have directed our experiences with this technology. For a more exhaustive description, the reader is directed to several excellent reviews (Pandey and Mann 2000; Aebersold and Goodlett 2001; Mann et al. 2001; Yarmush and Jayaraman 2002).

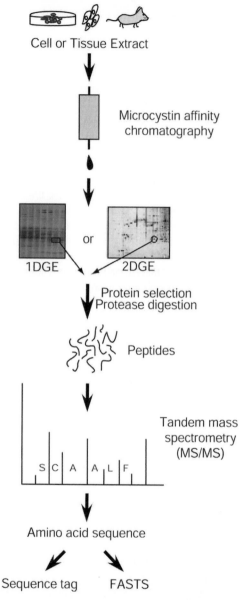

Fig. 1. A functional proteomics scheme for the identification of novel regulatory subunits of PP1 by mass spectrometry. A protein extract is prepared and applied to a matrix conjugated to microcystin-LR. The mixture of proteins is resolved by 1D- or 2DGE and visualized by staining. Individual proteins are then excised and subjected to proteolytic digestion. The amino acid sequence of an individual peptide is obtained by tandem mass spectrometry (MS/MS) and identified by database searching with sequence tags or the T/FASTS algorithms (http://fasta.bioch.virginia.edu).

The emergence of MS in life studies largely coincides with the development of appropriate ionization techniques. Because biomolecules such as proteins are large and polar, they are not easily transferred into the gaseous phase and ionized. However, the development of 'soft ionization methods' such as matrix-assisted laser desorption/ionization (MALDI) and electrospray (ESI) permitted the formation of ions without significant loss of sample and subsequently revolutionized this technique for the life sciences. In MALDI, the sample is incorporated into an energy-absorbing matrix and then subjected to laser irradiation, promoting the formation of singly charged ion species (Karas and Hillenkamp 1988). Often formatted as a 96-well plate, the entire process, including sample application, may be automated. Thus, as a high throughput operation, MALDI remains the optimal ionization method for the identification of multiple samples when speed is at a premium. In ESI ionization, an analyte solution is subjected to high voltage, resulting in the generation of a fine mist of highly charged droplets that are then introduced into the mass spectrometer (Fenn et al. 1989; Graves and Haystead 2002). ESI technology was further improved with the development of nanospray ionization (nano-) which allows the continuous spray of ions at a low flow rate, enabling reduced sample consumption and increased analysis time (Wilm et al. 1996).

Many mass spectrometers measure the mass-to-charge ratio (m/z) of peptides and proteins by the use of two main types of mass analyzers: time-of-flight (TOF-MS) and quadrupole electric fields (quadrupole-MS). TOF computes the time required for gas-phase ions to travel from the ionization source to the detector and, based on its ease of operation, represents the most common analyzer. Conversely, a quadrupole mass analyzer utilizes an electric field generated by successive metal rods to transmit all ions or acts as a mass filter to permit the transmission of only those ions with a certain m/z ratio. For amino acid sequencing, two steps of MS are often performed in tandem (tandem MS or MS/MS) by employing the same separation principle twice (e.g. TOF/TOF) or by combining two distinct mass separation principles (e.g. quad-TOF) (Mann et al. 2001). In MS/MS, peptide ions are separated based on their m/z ratio and a particular peptide ion is selectively passed into a collision chamber. There, the ions are further fragmented, generating a series of peptides that differ by a single amino acid (Hunt et al. 1986). The amino acid sequence can then be deduced by examination of mass differences between ion peaks. At the cost of speed, this format delivers quality sequence data of high confidence and sensitivity. Furthermore, because a particular peptide ion may be selected for analysis, components of complex mixtures may be identified.

Regardless of the method for obtaining the data, a crucial component of protein identification remains the selection of an appropriate database or search strategy for analysis. Herein the progress of proteomics critically depends on the advances of genome sequencing. In a method termed peptide mass fingerprint (PMF) database searching, data from MALDI-MS is searched against theoretical spectra obtained from primary sequence databases (Mann et al. 1993; Jensen et al. 1997; Graves and Haystead 2002). If a sufficient number of peptides in the analytical and theoretical spectrum coincide, then a match is considered for protein identification. Although very rapid, PMF often leads to ambiguous results because of its inability to decipher proteins of equal masses but differing sequence (mass redundancy) or those proteins that have undergone post-translational modifications. In addition, because the search is not error-tolerant, PMF is not productive with poorly annotated genomes. To circumvent these problems, one may exploit a more discriminating search method through the use of amino acid composition data derived from MS/MS. With this data in hand, two separate approaches can be enlisted: peptide mass tag or de novo sequence searching. In the former method, a partial amino acid sequence (sequence tag) from MS/MS spectra is searched against databases in conjunction with the combined mass of the peptide (Mann and Wilm 1994). With calculated mass information, this approach can unambiguously identify proteins with as few as two amino acids and consequently offers rapid analysis from poorly developed spectra. However, peptide mass sequence tag searching is also highly dependent on the strength of a given database and is less useful with non-annotated genomes. The most comprehensive yet time-consuming approach involves the acquisition of de novo sequence data from MS/MS. With this information, multiple peptide sequences of unknown order may be simultaneously searched against protein or even translated DNA databases with the T/FASTS and T/FASTF algorithms. Originally developed in a collaboration between our laboratory and that of Dr. William Pearson at the University of Virginia, these algorithms extend the capabilities of the traditional FASTA and BLAST programs by eliminating multiple independent database searches and maximizing search sensitivity through the use of probability-based scoring (Pearson and Lipman 1988; Altschul et al. 1990; Mackey et al. 2002). Due to this enhanced flexibility, this approach is useful regardless of genome annotation level. These programs may be accessed through http://fasta.bioch.virginia.edu.

In all proteomic studies, a final yet critical portion includes the validation of results through additional approaches. The identification of a protein by either Edman or MS does not guarantee a corresponding role in the biology of that experiment. In fact, the actual labor of a study is just beginning. In the context of microcystin affinity chromatography, subsequent experiments such as immunoprecipitation, far-Western, and colocalization studies must be performed in order to validate initial results.

9.2 PP1 research: Defining the regulation of regulatory subunits

Upon formation, PP1c/regulatory subunit complexes do not exist in a vacuum. Rather, extracellular and intracellular signals impinge on the integrity and function of this complex through multiple mechanisms, the most studied of which is reversible phosphorylation. This additional level of control provides yet another regulatory mechanism by which PP1 activity is modulated. However, relatively very little is understood concerning the implications of phosphorylation in the regulation of these subunits. Ongoing work on several targeting subunits including G_M, neurabin I, and NIPP1 have demonstrated that phosphorylation of serine residues within or near the conserved (R/K)(V/I)XF motif diminishes PP1c binding and subsequently subunit-directed dephosphorylation (Dent et al. 1989; Hubbard and Cohen 1989; Vulsteke et al. 1997; McAvoy et al. 1999; Cohen 2002). In contrast, phosphorylation of inhibitory subunits such as inhibitor-1, DARPP-32, and CPI-17 creates an additional binding site for PP1c, thereby strengthening their inhibitory properties (Hemmings et al. 1984; Eto et al. 1997).

In this section, we will describe current methods for identifying phosphorylation sites that remain imperative for phosphoprotein study. In addition, we will describe results from studies in our laboratory that examined the regulation of the myosin targeting subunit of myosin light chain phosphatase through the use of proteomics strategies (MacDonald et al. 2001; Borman et al. 2002).

9.2.1 Phosphorylation site analysis

The identification of relevant phosphorylation sites in a given protein promotes the further classification of responsible kinases as well as the analysis of the functional consequences of the modification. However, phosphoproteins often exist in vanishingly small amounts and therefore require sensitive yet specific methodology. As in protein identification, the two most applicable techniques remain Edman degradation and mass spectrometry.

Edman chemistry remains a practical method for phosphorylation site mapping due to its simplicity and application to a large variety of peptides (Aebersold et al. 1991; Boyle et al. 1991; Wettenhall et al. 1991; Graves and Haystead 2003). Briefly, a ^{32}P-labeled protein is cleaved by a protease, and the resulting peptides are resolved by high pressure liquid chromatography (HPLC) or thin layer chromatography. Following crosslinkage to an inert membrane, the peptides are then subjected to several cycles of Edman chemistry whereby each Ser, Thr, or Tyr residue may potentially release radioactive phosphate. One can then deduce putative phosphorylation sites by examining the correlation between radioactive counts and amino acid sequence. In addition, we recently developed the cleaved radioactive peptide program (CRP) that exploits bioinformatics to map sites of a known protein at the subfemtomolar level (MacDonald et al. 2002). This program assesses the number of Edman cycles required for the complete coverage of Ser, Thr, and Tyr residues in a given protein following protease cleavage. Within a single cleavage experiment, CRP may reduce the number of potential phosphorylation sites to 5-10 for most proteins. Further specificity is obtained by the use of a second protease as well as phosphoamino acid analysis, thereby unambiguously identifying relevant site(s). Developed in collaboration with Aaron Mackey and Dr. William Pearson at the University of Virginia, the CRP program is publicly available at http://fasta.bioch.virginia.edu/crp/crp2/.

As its instrumentation and sensitivity improves, MS is emerging as an ideal method for phosphorylation site mapping. This approach capitalizes on two properties of phosphoproteins: the mass incurred upon a peptide upon phosphate addition and the chemical lability of the phosphate ester bond within the collision chamber of the mass spectrometer. In principle, phosphopeptides can be distinguished from others by a net mass differential of 80 Da that results from the phosphorylation of a Ser, Thr, or Tyr residue. Upon careful comparison, one may then identify phosphopeptides from peptide mass maps and subsequently detect relevant residues within a peptide sequence by conventional MS/MS. Furthermore, the identity of phosphorylated residues can be confirmed by the presence of elimination products such as dehydroalanine (p-Ser) and dehydroamino-2-butyric acid (p-Thr) (Neubauer and Mann 1999). For a more detailed description of MS approaches in phosphorylation site mapping, the reader is again directed to several excellent reviews (Aebersold and Goodlett 2001; McLachlin and Chait 2001; Mann et al. 2002; Graves and Haystead 2003).

In our laboratory, we concluded that a sequential combination of HPLC, Edman degradation, and phosphopeptide sequencing by MS/MS provides ideal data for phosphorylation site determination. Following cleavage of a ^{32}P-labeled protein by a protease, the resulting peptides are resolved by HPLC. Only those peptide fractions that contain radioactivity are then submitted for analysis by MS, thereby decreasing the complexity of the analyte mixture and promoting site identification. As demonstrated in the following example, site mapping offers tremendous insight into the effect and subsequently the physiology of a given phosphorylation event.

9.2.2 Identification of the endogenous smooth muscle myosin phosphatase-associated kinase

Smooth muscle contraction is largely regulated through transient changes in free intracellular calcium and maintenance of myosin light chain (MLC20) phosphorylation, the latter of which is largely controlled by the opposing actions of myosin light chain kinase (MLCK) and myosin light chain phosphatase (SMPP-1M). Inhibition of SMPP-1M results in the hyperphosphorylation of MLC20, thereby promoting contraction. SMPP-1M is composed of three components including a catalytic PP1 subunit, a 110-130 kDa regulatory myosin targeting subunit (MYPT1), and a 21-kDa subunit of unknown function. Previous studies demonstrated that SMPP-1M activity is regulated through the inhibitory phosphorylation of MYPT1 at Thr[697] (Ichikawa et al. 1996). However, the endogenous SMPP-1M kinase remained unknown.

In order to identify the kinase responsible for the phosphorylation of MYPT1, we initially enriched for purine-binding proteins through the use γ-phosphate-linked adenosine triphosphate (ATP)-Sepharose. Originally developed in our laboratory, this resin links ATP in a conformation that is amenable to binding protein kinases in the active conformation. In addition, γ-ATP-Sepharose binds the large complement of purine-utilizing enzymes including dehydrogenases, heat shock proteins, ligases, and others that in sum account for approximately 5% of the expressed eukaryotic genome (Graves et al. 2002). Following the application of cow bladder homogenate to the resin and high stringency washes to remove non-specific binding, an enriched fraction of smooth muscle kinases was obtained. After further fractionation by anion exchange chromatography, both in vitro and in-gel kinase assays identified those fractions containing kinase activity against a peptide of MYPT encompassing the Thr[697] phosphorylation site (Fig. 2A,B). The

MYPT1 kinase was then purified to near homogeneity and identified by mixed peptide sequencing and FASTF analysis as most similar to HeLa zipper interacting protein kinase (ZIP-like kinase or MYPT1 kinase) (Fig. 2C). Additional experiments confirmed that the phosphorylation of MYPT1 in intact muscle results in the inhibition of SMPP-1M. Furthermore, the introduction of recombinant MYPT1 kinase into an intact ileal muscle strip results in Ca^{2+}-independent contraction, thereby validating our results and delineating a role for this protein in vivo. As MYPT-1 kinase is phosphorylated upon activation, ongoing studies are now examining the regulation of this protein (e.g. the identification of MYPT1-kinase kinase). Overall, using this proteomics strategy, we are rapidly delineating the components of this crucial signaling cascade.

9.3 PP1 Research: Identifying substrates of a PP1/regulatory subunit complex

Despite immense progress in the identification and characterization of novel regulatory subunits, our understanding of the physiological function and regulation of PP1 is still far from complete. This void mainly stems from a lack of known physiological PP1 substrates, and in particular, the effect of a given regulatory subunit on directed PP1 activity. According to a recent review, at least 15 bona fide regulatory subunits lack a physiological function, and numbers are likely to increase with the emerging identification of new PP1 binding proteins (Cohen 2002). Here, we describe a novel proteomics study from our laboratory that identified the physiological substrates of PP1/Glc7p-Reg1p in a yeast model system (Alms et al. 1999).

9.3.1 Reg1p targets PP1 to dephosphorylate hexokinase II in *Saccharomyces cerevisiae*

In yeast, Glc7 is the functional homolog of PP1c, and like its mammalian counterpart, is regulated by interaction with distinct targeting subunits. Among other functions as an essential protein, the Glc7 phosphatase is required for glucose repression (Tu and Carlson 1994). Depending on availability, yeast may thrive on a variety of carbon sources yet glucose and fructose are preferred. When glucose is highly abundant, the transcription of enzymes required for processes such as gluconeogenesis and the metabolism of alternate sugars are repressed. Biochemical and genetic studies identified Reg1 as a putative regulatory subunit of Glc7 that, upon deletion, impairs glucose repression (Tu and Carlson 1995; Huang et al. 1996). These results suggest that Reg1 potentially directs the participation of Glc7 in this process yet the physiological substrates of the Reg1/Glc7 complex remained entirely unknown.

In an approach that bridged both proteomics and genetics, we compared the phosphoproteome (the entire complement of phosphorylated proteins) of a wild

A.

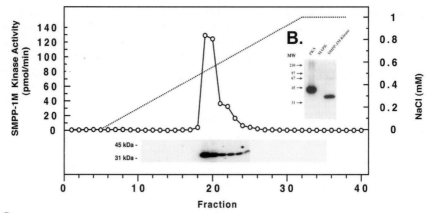

C.

FASTF Aligned Sequence	Called Protein
<pre>   MGEELGSGQFAIV---
          ::::::::::::::
MSTFRQEDVEDHYEMGEELGSGQFAIVRKCRQKGTGKEYAAKFIKKRRLPSSRRGVSREEIEREVNILREIRHPNIITLH
     10   20   30   40   50   60   70   80

--MLLDKXIFXRPIQ--
                                                        ::::: .   :
DIFENKTDVVLILELVSGGELFDFLAEKESLTEDEATQFLKQILDGVHYLHSKRIAHFDLKPENIMLLDKNVPNPRIKLI
     90   100   110   120   130   140   150   160

DFGIAHKIEAGNEFKNIFGTPEFVAPEIVNYEPLGLEADMWSIGVITYILLSGASPFLGETKQETLTNISAVNYDFDEEY
    170   180   190   200   210   220   230   240
----------------------MTIAQNLXYXXIX--------------------------------------
                      ::::: : . :
FSSTSELAKDFIRRLLVKDPKRRMTIAQSLEHSWIKVRRREDGARKPERRRLRAARLREYSLKSHSSMPRNTSYASFERF
    250   260   270   280   290   300   310   320</pre> | ZIP kinase |

D.

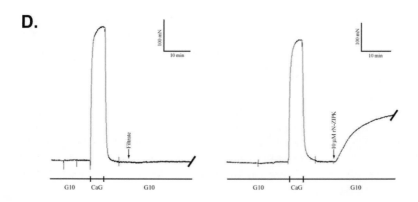

Fig. 2 (overleaf). A. Purification of SMPP-1M-associated kinase. Following elution by anion exchange chromatography, SMPP-1M kinase activity was identified by both in vitro and in-gel kinase assays. A peptide containing the Thr^{697} phosphorylation site of MYPT1 was used as substrate. B. An autoradiogram from an in-gel kinase assay demonstrates localized kinase activity to a discrete protein band at 32 kDa. C. Identification of SMPP-1M-associated kinase by mixed peptide sequencing. FASTF was used to search and match the mixed sequences to the National Center for Biotechnology Information (NCBI)/Human protein database. D. Recombinant MYPT1 kinase induces Ca^{2+}-independent contraction in intact muscle. Permeabilized rabbit ileal strips were contracted, relaxed, and then incubated in the absence or presence of 10 μM rMYPT1K. (Redrawn from (MacDonald et al. 2001; Borman et al. 2002)).

type yeast strain with a mutant strain that lacks Reg1 (*reg1△*). In the absence or disruption of REG1, relevant targets would potentially exist in a hyperphosphorylated state as a result of disrupted phosphatase activity. Indeed, following in vivo ^{32}P-labeling, resolution of cell lysates by 2DGE, and protein identification by mixed peptide sequencing, we demonstrated that hexokinase II (Hxk2) became hyperphosphorylated upon REG1 deletion without a concurrent change in expression. Furthermore, glucose depletion in wild type strains induced a similar shift in the phosphorylation of Hxk2. Additional experiments revealed an absence of phosphatase activity towards Hxk2 in *reg1△*-mutants, suggesting that Reg1 directs Glc7 to this substrate. Indeed, Hxk2 dephosphorylation is restored upon expression of normal Reg1 but not a mutant of Reg1 (F468R within the (R/K)(V/I)XF motif) that is unable to bind PP1.

Results from this study suggest that Glc7 mediates an effect on glucose repression in vivo through the Reg1-directed dephosphorylation of Hxk2, an enzyme involved in the glycolytic pathway. However, perhaps even more exciting than these findings is the strategy itself. On one level, this approach offers a unique tool for identifying novel substrates for PP1, thereby promoting the growth of the phosphatase field. Although, in addition, the juxtaposition of proteomics with genetics provides a phenomenal approach by which to delineate signaling pathways in vivo. Furthermore, in this context, the functional consequences of a particular genetic manipulation can be assessed directly on the proteome as a whole. In this study, we exploited the genetic ease of the yeast system, but a similar approach may certainly be adopted for higher eukaryotic systems through the use of RNA interference (RNAi) or gene knockouts. For example, a recent study used RNAi in a Drosophila cell system to delineate substrates of protein tyrosine kinases and phosphatases (Muda et al. 2002). By the simple introduction of double-stranded RNA into S2 cells, a 'knockout' phenotype was achieved within days. Through in vivo ^{32}P-labeling and immunoprecipitation experiments, the authors subsequently identified DPTP61F as the regulatory phosphatase of Dock protein. Likewise, the effects on the entire phosphoproteome may be examined through a functional proteomics strategy as described earlier. Clearly, the future for the merger of proteomics and genetics is exciting.

9.4 Conclusions

Functional proteomics has clearly emerged as a credible approach for interpreting genome sequence data. These novel technologies are now delineating and characterizing the components of entire signaling pathways at a rapid pace including the activity, regulation, and biology of phosphatases. Yeast two-hybrid assays and microcystin affinity chromatography followed by microsequencing allow for the rapid identification of novel regulatory subunits of PP1. In addition, regulatory kinases within a pathway may be rapidly identified through the use of γ-ATP-Sepharose and in vitro and in-gel kinase assays. Finally, genetic techniques including yeast deletions and RNAi facilitate the identification of in vivo substrates of a PP1/regulatory subunit complex. Future studies that incorporate these proteomics techniques should provide tremendous insight into phosphatase biology.

Most disease processes occur through aberrant abundance, activity, or regulation of proteins. Therefore, proteomic studies are likely to provide unique insight into the pathophysiology of a disease. A current aim of phosphatase research involves the development of drugs that may be applicable for the treatment of the wide array of human disorders for which PP1 is involved. Despite the participation of PP1 in a number of pathways, drug specificity may be obtained by specifically disrupting the interaction of a particular targeting subunit and PP1c, substrate targeting of the PP1 complex, or the regulation of targeting subunits. Indeed, a kinase inhibitor that decreases MYPT1 phosphorylation results in the relaxation of smooth muscle and thereby poses as a potential drug candidate for hypertension (Uehata et al. 1997). As our knowledge of the phosphatase biology expands, we are likely to see similar successes in drug targeting.

In just a few years, proteomics has made significant contributions to our understanding of individual proteins by using a battery of sophisticated genome-wide approaches. As our methodology and technology improve, we will undoubtedly acquire the ability to analyze complex and disparate data sets and integrate this information into meaningful biology.

Acknowledgments

We are very grateful to Elizabeth Herrick for figure making and design. We also thank Dr. Paul Graves, Dr. Michael Datto, and Dr. Jesse Kwiek for critical review of the manuscript.

References

Aebersold R, Goodlett DR (2001) Mass spectrometry in proteomics. Chem Rev 101:269-295

Aebersold R, Watts JD, Morrison HD, Bures EJ (1991) Determination of the site of tyrosine phosphorylation at the low picomole level by automated solid-phase sequence analysis. Anal Biochem 199:51-60

Allen PB, Kwon YG, Nairn AC, Greengard P (1998) Isolation and characterization of PNUTS, a putative protein phosphatase 1 nuclear targeting subunit. J Biol Chem 273:4089-4095

Alms GR, Sanz P, Carlson M, Haystead TA (1999) Reg1p targets protein phosphatase 1 to dephosphorylate hexokinase II in Saccharomyces cerevisiae: characterizing the effects of a phosphatase subunit on the yeast proteome. Embo J 18:4157-4168

Altschul SF, Gish W, Miller W, Meyers EW, Lipman DJ (1990) Basic Local Alignment Search Tool. Journal of Molecular Biology 215:403-410

Bennett D, Alphey L (2002) PP1 binds Sara and negatively regulates Dpp signaling in Drosophila melanogaster. Nat Genet 31:419-423

Bollen M (2001) Combinatorial control of protein phosphatase-1. Trends Biochem Sci 26:426-431

Borman MA, MacDonald JA, Muranyi A, Hartshorne DJ, Haystead TA (2002) Smooth muscle myosin phosphatase-associated kinase induces Ca2+ sensitization via myosin phosphatase inhibition. J Biol Chem 277:23441-23446

Boyle WJ, van der Geer P, Hunter T (1991) Phosphopeptide mapping and phosphoamino acid analysis by two-dimensional separation on thin-layer cellulose plates. Methods Enzymol 201:110-149

Campos M, Fadden P, Alms G, Qian Z, Haystead TA (1996) Identification of protein phosphatase-1-binding proteins by microcystin-biotin affinity chromatography. J Biol Chem 271:28478-28484

Ceulemans H, Stalmans W, Bollen M (2002) Regulator-driven functional diversification of protein phosphatase-1 in eukaryotic evolution. Bioessays 24:371-381

Cohen PT (2002) Protein phosphatase 1--targeted in many directions. J Cell Sci 115:241-256

Damer CK, Partridge J, Pearson WR, Haystead TA (1998) Rapid identification of protein phosphatase 1-binding proteins by mixed peptide sequencing and data base searching. Characterization of a novel holoenzymic form of protein phosphatase 1. J Biol Chem 273:24396-24405

Dawson JF, Holmes CF (1999) Molecular mechanisms underlying inhibition of protein phosphatases by marine toxins. Front Biosci 4:D646-658

Dent P, Campbell DG, Hubbard MJ, Cohen P (1989) Multisite phosphorylation of the glycogen-binding subunit of protein phosphatase-1G by cyclic AMP-dependent protein kinase and glycogen synthase kinase-3. FEBS Lett 248:67-72

Dunaief JL, King A, Esumi N, Eagen M, Dentchev T, Sung CH, Chen S, Zack DJ (2002) Protein Phosphatase 1 binds strongly to the retinoblastoma protein but not to p107 or p130 in vitro and in vivo. Curr Eye Res 24:392-396

Edman P (1949) A method for the determination of the amino acid sequence of peptides. Arch. Biochem. Biophys. 22:475-483

Egloff MP, Johnson DF, Moorhead G, Cohen PT, Cohen P, Barford D (1997) Structural basis for the recognition of regulatory subunits by the catalytic subunit of protein phosphatase 1. EMBO J 16:1876-1887

Eriksson JE, Toivola D, Meriluoto JA, Karaki H, Han YG, Hartshorne D (1990) Hepatocyte deformation induced by cyanobacterial toxins reflects inhibition of protein phosphatases. Biochem Biophys Res Commun 173:1347-1353

Eto M, Senba S, Morita F, Yazawa M (1997) Molecular cloning of a novel phosphorylation-dependent inhibitory protein of protein phosphatase-1 (CPI17) in smooth muscle: its specific localization in smooth muscle. FEBS Lett 410:356-360

Fenn JB, Mann M, Meng CK, Wong SF, Whitehouse CM (1989) Electrospray ionization for mass spectrometry of large biomolecules. Science 246:64-71

Fields S, Song O (1989) A novel genetic system to detect protein-protein interactions. Nature 340:245-246

Futcher B, Latter GI, Monardo P, McLaughlin CS, Garrels JI (1999) A sampling of the yeast proteome. Mol Cell Biol 19:7357-7368

Godovac-Zimmermann J, Brown LR (2001) Perspectives for mass spectrometry and functional proteomics. Mass Spectrom Rev 20:1-57

Graves PR, Haystead TA (2002) Molecular biologist's guide to proteomics. Microbiol Mol Biol Rev 66:39-63

Graves PR, Haystead TAJ (2003) A Functional Proteomics Approach to Signal Transduction. Recent Prog Horm Res 58:1-24

Graves PR, Kwiek JJ, Fadden P, Ray R, Hardeman K, Coley AM, Foley M, Haystead TA (2002) Discovery of novel targets of quinoline drugs in the human purine binding proteome. Mol Pharmacol 62:1364-1372

Gygi SP, Rochon Y, Franza BR, Aebersold R (1999) Correlation between protein and mRNA abundance in yeast. Mol Cell Biol 19:1720-1730

Hemmings HC, Jr., Greengard P, Tung HY, Cohen P (1984) DARPP-32, a dopamine-regulated neuronal phosphoprotein, is a potent inhibitor of protein phosphatase-1. Nature 310:503-505

Holmes CF, Maynes JT, Perreault KR, Dawson JF, James MN (2002) Molecular enzymology underlying regulation of protein phosphatase-1 by natural toxins. Curr Med Chem 9:1981-1989

Hsieh-Wilson LC, Allen PB, Watanabe T, Nairn AC, Greengard P (1999) Characterization of the neuronal targeting protein spinophilin and its interactions with protein phosphatase-1. Biochemistry 38:4365-4373

Huang D, Chun KT, Goebl MG, Roach PJ (1996) Genetic interactions between REG1/HEX2 and GLC7, the gene encoding the protein phosphatase type 1 catalytic subunit in Saccharomyces cerevisiae. Genetics 143:119-127

Hubbard MJ, Cohen P (1989) Regulation of protein phosphatase-1G from rabbit skeletal muscle. 1. Phosphorylation by cAMP-dependent protein kinase at site 2 releases catalytic subunit from the glycogen-bound holoenzyme. Eur J Biochem 186:701-709

Hunt DF, Yates JR, 3rd, Shabanowitz J, Winston S, Hauer CR (1986) Protein sequencing by tandem mass spectrometry. Proc Natl Acad Sci USA 83:6233-6237

Ichikawa K, Ito M, Hartshorne DJ (1996) Phosphorylation of the large subunit of myosin phosphatase and inhibition of phosphatase activity. J Biol Chem 271:4733-4740

Ito T, Ota K, Kubota H, Yamaguchi Y, Chiba T, Sakuraba K, Yoshida M (2002) Roles for the two-hybrid system in exploration of the yeast protein interactome. Mol Cell Proteomics 1:561-566

Jensen ON, Podtelejnikov AV, Mann M (1997) Identification of the components of simple protein mixtures by high-accuracy peptide mass mapping and database searching. Anal Chem 69:4741-4750

Karas M, Hillenkamp F (1988) Laser desorption ionization of proteins with molecular masses exceeding 10,000 daltons. Anal Chem 60:2299-2301

Klose J (1975) Protein mapping by combined isoelectric focusing and electrophoresis of mouse tissues. A novel approach to testing for induced point mutations in mammals. Humangenetik 26:231-243

MacDonald JA, Borman MA, Muranyi A, Somlyo AV, Hartshorne DJ, Haystead TA (2001) Identification of the endogenous smooth muscle myosin phosphatase-associated kinase. Proc Natl Acad Sci USA 98:2419-2424

MacDonald JA, Mackey AJ, Pearson WR, Haystead TA (2002) A strategy for the rapid identification of phosphorylation sites in the phosphoproteome. Mol Cell Proteomics 1:314-322

Mackey AJ, Haystead TAJ, Pearson WR (2002) Getting more from less: Algorithms for rapid protein identification with multiple short peptide sequences. Mol Cell Proteomics 1:139-147

Mann M, Hendrickson RC, Pandey A (2001) Analysis of proteins and proteomes by mass spectrometry. Annu Rev Biochem 70:437-473

Mann M, Hojrup P, Roepstorff P (1993) Use of mass spectrometric molecular weight information to identify proteins in sequence databases. Biol Mass Spectrom 22:338-345

Mann M, Ong SE, Gronborg M, Steen H, Jensen ON, Pandey A (2002) Analysis of protein phosphorylation using mass spectrometry: deciphering the phosphoproteome. Trends Biotechnol 20:261-268

Mann M, Wilm M (1994) Error-tolerant identification of peptides in sequence databases by peptide sequence tags. Anal Chem 66:4390-4399

Manning G, Whyte DB, Martinez R, Hunter T, Sudarsanam S (2002) The protein kinase complement of the human genome. Science 298:1912-1934

McAvoy T, Allen PB, Obaishi H, Nakanishi H, Takai Y, Greengard P, Nairn AC, Hemmings HC, Jr. (1999) Regulation of neurabin I interaction with protein phosphatase 1 by phosphorylation. Biochemistry 38:12943-12949

McLachlin DT, Chait BT (2001) Analysis of phosphorylated proteins and peptides by mass spectrometry. Curr Opin Chem Biol 5:591-602

Mikhailov A, Harmala-Brasken AS, Hellman J, Meriluoto J, Eriksson JE (2003) Identification of ATP-synthase as a novel intracellular target for microcystin-LR. Chem Biol Interact 142:223-237

Monshausen M, Rehbein M, Richter D, Kindler S (2002) The RNA-binding protein Staufen from rat brain interacts with protein phosphatase-1. J Neurochem 81:557-564

Moorhead G, MacKintosh RW, Morrice N, Gallagher T, MacKintosh C (1994) Purification of type 1 protein (serine/threonine) phosphatases by microcystin-Sepharose affinity chromatography. FEBS Lett 356:46-50

Muda M, Worby CA, Simonson-Leff N, Clemens JC, Dixon JE (2002) Use of double-stranded RNA-mediated interference to determine the substrates of protein tyrosine kinases and phosphatases. Biochem J 366:73-77

Neubauer G, Mann M (1999) Mapping of phosphorylation sites of gel-isolated proteins by nanoelectrospray tandem mass spectrometry: potentials and limitations. Anal Chem 71:235-242

O'Farrell PH (1975) High resolution two-dimensional electrophoresis of proteins. J Biol Chem 250:4007-4021

Pandey A, Mann M (2000) Proteomics to study genes and genomes. Nature 405:837-846

Pearson WR, Lipman DJ (1988) Improved tools for biological sequence comparison. Proc Natl Acad Sci USA 85:2444-2448

Rudenko A, Bennett D, Alphey L (2003) Trithorax interacts with type 1 serine/threonine protein phosphatase in *Drosophila*. EMBO Rep 4:59-63

Scheele GA (1975) Two-dimensional gel analysis of soluble proteins. Charaterization of guinea pig exocrine pancreatic proteins. J Biol Chem 250:5375-5385

Steen RL, Martins SB, Tasken K, Collas P (2000) Recruitment of protein phosphatase 1 to the nuclear envelope by A-kinase anchoring protein AKAP149 is a prerequisite for nuclear lamina assembly. J Cell Biol 150:1251-1262

Toivola DM, Eriksson JE, Brautigan DL (1994) Identification of protein phosphatase 2A as the primary target for microcystin-LR in rat liver homogenates. FEBS Lett 344:175-180

Topcu Z, Borden KL (2000) The yeast two-hybrid system and its pharmaceutical significance. Pharm Res 17:1049-1055

Tu J, Carlson M (1994) The GLC7 type 1 protein phosphatase is required for glucose repression in Saccharomyces cerevisiae. Mol Cell Biol 14:6789-6796

Tu J, Carlson M (1995) REG1 binds to protein phosphatase type 1 and regulates glucose repression in Saccharomyces cerevisiae. EMBO J 14:5939-5946

Uehata M, Ishizaki T, Satoh H, Ono T, Kawahara T, Morishita T, Tamakawa H, Yamagami K, Inui J, Maekawa M, Narumiya S (1997) Calcium sensitization of smooth muscle mediated by a Rho-associated protein kinase in hypertension. Nature 389:990-994

Vulsteke V, Beullens M, Waelkens E, Stalmans W, Bollen M (1997) Properties and phosphorylation sites of baculovirus-expressed nuclear inhibitor of protein phosphatase-1 (NIPP-1). J Biol Chem 272:32972-32978

Wasinger VC, Cordwell SJ, Cerpa-Poljak A, Yan JX, Gooley AA, Wilkins MR, Duncan MW, Harris R, Williams KL, Humphery-Smith I (1995) Progress with gene-product mapping of the Mollicutes: Mycoplasma genitalium. Electrophoresis 16:1090-1094

Wettenhall RE, Aebersold RH, Hood LE (1991) Solid-phase sequencing of 32P-labeled phosphopeptides at picomole and subpicomole levels. Methods Enzymol 201:186-199

Wilkins MR, Sanchez JC, Williams KL, Hochstrasser DF (1996) Current challenges and future applications for protein maps and post-translational vector maps in proteome projects. Electrophoresis 17:830-838

Wilm M, Shevchenko A, Houthaeve T, Breit S, Schweigerer L, Fotsis T, Mann M (1996) Femtomole sequencing of proteins from polyacrylamide gels by nano-electrospray mass spectrometry. Nature 379:466-469

Yan Z, Hsieh-Wilson L, Feng J, Tomizawa K, Allen PB, Fienberg AA, Nairn AC, Greengard P (1999) Protein phosphatase 1 modulation of neostriatal AMPA channels: regulation by DARPP-32 and spinophilin. Nat Neurosci 2:13-17

Yarmush ML, Jayaraman A (2002) Advances in proteomic technologies. Annu Rev Biomed Eng 4:349-373

10 Structure and function of the T-cell protein tyrosine phosphatase

Annie Bourdeau, Krista M. Heinonen, Daniel V. Brunet, Pankaj Tailor, Wayne S. Lapp, Michel L. Tremblay

Abstract

Protein tyrosine phosphatases (PTP) have gained recognition as important regulators of mammalian cell signaling. Among these, T-cell protein tyrosine phosphatase (TC-PTP) participates in the negative regulation of surface receptor signaling. Indeed, several members of the Jak/Stat family of molecules involved in cytokine and hormone receptor signaling have now been identified as substrates for this phosphatase. In addition, TC-PTP has recently been shown to exert a positive regulatory role on cell proliferation through the NF-κB pathway. The analysis of TC-PTP null mice has revealed an important function for this enzyme in hematopoiesis and immune regulation, as demonstrated by the impaired lymphocyte response to mitogenic stimuli. In addition, these mice display an inflammatory phenotype characterized by elevated levels of IFN-γ. The recent description of the three-dimensional structure and functional domains of TC-PTP provides an opportunity for the design of specific inhibitors of this phosphatase with potential therapeutic implications.

10.1 Cloning and characterization of T-cell protein tyrosine phosphatase (TC-PTP)

10.1.1 The gene

In the search for novel PTPs, human TC-PTP was identified by the screening of a human peripheral T-cell cDNA library with labeled oligonucleotides derived from the catalytic domain of the PTP1B protein tyrosine phosphatase. The cDNA sequence of human TC-PTP contains a full open reading frame of 1305 base pairs (bp), and a 978 bp 3'-untranslated region (Cool et al. 1989). Work describing the mapping of human TC-PTP to chromosome 18p11.2-p11.3 also identified two TC-PTP pseudogenes that were localized on chromosomes 1 and 13. Even if they had considerable sequence similarity with TC-PTP, neither of these pseudogenes could be translated in a single reading frame in order to generate a complete TC-PTP protein or any other known PTPs (Johnson et al. 1993).

Topics in Current Genetics, Vol. 5
J. Arino, D.R. Alexander (Eds.) Protein phosphatases
© Springer-Verlag Berlin Heidelberg 2004

Subsequently, the mouse homologue of TC-PTP, termed PTP-2 or MPTP was described (Miyasaka and Li 1992; Mosinger et al. 1992). The mouse gene is 88.8% identical to human TCPTP at the nucleotide level, and maps to chromosome 18, a region of synteny with the human TC-PTP locus. The murine cDNA sequence comprises 1570 bp and contains a 5'-untranslated region of 61 bp, a single full open reading frame of 1146 bp and a 3'-untranslated region of 61 bp that includes a polyadenylation signal located 90 bp upstream of the polyadenylation site (Mosinger et al. 1992).

A rat homologue of TC-PTP, termed PTP-S (for protein tyrosine phosphatase of spleen), was also cloned. Sequence analysis of rat TC-PTP revealed high sequence conservation with both mouse and human TC-PTPs (Reddy and Swarup 1995).

10.1.2 The promoter

To better understand the elements controlling and regulating TC-PTP, the promoter region of the murine TC-PTP gene was cloned and characterized using primer extension assays as well as S1 nuclease mapping techniques (Wee et al. 1999). Transfection of NIH-3T3 cells using various truncated promoter constructs fused to the CAT reporter gene revealed that the majority of the promoter activity is located within the first 147 bp of the TC-PTP promoter. This minimal portion of the promoter is required for functional gene expression. The 5' flanking region of the murine TC-PTP gene lacks the consensus TATAA and CAAT boxes, but possesses an initiator sequence typical of many housekeeping gene promoters. Indeed multiple transcriptional start sites were found located at the initiator sequence at 197-198 bp, 217 bp, 220 bp and 223 bp upstream from the translation initiation codon. Other cis-acting motifs included three GC-rich clusters that correlate with the presence of binding sites for the Sp1 and AP2 transcription factors. A computer-based analysis of the promoter sequence also identified putative binding sites for NF-κB, an APF binding site, two PEA3 binding sites and two c-myc recognition sequences. Therefore, this protooncogene might modulate transcription of TC-PTP and implicate TC-PTP in cell cycle progression. This hypothesis is further substantiated by the observation that peak expression of TC-PTP in late G1 phase is followed by marked repression. This pattern of expression was shown to require a repressor domain located between 492 and 1976 bp upstream from the transcriptional start site (Wee et al. 1999).

10.1.3 The protein

The cloning of the cDNA for human TC-PTP described an intracellular protein comprising 418 amino acid (aa) residues (Cool et al. 1989). Sequence comparison with the murine TC-PTP cDNA revealed 95% identity at the protein level, with the highest homology in the single catalytic domain (Mosinger et al. 1992). However, the human and murine TC-PTP open reading frames differed markedly in their 3' ends. This observation led to the recognition of a splice donor site (ACGT) in the human TC-PTP gene at the position where the sequence diverges from that of its murine counterpart. This splicing event generated two distinct mRNAs, termed TCPTPa and TC-PTPb of predicted protein size of 45 kDa and 48.5 kDa respectively (Champion-Arnaud et al. 1991). TC-PTPa mRNA contains a unique exon originating from the 3' end of the TC-PTP gene, which encodes a segment of 12 hydrophilic aa residues (Mosinger et al. 1992; Lorenzen et al. 1995). This transcript is the major gene product found in most human and mouse tissues (Kamatkar et al. 1996). TC-PTPb mRNA is less abundant in human cells and is absent in the mouse. The predicted protein contains an extra hydrophobic segment of 36 aa (residues 382- 418) at its carboxyl terminus (Mosinger et al. 1992; Lorenzen et al. 1995). In the rat TC-PTP there is also evidence for alternative splicing perhaps giving rise to more than two isoforms (Reddy and Swarup 1995).

The search for specific inhibitors of PTP1B for the treatment of type II diabetes and obesity, has provided the impetus for determining the three dimensional structure of TC-PTP, its closest homologue, as well as identifying key aa residues involved in binding of cognate substrates. Unfortunately, attempts at obtaining a crystal structure for the human TC-PTP have been unsuccessful, owing to an innate ability of human TC-PTP to homodimerize through interaction of residues 130–132, a region that has been termed the "DDQ loop" (Iversen et al. 2002). Interestingly, the mouse TC-PTP doesn't contain a 'DDQ' loop, but rather a 'DDR' loop in the same location. It remains to be tested if the mouse TC-PTP is more amenable to crystallization. Nevertheless, the structure for the first 321 aa residues of TC-PTP was obtained and compared to that of PTP1B. This comparison is justified by the high sequence identity between the two PTPs (65% overall identity and 74% identity in the catalytic domain) as well as nearly identical catalytic activity for p-nitrophenyl phosphate, suggesting a high degree of similarity between the two PTPs at the structural and functional levels (Iversen et al. 2002). The authors identified two areas in proximity to the active site pocket that differ between TC-PTP and PTP1B. One of these areas is found at these specific areas could account for the different substrate specificity of these PTPs, and might thus be used to develop selective inhibitors for each of these two enzymes. Another study by Asante-Appiah et al. (2001) examined the role of residues Tyr48, Arg49 and Asp50, termed the YRD motif, in determining substrate and inhibitor specificity in TC-PTP. The authors compared wild type TC-PTP with mutants bearing point mutations at position 49 or 50 in kinetics assays, and were thus able to show decreased substrate hydrolysis and affinity for the potent inhibitor N-benzoyl- L-glutamyl-[4-phosphono(difluoromethyl)]-L-phenylalanyl-[4-phosphono(difluoro methyl)]-phenylalanineamide in TC-PTP mutants. These findings complement a previous report by Jia et al. (1995) showing that the corresponding residues in PTP1B play an important role in substrate binding and orientation within the enzyme's active site. Together, these studies identify the YRD motif as a target for the design of specific inhibitors of TC-PTP and PTP1B.

10.2 Expression and localization of TC-PTP

10.2.1 TC-PTP tissue expression

Few studies have been reported on the regulation of TC-PTP expression. TC-PTP is ubiquitously expressed in embryonic and adult tissues. Northern blot analysis of total RNA isolated from different tissues revealed that mouse TC-PTP mRNA is ~1.9 kb in size. Its highest expression is found in lymphoid cells (You-Ten et al. 1997). TC-PTP is also expressed abundantly in the ovaries, testes, thymus and kidneys (Mosinger et al. 1992), and at lower levels in spleen, muscle, liver, heart, and brain (Miyasaka and Li 1992). The expression of mouse TC-PTP is low in embryonic stem cells and increases during the later stages of development (Mosinger et al. 1992). A second transcript of ~1.3 kb was detected exclusively in the testes (Mosinger et al. 1992).

Expression of intracellular TC-PTP was also measured by PCR analysis in various human lymphoid organs and cells. TC-PTP was expressed at high levels in tonsil tissue and in Jurkat T cells; moderate levels were found in bone marrow, lymph nodes and spleen; low levels were detected in fetal liver, peripheral blood lymphocytes and in the thymus (Gjorloff-Wingren et al. 2000). Mitogenic stimulation of T cells with Concanavalin-A increased TC-PTP mRNA level by three-fold. Moreover, the half-life of TC-PTP mRNA in resting T cells is approximately 25 min, but is increased to 5 h upon mitogenic stimulation (Rajendrakumar et al. 1993).

As suggested by the presence of c-myc binding sites within the TC-PTP promoter (Wee et al. 1999), TC-PTP might be implicated in cell cycle regulation. In COS cells, steady-state levels of TC-PTP transcripts were observed to fluctuate in a cell-cycle-dependent manner with peak levels in late G1 and marked repression thereafter (Tillmann et al. 1994). Similar results were obtained using rat fibroblasts and HeLa cells, where TC-PTP mRNA levels were increased three to four fold compared with serum-starved cells. TC-PTP transcript levels peaked 8 h after serum addition, corresponding to the late G1 phase of the cell cycle and decreased in the S phase (Radha et al. 1997). Together, promoter analysis and studies of mRNA expression suggest that TC-PTP is important for cell cycle regulation.

10.2.2 Intracellular localization of TC-PTP

Several groups suggested that localization of TC-PTP is one of the key elements leading to substrate specificity. As a nuclear localization signal (NLS) has been found in the non-catalytic C-terminus domain of TC-PTP, nuclear localization has been studied in the last decade using a variety of biochemical and immunofluorescence approaches. The NLS of the murine 45 kDa TC-PTP is composed of three basic aa clusters: 345 RKR 347, 352 RK 353, and 372 RKRKR 376. In overexpression studies, several laboratories have localized the 45 kDa form to the nucleus by immunofluorescence, whereas the 48.5 kDa TC-PTP localized to the endoplasmic reticulum (Cool et al. 1989; Tillmann et al. 1994; Lorenzen et al. 1995).

Our group has also confirmed the nuclear localization of the 45 kDa isoform using monoclonal antibodies (mAbs) raised against TC-PTP (Tailor et al. 2003 submitted). All three basic clusters of the NLS were required for nuclear localization (Tillmann et al. 1994) and the presence of a typical casein kinase recognition site at the threonine 335 of the TXXE motif was also recognized. This sequence is often found in front of a NLS and its phosphorylation is believed to modulate the rate of nuclear transport. However, in similar experiments, Lorenzen et al. (1995) identified the core NLS of human TCPTP as 377 RKRKR 381, also known as basic cluster III. Point mutations of K380Q and R381Q were found to abolish nuclear transport of the 45 kDa form of TC-PTP. The study by Tiganis et al. (1997) reaffirmed the importance of basic clusters I and II for nuclear localization, but also concluded that cluster III is necessary for efficient localization of the 45 kDa.

The report by Tillman et al. (1994) also noted that deletion of basic cluster III resulted in nucleolar localization of the 45 kDa TC-PTP. Our unpublished results using anti-TC-PTP mAbs also suggested the nucleolar localization of this PTP. Using overexpression studies and photobleaching experiments, we also observed that the 45 kDa was dynamically shuttling between nucleoplasm and nucleolus in COS7, MCF7, and HeLa cells. Interestingly, nucleolar localization was regulated by phosphorylation in the RNRNRYRVDSP motif located at the amino terminus of the 45 kDa by AKT kinase. Active nucleolar shuttling of the 45 kDa isoform may serve to regulate its access to potential substrates within the nucleoplasm and the nucleolus (Tailor et al. 2003 submitted).

Intracellular distribution of TC-PTP is altered in response to certain cellular stresses as well as mitogenic stimuli. Whereas in resting cells the 45 kDa isoform is localized almost exclusively within the nucleus, it readily shuttles to the cytoplasm in response to hyperosmotic stress (Lam et al. 2001). This process is achieved by passive diffusion and perhaps also by inhibition of its nuclear import. Remarkably, only stress-activated signaling pathways involving the AMP-activated protein kinase caused cytoplasmic accumulation of the 45 kDa (Lam et al. 2001). Work by Tiganis et al. (1998, 1999) has shown that the 45 kDa form can also translocate to the cytoplasm upon mitogenic stimulation. Through the use of immunofluorescence and heterokaryon studies, an significant proportion of the 45 kDa isoform was found to migrate from the nucleus to the cytoplasm when cells were treated with EGF. This relocalization was shown to be dependent on the EGFR tyrosine kinase activity, since DAPH (a specific inhibitor of the EGFR-PTK) inhibited both EGFR autophosphorylation as well as 45 kDa relocalization. Interestingly, an activated Src mutant (Src Y527F) did not trigger the relocalization of the 45 kDa isoform, indicating that this mechanism is tyrosine phosphorylation independent and linked to specific signaling pathways (Tiganis et al. 1998; Tiganis et al. 1999). Relocalization to the cytoplasm might allow the 45 kDa isoform to target proteins associated with the cytoplasmic membrane.

10.3 Analysis of TC-PTP function in vitro

10.3.1 Identification of downstream substrates

Substrate identification is a crucial step in delineating the signaling pathways regulated by TCPTP in vivo. To attain this goal, a substrate trapping approach was developed by the Tonks laboratory (Cold Spring Harbor) and employed by many groups to identify physiological targets of TC-PTP. Catalytic domain mutants of TC-PTP have been generated by mutation of Asp→Ala (DA) or Cys→Ser (CS). These mutations ablate the ability of TC-PTP to dephosphorylate target substrates but do not affect substrate binding. These mutants can form stable enzyme-substrate complexes, allowing identification of the substrate by co-immunoprecipitation with TC-PTP.

The majority of TC-PTP substrates identified to date belong to the Janus kinase (Jak) family or the signal transducer and activator of transcription (Stat) family of signal transduction molecules (Fig. 1). Simoncic et al. (2002) transfected CTLL-1 cells with the substrate-trapping mutant TC-PTP/DA. These cells also express the IL-2R, which signals through Jak1 and Jak3. Following incubation in the presence of IL-2, both Jak1 and Jak3 could be coimmunoprecipitated with TC-PTP/DA. This interaction was specific to TC-PTP, since substrate-trapping mutants of the highly homologous PTP-PEST and PTP1B did not co-immunoprecipitate either Jak1 or Jak3 in similar experiments. The association of TC-PTP with Jak1 required contact with the catalytic domain of TC-PTP since addition of orthovanadate completely abolished this interaction. The ability of TC-PTP to bind Jak1 following cytokine receptor activation was verified in human 293T cells (Simoncic et al. 2002). These cells express the IFN-γ-R, which signals through Jak1 and Jak2. Following treatment with IFN-γ, Jak1 but not Jak2 coimmunoprecipitated with TC-PTP/DA, although both Jak kinases were activated. Propagation of signals downstream of cytokine receptors involves Stat phosphorylation by Jak kinases (Aaronson and Horvath 2002). In addition, several papers have demonstrated that Stat1, Stat3, and Stat5a/5b are also substrates for TC-PTP. This indicates that TC-PTP functions as a major negative regulator of cytokine signaling (Haspel and Darnell 1999; McBride et al. 2000; Ibarra-Sanchez et al. 2001; ten Hoeve et al. 2002; Zhu et al. 2002; Yamamoto et al. 2002; Simoncic et al. 2002; Aoki et al. 2002; Aoki and Matsuda 2000).

Previous studies demonstrated the existence of a nuclear phosphatase activity that could inactivate Stat1 (Haspel and Darnell 1999; McBride et al. 2000; Ibarra-Sanchez et al. 2001). To determine whether TC-PTP might be responsible for the dephosphorylation of Stat1 in the nucleus, ten Hoeve et al. (2002) prepared nuclear extracts from HeLa cells and then depleted them of TC-PTP using a TC-PTP mAb. Phosphorylated Stat1 was not inactivated when incubated with the TC-PTP depleted nuclear extracts. Significantly, no other phosphatase could substitute for TC-PTP in regulating the activity of Stat1 within the nucleus. The ability of TCPTP to regulate Stat1 was subsequently examined in 293T cells expressing increasing amounts of TC-PTP. These experiments showed that IFN-γ-induced phosphorylation of Stat1 was inhibited by TC-PTP in a dose-dependent manner

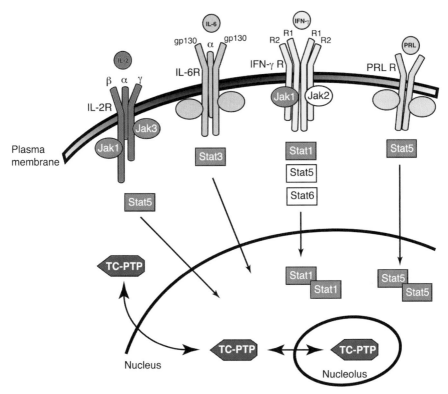

Fig. 1. Downstream substrates of TC-PTP. TC-PTP has been shown to negatively regulate several signaling molecules belonging to the cytokine pathway. Upon ligand binding, IL-2R (blue), IL-6R (green), IFN-γR (yellow) and PRLR (orange) associated Jaks become activated. These Jaks to its cognate mediate phosphorylation of specific tyrosine residues and recruit Stats. Activated Stats are released from the receptor and translocate to the nucleus. The 45 kDa form of TC-PTP was shown to dephosphorylate Jak1 and Jak3 as well as Stat1, Stat3, and Stat5 (molecules illustrated in red). To gain access to its substrates (red), TC-PTP shuttling has been demonstrated from the cytoplasm to the nucleus and to the nucleolus.

(ten Hoeve et al. 2002). Moreover, transfection of TC-PTP in a human osteosarcoma cell line caused a decrease in IFN-γ-induced phosphorylation of Stat1 compared to the parental cell line (ten Hoeve et al. 2002). Corroborating evidence was also obtained using TC-PTP-/- murine embryonic fibroblasts (MEF) and primary thymocytes wherein IFN-γ-induced phosphorylation of Stat1 was increased compared to their wild type counterparts. This phenotype was rescued with transfection of the 45 kDa form of TC-PTP but not the 48.5 kDa. The activity of TC-PTP was specific for Stat1, as the activity of Stat5 and Stat6 was not affected by the presence or absence of TC-PTP (ten Hoeve et al. 2002). Zhu et al. (2002) investigated the nature of the protein-protein interaction in the dephosphorylation of Stat1 by TC-PTP. Specifically, the authors examined the role of Arg methylation by performing pulse-chase experiments in U266 cells or human primary fibro-blasts, in the absence or presence of methylation inhibitors. Lack of Arg methyla-tion resulted in a significant delay in the dephosphorylation of Stat1 on Tyr701. The authors showed that lack of Arg methylation favored association of Stat1 with TC-PTP over PIAS1 (an inhibitor of Stat1). Thus, Arg methylation of Stat1 nega-tively regulates its interaction with PIAS1 and TC-PTP, thereby increasing the du-ration of Stat1 tyrosine phosphorylation.

Another member of the Stat family, Stat3, was also shown to be a substrate for TC-PTP. Yamamoto et al. (2002) studied the activity of Stat3 in 293T cells trans-fected with WT TCPTPa, or inactive TC-PTP/DA or TC-PTP/CS mutants. Stat3 activation was achieved by stimulation with IL-6 and soluble IL-6Rα. Whereas Stat3 was dephosphorylated in cells expressing WT TC-PTP, Stat3 remained phosphorylated in cells expressing either of the inactive TC-PTP mutants. These results suggest that TC-PTP is essential for dephosphorylation of Stat3. In addi-tion, the authors showed that retrovirus-mediated overexpression of TC-PTPa suppressed IL-6-induced growth arrest in M1 myeloid leukemia cells. Further-more, TC-PTP/Stat3 complexes were observed in the nucleus of M1 cells. To-gether, these results indicate that TCPTP inhibits IL-6 signaling through dephos-phorylation of Stat3.

Regulation of Stat5 by TC-PTP following IL-2R signaling in CTLL-1 cells was examined by Simoncic et al. (2002). Following IL-2 stimulation, Stat5 was hyper-phosphorylated in TCPTP-/- cells, whereas hypophosphorylation was observed in cells overexpressing TC-PTP. Hormonal receptors are also able to activate the Jak/Stat pathway. For instance, the binding of prolactin (PRL) to its receptor re-sults in activation of Jak2 and Stat5. Aoki et al. (2002) demonstrated that TC-PTP was a negative regulator of PRL-mediated signal transduction by specifically dephosphorylating and deactivating Stat5 in the nucleus. The authors had shown in a previous study that PTP1B could dephosphorylate Stat5a and Stat5b in trans-fected COS-7 cells as well as in mammary epithelial COMMA-1D cells, thereby regulating the PRL signaling pathway (Aoki and Matsuda 2000). Hence, it is likely that negative regulation of several signaling pathways is mediated by TC-PTP but in combination with other PTPs.

Other downstream substrates of TC-PTP have also been identified but less characterized. TC-PTP inhibited Epidermal growth factor (EGF)-induced associa-tion of p52Shc with Grb2 (Tiganis et al. 1998). More recently, it was demon-strated using 293 cells, that TC-PTP could dephosphorylate the insulin-receptor β subunit and downregulate insulin-induced signaling (Galic et al. 2003).

10.3.2 Cell cycle regulation and TC-PTP

The ubiquitous nature of TC-PTP, combined with its preference for nuclear local-ization, suggests a possible role in cell cycle regulation. In support of this notion, TC-PTP mRNA levels have been shown to fluctuate during the cell cycle, begin-ning with an initial increase in G0 and early G1, followed by a reduction for the remainder of the cycle (Tillmann et al. 1994). In addition, overexpression studies of TC-PTP (Radha et al. 1997), as well as analyses of knockout cell lines and murine embryonic fibroblasts (MEFs) (Ibarra-Sanchez et al. 2001) all point to a positive role for TC-PTP in cellular proliferation.

Our group compared the proliferation of TC-PTP-/- MEFs with that of their WT counterparts (Ibarra-Sanchez et al. 2001). We observed that MEFs lacking TC-PTP proliferated slower compared to controls. Cell cycle analysis by flow cytometry uncovered a slower progression through the G1 phase in cells lacking TC-PTP. This was shown to result from delayed induction of cyclin D1, and prolonged expression at high levels of the cyclin-dependent kinase (cdk) inhibitor p27KIP1, whose degradation is essential for progression through G1. Slower induction of cyclin D1 also delayed the phosphorylation of the retinoblastoma protein (Rb) as well as cdk2, two events that prime the cell for continuation through the cell cycle.

A delayed activation of NFκB and a reduction in the activity of IκB kinase (IKKβ) were observed in TC-PTP-/- MEFs following PDGF signaling, as compared to wild type cells (Ibarra- Sanchez et al. 2001). Since NFκB is known to be involved in the activation of cyclin D1 (Guttridge et al. 1999; Bharti et al. 2003), this finding suggested that TC-PTP might exert a positive regulatory effect on the cell cycle by enhancing signaling through the NF-κB pathway. In agreement with this possibility, reintroduction of WT TC-PTP in TC-PTP-/- MEFs rescued the defective proliferation, cyclin D1 expression, NF-κB activation, as well as IκB phosphorylation (Ibarra-Sanchez et al. 2001).

10.4 Analysis of TC-PTP function in vivo

10.4.1 TC-PTP knockout mice

To better elucidate the physiological role of TC-PTP, we generated TC-PTP null mutant mice using gene targeting technology. The mutant allele contains a deletion of the catalytic domain and results in a loss of expression of the gene product (You-Ten et al. 1997). Offspring from heterozygous matings fall into the expected Mendelian distribution (1 : 2 :1), which indicates that the null mutation is not embryonic lethal. Moreover, the TC-PTP-/- mice appear physically normal until 10-14 days of age, at which time they begin to show signs of growth retardation. At three weeks, they exhibit weight loss, piloerection, hunched posture, and diarrhea. They also display defective hematopoiesis and immune function, characterized by splenomegaly, lymphadenopathy, anemia, and thymic atrophy. All TC-PTP-/- mice die at 3-5 weeks after birth (You-Ten et al. 1997).

10.4.2 TC-PTP and hematopoiesis

The higher level of TC-PTP expression in the hematopoietic system led us to hypothesize that TC-PTP may have a role in either hematopoiesis or immune function. Three-week-old TC-PTP-/- mice show a three to four-fold increase in spleen and lymph node cellularity. In the lymph nodes, this could mostly be attributed to a steady increase in the percentage of B220+ cells seen as early as 13 days after birth (You-Ten et al. 1997). Conversely, there was a decrease in the percentage of B cells in the spleen, although the total number of B220+ cells was not changed due to the increase in overall cellularity. The red pulp of the TC-PTP-/- spleen was expanded, consistent with sequestration of red blood cells and correlating with the decreased hematocrit. The bone marrow cellularity decreased strikingly between days 17 and 21, accompanied by depletion of B220+ cells. Therefore, it seems that there is a switch toward extramedullary hematopoiesis, which cannot fully populate the various compartments. In terms of T cell numbers, there was a substantial decrease in thymic double positive cells on day 21. However, the decrease in the absolute numbers of peripheral T cells was modest and the CD4+/CD8+ ratio was essentially unchanged (You-Ten et al. 1997).

Bone marrow transplantation experiments suggested that the hematopoietic defect was not due to a problem with the stem cells but rather with the bone marrow stroma (You-Ten et al. 1997). Irradiated +/+ recipients were fully reconstituted with -/- bone marrow, whereas +/+ bone marrow graft did not rescue -/- recipients. Two months after the graft, there was no difference in T and B cell numbers in different organs between irradiated +/+ mice that had received +/+ or -/- bone marrow. Their hematocrit was normal, and there was no increase in the size of either spleen or lymph nodes. However, the functional defects seen in -/- mice were still present in the +/+ mice reconstituted with -/- bone marrow: mature T and B cells from the spleen of the animals failed to proliferate in response to mitogenic stimuli (ConA or anti-CD3ε for T cells and LPS for B cells).

There is no overt defect in TC-PTP+/- animals: they are generally healthy and fertile past 12 months of age. The percentages of various lymphoid and myeloid populations follow closely those seen in +/+ mice, and the mitogen and plaque-forming cell responses are normal (You-Ten et al. 1997). This indicates that one normal allele is sufficient to sustain hematopoiesis, proliferative T and B cell responses as well as T cell-dependent B cell responses. However, it is important to note that these animals are kept in a pathogen free environment. It remains to be evaluated if they would respond equally well as wild type animals to challenge by an infectious agent.

10.4.3 TC-PTP and immune function

In addition to lymphopoietic abnormalities, there are functional defects in the T and B cell compartments. The mitogen response of TC-PTP-/- splenocytes to ConA and LPS is severely impaired. There is also a defect in T cell-dependent B cell responses as shown by decreased number of plaque-forming colonies from TC-PTP-/- spleens after immunization with sheep RBCs (You-Ten et al. 1997). Part of the defect in mitogen assays may stem from the inherent proliferative defect of -/- cells (Ibarra-Sanchez et al. 2001). Splenic T cells from -/- mice produce less IL-2 than their +/+ counterparts (Ibarra-Sanchez et al. 2000); however, the addition of exogenous IL-2 does not rescue the phenotype. Stimulation with anti-CD3ε is followed by normal tyrosine phosphorylation patterns and calcium flux, providing evidence that the immediate events downstream of the T cell receptor are not affected. Nevertheless, stimulating purified populations of TC-PTP-/- T or B cells partially rescues the phenotype, suggesting that although TC-PTP is necessary for rapid progression through the cell cycle, there may be other factors contributing to the lack of proliferation in mixed cultures (Dupuis et al. 2003).

Recent data from our laboratory provide an explanation for the above findings. Culturing purified populations of Gr1+ TC-PTP-/- cells with TC-PTP+/+ T cells produced a similar defect in proliferation in response to anti-CD3ε as that seen in -/- spleen cultures (Dupuis et al. 2003). This was shown to correlate with NO production and was preventable by adding iNOS inhibitors or anti-IFN-γ to the culture medium. Induction of NO production was also contact-dependent, suggesting that other signals in addition to IFN-γ was necessary to induce this response in the Gr-1+ cells. One example would be the engagement of co-stimulatory molecules, as CD80 is 14 upregulated on TC-PTP-/- Gr-1+ cells (Dupuis et al. 2003). It must be noted that TC-PTP+/+ Gr-1+ cells did not have the same effect on T cell proliferation in vitro, correlating with the hypothesis that the -/- cells have a higher activation status.

Our in vivo studies support these findings, showing increased production of proinflammatory and inflammatory cytokines in TC-PTP-/- mice (Heinonen et al. 2003 submitted). This increase could be seen as early as three days after birth in the liver and spread from the liver to the periphery, including non-lymphoid organs. The increased production of IFN-γ in vivo accounts for the activated phenotype of splenic macrophages and Gr-1+ cells (Gifford and Lohmann-Matthes 1987). Moreover, TC-PTP-/- mice displayed augmented LPS sensitivity, showing symptoms of septic shock after receiving much lower doses of LPS than +/+ mice. The production of inflammatory cytokines would also explain the lack of T cell-dependent B cell responses, which are mainly of the Th2-type, and in vivo NO would contribute to the immunosuppression and thymic atrophy (Bobe et al. 1999; Angulo et al. 2000).

In summary, data from our laboratory clearly show that TC-PTP plays an important role both in immune development and in immune function. This is most likely mediated via regulation of cytokine receptor signaling and, to a certain extent, through cell cycle control.

10.4.4 p53 and TC-PTP

p53 is a tumor-suppressor protein whose mechanism of action remains unclear at present (Sharpless and DePinho 2002) In p53-/- mice, tumors develop spontaneously by 6 months of age (Donehower et al. 1992). To explain this phenotype, several groups have demonstrated that the loss of p53 causes genomic instability (Bender et al. 2002; Dixon and Norbury 2002). Generally, it is postulated that p53 acts as a regulatory molecule, which determines whether a cell containing damaged DNA undergoes cell cycle arrest or apoptosis (Lane 1992). In support of this hypothesis, transgenic mice containing 3 or 4 functional copies of p53 show a strong response to DNA damage, and increased resistance to cancer without accelerating the normal aging process (Garcia-Cao et al. 2002).

A possible functional link between p53 and TC-PTP was suggested by the observation that overexpression of TC-PTP in p53+/+ COS cells induced apoptosis, whereas p53-/- cells were resistant (Radha et al. 1999). Furthermore, TC-PTP overexpression increased both transcriptional activity of the p53 gene and p53 protein levels, as evidenced by enhanced expression of the 15 caspase-1 gene (Gupta et al. 2002). Together, these results show that TC-PTP can induce p53 expression, leading to increased apoptosis. To verify this hypothesis, our group has generated mice deficient for both TC-PTP and p53. These animals survived on average one week longer than the TC-PTP-/- mice (Ibarra-Sanchez and Tremblay unpublished data). Thus, lack of p53 partially mitigates the phenotype of TC-PTP-/- mice. This result also implies that absence of TC-PTP results in cellular anomalies that cannot be compensated by abrogation of p53 function alone. Additional studies are needed to characterize the function of TC-PTP *in vivo*.

10.5 Conclusion

Recent work in cell lines has uncovered an important role for TC-PTP as a negative regulator of cytokine and hormone receptor signaling. Moreover, TC-PTP has been implicated in cell cycle regulation, and the NFκB signaling cascade has been identified as one pathway through which TC-PTP may exert its regulatory effect. Additional work is required to identify other mitogen signaling pathways wherein TC-PTP may also participate. Characterization of the protein structure has identified key aa residues involved in determining substrate specificity and catalytic activity of TC-PTP. This work opens the door to the rational design of specific inhibitors of TCPTP, providing a novel therapeutic approach for disorders in the field of oncology, infection and cardiovascular diseases. For example, specific inhibitors of TC-PTP could delay the activation of NFκB by decreasing IκB activity thus reducing the possibility of skin cancer (Dajee et al 2003). Secondary mycotic infections especially seen in AIDS patients could also be treated with inhibitors of TC-PTP. It could be envisaged that the decreased expression of TC-PTP would increase IFN-γ activation thus modulating fungicidal macrophages and reduce the formation of granulomas (Sisto et al 2003). Another application would use TC-PTP inhibitors to enhance the Jak-Stat pathway that has been reported deficient in patients with end-stage dilated cardiomyopathies (Podewshi et al 2003). Finally, insight into the role played by TC-PTP in vivo has been gained through the study of TC-PTP null mice. It is now clear that this phosphatase is crucial for hematopoietic development as well as maintaining the immune system in homeostatic balance. Future studies will aim to determine the relative contribution of hematopoietic and stromal cell defects to the phenotype of TC-PTP null mice.

Acknowledgments

This research has been supported by an operating grant from the National Cancer Institute of Canada to MLT. AB is supported by a cancer immunology postdoctoral fellowship from the Canadian Research Institute of New York, KMH received a Canadian Institutes of Health Research (CIHR) Cancer Consortium training award and MLT is a Scientist from the CIHR. We thank Dr. Sebastien Trop for critical review of this manuscript.

References

Aaronson DS, Horvath CM (2002) A road map for those who know JAK-STAT. Science 296:1653-1655

Angulo I, de las Heras FG, Garcia-Bustos JF, Gargallo D, Munoz-Fernandez MA, Fresno M (2000) Nitric oxide-producing CD11b(+)Ly-6G(Gr-1)(+)CD31(ER-MP12)(+) cells in the spleen of cyclophosphamide-treated mice: implications for T-cell responses in immunosuppressed mice. Blood 95:212-220

Aoki N, Matsuda T (2000) A cytosolic protein-tyrosine phosphatase PTP1B specifically dephosphorylates and deactivates prolactin-activated STAT5a and STAT5b. J Biol Chem 275:39718-39726

Aoki N, Matsuda T (2002) A nuclear protein tyrosine phosphatase TC-PTP is a potential negative regulator of the PRL-mediated signaling pathway: dephosphorylation and de-activation of signal transducer and activator of transcription 5a and 5b by TC-PTP in nucleus. Mol Endocrinol 16:58-69

Asante-Appiah E, Ball K, Bateman K, Skorey K, Friesen R, Desponts C, Payette P, Bayly C, Zamboni R, Scapin G, Ramachandran C, Kennedy BP (2001) The YRD motif is a major determinant of substrate and inhibitor specificity in T-cell protein-tyrosine phosphatase. J Biol Chem 276:26036-26043

Bender CF, Sikes ML, Sullivan R, Huye LE, Le Beau MM, Roth DB, Mirzoeva OK, Oltz EM, Petrini JH (2002) Cancer predisposition and hematopoietic failure in Rad50(S/S) mice. Genes Dev 16:2237-2251

Bharti AC, Donato N, Singh S, Aggarwal BB (2003) Curcumin (diferuloylmethane) down-regulates the constitutive activation of nuclear factor-kappa B and Ikappa Balpha kinase in human multiple myeloma cells, leading to suppression of proliferation and induction of apoptosis. Blood 101:1053-1062

Bobe P, Benihoud K, Grandjon D, Opolon P, Pritchard LL, Huchet R (1999) Nitric oxide mediation of active immunosuppression associated with graft-versus-host reaction. Blood 94:1028-1037

Champion-Arnaud P, Gesnel MC, Foulkes N, Ronsin C, Sassone-Corsi P, Breathnach R (1991) Activation of transcription via AP-1 or CREB regulatory sites is blocked by protein tyrosine phosphatases. Oncogene 6:1203-1209

Cool DE, Tonks NK, Charbonneau H, Walsh KA, Fischer EH, Krebs EG (1989) cDNA isolated from a human T-cell library encodes a member of the protein-tyrosine-phosphatase family. Proc Natl Acad Sci USA 86:5257-5261

Dajee M, Lazaeov M, Zhang JY, Cai T, Green CL, Russell AJ, Marinkovich MP, Tao S, Lin S, Lin Q, Kubo Y, Khavari PA (2003)NF-kappaB blockade and oncogenic Ras trigger invasive human epidermal neoplasis. Nature 421:639-643

Dixon H, Norbury CJ (2002) Therapeutic Exploitation of Checkpoint Defects in Cancer Cells Lacking p53 Function. Cell Cycle 1:362-368

Donehower LA, Harvey M, Slagle BL, McArthur MJ, Montgomery CA, Jr., Butel JS, Bradley A (1992) Mice deficient for p53 are developmentally normal but susceptible to spontaneous tumours. Nature 356:215-221

Dupuis M, Ibarra-Sanchez MJ, Tremblay ML, Duplay P (2003) Gr-1+ myeloid cells lacking TC-PTP inhibit lymphocyte proliferation by IFN-gamma and NO-dependent mechanism. J Immunol 171:726-732

Galic S, Klingler-Hoffmann M, Fodero-Tavoletti MT, Puryer MA, Meng TC, Tonks NK, Tiganis T (2003) Regulation of insulin receptor signalin by the protein tyrosine phosphatase TCPTP. Mol Cell Biol 23:2096-2108

Garcia-Cao I, Garcia-Cao M, Martin-Caballero J, Criado LM, Klatt P, Flores JM, Weill JC, Blasco MA, Serrano M (2002) "Super p53" mice exhibit enhanced DNA damage response, are tumor resistant and age normally. EMBO J 21:6225-6235

Gifford GE, Lohmann-Matthes ML (1987) Gamma interferon priming of mouse and human macrophages for induction of tumor necrosis factor production by bacterial lipopolysaccharide. J Natl Cancer Inst 78:121-124

Gjorloff-Wingren A, Saxena M, Han S, Wang X, Alonso A, Renedo M, Oh P, Williams S, Schnitzer J, Mustelin T (2000) Subcellular localization of intracellular protein tyrosine phosphatases in T cells. Eur J Immunol 30:2412-2421

Gupta S, Radha V, Sudhakar C, Swarup G (2002) A nuclear protein tyrosine phosphatase activates p53 and induces caspase-1-dependent apoptosis. FEBS Lett 532:61-66

Guttridge DC, Albanese C, Reuther JY, Pestell RG, Baldwin AS, Jr. (1999) NF-kappaB controls cell growth and differentiation through transcriptional regulation of cyclin D1. Mol Cell Biol 19:5785-5799

Haspel RL, Darnell JE, Jr. (1999) A nuclear protein tyrosine phosphatase is required for the inactivation of Stat1. Proc Natl Acad Sci USA 96:10188-10193

Ibarra-Sanchez MJ, Simoncic PD, Nestel FR, Duplay P, Lapp WS, Tremblay ML (2000) The T-cell protein tyrosine phosphatase. Semin Immunol 12:379-386

Ibarra-Sanchez MJ, Wagner J, Ong MT, Lampron C, Tremblay ML (2001) Murine embryonic fibroblasts lacking TC-PTP display delayed G1 phase through defective NF-kappaB activation. Oncogene 20:4728-4739

Iversen LF, Andersen HS, Moller KB, Olsen OH, Peters GH, Branner S, Mortensen SB, Hansen TK, Lau J, Ge Y, Holsworth DD, Newman MJ, Hundahl Moller NP (2001) Steric hindrance as a basis for structure-based design of selective inhibitors of protein-tyrosine phosphatases. Biochemistry 40:14812-14820

Iversen LF, Moller KB, Pedersen AK, Peters GH, Petersen AS, Andersen HS, Branner S, Mortensen SB, Moller NP (2002) Structure determination of T cell protein-tyrosine phosphatase. J Biol Chem 277:19982-19990

Jia Z, Barford D, Flint AJ, Tonks NK (1995) Structural basis for phosphotyrosine peptide recognition by protein tyrosine phosphatase 1B. Science 268:1754-1758

Johnson CV, Cool DE, Glaccum MB, Green N, Fischer EH, Bruskin A, Hill DE, Lawrence JB (1993) Isolation and mapping of human T-cell protein tyrosine phosphatase sequences: localization of genes and pseudogenes discriminated using fluorescence hybridization with genomic versus cDNA probes. Genomics 16:619-629

Kamatkar S, Radha V, Nambirajan S, Reddy RS, Swarup G (1996) Two splice variants of a tyrosine phosphatase differ in substrate specificity, DNA binding, and subcellular location. J Biol Chem 271:26755-26761

Lam MH, Michell BJ, Fodero-Tavoletti MT, Kemp BE, Tonks NK, Tiganis T (2001) Cellular stress regulates the nucleocytoplasmic distribution of the protein-tyrosine phosphatase TCPTP. J Biol Chem 276:37700-37707

Lane DP (1992) Cancer p53, guardian of the genome. Nature 358:15-16

Lorenzen JA, Dadabay CY, Fischer EH (1995) COOH-terminal sequence motifs target the T cell protein tyrosine phosphatase to the ER and nucleus. J Cell Biol 131:631-643

McBride KM, McDonald C, Reich NC (2000) Nuclear export signal located within theDNA-binding domain of the STAT1transcription factor. EMBO J 19:6196-6206

Miyasaka H, Li SS (1992) Molecular cloning, nucleotide sequence and expression of a cDNA encoding an intracellular protein tyrosine phosphatase, PTPase-2, from mouse testis and T-cells. Mol Cell Biochem 118:91-98

Mosinger B, Jr., Tillmann U, Westphal H, Tremblay ML (1992) Cloning and characterization of a mouse cDNA encoding a cytoplasmic protein-tyrosine-phosphatase. Proc Natl Acad Sci USA 89:499-503

Podewski EK, Hilfiker-Kleiner D, Hilfiker A, Morawietz H, Lichtenberg A, Wollert KC, Drexler H (2003) Alterations in Janus kinase (JAK)-signal transducers and activators of transcription (STAT) signaling in patients with end-stage dilated cardiomyopathy. Circulation 107:798-802

Radha V, Nambirajan S, Swarup G (1997) Overexpression of a nuclear protein tyrosine phosphatase increases cell proliferation. FEBS Lett 409:33-36

Radha V, Sudhakar C, Swarup G (1999) Induction of p53 dependent apoptosis upon overexpression of a nuclear protein tyrosine phosphatase. FEBS Lett 453:308-312

Rajendrakumar GV, Radha V, Swarup G (1993) Stabilization of a protein-tyrosine phosphatase mRNA upon mitogenic stimulation of T-lymphocytes. Biochim Biophys Acta 1216:205-212

Reddy RS, Swarup G (1995) Alternative splicing generates four different forms of a nontransmembrane protein tyrosine phosphatase mRNA. DNA Cell Biol 14:1007-1015

Sharpless NE, DePinho RA (2002) p53: good cop/bad cop. Cell 110:9-12

Simoncic PD, Lee-Loy A, Barber DL, Tremblay ML, McGlade CJ (2002) The T cell protein tyrosine phosphatase is a negative regulator of janus family kinases 1 and 3. Curr Biol 12:446-453

Sisto F, Miluzio A, Leopardi O, Mirra M, Boelaert JR, Taramelli D (2003) Differential cytokine pattern in the spleens and livers of BALB/c mice infected with Penicillium marneffei:protective role of gamma interferon. Infect Immun 71:465-473

ten Hoeve J, de Jesus Ibarra-Sanchez M, Fu Y, Zhu W, Tremblay M, David M, Shuai K (2002) Identification of a nuclear Stat1 protein tyrosine phosphatase. Mol Cell Biol 22:5662-5668

Tiganis T, Bennett AM, Ravichandran KS, Tonks NK (1998) Epidermal growth factor receptor and the adaptor protein p52Shc are specific substrates of T-cell protein tyrosine phosphatase. Mol Cell Biol 18:1622-1634

Tiganis T, Flint AJ, Adam SA, Tonks NK (1997) Association of the T-cell protein tyrosine phosphatase with nuclear import factor p97. J Biol Chem 272:21548-21557

Tiganis T, Kemp BE, Tonks NK (1999) The protein-tyrosine phosphatase TCPTP regulates epidermal growth factor receptor-mediated and phosphatidylinositol 3-kinase-dependent signaling. J Biol Chem 274:27768-27775

Tillmann U, Wagner J, Boerboom D, Westphal H, Tremblay ML (1994) Nuclear localization and cell cycle regulation of a murine protein tyrosine phosphatase. Mol Cell Biol 14:3030-3040

Wee C, Muise ES, Coquelet O, Ennis M, Wagner J, Lemieux N, Branton PE, Nepveu A, Tremblay ML (1999) Promoter analysis of the murine T-cell protein tyrosine phosphatase gene. Gene 237:351-360

Yamamoto T, Sekine Y, Kashima K, Kubota A, Sato N, Aoki N, Matsuda T (2002) The nuclear isoform of protein-tyrosine phosphatase TC-PTP regulates interleukin-6-mediated signaling pathway through STAT3 dephosphorylation. Biochem Biophys Res Commun 297:811-817

You-Ten KE, Muise ES, Itie A, Michaliszyn E, Wagner J, Jothy S, Lapp WS, Tremblay ML (1997) Impaired bone marrow microenvironment and immune function in T cell protein tyrosine phosphatase-deficient mice. J Exp Med 186:683-693

Zhu W, Mustelin T, David M (2002) Arginine methylation of STAT1 regulates its dephosphorylation by T cell protein tyrosine phosphatase. J Biol Chem 277:35787-35790

11 Protein tyrosine phosphatase-based therapeutics: lessons from PTP1B

Jannik N. Andersen and Nicholas K. Tonks

Abstract

The family of Protein Tyrosine Phosphatases (PTPs), which is encoded by ~100 genes in humans, plays a critical role in the regulation of signal transduction. Recently a variety of links between aberrant PTP function and human disease have been defined and it has become apparent that generation of PTP inhibitors may present novel avenues for therapeutic intervention in several, major human diseases. The most clearly defined example is the potential of PTP1B as a target for treatment of diabetes and obesity. Nevertheless, it has also become apparent that the properties of the PTPs present significant challenges to drug development. In this review we focus on PTP1B. We describe its structure, catalytic mechanism and biological function, together with a review of the specialist literature that describes the generation of inhibitors of PTP1B. In doing so we have attempted to present an overview, aimed at those with a general interest in signal transduction, that illustrates both the progress that has been made and the challenges that are associated with the quest for PTP inhibitors as drugs.

11.1 Introduction

Cell surface receptors have been a focus of attention for drug development, with the G protein coupled receptors representing the major class of pharmaceutical targets (Chalmers and Behan 2002). More recently, industry has begun to explore other aspects of signal transduction downstream of ligand-receptor interactions, in particular the importance of reversible phosphorylation, mediated by protein kinases and phosphatases, in the regulation of protein function. Disruption of the normal patterns of phoshorylation of tyrosyl residues in proteins, resulting in aberrant regulation of signal transduction, has been implicated in the etiology of a variety of human diseases, including diabetes and cancer. The ability to modulate selectively signalling pathways regulated by protein tyrosine phosphorylation holds enormous therapeutic potential. The first drugs directed against protein tyrosine kinases (PTKs) have now entered the market and represent breakthroughs in cancer therapy (reviewed in Traxler 2003 and Drevs et al. 2003). For example, Herceptin, a monoclonal antibody targeting the transmembrane PTK HER-2, is used for the treatment of breast cancer and Gleevec/STI-571, a small molecule inhibitor of the p210 Bcr-Abl oncoprotein PTK, is a new treatment for chronic myeloge-

Topics in Current Genetics, Vol. 5
J. Arino, D.R. Alexander (Eds.) Protein phosphatases
© Springer-Verlag Berlin Heidelberg 2004

nous leukaemia. In addition, more than 20 low molecular weight PTK inhibitors, including molecules directed to the ATP binding site, are in various phases of clinical evaluation (Traxler 2003). Nevertheless, one must not lose sight of the fact that tyrosine phosphorylation is reversible and, therefore, there is the potential to manipulate signal transduction pathways at the level of both PTKs and the protein tyrosine phosphatases (PTPs). Although initially dismissed as housekeeping enzymes, the PTPs are now recognized as regulators of signalling in their own right and have recently garnered attention as potential therapeutic targets (Johnson et al. 2002; Zhang and Lee 2003; Huijsduijnen al. 2002; Tobin and Tam 2002; Ramachandran and Kennedy 2003; Taylor 2003). In this review, we will focus on the prototypic PTP, PTP1B, discussing recent developments in our understanding of its biological function and progress in the identification of therapeutic inhibitors of the enzyme.

11.2 PTP1B structure and mechanism

PTP1B was the first PTP to be purified to homogeneity, originally isolated from human placenta as a 37kDa catalytic domain (Tonks et al. 1988b). The cloning of cDNA encoding the enzyme revealed a full-length form of the protein that also contains a regulatory segment of ~115 residues on the C-terminal side of the catalytic domain (Chernoff et al. 1990). The C-terminal 35 residues are predominantly hydrophobic in nature and function in targeting the enzyme to the cytoplasmic face of membranes of the endoplasmic reticulum (ER) (Frangioni et al. 1992). Such sub-cellular targeting is now recognized as an important component of the regulation of substrate specificity, and thus physiological function, of members of the PTP family. The C-terminal segment also contains sites of phosphorylation by Ser/Thr kinases (Flint et al. 1993; Moeslein et al. 1999) and a site of proteolytic cleavage by calpain, which generates a truncated, soluble, activated form of the enzyme (Frangioni et al. 1993), suggesting a role for this segment in the regulation of PTP1B activity.

The first crystal structure to be solved for a member of the PTP family was also that of PTP1B - a construct comprising the 37 kDa catalytic domain (Barford et al. 1994). This structure and those of several mutant forms of PTP1B, coupled with extensive enzymatic and kinetic analyses from several laboratories, has revealed the mechanism of PTP1B-mediated substrate recognition and catalysis and facilitated important insights into the function of the PTP family as a whole (reviewed in Barford et al. 1998; Zhang 2002). In fact, we have now visualized each of the reaction steps in PTP-mediated catalysis (Pannifer et al. 1998). All members of the PTP family are characterized by the presence of a signature motif, [I/V]HCXXGXXR[S/T], which recognizes the dianionic phosphate moiety of the target substrate and contains the cysteinyl residue (Cys 215 in PTP1B) that is essential for catalysis. This signature motif is located at the base of a pronounced cleft on the surface of the protein, the sides of which are formed by three motifs: the WPD loop, the Q loop and the pTyr loop (Andersen et al. 2001). Of these, the pTyr loop contains a tyrosine residue (Tyr 46 in PTP1B), which defines the depth of the cleft and contributes to the absolute specificity that PTP1B displays for phosphotyrosine-containing substrates, since the smaller phosphoserine and phosphothreonine residues would not reach down to the phosphate binding site at the base of the cleft.

PTP mediated catalysis proceeds via a 2-step mechanism, involving the production of a cysteinyl-phosphate catalytic intermediate. In the first step, substrate binding is accompanied by a large conformational change in the active site in which the WPD loop closes around the side chain of the pTyr residue of the substrate (Fig. 1). In fact, PTP1B represents an example of the concept of "induced fit", in which substrate binding induces a conformational change that creates the catalytically competent form of the enzyme (Jia et al. 1995). Following substrate binding, there is nucleophilic attack on the substrate phosphate by the sulphur atom of the thiolate side chain of the essential Cys residue. This is coupled with protonation of the tyrosyl leaving group of the substrate by the side chain of a conserved acidic residue in the WPD loop (Asp 181 in PTP1B) which, after closure of the active site, is positioned to function as a general acid (Barford et al. 1995). The second step of catalysis involves hydrolysis of the catalytic intermediate, mediated by Gln 262 from the Q loop, which coordinates a water molecule, and Asp 181, which now functions as a general base, culminating in release of phosphate. This was revealed in the structure of a PTP1B-orthovanadate complex, which is a mimic of the pentavalent phosphorus transition state, and the structure of a Q262A mutant form of PTP1B, which allowed trapping and visualization of the catalytic intermediate in a crystal of the mutant protein because its hydrolysis is impaired (Pannifer et al. 1998). These structures also revealed that the WPD loop is closed over the entrance to the active site, thereby sequestering the phosphocysteine intermediate with water molecules at the catalytic centre, promoting its hydrolysis and preventing the transfer of phosphate to extraneous phosphoryl acceptors. This explains why the PTPs do not function as 'kinases in reverse', being unable to phosphorylate other phosphate acceptors.

11.3 Substrate-trapping mutant PTPs

The critical step in the purification of PTP1B was affinity chromatography on an immobilized thiophosphorylated substrate. The principle underlying this step was that thiophosphorylated substrates retain the ability to bind to the PTP, but tend to be resistant to dephosphorylation. Therefore, the PTP recognizes the thiophosphorylated substrate and forms a stable complex with it, which subsequently can be disrupted to elute the enzyme, for example by application of a salt gradient (Tonks et al. 1988b). The ability to define the substrate specificity of members of the PTP family is an essential aspect of gaining an understanding of the function of these enzymes. Using insights from the crystal structure, a mutational analysis was undertaken in an attempt to generate the converse of the affinity purification step, i.e. to produce a form of PTP1B that maintains its high affinity for substrate

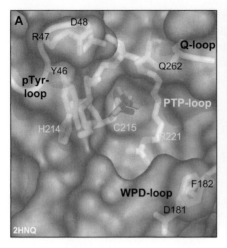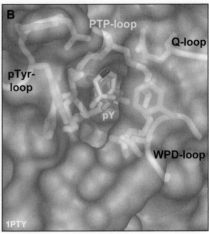

Fig. 1. Substrate binding induces a conformational change at the active site of PTP1B. The figure illustrates surface representations of the structure of (A) PTP1B in complex with tungstate (apo-structure) and (B) the inactive PTP1B C215S mutant bound to phosphotyrosine. Surface loops containing catalytic residues are shown beneath the semi-transparent surface. Atoms are coloured by type (oxygen in red; nitrogen in blue; carbon, green for PTP1B and pink for substrates). The Cα-backbone of the PTP-loop is coloured yellow to highlight its unique architecture; it coordinates the binding of phosphate, forming a semi-circle with Cys215 (yellow surface) in the centre and the main chain amides (backbone nitrogens shown in blue) oriented towards Cys215 to accommodate H-bonding interactions with the dianionic phosphate group of the pTyr substrate. The WPD-loop, which moves upon substrate binding to close the active site cleft and create the catalytically-competent form of the enzyme, is highlighted in green (Phe182). For comparison purposes, we have used the same view and orientation of PTP1B in each of the subsequent figures. The accession number for the structure file in the Protein Database bank (PDB file) is shown in white in the lower left corner of the figure.

but does not catalyze dephosphorylation effectively, thereby to convert this extremely active enzyme into a "substrate trap". Mutation of the invariant catalytic acid (Asp181 in PTP1B), a residue that is conserved in all members of the PTP family, yielded such a substrate-trapping mutant (Flint et al. 1997). This has afforded us a unique approach to identification of physiological substrates of PTPs in general. Following expression, the mutant PTP binds to its physiological substrates in the cell but, because it is unable to dephosphorylate the target efficiently, the mutant and substrate become locked in a stable, "dead-end" complex. These complexes can then be isolated and the substrates identified. Application of this strategy has revealed that members of the PTP family display exquisite substrate specificity in a cellular context, consistent with a function as highly selective regulators of signal transduction.

Characterization of the PTP1B substrate trapping mutant (PTP1B-D181A) indicated a role for the enzyme in dephosphorylation of the EGF receptor PTK

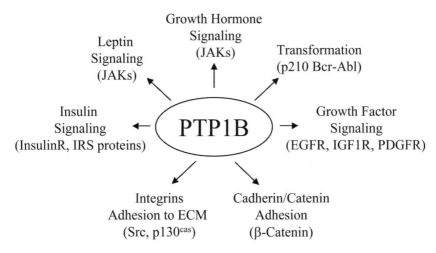

Fig. 2. Signalling events regulated by PTP1B. PTP1B regulates various signalling pathways associated with growth and proliferation, adhesion and metabolism. In each case, the substrates that have been identified are highlighted in parentheses.

(Flint et al. 1997). More recently, an elegant study using fluorescence resonance energy transfer (FRET) in PTP1B-/- fibroblasts reconstituted with PTP1B substrate trapping mutant demonstrated that dephosphorylation of EGF and PDGF receptors occurred at specific sites in the ER following endocytosis of the PTK (Haj et al. 2002). Furthermore, an alternative strategy using bioluminescence resonance energy transfer (BRET) illustrated both a basal interaction between PTP1B and nascent insulin receptors, as well as an insulin-induced association between PTP1B substrate trapping mutant and the internalized receptor (Boute et al. 2003). The localization of PTP1B to the ER has prompted questions regarding its importance as a regulator of PTK signalling. Nevertheless, these data indicate that PTP1B has the potential to attenuate the activity of newly synthesized receptor PTKs, as well as functioning in a "dephosphorylation compartment" that is encountered by down-regulated PTKs before being directed either to the lysosome or recycled to the plasma membrane. This suggests that PTP1B may play an important role in terminating receptor-PTK signalling.

11.4 The biological function of PTP1B

Several physiological substrates of PTP1B have now been identified, shedding light on the function of the enzyme as a regulator of signal transduction (Fig. 2). Early studies demonstrated its potential to suppress transformation by oncoprotein PTKs (Brown-Shimer et al. 1992). Interestingly, changes in the level of expression of PTP1B have been noted in several human diseases, particularly those associated with disruption of the normal patterns of tyrosine phosphorylation, including certain cancers. For example, the expression of PTP1B is induced specifically by the p210 Bcr-Abl oncoprotein, a PTK that is directly responsible for the initial manifestations of chronic myelogenous leukaemia (CML) (LaMontagne et al. 1998). Furthermore, PTP1B has the ability to antagonize p210 Bcr:Abl-induced transformation (LaMontagne et al. 1998), suggesting that this change in expression of the phosphatase may represent a means by which the cell attempts to attenuate the action of the oncoprotein PTK, particularly in the early chronic phase of the disease. Characterization of the *PTP1B* promoter (Forsell et al. 2000; Fukada and Tonks 2001) has now illustrated mechanisms by which expression of the enzyme is regulated in both physiological and pathophysiological conditions (Fukada and Tonks 2003).

PTP1B has also been implicated in the control of cell adhesion mediated by both cadherin-catenin complexes, which regulate cell-cell contact at adherens junctions, and integrins, which regulate interactions with the extracellular matrix at focal adhesion complexes (Lilien et al. 2002; Pathre et al. 2001; Arregui et al. 1998). The cadherins are transmembrane proteins that participate in Ca^{2+}-dependent, homophilic binding interactions via their extracellular segment. The intracellular segment is also essential for adhesion and serves as a binding site for proteins termed catenins, which link the cadherin to the actin cytoskeleton, and are subject to tyrosine phosphorylation (Lilien et al. 2002). PTP1B associates with the cytoplasmic domain of N-cadherin at a site that partially overlaps with the β-catenin binding site (Balsamo et al. 1998; Xu et al. 2002). Disruption of this association increases the pool of free tyrosine phosphorylated β-catenin, which impairs cadherin-mediated adhesion (Balsamo et al. 1998; Xu et al. 2002). Furthermore, upon expression in L-cells, the PTP1B-C215S inactive mutant was observed to interact with N-cadherin, apparently displacing the endogenous wild type enzyme and functioning in a dominant-negative manner. Expression of the inactive PTP1B mutant enhanced tyrosine phosphorylation of β-catenin, uncoupled cadherin from the actin cytoskeleton and disrupted cadherin-mediated adhesion (Balsamo et al. 1998). Thus, PTP1B appears to regulate cadherin-mediated adhesion by controlling the phosphorylation status of β-catenin. PTP1B has also been shown to associate with integrins and, once again, the inactive PTP1B-C215S mutant was observed to inhibit integrin-mediated adhesion and spreading on fibronectin, disrupting focal adhesions and stress fibres and resulting in morphological changes (Arregui et al. 1998). The effects of PTP1B on integrin signalling appear to be exerted at the level of dephosphorylation and activation of the PTK Src (Pathre et al. 2001; Bjorge et al. 2000). PTP1B has now been identified as a major PTP for the dephosphorylation of Src in various cell lines (Bjorge et al. 2000) and fibroblasts derived from PTP1B knockout mice display attenuated integrin signalling and cell spreading that is similar to the phenotype of Src-deficient fibroblasts (Cheng et al. 2001). There have been suggestions that PTP1B may function to inhibit integrin signalling, with the scaffold protein p130[cas] a substrate of PTP1B in this context (Balsamo et al. 1998); however, there appears to be some controversy around this point (Cheng et al. 2001). Interestingly, PTP1B has also recently been implicated in the mechanism by which the Sprouty proteins inhibit cell migration (Yigzaw et al. 2003).

Among the most exciting recent developments has been the establishment of PTP1B as a regulator of the signalling pathways associated with diabetes and obesity. A variety of studies have explored regulatory links between PTP1B and the insulin receptor, including the use of neutralizing antibodies, overexpression, and substrate trapping strategies. Furthermore, recent analyses of quantitative trait loci and mutations in the *PTP1B* gene in humans support the concept that aberrant expression of PTP1B may contribute to diabetes and obesity (Lee et al. 1999; Di Paola et al. 2002; Mok et al. 2002; Echwald et al. 2002). Perhaps the most exciting observations, however, have come from the development of *PTP1B* knockout mice (Elchebly et al. 1999; Klaman et al. 2000). Despite data linking PTP1B to the down-regulation of growth factor receptor PTKs, and evidence to indicate hyperphosphorylation of these PTKs following ablation of PTP1B expression, the mice did not show a predisposition to cancer. It appears that compensatory mechanisms are induced to prevent hyperactivation of the signalling pathways triggered by these receptors (Haj et al. 2003). In contrast, the PTP1B-/- mice display a tissue-specific increased sensitivity to insulin (Elchebly et al. 1999; Klaman et al. 2000). In the fed state, the levels of serum glucose and insulin were lower in the knockouts than the wild type littermates and they showed enhanced sensitivity in glucose and insulin tolerance tests. These effects coincided with enhanced tyrosine phosphorylation of the insulin receptor (IR) in muscle and liver, suggesting that the receptor may be a direct substrate of PTP1B. More surprising was the observation that ablation of PTP1B conferred resistance to obesity induced by a high fat diet (Elchebly et al. 1999; Klaman et al. 2000), an effect related to increases in both basal metabolic rate and total energy expenditure (Klaman et al. 2000). As a further indication of the specificity of these effects it is interesting to note that despite the close structural relationship between insulin and IGF-1 receptor PTKs, and data to indicate enhanced phosphorylation and activity of the IGF-1 receptor in PTP1B-deficient cell lines (Buckley et al. 2002), the PTP1B-deficient mice do not display phenotypes associated with aberrant IGF-1 induced signalling.

11.5 Insights from the structure of PTP1B-substrate complexes

The catalytic domain of the classical PTPs, such as PTP1B, comprises 280 residues and can be defined by 10 conserved sequence motifs (Andersen et al. 2001). The construction of the active site ensures specificity for pTyr residues in substrate proteins, although the affinity for free pTyr is some 10,000-fold lower than that of optimal peptide substrates (Zhang et al. 1994). Although there is extensive structural similarity within the catalytic centre of members of the PTP family, the surface residues surrounding the active site are more variable (Andersen et al. 2001). Interaction between residues flanking the pTyr of the substrate and the residues surrounding the PTP active site not only confer the higher affinity of interactions with protein substrates but also offer the potential for specificity in such

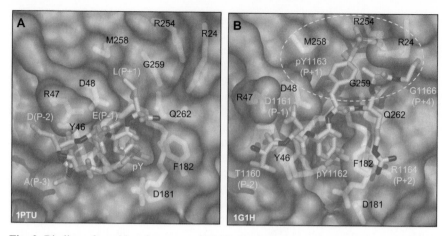

Fig. 3. Binding of peptide substrates to PTP1B. Surface representations of the structure of (A) PTP1B C215S in a complex with the EGFR-derived peptide DADEpYL and (B) PTP1B C215A in a complex with the bis-phosphorylated insulin receptor activation loop peptide ETDpYpYRKGGKGL. Note the closed conformation of the WPD-loop and the extensive interaction between residues in the peptide substrate (pink) and residues on the surface of PTP1B (black). The second pTyr binding site, which is characterized by salt bridge interactions between the phosphate group of a pTyr residue and the side chains of Arg 24 and Arg 254, is highlighted by the yellow oval.

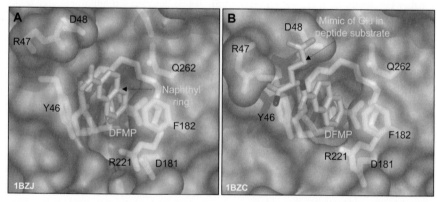

Fig. 4. Interaction of PTP1B with aryl difluoromethylene phoshonate-based inhibitors. Surface representations of the structure of PTP1B in a complex with naphthyl compounds bearing difluoromethylene phosphonate (DFMP) groups and either a carboxylic acid (A) or glutamic acid (B) functionality. The DFMP substituted naphthyl rings, which functioned as pTyr mimetics, were buried within the active site pocket, displaying hydrophobic interactions with the phenyl ring of Tyr46 and Phe182, as well as with Val 49, Ala 217 and Ile219 (residues not shown). The glutamic acid substitution (B) was introduced to mimic the acidic residue present in the EGFR peptide at position (P-1). This semi-rational design enhanced the potency of the compound compared to the non-substituted naphtyl compound by interacting with the side chain of Arg47.

interactions (Jia et al. 1995; Sarmiento et al. 2000; Salmeen et al. 2000). Crystallographic studies of the recognition of substrates by PTP1B have revealed how specificity may be achieved (Fig. 3). For example, the structure of a complex between PTP1B and the pTyr peptide (DADEpYL) from the EGF receptor, revealed an interaction between acidic residues in the peptide and the side chain of Arg 47 in PTP1B (Jia et al. 1995) (Fig. 3A). Furthermore, the molecular basis of specificity in the interaction between PTP1B and activation loop of the IR has also been defined (Fig. 3B). In this case, interactions between PTP1B and residues on both N- and C-terminal sides of the substrate pTyr were explored and the sequence E/D-pY-pY-R/K was identified as important for optimal recognition of the IR peptide as a substrate (Salmeen et al. 2000). The pTyr 1162 of the IR binds to the catalytic centre, while the adjacent pTyr 1163 is located within a shallow groove on the surface of PTP1B that is connected to the active site by a channel. In this groove, previously identified as "a second aryl-phosphate binding site" (Puius et al. 1997), the phosphate group of pTyr 1163 forms salt bridges with the side chains of Arg 24 and Arg 254 of PTP1B, with specificity for pTyr binding to this site being determined by the length of the phosphorylated residue and the positioning of these arginines (Salmeen et al. 2000). These interactions explain the observation that PTP1B displays 70 fold higher affinity for tandem pTyr-containing peptides compared to mono-pTyr derivatives (Salmeen et al. 2000).

There are 2 corollaries of these structural analyses. Firstly, it may be possible to use the presence of the consensus motif, E/D-pY-pY-R/K, to predict other physiological substrates of PTP1B. On this basis, we were able to demonstrate that the JAK subfamily of PTKs, in particular JAK2 and TYK2, are physiological substrates of PTP1B, consistent with an important role for the phosphatase as a regulator of cytokine signalling (Myers et al. 2001). Interestingly, the satiety hormone leptin exerts its effects through a receptor in the hypothalamus that displays the hallmarks of a cytokine receptor and signals through the JAK PTKs. On the basis of our observation, we suggested that PTP1B may normally function as a negative regulator of leptin signalling in the brain and that effects on leptin signalling resulting from aberrant down-regulation of the JAK PTKs may contribute to the resistance of PTP1B-knockout mice to high fat diet-induced obesity (Salmeen et al. 2000; Myers et al. 2001). In fact, experiments in the knockout mice have now illustrated such a role for PTP1B (Zabolotny et al. 2002; Cheng et al. 2002). Interestingly, dephosphorylation of the JAKs has also been implicated in the mechanism by which PTP1B regulates the signalling response to Growth Hormone (Gu et al. 2003). Secondly, the observation that there is specificity in the interaction between PTP1B and its target substrates suggests that it may be possible to develop specific inhibitors of such interactions as the basis for novel therapeutic strategies for the treatment of diabetes and obesity.

11.6 The development of PTP1B as a therapeutic target

The functional characterization of PTP1B suggests that an inhibitor of the enzyme will address both obesity and insulin resistance, thereby presenting a unique therapeutic opportunity in one of the major healthcare challenges of the 21st century. The importance of PTP1B as a target can be seen from the fact that the scientific and patent literature contains disclosures from 16 different pharmaceutical and biotechnology companies that have worked in this area (Blaskkovitch and Kim 2002)! Although great progress has been made, at this time, a drug is still a long way off.

Several PTP inhibitors have been identified in the scientific literature that display minimal specificity. These include zinc, molybdate, tungstate, phenyl arsine oxide, and acidic polymers such as heparin (Tonks et al. 1988a). Perhaps the best known is vanadate, which acts at the PTP active site as a transition state analogue of phosphate and modifies the active site Cys residue (Pannifer et al. 1996), and pervanadate, which leads to oxidation of the active site Cys (Huyer et al. 1997). Interestingly, vanadium-containing compounds have long been known to display insulin-mimetic properties and have been used in the treatment of diabetes; however, their utility is severely limited (Heffetz et al. 1990; Goldfine et al. 1995; Cohen et al. 1995). Although this has helped to highlight the potential of PTPs as therapeutic targets, it has also emphasized the challenges. For example, the observation that treatment of NRK cells with vanadate induced a transformed phenotype (Klarlund 1985) illustrated the critical role of PTPs in attenuating cellular tyrosine phosphorylation and drew attention to the importance of specificity in the action of a potential PTP-based therapeutic. In order to avoid inducing global changes in tyrosine phosphorylation, any such inhibitor must act selectively on the PTP(s) that regulate the signalling pathways associated with the disease state.

Attempts have been made to identify drug candidates that target members of the PTP family using standard activity-based, high-throughput screening (HTS) of libraries of small molecules. Various assay formats have been used including dephosphorylation of ^{32}P-labelled phophopeptides (Andersen et al. 2000), as well as fluorescence (Huang et al. 1999) or colorimetric (Urbanek et al. 2001; Andersen et al. 2002) detection of the dephosphorylation of low Mr phosphate esters. Due to the unique environment of the PTP active site, the essential Cys residue of the signature motif displays an unusually low pKa and is present predominantly as the thiolate anion at neutral pH. This renders the PTPs susceptible to oxidation with concomitant inhibition of activity, since the oxidized Cys can no longer function as a nucleophile. Interestingly, this property has been harnessed by nature as a mechanism for transient inhibition of PTP function and fine-tuning of tyrosine phosphorylation dependent signalling pathways (Finkel 2003). Reactive oxygen species, such as H_2O_2, are produced in response to a variety of physiological stimuli and promote oxidation of PTPs to a stable, singly oxidized sulfenic acid (-SOH) (Claiborne et al. 1999), which can be readily reduced to the active form of the enzyme via a sulfenyl-amide intermediate (Salmeen et al. 2003; Van Montfort et al. 2003). For example, PTP1B is oxidized reversibly in response to EGF (Lee et al. 1998) and insulin (Mahadev et al. 2001). Consequently, PTP1B is exquisitely sensitive to oxidizing and alkylating agents in HTS activity assays. Pervanadate and alendronate inhibit PTP1B by oxidation of the active site Cys to sulfonic acid ($-SO_3H$) (Huyer et al. 1997) and sulfinic acid ($-SO_2H$) (Skorey et al. 1997) respectively, with the production of such higher oxidized forms representing an irreversible modification. A recent report of a drug-screening program showed that of 25 HTS hits, 21 were found to inhibit PTP1B by time-dependent or non-competitive mechanisms (Doman et al. 2002), highlighting the fact that great care must be taken in characterizing the mechanisms by which such compounds inactivate the enzyme so as to avoid non-specific agents that may inhibit the enzyme by covalent modification.

Alternative assay protocols are being developed to circumvent this problem. For example, mutant forms of PTP1B in which Cys 215 has been altered to Ser or Ala display properties of a substrate trapping mutant and retain some of the enzyme's ability to bind substrate (Milarski et al. 1993). The PTP1B-C215S mutant has been used for development of high-throughput binding assays in which the enzyme (now converted to a binding protein) is no longer susceptible to oxidation or alkylation at the active site (Skorey et al. 2001; Wang et al. 2001; Shen et al. 2001). For example, a scintillation proximity assay, which measures the ability of compounds to disrupt the association between the tagged mutant PTP bound to protein A scintillation beads and a [^{3}H]-labelled tripeptide inhibitor (Skorey et al. 2001) and a non-radioactive ELISA-based assay, which measures the ability of potential inhibitors to compete with an immobilized biotinylated phosphopeptide substrate for association with the mutant PTP (Shen et al. 2001). Nevertheless, one must not forget that the mutant PTP may exhibit subtle differences in properties when compared to the wild type enzyme (Scapin et al. 2001).

11.7 Structure-based design of PTP1B inhibitors

In addition to its extensive biological validation, a major asset in the development of inhibitors of PTP1B is the availability of crystal structures of the enzyme (Table 1), including complexes with substrates (Fig. 1 and 3). These structures, and their insights into the mechanism of substrate recognition and catalysis mediated by PTP1B, have afforded investigators the opportunity to explore a variety of strategies in their quest for inhibitors of the PTP that display both high affinity and selectivity.

An early strategy for the design of PTP1B antagonists was to begin with peptide substrates and generate inhibitors from them by substituting pTyr with non-hydrolyzable analogues (Burke et al. 1994; Groves et al. 1998; Desmarais et al. 1999). The best such analogue was based upon the EGF receptor peptide DADEpYL, in which the pTyr residue was substituted by F_2PMP (phosphonodifluoromethylene Phe), generating an inhibitor of 100nM potency (Burke et al. 1994). Interestingly, the presence of the F atoms enhanced affinity by 3 orders of

Table 1. Published X-ray crystal structures of PTP1B

Description	PDB ID	References
Apo structures		
Apo form (with & without tungstate)	2HNP,2HNQ	(Barford et al. 1994)
Apo form (C215S mutant)	1I57	(Scapin et al. 2001)
Phosphocysteine interme-diate	1A5Y	(Pannifer et al. 1998)
Substrate complexes		
Phosphotyrosine	1PTV,1PTY	(Jia et al. 1995; Puius et al. 1997)
Bis(para-phosphophenyl) methane (BPPM)	1AAX	(Puius et al. 1997)
EGFR-derived peptides	1PTT,1PTU	(Jia et al. 1995)
Insulin receptor derived peptides	1G1F,1G1G,1G1H	(Salmeen et al. 2000)
Designed phosphopeptides	1EEN,1EEO	(Sarmiento et al. 2000)
Inhibitor complexes		
Cyclic EGFR peptide	1BZH	(Groves et al. 1998)
Difluorophosphonate-based small molecules	1BZC,1BZJ	(Groves et al. 1998)
Compounds with two pTyr mimetic groups	1KAK,1KAV,1LQF, 1N6W	(Jia et al. 2001;Asante-Appiah et al. 2002; Sun et al. 2003)
2-(oxalylamino)-benzoic acid based	1C83,1C84,1C85,1C86, 1C87,1C88,1ECV, 1GFY,1L8G	(Andersen et al. 2000; Andersen et al. 2002; Iversen et al. 200; Peters et al. 2000
Peptidomimetics (CCK-based)	1G7F,1G7G,1JF7	(Bleasdale et al. 2001; Larsen et al. 2002)
SAR-by-NMR and modular design approach	1NL9,1NNY,1NO6, 1ONZ, 1ONY,1NZ7	(Szczepankiewicz et al. 2003; Xin et al. 2003)
Breakaway tethering	1NWL	(Erlanson et al. 2003)
Oxidized enzyme forms		
Reversible and irreversible inactivated	1OEO,1OEM,1OES, 1OET,1OEU,1OEV	(Salmeen et al. 2003; Montfort et al. 2003)

magnitude relative to the phosphonomethylene Phe derivative (Burke et al. 1994), by contributing interactions between the F atoms and the PTP active site (Burke et al. 1996; Murthy and Kulkarni 2002). Although these studies illustrated an effective pTyr-mimetic, the problems of poor bioavailability of such peptide inhibitors encouraged the investigators to generate aryl difluoromethylene-phosphonate (DFMP) substitutions of small molecule scaffolds. Phenyl derivatives, as close mimics of pTyr itself, were essentially inactive, as might be anticipated in light of the low affinity of PTP1B for free pTyr. Nevertheless, incorporation of DFMP into a naphthyl ring dramatically improved inhibitor potency by enhancing hydrophobic interactions between the compound and the PTP1B active site (Yao et al. 1998). This compound, which induced closure of the WPD loop of the active site in a similar manner to substrate binding, was further derivatized in the naphthyl ring to generate additional inhibitor-protein interactions (Fig. 4) (Groves et al.

1998). For example, introduction of a carboxylate group in the naphthyl ring (inhibitor INH), promoted indirect associations with R47 and D48 of PTP1B via a water molecule (Fig. 4A). Further substitution by including a carboxamidoglutamate group, to mimic the Glu residue in the EGF receptor peptide substrate (inhibitor TPI), facilitated direct H-bonding interactions with the side chains of R47 and D48, which are flexible and moved to accommodate interaction with the inhibitor (Fig. 4B) (Groves et al. 1998). These observations are important because they provided the first illustrations of the potential to build both enhanced affinity and specificity into PTP1B-inhibitor interactions, while maintaining a low molecular weight for the inhibitor.

One of the few reported successes in the search for inhibitors of PTP1B using HTS screens arises from the identification of 2-(oxalylamino)-benzoic acid (OBA) as a broad specificity, reversible inhibitor that is directed to the active site and displays classical competitive kinetics of inhibition (Andersen et al. 2000). As such, OBA has been used as a template to create derivatives with greater potency and enhanced specificity (Andersen et al. 2002; Peters et al. 2000). OBA binds to the active site of PTP1B as a pTyr mimetic (Fig. 5A). Following the lessons learned from the enhanced hydrophobic interactions with naphthyl fluorophosphonates, compared to the phenyl derivatives, it was found that substitution of the phenyl ring of OBA with a thiophene ring facilitated derivitization to generate inhibitors with an additional saturated ring structure, which not only enhanced potency but also facilitated introduction of substituents that would generate specificity (Andersen et al. 2002; Iversen et al. 2000). Analysis of the sequences and structures of the classical PTPs illustrated that the arrangement of residues Arg 47/Asp 48 and Met 258/Gly 259 at the active site was unique to PTP1B and its closest relative TC-PTP (Iversen et al. 2000). The structure of PTP1B-substrate complexes have illustrated the importance of these residues in determining specificity in enzyme-substrate interactions (Jia et al. 2000; Sarmiento et al. 2000). Arg 47 binds to the side chains of acidic residues at the P-1 and P-2 positions, N-terminal to the pTyr in a substrate, although conformational flexibility in this residue allows PTP1B also to accommodate hydrophobic side chains at the P-1 position (Sarmiento et al. 2000). Asp 48 is also important for substrate binding, its side chain participating in H-bonding interactions with main chain N atoms in the peptide substrate (Jia et al. 1995; Salmeen et al. 2000; Sarmiento et al. 2000). Gly 259 is a key determinant of substrate specificity (Peters et al. 2000). In the structure of the complex between PTP1B and a peptide substrate derived from the insulin receptor activation loop, it is apparent that as a consequence of the presence of Gly at this position in the enzyme the substrate has access to a second binding site for pTyr on the surface of PTP1B, contributing to the preference shown by the enzyme for substrates with tandem pTyr residues (Salmeen et al. 2000). In contrast, many PTPs have bulky hydrophobic residues at this position, which would sterically hinder access to this binding site (Peters et al. 2000). In an elegant series of studies, derivatives of OBA have been generated that engage these unique features of the active site of PTP1B. For example, introduction of a basic nitrogen into the core structure of the inhibitor was observed to enhance specificity for PTP1B, by promoting a salt bridge with Asp 48, while simultaneously causing repulsion in

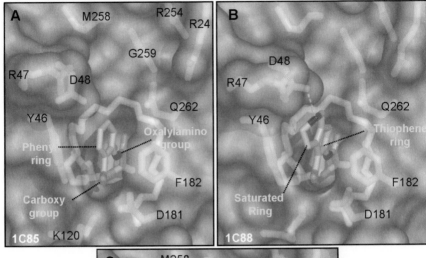

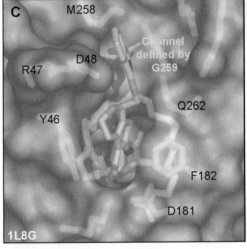

other PTPs that have an Asn residue at the equivalent position (Fig. 5B) (Iversen et al. 2000). Interestingly mutation of Asp 48 in PTP1B to Asn markedly reduced affinity, whereas introduction of Asp for Asn at the equivalent position in PTPα increased its affinity for the inhibitor. In a separate study, OBA was derivatized with a bulky group, in such a way as to engage the channel created by the presence of Gly 259, thereby promoting binding to PTP1B while simultaneously further enhancing specificity by steric hindrance between the compound and the large hydrophobic residues that block the channel in the majority of PTPs and, therefore, exclude the modified inhibitors (Fig. 5C) (Iversen et al. 2001). These observations provide a clear structural explanation of how potential inhibitors may be designed around the features of the active site of PTP1B, not only to enhance affinity but also to improve specificity.

Fig. 5 (overleaf). Structure-based optimization of a general, reversible and competitive PTP inhibitor identified from a high throughput screen. (A) Structure of PTP1B in a complex with 2-(oxalylamino)-benzoic acid (OBA). OBA binds to PTP1B with the WPD loop in the closed conformation. The oxalylamino group binds to the PTP loop in similar manner to the phosphate of pTyr substrates. The o-carboxylic acid group hydrogen bonds with Tyr46 and Asp181 and forms a unique salt bridge with Lys120. (B & C) OBA has been used as a lead compound for further optimization; the phenyl ring in OBA has been substituted with a thiophene ring, which in turn may be fused to an additional saturated ring, both to enhance potency through hydrophobic interactions and introduce specificity through interaction with additional surface residues. Selectivity towards PTP1B has been achieved via two different strategies (i) attraction-repulsion and (ii) steric hindrance. (B) Introduction of a basic nitrogen, which forms a salt-bridge with Asp48 (highlighted by the yellow dashed line), both enhances the interaction with PTP1B and causes repulsion in other PTPs that contain a Asn residue at this position. (C) The presence of a glycine at position 259 allows compounds to be derivatized with large groups that engage the channel created in PTP1B, but that cannot be accommodated by the majority of PTPs, which contain a bulky residue at the equivalent position.

The identification of a second binding site for pTyr on the surface of PTP1B (Puius et al. 1997) and the demonstration that engagement of this site by substrates containing tandem pTyr residues enhances binding affinity (Salmeen et al. 2000) has been influential in approaches to design inhibitors of the enzyme. Early in the process it was shown that the tripeptide inhibitor Glu-F_2PMP-F_2PMP, containing two adjacent residues that are non-hydrolyzable analogues of pTyr, was a potent inhibitor of PTP1B (IC_{50} 40nM) that displayed >100 fold selectivity compared to the other PTPs that were tested (Desmarais et al. 1999). Such observations raised expectations that combinatorial strategies generating molecules that engaged these two sites on PTP1B simultaneously would enhance the affinity and specificity of potential inhibitors. In developing the above inhibitor, it was found that addition of a benzoyl group to the N-terminus of the tripeptide enhanced affinity still further (IC_{50} 6nM); however, the structure of a complex of the inhibitor and PTP1B revealed a binding mode that did not involve the second site (Asante-Appiah et al. 2002) (Fig. 6A). In fact, while one F_2PMP residue was engaged at the active site, the second interacted with Arg 47 and Asp 48, with mutation of Asp 48 to Ala resulting in a 100 fold loss of potency (Asante-Appiah et al. 2002). A similar, highly potent bidentate inhibitor that is predominantly peptidic in character was generated independently (Shen et al. 2001; Sun et al. 2003). In this case, a small library of compounds was constructed in which pTyr was connected to various aryl acids by different linker sequences. High affinity substrates were identified in the library and converted to inhibitors by replacement of pTyr with F_2PMP. A substituted dipeptide inhibitor was generated, containing F_2PMP separated from an aryl DFMP moiety by an Asp residue, which was a potent inhibitor of PTP1B (K_i 2.4 nM) and even displayed 10 fold specificity over TC-PTP (Shen et al. 2001) (Fig. 6B). In a separate study, linkage of two aryl DFMPs by an alkyl chain reduced the peptidic character of the inhibitor, as well as introducing flexibility and sufficient separation of the charged groups to be consistent with engagement of the second binding site (Jia et al. 2001) (Fig. 6C). Nevertheless, as observed for the

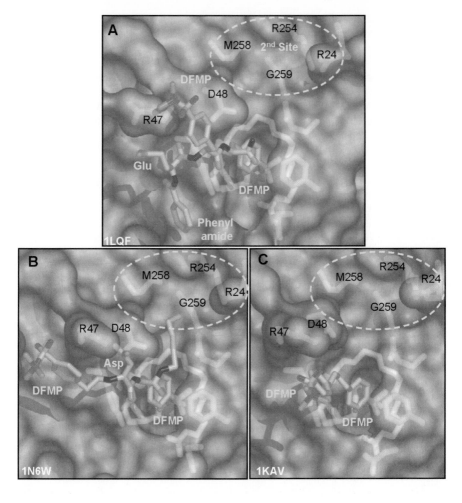

Fig. 6. Inhibitors of PTP1B bearing two phosphonate groups. (A – C) Each panel illustrates the structure of PTP1B in a complex with an inhibitor containing adjacent difluoromethylene phosphonate (DFMP) groups that were designed to mimic the tandem arrangement of pTyr residues in certain substrates of PTP1B. These inhibitors were expected to interact simultaneously with the catalytic site and the second phosphate binding site (yellow circle). Surprisingly, none of the distal DFMP groups bound to the second site, but instead engaged Arg 47 and Asp 48.

F_2PMP-containing tripeptide, X ray crystal structures of complexes of PTP1B and these inhibitors revealed, unexpectedly, that the second pTyr-binding site was not engaged and that the primary interactions of the distal phosphonate group involved Arg 47 and Asp 48 (Sun et al. 2003; Jia et al. 2001) (Fig. 6B & 6C). The structure of the complex between PTP1B and the insulin receptor activation loop peptide revealed that the motif E/D-pY-pY-K/R was important for optimal substrate recognition, with interactions between residues flanking the pTyr in the substrate and residues on the surface of PTP1B positioning the peptide in the active site of the phosphatase. In particular, the guanidinium side chain of Arg 1164 at the P+2 position, on the C-terminal side of the substrate pTyr, forms a classical π-cation interaction with Phe 182 of PTP1B following closure of the WPD loop (Salmeen et al. 2000). It is likely that such additional interactions, which are missing in these inhibitors, are important to guide substrates, or potential inhibitors, to the second pTyr binding site.

Different approaches have now been tested in an attempt to engage directly the second pTyr binding site, including the "Linked-Fragment" strategy, also referred to as "SAR by NMR" (Shuker et al. 1996). In this strategy, the binding of small molecule fragments to distinct sites on a target protein is assessed by NMR; such fragments then serve as building blocks, subsequently to be optimized and linked together to create novel inhibitors that display both high affinity and specificity for their targets. In the case of PTP1B, an NMR screen identified diaryloxamic acid as a pTyr-mimetic that bound in the active site of the enzyme (Szczepankiewicz, Liu et al. 2003). Further optimization, to enhance interactions at the active site by including a bulky hydrophobic group, led to the generation of naphthyloxamic acid, which inhibited PTP1B competitively and reversibly with a K_i of ~39 μM (Szczepankiewicz et al. 2003) (Fig. 7A). Interestingly, this inhibitor binds to PTP1B with the WPD loop in the open conformation, the form of the enzyme that occurs in the absence of substrate, which presents a larger binding surface to the inhibitor than is encountered when the WPD loop is closed. By building outward from the active site, to include a diamido chain as a linker segment, the positioning of the compound at the active site was altered so that it now also engaged Asp 48, thereby enhancing affinity. Using a separate NMR screen, fragments that recognized the second pTyr binding site were also identified. These were primarily small, fused-ring aromatic acids, of which naphthoic acid was chosen. By linking these fragments together an inhibitor of nanomolar potency was created (K_i ~22 nM), which engaged simultaneously the active site and both Arg 254 and Arg 24 at the second binding site (Fig. 7B) (Szczepankiewicz et al. 2003). Variations on this theme were used to create a distinct inhibitor by exploring different linker and Site 2 ligands (Xin et al. 2003). In testing amino acids as Site 2 ligands, an inhibitor was produced that not only displayed high potency for PTP1B (K_i ~79 nM) but also 5 fold selectively compared to inhibition of the closest relative, TC-PTP (Fig. 7C) (Xin et al. 2003). Although these studies illustrate the potential for generating potent and specific inhibitors of PTP1B that engage both Site 2 and the catalytic site simultaneously, the resulting compounds are not without their problems. The PTP active site is highly charged, as might be expected from our understanding of the catalytic mechanism and the importance of the dianionic phosphate group in substrate binding. Therefore, many of the active site-directed inhibitors described to date are themselves highly charged. Furthermore, interactions with this "second site", which is located in a shallow groove on the surface of PTP1B and, unlike the active site, is not a clearly defined pocket, are dominated by salt bridges with Arg 254 and Arg 24. As a result, these strategies have generated highly charged

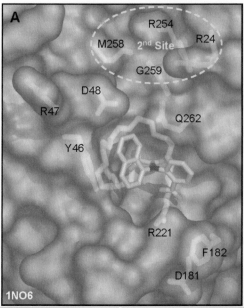

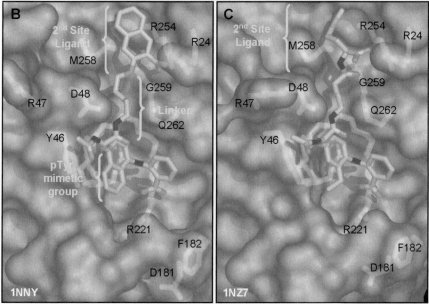

compounds, in which the problems already experienced at the active site are exacerbated and the capacity to cross the plasma membrane is limited.

In light of the importance of pTyr in substrate binding to PTP1B, various labs have searched for novel pTyr mimetics, also referred to as bioisosteres – structural analogues that mimic the phosphorylated residue but present properties more

Fig. 7 (overleaf). The "Linked-Fragment" or "SAR-by-NMR" strategy for the identification of a second site ligand. (A) Structure of PTP1B bound with the pTyr surrogate naphthyloxamic acid, in which the inhibitor binds to the open form of the enzyme, where the WPD-loop retains a conformation similar to the apo-structure seen in Figure 1A. (B) Structure of a more potent inhibitor obtained by linking naphthyloxamic acid to a second site ligand (identified by NMR screening) to the catalytic site ligand. Specifically, the carboxyl group of the second site ligand naphthoic acid is positioned 2.6 Å from Arg254 and 2.4 Å from Arg24, consistent with salt bridge or H-bond interactions. (C) Structure of PTP1B bound to distinct compound, obtained from the same Linked-Fragment/SAR-by-NMR program, in which the second site ligand is the amino acid methionine. This compound also engages Arg24 and Arg254 via the carboxylic acid of methionine, suggesting again that a negatively charged acidic group might be preferred as part of a site 2 ligand.

consistent with drug development. One such analogue was identified in a study in which derivatives of the peptide cholecystokinin (CCK-8), in particular the tripeptide motif N-acetyl Asp-Tyr (SO$_3$H)-Nle, were generated as inhibitors of PTP1B (Bleasdale et al. 2001). This sulfotyrosyl peptide is a competitive inhibitor of PTP1B (IC$_{50}$ 5 μM), with the presence of the sulphated tyrosine, an analogue of pTyr, being essential for inhibition. A variety of derivatives were generated with the goal of reducing the peptidic character and finding more stable analogues of sulfotyrosine (Liljebris et al. 2002; Larsen et al. 2002). Replacement of O-sulfotyrosine with an O-malonytyrosyl side chain, a known pTyr bioisostere (Roller et al. 1998), enhanced potency and yielded an inhibitor that bound to PTP1B with the WPD loop in the open conformation (Fig. 8A). In contrast, when sulfotyrosine was replaced by the novel pTyr bioisostere, 2 carboxymethoxybenzoic acid, binding of the inhibitor (IC$_{50}$ 2.8 μM), like binding of substrate, induced the WPD loop to close (Larsen et al. 2002) (Fig 8B). This modification, coupled with substitution of Phe at the N-terminus (Larsen et al. 2003), generated a sub-micromolar inhibitor (IC$_{50}$ 0.25 μM) that was able to inhibit dephosphorylation of the insulin receptor by PTP1B and disrupt the interaction between the insulin receptor and the PTP1B C215S mutant *in vitro* (Larsen et al. 2002). Most importantly, derivatives of these compounds displayed efficacy in cell based models of insulin action (Larsen et al. 2002). For example, esterification of the 3 free carboxylate groups in this inhibitor generated a compound that augmented insulin induced glucose uptake in 3T3 L1 cells. Therefore, these studies demonstrate that it is feasible to antagonize PTP1B function with small molecule inhibitors in a cellular environment.

11.8 Conclusions and perspectives

Despite major breakthroughs in the characterization of the structure, regulation and function of members of the PTP family, and the exciting progress to date in the development of inhibitors of these enzymes, particularly PTP1B, they remain challenging targets for drug development. Nevertheless, our understanding of the

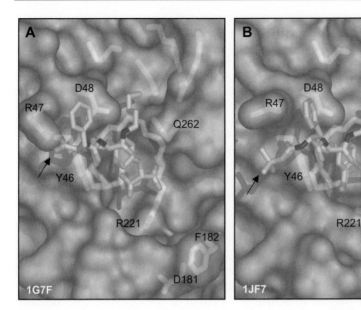

Fig. 8. From peptide to peptoid – optimization of a simple tripeptide lead. Derivatives of the sulfotyrosyl tripetide N-acetyl-Asp-Tyr(SO₃H)-Nle were generated, displaying attenuated peptidic character and improved physiochemical properties, and were co-crystallized with PTP1B. The O-sulfotyrosine was replaced by either (A) O-malonyl-tyrosine or (B) the novel pTyr bioisostere O-carboxylmethyl salicylic acid. The malonyl tyrosine group binds with the WPD-loop in the open conformation, whereas the template harbouring the O-carboxylmethyl salicylic acid pTyr surrogate binds with the WPD-loop closed. Analogues with activity in cell based assays of insulin action have been generated that included substitutions at the N-terminus of the peptide (arrow) (Larsen et al 2003).

physiological function of PTP1B reinforces its unique potential as the basis for a novel strategy for therapeutic intervention in diabetes and obesity and emphasizes the value of the prize. The challenges are illustrated by the fact that to date there has been only one report of a PTP1B inhibitor as a clinical candidate. This compound, termed PTP-112, which demonstrated efficacy as an anti-hyperglycemic agent in *ob/ob* mice, was withdrawn from Phase II trials due to unsatisfactory clinical efficacy and dose-limiting side effects (Tobin and Tam 2002; Taylor 2003). Compounds similar to PTP-112 activated PPAR-γ, raising the possibility that it may recognize targets other than PTP1B in the cell. The development of further clinical candidates will take perseverance and will most likely benefit from novel approaches to the inhibition of PTP1B function. Strategies will have to be developed to overcome the major issue of combining potency and specificity with bioavailability in any therapeutic inhibitor of PTP1B. Structure-based docking methods and molecular dynamics simulations are being used to identify compounds that display complementarity with the PTP1B active site (Sarmiento et al. 2000; Murthy and Kulkarni 2002; Doman et al. 2003). Top scoring compounds from *in silico* screens are then selected and examined in kinetic assays. In one report, a virtual PTP1B screen using the DOCK algorithm enhanced the hit rate by 1700 fold over the experimental HTS screening process (Doman et al. 2002). Docking analysis from another group identified 25 compounds of which 7 exhibited measurable inhibition of PTP1B at 100 μM (Sarmiento et al. 2000). However, in both of these molecular docking studies, the correlation between IC_{50} values and docking scores were poor, illustrating the challenges of predicting ligand binding affinities and binding modes reliably. In particular, failure to consider conformational changes in the target protein contributes to this problem. The conformational changes in the WPD loop that accompany pTyr binding at the active site and the flexibility of Arg 47 and Asp 48 (now referred to as the YRD motif in TC-PTP (Asante-Appiah et al. 2001; Sarmiento et al. 2000)), highlight the importance of this issue in the context of PTP1B. With the current access to a large number of 3D protein structures (Table 1), molecular mechanics-based methods are being developed that take induced fit and enzyme plasticity into consideration (Frimurer et al. 2003), which should provide an important complementary technique to HTS for lead discovery.

As illustrated by the "2nd pTyr binding site", exploration of additional, novel binding sites on the surface of PTP1B will also be important in the design of potent and specific inhibitors of the enzyme. One such approach, referred to as "Breakaway Tethering", introduces a Cys residue on the surface of the enzyme, which acts as a tethering point. The modified enzyme is incubated with a library of disulphide-containing small molecule fragments under reducing conditions, which promote thiol interchange and formation of an S-S bond between the ligand and the surface Cys residue. This tethering allows the capture of compounds that display even a weak affinity for the target protein, which can be identified by mass spectrometry (Erlanson et al. 2000). Using this approach, the authors identified a compound that bound at the active site of PTP1B, but which only contained a single carboxylate (Erlanson et al. 2003). Such compounds would show enhanced bioavailability compared to the inhibitors described to date, which contain multiple carboxylates. Although currently this inhibitor displays low affinity, in the millimolar range, it can now be used as a template for addition of further molecular fragments that explore novel binding sites on the enzyme and enhance the potency of the resulting inhibitor. Binding sites remote from the active site of PTP1B have already been implicated in protein-protein interactions. For example, the N-terminal segment of the catalytic domain contains an unidentified site that contributes to binding of the insulin receptor (Dadke et al. 2000). The interaction of PTP1B with N-cadherin requires the presence of Tyr at position 152 in the phosphatase (Rhee et al. 2001). Furthermore, a Pro-rich sequence in PTP1B has been implicated in binding to p130^{cas} (Liu et al. 1996). Interestingly, compounds have been identified (Fig. 7 & 8A), including the one described above, that recognize the inactive form of PTP1B in which the WPD loop is in the open conformation, highlighting two different philosophies of inhibitor design. The active conformation of PTP1B, in which the WPD loop is closed, presents an attractive hydrophobic binding pocket; however, some of the binding energy will be sacrificed as a compound binding to this form of the enzyme would have to trigger the same conformational change that is induced by substrate. In contrast, the inactive, open conformation of the WPD loop presents a broad pocket that is not as clearly defined as in the active conformation, but may be more amenable to exploitation of unique structural features that would enhance specificity. Of course, it is important to remember that the studies to date have focused on PTP1B constructs that comprise the catalytic domain and are based on the truncated form of the enzyme originally isolated from human placenta (Tonks et al. 1988b). It will be interesting also to pursue the role of the regulatory C-terminal segment of PTP1B, to test whether it interacts with the active site. As this segment is less conserved than the catalytic domain, it may offer additional opportunities to exploit specificity in inhibitor design.

The reversible oxidation of PTPs is a novel mechanism for the fine-tuning of tyrosine phosphorylation dependent signal transduction pathways. It is one physiological mechanism by which the activity of the PTPs that normally down-regulate a signal transduction pathway is attenuated, thereby facilitating initiation of the signalling response. Upon encountering Reactive Oxygen Species, such as H_2O_2, the nucleophilic Cys residue at the active site of PTP1B is oxidized to the sulfenic acid intermediate, which is rapidly converted to a sulfenamide (isothiazolidine) species (Salmeen et al. 2003; Van Montfort et al. 2003). A covalent bond is formed between the sulphur atom of the active site Cys and the main chain nitrogen of the adjacent residue, Ser 216, leading to formation of a novel 5-atom ring structure, which is accompanied by profound, but reversible, changes in the architecture of the active site (Salmeen et al. 2003; Van Montfort et al. 2003). With this conformational change, the oxidized Cys adopts a solvent exposed position that would facilitate reduction to the active form of the enzyme. In addition, Tyr 46, from the pTyr loop, also becomes solvent exposed and susceptible to phosphorylation. Perhaps we can learn from biology and harness this regulatory process as a mechanism for therapeutic inhibition of PTP function. In this context, it is interesting that small molecule NO-donors and covalent modifiers of other thiol-dependent enzymes (i.e. cysteine proteases and caspases) are being investigated as novel therapeutics (Leung-Toung et al. 2002).

Despite the encouraging progress to date, ultimately it may not be possible to generate small molecule inhibitors of PTP1B with the appropriate combination of affinity, specificity and bioavailability. Antisense strategies for inhibition of PTP1B function have shown promise. In one particular series of experiments, the effects of the PTP1B-directed antisense oligonucleotide (ASO) ISIS-113715 were tested in the *ob/ob* mouse model of obesity and diabetes (Zinker et al. 2002). The levels of PTP1B protein (and mRNA) were reduced in liver and fat, but not skeletal muscle, and this was accompanied by enhanced insulin sensitivity, in terms of tyrosine phosphorylation of the insulin receptor and IRS proteins as well as activation of downstream signalling pathways (Zinker et al. 2002;Gum et al. 2003b). When viewed in conjunction with the data from PTP1B knockout mice, which display enhanced insulin sensitivity in muscle and liver, but not adipose tissue, these studies highlight the liver as a potentially important site for the action of PTP1B (Gum et al. 2003a). Interestingly, insulin signalling was also enhanced in the muscle of PTP1B ASO-treated *ob/ob* mice, despite the fact that the levels of PTP1B protein in muscle were apparently unchanged (Gum et al. 2003b). Exciting developments in powerful new technologies, such as RNA interference, may also offer alternative strategies for inhibition of PTP function in a therapeutic context.

Finally, it is important to remember that there is more to the PTP family than just PTP1B! Access to the essentially complete human genome sequence has allowed the identification of ~100 genes classified as tyrosine specific and dual specificity phosphatases – a number comparable to the family of tyrosine-specific and dual specificity protein kinases (see the website http://ptp.cshl.edu, and the parallel site http://science.novonordisk.com/ptp, for a detailed analysis of the PTP family). The diversity of the family is further enhanced by such mechanisms as alternative splicing, alternative promoter usage and various post translational modifications. Various approaches, from the development of substrate-trapping mutants to gene ablation, have highlighted the potential for PTPs to display exquisite substrate specificity *in vivo* and to function as regulators of fundamentally important signal transduction events. Additional links between PTPs and human diseases, including cancer, inflammation and infectious disease, have already been established and it is likely that more will be defined. In fact, certain dual specificity phosphatases, such as the cell cycle control enzyme cdc25, are already recognized as therapeutic targets in cancer (Lyon et al. 2002). Hopefully, further characterization will establish new functions for these enzymes and perhaps the PTPs may ultimately be recognized as a platform for the development of novel therapeutics in a variety of disease states.

Acknowledgments

Work in the authors' laboratory is supported by grants R37-CA53840 and R01-GM55989 from the National Institutes of Health.

References

Andersen HS, Iversen LF, Jeppesen CB, Branner S, Norris K, Rasmussen HB, Moller KB, Moller NP (2000) 2-(oxalylamino)-benzoic acid is a general, competitive inhibitor of protein-tyrosine phosphatases. J Biol Chem 275:7101-7108

Andersen HS, Olsen OH, Iversen LF, Sorensen AL, Mortensen SB, Christensen MS, Branner S, Hansen TK, Lau JF, Jeppesen L, Moran EJ, Su J, Bakir F, Judge L, Shahbaz M, Collins T, Vo T, Newman MJ, Ripka WC, Moller NP (2002) Discovery and SAR of a novel selective and orally bioavailable nonpeptide classical competitive inhibitor class of protein-tyrosine phosphatase 1B. J Med Chem 45:4443-4459

Andersen JN, Mortensen OH, Peters GH, Drake PG, Iversen LF, Olsen OH, Jansen PG, Andersen HS, Tonks NK, Moller NP (2001) Structural and evolutionary relationships among protein tyrosine phosphatase domains. Mol Cell Biol 21:7117-7136

Arregui CO, Balsamo J, Lilien J (1998) Impaired integrin-mediated adhesion and signaling in fibroblasts expressing a dominant-negative mutant PTP1B. J Cell Biol 143:861-873

Asante-Appiah E, Ball K, Bateman K, Skorey K, Friesen R, Desponts C, Payette P, Bayly C, Zamboni R, Scapin G, Ramachandran C, Kennedy BP (2001) The YRD motif is a major determinant of substrate and inhibitor specificity in T-cell protein-tyrosine phosphatase. J Biol Chem 276:26036-26043

Asante-Appiah E, Patel S, Dufresne C, Roy P, Wang Q, Patel V, Friesen RW, Ramachandran C, Becker JW, Leblanc Y, Kennedy BP, Scapin G. (2002) The structure of PTP-1B in complex with a peptide inhibitor reveals an alternative binding mode for bisphosphonates. Biochemistry 41:9043-9051

Balsamo J, Arregui C, Leung T, Lilien J (1998) The nonreceptor protein tyrosine phosphatase PTP1B binds to the cytoplasmic domain of N-cadherin and regulates the cadherin-actin linkage. J Cell Biol 143:523-532

Barford D, Das AK, Egloff MP (1998) The structure and mechanism of protein phosphatases: insights into catalysis and regulation. Annu Rev Biophys Biomol Struct 27:133-164

Barford D, Flint AJ, Tonks NK (1994) Crystal structure of human protein tyrosine phosphatase 1B. Science 263:1397-1404

Barford D, Jia Z, Tonks NK (1995) Protein tyrosine phosphatases take off. Nat Struct Biol 2:1043-1053

Bjorge JD, Pang A, Fujita DJ (2000) Identification of protein-tyrosine phosphatase 1B as the major tyrosine phosphatase activity capable of dephosphorylating and activating c-Src in several human breast cancer cell lines. J Biol Chem 275:41439-41446

Blaskkovitch MA, Kim OH (2002) Recent advances in the discovery and development of Protein Tyrosine Phosphatase Inhibitors. Expert Opin Ther Patents 12:871-905

Bleasdale JE, Ogg D, Palazuk BJ, Jacob CS, Swanson ML, Wang XY, Thompson DP, Conradi RA, Mathews WR, Laborde AL, Stuchly CW, Heijbel A, Bergdahl K, Bannow CA, Smith CW, Svensson C, Liljebris C, Schostarez HJ, May PD, Stevens FC, Larsen SD (2001) Small molecule peptidomimetics containing a novel phosphotyrosine bioisostere inhibit protein tyrosine phosphatase 1B and augment insulin action. Biochemistry 40:5642-5654

Boute N, Boubekeur S, Lacasa D, Issad T (2003) Dynamics of the interaction between the insulin receptor and protein tyrosine-phosphatase 1B in living cells. EMBO Rep 4:313-319

Brown-Shimer S, Johnson KA, Hill DE, Bruskin AM (1992) Effect of protein tyrosine phosphatase 1B expression on transformation by the human neu oncogene. Cancer Res 52:478-482

Buckley DA, Cheng A, Kiely PA, Tremblay ML, O'Connor R (2002) Regulation of insulin-like growth factor type I (IGF-I) receptor kinase activity by protein tyrosine phosphatase 1B (PTP-1B) and enhanced IGF-I-mediated suppression of apoptosis and motility in PTP-1B-deficient fibroblasts. Mol Cell Biol 22:1998-2010

Burke TR Jr, Kole HK, Roller PP (1994) Potent inhibition of insulin receptor dephosphorylation by a hexamer peptide containing the phosphotyrosyl mimetic F2Pmp. Biochem Biophys Res Commun 204:129-134

BurkeTR Jr, Ye B, Yan X, Wang S, Jia Z, Chen L, Zhang ZY, Barford D. (1996) Small molecule interactions with protein-tyrosine phosphatase PTP1B and their use in inhibitor design. Biochemistry 35:15989-15996

Chalmers DT, Behan DP (2002) The use of constitutively active GPCRs in drug discovery and functional genomics. Nat Rev Drug Discov 1:599-608

Cheng A, Bal GS, Kennedy BP, Tremblay ML (2001) Attenuation of adhesion-dependent signaling and cell spreading in transformed fibroblasts lacking protein tyrosine phosphatase-1B. J Biol Chem 276:25848-25855

Cheng A, Uetani N, Simoncic PD, Chaubey VP, Lee-Loy A, McGlade CJ, Kennedy BP, Tremblay ML (2002) Attenuation of leptin action and regulation of obesity by protein tyrosine phosphatase 1B. Dev Cell 2:497-503

Chernoff J, Schievella AR, Jost CA, Erikson RL, Neel BG (1990) Cloning of a cDNA for a major human protein-tyrosine-phosphatase. Proc Natl Acad Sci USA 87:2735-2739

Claiborne,A, Yeh JI, Mallett TC, Luba J, Crane EJ, Charrier V, Parsonage D (1999) Protein-sulfenic acids: diverse roles for an unlikely player in enzyme catalysis and redox regulation. Biochemistry 38:15407-15416

Cohen N, Halberstam M, Shlimovich P, Chang CJ, Shamoon H, Rossetti L (1995) Oral vanadyl sulfate improves hepatic and peripheral insulin sensitivity in patients with non-insulin-dependent diabetes mellitus. J Clin Invest 95:2501-2509

Dadke S, Kusari J, Chernoff J (2000) Down-regulation of insulin signaling by protein-tyrosine phosphatase 1B is mediated by an N-terminal binding region. J Biol Chem 275:23642-23647

Denu JM, Lohse DL, Vijayalakshmi J, Saper MA, Dixon JE (1996) Visualization of intermediate and transition-state structures in protein-tyrosine phosphatase catalysis. Proc Natl Acad Sci USA 93:2493-2498

Desmarais S, Friesen RW, Zamboni R, Ramachandran C (1999) [Difluro(phosphono)methyl]phenylalanine-containing peptide inhibitors of protein tyrosine phosphatases. Biochem J 337:219-223

Di Paola R, Frittitta L, Miscio G, Bozzali M, Baratta R, Centra M, Spampinato D, Santagati MG, Ercolino T, Cisternino C, Soccio T, Mastroianno S, Tassi V, Almgren P, Pizzuti A, Vigneri R, Trischitta V (2002) A variation in 3' UTR of hPTP1B increases specific gene expression and associates with insulin resistance. Am J Hum Genet 70:806-812

Doman TN, McGovern SL, Witherbee BJ, Kasten TP, Kurumbail R, Stallings WC, Connolly DT, Shoichet BK (2002) Molecular docking and high-throughput screening for novel inhibitors of protein tyrosine phosphatase-1B. J Med Chem 45:2213-2221

Drevs J, Medinger M, Schmidt-Gersbach C, Weber R, Unger C (2003) Receptor tyrosine kinases: the main targets for new anticancer therapy. Curr Drug Targets 4:113-121

Echwald SM, Bach H, Vestergaard H, Richelsen B, Kristensen K, Drivsholm T, Borch-Johnsen K, Hansen T, Pedersen O (2002) A P387L variant in protein tyrosine phosphatase-1B (PTP-1B) is associated with type 2 diabetes and impaired serine phosphorylation of PTP-1B in vitro. Diabetes 51:1-6

Elchebly M, Payette P, Michaliszyn E, Cromlish W, Collins S, Loy AL, Normandin D, Cheng A, Himms-Hagen J, Chan CC, Ramachandran C, Gresser MJ, Tremblay ML, Kennedy BP (1999) Increased insulin sensitivity and obesity resistance in mice lacking the protein tyrosine phosphatase-1B gene. Science 283:1544-1548

Erlanson DA, Braisted AC, Raphael DR, Randal M, Stroud RM, Gordon EM, Wells JA (2000) Site-directed ligand discovery. Proc Natl Acad Sci USA 97:9367-9372

Erlanson DA, McDowell RS, He MM, Randal M, Simmons RL, Kung J, Waight A, Hansen SK (2003) Discovery of a New Phosphotyrosine Mimetic for PTP1B Using Breakaway Tethering. J Am Chem Soc 125:5602-5603

Finkel T (2003) Oxidant signals and oxidative stress. Curr Opin Cell Biol 15:247-254

Flint AJ, Gebbink MF, Franza BR Jr, Hill DE, Tonks NK (1993) Multi-site phosphorylation of the protein tyrosine phosphatase, PTP1B: identification of cell cycle regulated and phorbol ester stimulated sites of phosphorylation. EMBO J 12:1937-1946

Flint AJ, Tiganis T, Barford D, Tonks NK (1997) Development of "substrate-trapping" mutants to identify physiological substrates of protein tyrosine phosphatases. Proc Natl Acad Sci USA 94:1680-1685

Forsell PA, Boie Y, Montalibet J, Collins S, Kennedy BP (2000) Genomic characterization of the human and mouse protein tyrosine phosphatase-1B genes. Gene 260:145-153

Frangioni JV, Beahm PH, Shifrin V, Jost CA, Neel BG (1992) The nontransmembrane tyrosine phosphatase PTP-1B localizes to the endoplasmic reticulum via its 35 amino acid C-terminal sequence. Cell 68:545-560

Frangioni JV, Oda A, Smith M, Salzman EW, Neel BG (1993) Calpain-catalyzed cleavage and subcellular relocation of protein phosphotyrosine phosphatase 1B (PTP-1B) in human platelets. EMBO J 12:4843-4856

Frimurer TM, Peters GH, Iversen LF, Andersen HS, Moller NP, Olsen OH (2003) Ligand-induced conformational changes: improved predictions of ligand binding conformations and affinities. Biophys J 84:2273-2281

Fukada T, Tonks NK (2001) The reciprocal role of Egr-1 and Sp family proteins in regulation of the PTP1B promoter in response to the p210 Bcr-Abl oncoprotein-tyrosine kinase. J Biol Chem 276:25512-25519

Fukada T, Tonks NK (2003) Identification of YB-1 as a regulator of PTP1B expression: implications for regulation of insulin and cytokine signaling. EMBO J 22:479-493

Goldfine AB, Simonson DC, Folli F, Patti ME, Kahn CR (1995) Metabolic effects of sodium metavanadate in humans with insulin-dependent and noninsulin-dependent diabetes mellitus in vivo and in vitro studies. J Clin Endocrinol Metab 80:3311-3320

Groves MR, Yao ZJ, Roller PP, Burke TR Jr, Barford D (1998) Structural basis for inhibition of the protein tyrosine phosphatase 1B by phosphotyrosine peptide mimetics. Biochemistry 37:17773-17783

Gu F, Dube N, Kim JW, Cheng A, Ibarra-Sanchez MJ, Tremblay ML, Boisclair YR (2003) Protein tyrosine phosphatase 1B attenuates growth hormone-mediated JAK2-STAT signaling. Mol Cell Biol 23:3753-3762

Gum RJ, Gaede LL, Heindel MA, Waring JF, Trevillyan JM, Zinker BA, Stark ME, Wilcox D, Jirousek MR, Rondinone CM, Ulrich RG (2003a) Antisense Protein Tyrosine Phosphatase 1B (PTP1B) Reverses Activation of p38 Mitogen-Activated Protein Kinase in Liver of ob/ob Mice. Mol Endocrinol 17:1131-1143

Gum RJ, Gaede LL, Koterski SL, Heindel M, Clampit JE, Zinker BA, Trevillyan JM, Ulrich RG, Jirousek MR, Rondinone CM (2003b) Reduction of protein tyrosine phosphatase 1B increases insulin-dependent signaling in ob/ob mice. Diabetes 52:21-28

Haj FG, Markova B, Klaman LD, Bohmer FD, Neel BG (2003) Regulation of receptor tyrosine kinase signaling by protein tyrosine phosphatase-1B. J Biol Chem 278:739-744

Haj FG, Verveer PJ, Squire A, Neel BG, Bastiaens PI (2002) Imaging sites of receptor dephosphorylation by PTP1B on the surface of the endoplasmic reticulum. Science 295:1708-1711

Heffetz D, Bushkin I, Dror R, Zick Y (1990) The insulinomimetic agents H_2O_2 and vanadate stimulate protein tyrosine phosphorylation in intact cells. J Biol Chem 265:2896-2902

Huang Z, Wang Q, Ly HD, Gorvindarajan A, Scheigetz J, Zamboni R, Desmarais S, Ramachandran C (1999) 3,6-Fluorescein Diphosphate: A Sensitive Fluorogenic and Chromogenic Substrate for Protein Tyrosine Phosphatases. J Biomol Screen 4:327-334

Huijsduijnen RH, Bombrun A, Swinnen D (2002) Selecting protein tyrosine phosphatases as drug targets. Drug Discov Today 7:1013-1019

Huyer G, Liu S, Kelly J, Moffat J, Payette P, Kennedy B, Tsaprailis G, Gresser MJ, Ramachandran C (1997) Mechanism of inhibition of protein-tyrosine phosphatases by vanadate and pervanadate. J Biol Chem 272:843-851

Iversen LF, Andersen HS, Branner S, Mortensen SB, Peters GH, Norris K, Olsen OH, Jeppesen CB, Lundt BF, Ripka W, Moller KB, Moller NP (2000) Structure-based design of a low molecular weight, nonphosphorus, nonpeptide, and highly selective inhibitor of protein-tyrosine phosphatase 1B. J Biol Chem 275:10300-10307

Iversen LF, Andersen HS, Moller KB, Olsen OH, Peters GH, Branner S, Mortensen SB, Hansen TK, Lau J, Ge Y, Holsworth DD, Newman MJ, Moller NP (2001) Steric hindrance as a basis for structure-based design of selective inhibitors of protein-tyrosine phosphatases Biochemistry 40:14812-14820

Jia Z, Barford D, Flint AJ, Tonks NK (1995) Structural basis for phosphotyrosine peptide recognition by protein tyrosine phosphatase 1B. Science 268:1754-1758

Jia,Z, Ye Q, Dinaut AN, Wang Q, Waddleton D, Payette P, Ramachandran C, Kennedy B, Hum G, Taylor SD (2001) Structure of protein tyrosine phosphatase 1B in complex with inhibitors bearing two phosphotyrosine mimetics. J Med Chem 44:4584-4594

Johnson TO, Ermolieff J, Jirousek MR (2002) Protein tyrosine phosphatase 1B inhibitors for diabetes. Nat Rev Drug Discov 1:696-709

Klaman LD, Boss O, Peroni OD, Kim JK, Martino JL, Zabolotny JM, Moghal N, Lubkin M, Kim YB, Sharpe AH, Stricker-Krongrad A, Shulman GI, Neel BG, Kahn BB (2000) Increased energy expenditure, decreased adiposity, and tissue-specific insulin sensitivity in protein-tyrosine phosphatase 1B-deficient mice 1968. Mol Cell Biol 20:5479-5489

Klarlund JK (1985) Transformation of cells by an inhibitor of phosphatases acting on phosphotyrosine in proteins. Cell 41:707-717

LaMontagne KR Jr, Flint AJ, Franza BR Jr, Pandergast AM, Tonks NK (1998) Protein tyrosine phosphatase 1B antagonizes signalling by oncoprotein tyrosine kinase p210 bcr-abl in vivo. Mol Cell Biol 18:2965-2975

LaMontagne KR Jr, Hannon G, Tonks NK (1998) Protein tyrosine phosphatase PTP1B suppresses p210 bcr-abl-induced transformation of rat-1 fibroblasts and promotes differentiation of K562 cells. Proc Natl Acad Sci USA 95:14094-14099

Larsen SD, Barf T, Liljebris C, May PD, Ogg D, O'Sullivan TJ, Palazuk BJ, Schostarez HJ, Stevens FC, Bleasdale JE (2002) Synthesis and biological activity of a novel class of small molecular weight peptidomimetic competitive inhibitors of protein tyrosine phosphatase 1B. J Med Chem 45:598-622

Larsen SD, Stevens FC, Lindberg TJ, Bodnar PM, O'Sullivan TJ, Schostarez HJ, Palazuk BJ, Bleasdale JE (2003) Modification of the N-terminus of peptidomimetic protein tyrosine phosphatase 1B (PTP1B) inhibitors: identification of analogues with cellular activity. Bioorg Med Chem Lett 13:971-975

Lee JH, Reed DR, Li WD, Xu W, Joo EJ, Kilker RL, Nanthakumar E, North M, Sakul H, Bell C, Price RA (1999) Genome scan for human obesity and linkage to markers in 20q13. Am J Hum Genet 64:196-209

Lee SR, Kwon KS, Kim SR, Rhee SG (1998) Reversible inactivation of protein-tyrosine phosphatase 1B in A431 cells stimulated with epidermal growth factor. J Biol Chem 273:15366-15372

Leung-Toung R, Li W, Tam TF, Karimian K (2002) Thiol-dependent enzymes and their inhibitors: a review. Curr Med Chem 9:979-1002

Lilien J, Balsamo J, Arregui C, Xu G (2002) Turn-off, drop-out: functional state switching of cadherins. Dev Dyn 224:18-29

Liljebris C, Larsen SD, Ogg D, Palazuk BJ, Bleasdale JE (2002) Investigation of potential bioisosteric replacements for the carboxyl groups of peptidomimetic inhibitors of protein tyrosine phosphatase 1B: identification of a tetrazole-containing inhibitor with cellular activity. J Med Chem 45:1785-1798

Liu F, Hill DE, Chernoff J (1996) Direct binding of the proline-rich region of protein tyrosine phosphatase 1B to the Src homology 3 domain of p130(Cas). J Biol Chem 271:31290-31295

Lyon MA, Ducruet AP, Wipf P, Lazo JS (2002) Dual-specificity phosphatases as targets for antineoplastic agents. Nat Rev Drug Discov 1:961-976

Mahadev K, Zilbering A, Zhu L, Goldstein BJ (2001) Insulin-stimulated hydrogen peroxide reversibly inhibits protein-tyrosine phosphatase 1b in vivo and enhances the early insulin action cascade. J Biol Chem 276:21938-21942

Milarski,KL, Zhu G, Pearl CG, McNamara DJ, Dobrusin EM, MacLean D, Thieme-Sefler A, Zhang ZY, Sawyer T, Decker SJ (1993) Sequence specificity in recognition of the epidermal growth factor receptor by protein tyrosine phosphatase 1B. J Biol Chem 268:23634-23639

Moeslein FM, Myers MP, Landreth GE (1999) The CLK family kinases, CLK1 and CLK2, phosphorylate and activate the tyrosine phosphatase, PTP-1B. J Biol Chem 274:26697-26704

Mok A, Cao H, Zinman B, Hanley AJ, Harris SB, Kennedy BP, Hegele RA (2002) A single nucleotide polymorphism in protein tyrosine phosphatase PTP-1B is associated with protection from diabetes or impaired glucose tolerance in Oji-Cree. J Clin Endocrinol Metab 87:724-727

Murthy VS, Kulkarni VM (2002) Molecular modeling of protein tyrosine phosphatase 1B (PTP 1B) inhibitors. Bioorg Med Chem 10:897-906

Myers MP, Andersen JN, Cheng A, Tremblay ML, Horvath CM, Parisien JP, Salmeen A, Barford D, Tonks NK (2001) TYK2 and JAK2 are substrates of protein-tyrosine phosphatase 1B. J Biol Chem 276:47771-47774

Pannifer AD, Flint AJ, Tonks NK, Barford D (1998) Visualization of the cysteinyl-phosphate intermediate of a protein-tyrosine phosphatase by x-ray crystallography. J Biol Chem 273:10454-10462

Pathre P, Arregui C, Wampler T, Kue I, Leung TC, Lilien J, Balsamo J (2001) PTP1B regulates neurite extension mediated by cell-cell and cell-matrix adhesion molecules. J Neurosci Res 63:143-150

Peters GH, Iversen LF, Branner S, Andersen HS, Mortensen SB, Olsen OH, Moller KB, Moller NP (2000) Residue 259 is a key determinant of substrate specificity of protein-tyrosine phosphatases 1B and alpha. J Biol Chem 275:18201-18209

Puius YA, Zhao Y, Sullivan M, Lawrence DS, Almo SC, Zhang ZY (1997) Identification of a second aryl phosphate-binding site in protein-tyrosine phosphatase 1B: a paradigm for inhibitor design. Proc Natl Acad Sci USA 94:13420-13425

Ramachandran C, Kennedy BP (2003) Protein tyrosine phosphatase 1B: a novel target for type 2 diabetes and obesity. Curr Top Med Chem 3:749-757

Rhee J, Lilien J, Balsamo J (2001) Essential tyrosine residues for interaction of the non-receptor protein-tyrosine phosphatase PTP1B with N-cadherin. J Biol Chem 276:6640-6644

Roller PP, Wu L, Zhang ZY, Burke TR Jr (1998) Potent inhibition of protein-tyrosine phosphatase-1B using the phosphotyrosyl mimetic fluoro-O-malonyl tyrosine (FOMT). Bioorg Med Chem Lett 8:2149-2150

Salmeen A, Andersen JN, Myers MP, Meng TC, Hinks JA, Tonks NK, and Barford D (2003) Redox regulation of protein tyrosine phosphatase 1B involves a sulphenyl-amide intermediate. Nature 423:769-773

Salmeen A, Andersen JN, Myers MP, Tonks NK, Barford D (2000) Molecular basis for the dephosphorylation of the activation segment of the insulin receptor by protein tyrosine phosphatase 1B. Mol Cell 6:1401-1412

Sarmiento M, Puius YA, Vetter SW, Keng YF, Wu L, Zhao Y, Lawrence DS, Almo SC, Zhang ZY. (2000) Structural basis of plasticity in protein tyrosine phosphatase 1B substrate recognition. Biochemistry 39:8171-8179

Sarmiento M, Wu L, Keng YF, Song L, Luo Z, Huang Z, Wu GZ, Yuan AK, Zhang ZY (2000) Structure-based discovery of small molecule inhibitors targeted to protein tyrosine phosphatase 1B. J Med Chem 43:146-155

Scapin G, Patel S, Patel V, Kennedy B, Asante-Appiah E (2001) The structure of apo protein-tyrosine phosphatase 1B C215S mutant: more than just an S --> O change. Protein Sci 10:1596-1605

Shen K, Keng YF, Wu L, Guo XL, Lawrence DS, Zhang ZY (2001) Acquisition of a specific and potent PTP1B inhibitor from a novel combinatorial library and screening procedure. J Biol Chem 276:47311-47319

Shuker SB, Hajduk PJ, Meadows RP, Fesik SW (1996) Discovering high-affinity ligands for proteins: SAR by NMR. Science 274:1531-1534

Skorey K, Ly HD, Kelly J, Hammond M, Ramachandran C, Huang Z, Gresser MJ, Wang Q (1997) How does alendronate inhibit protein-tyrosine phosphatases? J Biol Chem 272:22472-22480

Skorey KI, Kennedy BP, Friesen RW, Ramachandran C (2001) Development of a robust scintillation proximity assay for protein tyrosine phosphatase 1B using the catalytically inactive (C215S) mutant. Anal Biochem 291:269-278

Sun JP, Fedorov AA, Lee SY, Guo XL, Shen K, Lawrence DS, Almo SC, and Zhang ZY (2003) Crystal structure of PTP1B complexed with a potent and selective bidentate inhibitor. J Biol Chem 278:12406-12414

Szczepankiewicz BG, Liu G, Hajduk PJ, Abad-Zapatero C, Pei Z, Xin Z, Lubben TH, Trevillyan JM, Stashko MA, Ballaron SJ, Liang H, Huang F, Hutchins CW, Fesik SW, Jirousek MR (2003) Discovery of a potent, selective protein tyrosine phosphatase 1B inhibitor using a linked-fragment strategy. J Am Chem Soc 125:4087-4096

Taylor SD (2003) Inhibitors of protein tyrosine phosphatase 1B (PTP1B). Curr Top Med Chem 3:759-782

Tobin JF, Tam S (2002) Recent advances in the development of small molecule inhibitors of PTP1B for the treatment of insulin resistance and type 2 diabetes. Curr Opin Drug Discov Devel 5:500-512

Tonks NK, Diltz CD, Fischer EH (1988a) Characterization of the major protein-tyrosine-phosphatases of human placenta. J Biol Chem 263:6731-6737

Tonks NK, Diltz CD, Fischer EH (1988b) Purification of the major protein-tyrosine-phosphatases of human placenta. J Biol Chem 263:6722-6730

Traxler P (2003) Tyrosine kinases as targets in cancer therapy - successes and failures. Expert Opin Ther Targets 7:215-234

Urbanek RA, Suchard SJ, Steelman GB, Knappenberger KS, Sygowski LA, Veale CA, Chapdelaine MJ (2001) Potent reversible inhibitors of the protein tyrosine phosphatase CD45. J Med Chem 44:1777-1793

Van Montfort RL, Congreve M, Tisi D, Carr R, Jhoti H (2003) Oxidation state of the active-site cysteine in protein tyrosine phosphatase 1B. Nature 423:773-777

Wang XY, Bergdahl K, Heijbel A, Liljebris C, Bleasdale JE (2001) Analysis of in vitro interactions of protein tyrosine phosphatase 1B with insulin receptors. Mol Cell Endocrinol 173:109-120

Xin Z, Oost TK, Abad-Zapatero C, Hajduk PJ, Pei Z, Szczepankiewicz BG, Hutchins CW, Ballaron SJ, Stashko MA, Lubben T, Trevillyan JM, Jirousek MR, Liu G (2003) Potent, selective inhibitors of protein tyrosine phosphatase 1B. Bioorg Med Chem Lett 13:1887-1890

Xu G, Arregui C, Lilien J, Balsamo J (2002) PTP1B modulates the association of beta-catenin with N-cadherin through binding to an adjacent and partially overlapping target site. J Biol Chem 277:49989-49997

Yao ZJ, Ye B, Wu XW, Wang S, Wu L, Zhang ZY, Burke TR Jr (1998) Structure-based design and synthesis of small molecule protein-tyrosine phosphatase 1B inhibitors. Bioorg Med Chem 6:1799-1810

Yigzaw Y, Poppleton HM, Sreejayan N, Hassid A, Patel TB (2003) Protein-tyrosine phosphatase-1B (PTP1B) mediates the anti-migratory actions of Sprouty. J Biol Chem 278:284-288

Zabolotny JM, Bence-Hanulec KK, Stricker-Krongrad A, Haj F, Wang Y, Minokoshi Y, Kim YB, Elmquist JK, Tartaglia LA, Kahn BB, Neel BG (2002) PTP1B regulates leptin signal transduction in vivo 6482. Dev Cell 2:489-495

Zhang ZY (2002) Protein tyrosine phosphatases: structure and function, substrate specificity, and inhibitor development Annu Rev Pharmacol Toxicol 42:209-234

Zhang ZY, Lee SY (2003) PTP1B inhibitors as potential therapeutics in the treatment of type 2 diabetes and obesity. Expert Opin Investig Drugs 12:223-233

Zhang ZY, MacLean D, McNamara DJ, Sawyer TK, Dixon JE (1994) Protein tyrosine phosphatase substrate specificity: size and phosphotyrosine positioning requirements in peptide substrates. Biochemistry 33:2285-2290

Zinker BA, Rondinone CM, Trevillyan JM, Gum RJ, Clampit JE, Waring JF, Xie N, Wilcox D, Jacobson P, Frost L, Kroeger PE, Reilly RM, Koterski S, Opgenorth TJ, Ulrich RG, Crosby S, Butler M, Murray SF, McKay RA, Bhanot S, Monia BP, Jirousek MR (2002) PTP1B antisense oligonucleotide lowers PTP1B protein, normalizes blood glucose, and improves insulin sensitivity in diabetic mice. Proc Natl Acad Sci USA 99:11357-11362

12 The CD45 phosphotyrosine phosphatase

Denis R. Alexander

Abstract

The CD45 transmembrane tyrosine phosphatase is an abundant glycoprotein expressed on all nucleated haematopoietic cells. It plays a critical role in regulating the threshold for signalling via the T and B cell antigen receptors and can also exert positive or negative effects on other receptors in immune cells. Mice and humans lacking CD45 have a severe combined immunodeficiency syndrome characterised by dysfunctional thymic development and few mature peripheral T cells. The key substrates for CD45 include members of the Src family of tyrosine kinases such as p56^{lck} and p59^{fyn}. The actions of CD45 are regulated by its subcellular localisation with reference to substrates, by interactions of motifs within its cytoplasmic tail and possibly by homodimerisation. Alternative splicing generates multiple CD45 isoforms that are differentially expressed on the surface of immune cells during development and activation. Four different models to explain the differential functions of isoforms are considered.

12.1 Introduction

The CD45 phosphotyrosine phosphatase (PTPase) is an abundant transmembrane glycoprotein expressed on the surface of all nucleated haematopoietic cells. In the earlier literature, CD45 was also called T200 or the leucocyte common antigen. The molecule has a large external domain (391-552 amino acids) and an extensive cytoplasmic tail (700 amino acids) containing tandem repeat sequences known as Domain 1 and Domain 2. Only Domain 1 expresses PTPase activity and Domain 2 has a regulatory function. The main role of CD45 is to regulate the signalling thresholds of receptors expressed on immune cells. A series of extensive reviews are available describing the properties, regulation and functions of CD45 (Thomas 1989; Trowbridge and Thomas 1994; Alexander 1997; Thomas and Brown 1999; Ashwell and Doro 1999; Alexander 2000; Justement 2001; Hermiston et al. 2003), and the reader is referred to these publications for discussions of earlier findings. This Chapter will focus on more recent results and will emphasise the different model systems used to investigate the functions of CD45, together with a discussion of the outstanding questions that continue to intrigue investigators in this research field.

Topics in Current Genetics, Vol. 5
J. Arino, D.R. Alexander (Eds.) Protein phosphatases
© Springer-Verlag Berlin Heidelberg 2004

232 Denis R. Alexander

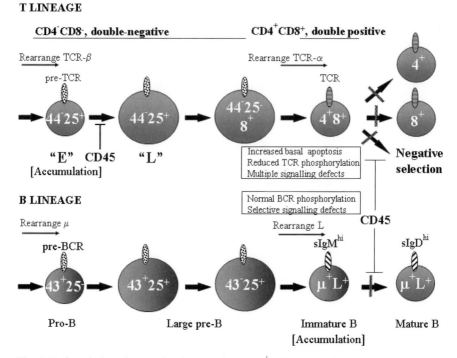

Fig. 1. Defects in lymphocyte development in CD45^{-/-} mice. Note that whereas positive selection in T cells is severely affected, negative selection and the maturation of CD4-CD8-
to CD4+CD8+ thymocytes are only partially inhibited. "E" = "expected size", referring to
CD44-CD25+ cells which have not yet entered the cell-cycle, whereas "L" refers to "larger
than expected" blast cells which are receiving pre-TCR induced proliferation signals.

12.2 CD45 regulates lymphocyte development

The generation of three independent mouse lines in which the CD45 gene has
been deleted by targeting exon-6 (Kishihara et al. 1993), exon-9 (Byth et al. 1996)
or exon-12 (Mee et al. 1999) has established a critical role for the molecule in
regulating the development of both the T and B cell lineages. Lymphoid precursors develop into mature T cell antigen receptor expressing (TCRαβ+) T cells
through a series of discrete maturation stages characterised by expression of CD4
and CD8 co-receptors. Successful TCRβ rearrangements lead to expression of the
immature form of the TCR, known as the pre-TCR, on CD4⁻CD8⁻ (double-
negative, DN) thymocytes and the subsequent mediation of survival and proliferative signals in a process known as β-selection. In CD45^{-/-} mice, as Figure 1 illustrates, there are two distinct defects in thymic development: a two-fold reduction
in the transition from DN to CD4⁺CD8⁺ (double-positive, DP) cells, and a five-
fold reduction in the further maturation of DP to single-positive CD4⁺ and CD8⁺

thymocytes (Byth et al. 1996; Kishihara et al. 1993). Partial defects in pre-TCR signalling are observed in the absence of CD45, readily explaining the partial block in DN thymocyte maturation (Pingel et al. 1999). Increased apoptosis in CD45$^{-/-}$ DP thymocytes also contributes to the reduced size of the CD45-deficient thymus (Byth et al. 1996). The marked increase in the TCR signalling threshold of CD45$^{-/-}$ DP thymocytes results in a major defect in positive selection, so explaining the small numbers of mature SP thymocytes, whereas negative selection is only partially defective (Conroy et al. 1996; Wallace et al. 1997; Mee et al. 1999). As a result of these shifts in the thresholds of selection events, self MHC-peptides that would normally have caused deletion of self-reactive TCRs now result instead in positive selection, leading to a high proportion of T cells expressing autoreactive TCRs (Trop et al. 2000). Nevertheless the development of autoimmunity in CD45$^{-/-}$ mice is not as aggressive as this finding might suggest, presumably because the few T cells that exit to the periphery are markedly non-responsive to antigenic stimulation (Stone et al. 1997).

Defects in B cell development in the absence of CD45 are less marked than those observed in the T lineage. The number of B cells produced in the bone marrow and their migration to the periphery appear largely normal, but their further maturation from IgM^{hi} IgD^{hi} (T2) cells into the IgM^{lo} IgD^{hi} phenotype typical of follicular B cells is impaired (Byth et al. 1996). The thresholds for B cell selection events are altered in CD45$^{-/-}$ mice in a manner analogous to the changes observed in the T lineage. In CD45$^{-/-}$ mice back-crossed to mice carrying immunoglobulin genes specific for hen egg lysozyme (HEL), the circulating HEL autoantigen which mediates negative selection in wild type mice now instead positively selects HEL-binding B cells, leading to their accumulation as long-lived IgD^{hi} cells (Cyster et al. 1996).

Patients with CD45$^-$deficiency have been described resulting, as in mice, in a severe-combined immunodeficiency (Kung et al. 2000; Tchilian et al. 2001). Overall, the defects observed in lymphocyte development in both CD45$^{-/-}$ mice and humans point to a dominantly positive role for CD45 in regulating signalling via the T and B cell antigen receptors. In the absence of CD45, the signal transduction thresholds are higher. During T cell development in a CD45$^{+/+}$ wild type mouse the threshold for signalling increases as maturation proceeds. The pre-TCR has the lowest threshold, apparently requiring no ligand, signalling spontaneously following its correct assembly; the threshold for the selection events that occur at the DP stage of thymic differentiation are at an intermediate level; finally the TCR in naive mature peripheral T cells is set at an even higher level. In the absence of CD45, the threshold for receptor signal transduction is increased at each of these maturation stages.

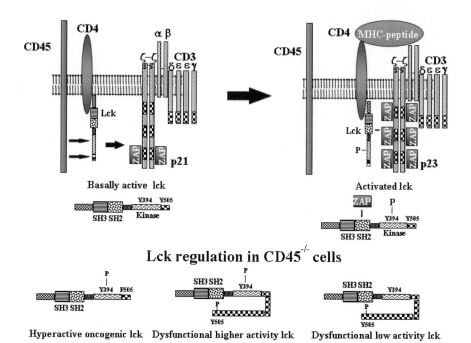

Fig. 2. A model explaining the role of CD45-regulated p56^{lck} in T cell antigen receptor signal transduction. The upper panel illustrates the molecular changes thought to result from ligation of the CD4/TCR complex by MHC-peptides. The lower panel shows the way in which CD45 can act at either of the two major regulatory p56^{lck} phosphorylation sites

12.3 CD45 regulates the Src family tyrosine kinases

12.3.1 In T cell antigen receptor signalling

The lymphocyte antigen receptors do not contain intrinsic kinase activity but instead couple to intracellular signalling pathways by means of tyrosine kinases belonging to the Src family. In T cells, the p56^{lck} kinase, associated with the CD4 and CD8 co-receptors, plays the main role in this process, although p59^{fyn} is also involved. Upon engagement of the TCR by MHC-peptide ligands expressed on antigen presenting cells, tyrosine containing (ITAM) motifs in the invariant CD3/TCR polypeptide cytoplasmic tails are phosphorylated leading to the recruitment of SH2-domain containing proteins such as the ZAP-70 tyrosine kinase (Fig. 2). Activation of ZAP-70 by p56^{lck} results in further ZAP-70 phosphorylation events resulting in turn in the recruitment of adapter proteins that couple the receptor to intracellular signalling pathways.

The consequences of CD45-deficiency in primary T cells are consistent with this model of receptor coupling. In DN thymocytes, the invariant CD3 polypeptide components of the pre-TCR are defective in their ability to induce normal tyrosine phosphorylation signals and to cause the further maturation of cells to the DP stage (Pingel et al. 1999). In DP thymocytes the tyrosine phosphorylation of the TCR-ζ chain and subsequent recruitment of the ZAP-70 kinase are severely defective, correlating with hyper-phosphorylation of p56^{lck}, and leading to a defect in calcium signalling (Stone et al. 1997). In mature peripheral CD45$^{-/-}$ T cells, TCR-induced proliferation is essentially ablated (Stone et al. 1997) and the production of cytokines such as IL-2, IL-4 and IL-10 is also much reduced (Perry and Alexander, unpublished). The extent of signalling defects in each case correlates with the T cell maturation stage, milder dysfunction characterising the CD3/pre-TCR on DN thymocytes, in contrast to the TCR on mature T cells that appears to be largely uncoupled from intracellular activation signals.

The p56^{lck} kinase, as with other members of the Src kinase family, is regulated by a wide range of factors. These include: localisation of the kinase in reference to the plasma membrane; binding of proteins containing Pro-rich motifs to the p56^{lck} SH3 domain; engagement of its SH2 domain by pTyr residue-containing proteins; autophosphorylation of the activatory Tyr-394 residue; and phosphorylation by the C-terminal Src kinase (Csk) of the inhibitory Tyr-505 residue. In the absence of CD45 in either T cell lines (Ostergaard et al. 1989; McFarland et al. 1993; Sieh et al. 1993) or primary T cells (Stone et al. 1997), p56^{lck} becomes hyper-phosphorylated at Tyr-505. This causes the kinase to adopt a 'closed conformation' in which the pTyr-505 residue forms an intramolecular association with its SH2 domain (Sieh et al. 1993; Stone et al. 1997). This version of the kinase lacks optimal activity and, perhaps more importantly, could render the p56^{lck} SH2 and SH3 domains less available to engage exogenous proteins. One candidate protein for such interactions is pTyr-319 in the linker region of ZAP-70, which has been proposed to engage with the p56^{lck} SH2 domain (Pelosi et al. 1999). Upon dephosphorylation of pTyr-505 by CD45, p56^{lck} then adopts an 'open conformation', with a modest increase in kinase activity, and with its SH2 and SH3 domains fully available for exogenous liaisons. It should be noted that dephosphorylation of pTyr-505 is not the only way in which such an open conformation is promoted. For example, a proline motif containing peptide in the CD28 cytoplasmic tail can bind to the p56^{lck} SH3 domain (Holdorf et al. 1999), and the Herpes Saimiri viral Tip protein can disrupt the relatively weak interaction between the p56^{lck} SH2 domain and pTyr-505 (Hartley et al. 1999), in both cases leading to increased p56^{lck} functionality. The idea that dephosphorylation of p56^{lck} pTyr-505 by CD45 is a physiologically relevant action of this PTPase is supported by the finding that in CD45$^{-/-}$ mice back-crossed to mice expressing a mutant lckY505F transgene, there is significant restoration of T cell development and reversal of the increased apoptosis that characterises CD45$^{-/-}$ DP thymocytes (Pingel et al. 1999; Seavitt et al. 1999).

Mutational analysis of p56^{lck} has shown that autophosphorylation of the Tyr-394 residue plays a stronger positive role in promoting the enzyme activity of the kinase in comparison with the relatively weak inhibitory effect of Tyr-505 phosphorylation (Doro et al. 1996). In some CD45-deficient transformed cell lines (Doro et al. 1996), as well as murine CD45$^{-/-}$ thymocytes (Baker et al. 2000), increased phosphorylation at Tyr-394 has been observed, suggesting that CD45 also has the capability to dephosphorylate this residue. Indeed, the overall $p56^{lck}$ kinase activity in the CD45$^{-/-}$ thymus is higher than in the wild type (Doro and Ashwell 1999), consistent with the idea that it is the pTyr-394 site that is most potent in regulating activity. Dramatic confirmation that CD45 could dephosphorylate $p56^{lck}$ pTyr-394 in vivo came from the observation that aggressive T cell lymphomas develop in CD45$^{-/-}p56^{lck}$-Y505F mice, caused by the conversion of the mutant $p56^{lck}$-Y505F into an oncogene by increased Tyr-394 phosphorylation in the absence of CD45 (Baker et al. 2000). It therefore appears that in the CD45-deficient T cell, pools of $p56^{lck}$ may exist that are either hyper-phosphorylated at Tyr-505, or at Tyr-394, and perhaps at both sites simultaneously. How CD45 might regulate both these sites in situ will be discussed further below in the context of its subcellular localisation in relation to its kinase substrates.

In addition to $p56^{lck}$, the $p59^{fyn}$ tyrosine kinase has also been shown to be regulated by CD45 in vivo in both cell lines (Shiroo et al. 1992; McFarland et al. 1993) and in primary thymocytes (Stone et al. 1997). The adapter protein SKAP-55 may play a role in binding to CD45 and mediating its effects on $p59^{fyn}$ (Wu et al. 2002). The precise function of CD45-regulated $p59^{fyn}$ in coupling the TCR or other receptors to specific signalling pathways has not as yet been completely defined.

12.3.2 In B cell antigen receptor signalling

The BCR is almost completely uncoupled from proliferative signals in CD45$^{-/-}$ B cells (Benatar et al. 1996; Byth et al. 1996), correlating with defects in calcium signalling and in activation of the Erk-2/pp90-Rsk pathways (Benatar et al. 1996; Cyster et al. 1996). Unlike the situation in T cells, however, it has proved much more difficult to establish unambiguous CD45 substrates in B cells. Indeed, BCR-stimulated tyrosine phosphorylation events appear normal in CD45$^{-/-}$ B cells. The reason for this may be the greater degree of redundancy in the utilisation of tyrosine kinases by the BCR in comparison with the TCR. Furthermore $p59^{lyn}$, a further member of the Src family of tyrosine kinases, is involved not only in coupling the BCR to intracellular signals, but also in mediating negative signals via the CD22 co-receptor, so there is scope for compensation between different receptors expressed on the same B cell. In fact, recruitment of the Syk tyrosine kinase to the BCR proceeds normally in a CD45-deficient cell-line, even though $p59^{lyn}$ is hyper-phosphorylated at its C-terminus inhibitory Tyr-508 regulatory site and is no longer itself recruited to the receptor (Pao et al. 1997).

Whether $p59^{lyn}$ is the key B cell substrate for CD45 under physiological conditions, explaining the defects observed in BCR signalling, remains controversial. In CD45-deficient B cell lines, there is evidence that the kinase is hyper-phosphorylated at both its autophosphorylation and C-terminus inhibitory regulatory Tyr residues, correlating with reduced activity of the kinase (Yanagi et al. 1996). But in a different CD45-deficient B cell line, basal $p59^{lyn}$ kinase activity is normal despite hyper-phosphorylation of $p59^{lyn}$ (Pao and Cambier 1997), whereas in a CD45-deficient sub-clone of the WEHI-231 B cell-line, overall p59^{lyn} activity is higher with increased phosphorylation at both regulatory Tyr residues (Katagiri et al. 1999). In contrast to these cell-line studies, dysfunctional regulation of p59^{lyn} in primary CD45$^{-/-}$ B cells has not yet been reported. One possible reason for these conflicting results may be that different pools of Src family kinases are differentially regulated by CD45 and by other tyrosine phosphatases depending on their subcellular localisation (Biffen et al. 1994). Measurements of the 'average' activities and phosphorylation states of kinases by using enzyme immunoprecipitates from whole cell lysates may not sufficiently discriminate between such pools. Therefore, in B cells physiological CD45 substrates have not yet been unambiguously established.

12.3.3 In integrin receptor signalling

CD45$^{-/-}$ macrophages display abnormally high integrin receptor-mediated adherence compared with wild type controls, correlating with the maturation of fewer bone marrow-derived macrophages when differentiated in vitro (Roach et al. 1997). The p56/p59^{hck} and p59^{lyn} tyrosine kinases are both hyperactive in CD45$^{-/-}$ macrophages, so in contrast to the situation with antigen receptors, CD45 appears to exert a predominantly negative effect on integrin signalling. Consistent with this observation, in a CD45-deficient T cell line enhanced adhesion to fibronectin via the $\alpha5\beta1$ (VLA-5) integrin but not via $\alpha4\beta1$ (VLA-4) has been noted (Shenoi et al. 1999).

12.3.4 In chemokine receptor signalling

The chemokine receptor CXCR4 and its ligand, stromal cell-derived factor-1α (CXCL12), regulate lymphocyte trafficking and play an important role in host immune surveillance. A reduction in CXCL12-induced chemotaxis has been noted in a CD45-deficient Jurkat T cell line, which could be restored upon CD45 re-expression (Fernandis et al. 2003). Whereas CXCL12 caused increased p56^{lck} kinase activity in the CD45-transfected cells, no activation was detected in the CD45-deficient cells in which p56^{lck} is hyper-phosphorylated at Tyr-505, correlating with defective tyrosine phosphorylation of downstream effectors such as ZAP-70 and Slp-76 (Fernandis et al. 2003). Therefore, in this cell line system, at least, CD45 exerts a positive regulatory effect on chemokine receptor signal transduction that appears to be p56^{lck}-dependent.

12.4. The subcellular localisation of CD45 in relation to its substrates

A critical factor in the regulation of CD45 is likely to be its precise subcellular localisation. The protein appears to reach the cell surface by both Golgi-dependent and -independent pathways (Baldwin and Ostergaard 2002). There is also evidence that TCR-stimulation results in a redistribution of CD45 from the Golgi (Minami et al. 1991) and that some of this CD45 relocates to the cell surface, facilitated by β1 spectrin and ankyrin cytoskeletal components (Pradhan and Morrow 2002). Ligation of CD45 itself with a monoclonal antibody also caused a marked depletion of intracellular CD45 from the Golgi region, although in this case cell-surface expression of CD45 was not increased (Shivnan et al. 1996). Therefore, CD45 subcellular localisation appears to be dynamically regulated and is influenced by receptor-mediated events. How is the location of CD45 regulated once it reaches the cell surface?

12.4.1 CD45 and lipid rafts

Lipid rafts are clusters of glycosphingolipids, cholesterol and other lipids, which form discrete assemblies distinct from the plasma membrane phospholipid bilayer. Co-localisation of signalling molecules and receptors within lipid rafts has been proposed as a mechanism for promoting receptor signal transduction coupling. It has been reported that CD45 is largely excluded from lipid rafts in both T cells (Rodgers and Rose 1996; Xavier et al. 1998) and in B cells (Cheng et al. 1999). Since the Src family tyrosine kinases are mainly located in lipid rafts, this raises the question as to how CD45 might regulate these kinases when located in a different membrane compartment. Indeed, the $p56^{lck}$ found in Triton-X100 insoluble lipid raft fractions isolated from T cells was found to be of lower activity in comparison to detergent-soluble $p56^{lck}$, as one might expect from the exclusion of CD45 from lipid rafts (Rodgers and Rose 1996). On the other hand, targeting of the CD45 cytoplasmic tail to lipid rafts in a T cell hybridoma caused downregulation of TCR signalling and a decrease in the raft-associated tyrosine kinase activity, presumably by dephosphorylating the autophosphorylation sites of kinases such as $p56^{lck}$ and $p59^{fyn}$ (He et al. 2002). Clearly, the precise amount of CD45 located in or near lipid rafts has the potential to be a critical regulator of Src kinase function. One possibility is that CD45, which is expressed abundantly at the cell surface, could be closely located around lipid rafts, albeit excluded, thereby making its phosphatase activity available for the regulation of the Src kinases (Thomas 1999).

However, there are also reports that small amounts of CD45 can be detected in lymphocyte lipid raft fractions in some experimental model systems. For example, microscopic visualisation of living B cells indicated that CD45 co-localises with rafts in about 10% of cells and this association increased upon BCR stimulation (Gupta and DeFranco 2003), although a different study on T cells that also employed imaging concluded that CD45 was excluded from lipid rafts (Janes et al. 1999). Raft fractions from murine thymocytes prepared with Brij-58 detergent rather than Triton-X100 contain small amounts of CD45 (Montixi et al. 1998) and the amount of CD45 found in such fractions is clearly detergent-dependent (Ilangumaran et al. 1999). As much as 5% of the total CD45 has been reported to be present in Triton-X100-insoluble lipid rafts fractions prepared from a T cell line (Edmonds and Ostergaard 2002). The importance of precisely titrating the amount of CD45 tyrosine phosphatase available in lipid rafts has also been highlighted by a study in which three chimaeric proteins, each containing the CD45 cytoplasmic tail but having small (Thy-1), intermediate (CD2) or large (CD43) ectodomains, were expressed in a CD45-deficient T cell line (Irles et al. 2003). The expectation in this investigation was that the large ectodomain would prevent superantigenmediated TCR stimulation by preventing the close apposition of the ligand. Unexpectedly, however, the CD43-CD45 chimaera was found to be most efficient at restoring TCR signalling, whereas the chimaeras with small ectodomains were poor in this respect. This correlated with the localisation of the CD43-CD45 chimaera in lipid rafts, whereas chimaeras with smaller ectodomains were excluded. In this study, some wild type CD45 was also found in raft fractions. Very small amounts of CD45 are efficient in regulating the Src kinases, so small amounts of the phosphatase could well be significant in terms of kinase regulation. It is possible that low amounts of CD45 tyrosine phosphatase activity targeted to rafts could exert a positive effect on Src kinases by dephosphorylating their inhibitory C-termini (Irles et al. 2003), whereas higher levels could, in addition, act at kinase autophosphorylation sites, thereby exerting a net negative effect (He et al. 2002).

In summary, the localisation of CD45 in relation to lipid rafts remains controversial and there are a number of studies reporting contradictory results. Resolution of this issue will most likely require more discriminatory imaging techniques that avoid the use of detergents.

12.4.2 CD45 and the immune synapse

The engagement of antigen receptors by ligand results in the formation of a highly structured ensemble of molecules at the receptor contact point known as 'the immune synapse', 'immunological synapse' or 'supramolecular activation cluster' (SMAC). All these terms refer to the same entity. It is thought that the immune synapse is a consequence of initial signalling events and provides a mechanism to facilitate a prolonged and carefully orchestrated transmission of intracellular signals during the period of time required for full lymphocyte activation. Selective repertoires of molecules characterise the central and peripheral regions of the immune synapse (cSMAC and pSMAC, respectively) (Monks et al. 1998). The immune synapse is a dynamic system in which molecules rapidly traffic between its various regions.

The importance question as to the precise relationship between CD45 localisation and the immune synapse has not yet been fully resolved. As with lipid rafts, the results obtained have varied somewhat depending on the model system utilised. A critical question is how CD45 localises in relation to its p56^{lck} and p59^{fyn} kinase substrates. Initially in T cells these kinases appear to cocluster in the cSMAC but then p56^{lck}, at least, moves to the periphery within a few minutes of receptor stimulation (Monks et al. 1998; Ehrlich et al. 2002; Holdorf et al. 2002). At the same time, the immune synapse appears to be replenished with more p56^{lck} from intracellular stores (Ehrlich et al. 2002). The question, then, is how CD45 relates to this moving target. Some imaging studies using either T cells (Leupin et al. 2000) or B cells (Batista et al. 2001) have suggested that CD45 is excluded from the immune synapse. However, in a careful kinetic study in which three-dimensional immunofluorescence microscopy was utilised to study the interaction between primary T cells and cognate peptide presented on an antigen presenting cell, both CD45 and p56^{lck} were initially identified in the cSMAC (Freiberg et al. 2002). Then, following 7 minutes of receptor stimulation, CD45 moved out of the cSMAC to a region peripheral to the pSMAC. Even in the absence of antigen presenting cells, using a system in which the MHC-peptide ligand was presented to T cells on beads, CD45 was still initially found in the cSMAC in close association with the TCR, showing that recruitment of CD45 to the immune synapse was not caused by an exogenous ligand (Freiberg et al. 2002). In a different study in which planar lipid bilayers containing MHC-peptide and the adhesion molecule ICAM-1 were used for antigen presentation, CD45 was at first excluded from the cSMAC, but within 10 minutes was then recruited back to the centre of the contact area (Johnson et al. 2000). The discrepancies between these results may be due to the different experimental systems utilised, in particular the use of lipid bilayers for antigen presentation in contrast to antigen presenting cells. A further intriguing possibility is that different CD45 isoforms (see below) might behave differently with respect to their localisation within the immune system and this question was not addressed in the cited studies.

Overall, although the data are not yet completely clear, a likely scenario is that both CD45 and p56^{lck} are present in the cSMAC at the earliest stage of its formation, enabling CD45 to dephosphorylate p56^{lck} at pTyr-505 and maintain the kinase in a basally functional state. Following p56^{lck} kinase activation, CD45 then moves to a peripheral region where it may not have continued contact with p56^{lck}, so preventing rapid dephosphorylation of pTyr-394 and preserving the kinase in an activated state to facilitate TCR coupling to intracellular signalling pathways (Alexander 2000). In this context it is of interest that a potent inhibitor of p56^{lck} kinase activity completely inhibited T cell proliferation even when added 1 hour after TCR stimulation, suggesting that on-going p56^{lck} activity is important for the T cell activation process (Ehrlich et al. 2002).

12.5 How are the actions of CD45 regulated?

As discussed in Section 12.4, one method whereby CD45 appears to be regulated is by movement of the transmembrane protein within the plane of the plasma membrane so that its dephosphorylating actions are brought into the vicinity of relevant substrates. A second method that has been proposed, homodimerisation of the protein, will be discussed in section 6 below in the context of CD45 isoforms. A third dimension of regulation is suggested by the structure of the CD45 cytoplasmic tail, which comprises two tandem PTPase-homology domains of about 240 residues, termed Domain 1 and Domain 2, separated by a spacer region of 45 residues and a C-terminus of 79 residues. A highly acidic region of 24 amino acids within Domain 2 close to the spacer region is not found within other transmembrane PTPases and provides multiple potential sites of serine phosphorylation. Only Domain 1 contains PTPase activity and in fact the putative PTPase active site in Domain 2 differs at 6 out of the 11 positions when compared to Domain 1. Substitution of 3 out of 5 of these altered amino acids from Domain 2 into the equivalent positions in Domain 1 was sufficient to ablate the PTPase activity of Domain 1 (Streuli et al. 1990; Desai et al. 1994).

Several studies point to an important regulatory role for Domain 2 in CD45 function. Mutation of various conserved residues surrounding the Cys1144 in the pseudo-active site of Domain 2 caused a 50% reduction in the PTPase activity of Domain 1 (Johnson et al. 1992), whereas mutation of Glu1180 to Gly in Domain 2 abrogated PTPase activity completely (Ng et al. 1995). Evidence using recombinant polypeptides in vitro suggests that Domain 2 stabilises Domain 1 and increases its PTPase activity: the destabilising Glu1180 to Gly point mutation affected this stabilisation (Felberg and Johnson 1998; Felberg and Johnson 2000). Furthermore, the spacer region between the two tandem repeats appears to interact with Domain 2 (Hayami-Noumi et al. 2000). In functional experiments in which TCR signalling was examined following reconstitution of a CD45-deficient cell line with mutant forms of CD45, substitution of Domain 2 with that from the LAR PTPase abrogated the production of Interleukin-2 (Kashio et al. 1998). This defect could not be explained by ablation of the PTPase activity, suggesting a specific role for Domain 2.

Biochemical investigation also indicates a regulatory role for the acidic 19 amino acid insert in Domain 2. This novel domain has been reported to bind to Casein Kinase 2 and its ablation significantly decreased the ability of CD45 to restore TCR signalling in a CD45-deficient cell line (Greer et al. 2001). Mutation of the four serine residues within the insert affected CD45 function to a similar extent compared with that of the deletion mutant. In fact, CD45 is phosphorylated at high stoichiometry by Casein Kinase 2 in vitro (Tonks et al. 1990) and this has been reported to regulate both its substrate specificity as well as its activity in vitro (Stover et al. 1991; Wang et al. 1999).

12.6 The role of CD45 isoforms

CD45 is expressed as multiple isoforms resulting from alternative RNA splicing of exons 4, 5 and 6 (Trowbridge and Thomas 1994; Alexander 1997). The tradition of naming exons 4-6 as A, B, and C gives rise to the widely used isoform nomenclature of CD45RABC, CD45RB etc: the letter 'R' denotes 'Restricted to' and the letters that follow refer to the exon-products included in that particular iso-form. The term 'CD45R0' refers to the isoform in which all three exon products have been excised, whereas 'B220' is the particular form of CD45RABC carrying a unique oligosaccharide-dependent epitope, which is expressed on B cells. CD45 exons 4-6 encode only about 50-60 amino acids each located at the ectodomain N-terminus, but the effects of splicing out these alternative exons is amplified by the loss of their multiple O-linked glycosylation sites. The cytoplasmic tail is identical in all CD45 isoforms. In theory, alternative splicing of exons 4-6 generates up to 8 different isoforms, but in practice only five of these isoforms have been reported as proteins expressed at significant levels in lymphocytes: CD45RABC, CD45RAB, CD45RBC, CD45RB, and CD45R0, varying in molecular weight from 180-240 kDa (Rogers et al. 1992; Fukuhara et al. 2002). Additional alternate use of exons 7, 8, and 10 in murine T cells and T cell lines has also been reported (Chang et al. 1991; Virts et al. 1998), although the possible significance of this finding at the protein expression level remains unclear. Alternative splicing (by mechanisms reviewed in Hermiston et al. 2003) has been most studied in the T lineage where it is under tight regulatory control during thymic development and in the activation of mature T cells. For example, CD45R0 is upregulated on CD4+CD8+ thymocytes, but then declines following positive selection with upregulation of higher molecular weight isoforms (such as CD45RB in the mouse) in CD4+ and CD8+ thymocytes (Fukuhara et al. 2002). Human cord blood T cells that have not previously been exposed to antigen express low CD45R0 levels relative to higher molecular weight isoforms containing the A-exon product. With increasing age, the proportion of peripheral T cells expressing CD45R0 gradually increases until it typically reaches 40-60% of the total repertoire in the adult (Hayward et al. 1989). Indeed, when CD45RA+ T cells are activated in vitro they upregulate CD45R0 and downregulate the CD45RA+ isoforms over a period of several days (Akbar et al. 1988). However, this conversion is never complete and all T cells express more than one CD45 isoform. It is not therefore possible to purify primary T cell subsets from wild type mouse or human and assess the properties of single CD45 isoforms in isolation. The upregulation of CD45R0 on antigen-experienced T cells, and upon T cells activated in vitro, has led to the idea that CD45R0[+] T cells comprise, or contain a population, of 'memory/effector' cells. However, this suggestion remains controversial and the only clear conclusion at present is that in the mature T cell repertoire CD45R0 is a marker for previously activated T cells. There is also evidence suggesting that conversion to a CD45R0[hi] phenotype is a reversible process, so in vivo it seems probable that CD45R0 is not only a marker for previously activated T cells, but also for T cells stimulated relatively recently.

The CD45 cytoplasmic tail, or the tail incorporated into a chimaeric protein with a non-CD45 ectodomain, has been shown to restore TCR signal transduction in CD45-deficient T cell lines, albeit less efficiently than wild type CD45 in some instances (Desai et al. 1993; Hovis et al. 1993; Volarevic et al. 1993). Such results might suggest that the CD45 ectodomain is unimportant in regulating the functions of the enzyme. However, given the carefully orchestrated regulation of CD45 isoform expression, a teleological perspective suggests that such changes

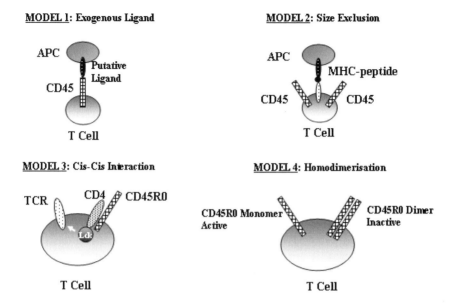

Fig. 3. Four possible models for explaining the functions of CD45 isoforms. See the text for further details

may have molecular and functional implications for the T cell. Nevertheless, the precise nature of these implications continues to remain elusive.

Four models (illustrated in Fig. 3) have been suggested to explain the different molecular actions of CD45 isoforms and these models will be briefly reviewed. None of the models are mutually exclusive.

12.6.1 The exogenous ligand model

It is a reasonable assumption that CD45 might have exogenous ligands, either in soluble form or expressed on other cells such as antigen presenting cells, and that such ligands could vary between different CD45 isoforms. However, until now, no physiologically relevant CD45 ligands have been described. Although CD45 binds effectively to lectins, not surprisingly in light of its abundant oligosaccharide branches, such interactions have not as yet been demonstrated to play a role in vivo in modulating the actions of CD45.

12.6.2 The size exclusion model

An equally reasonable assumption is that the larger CD45 isoform ectodomain might sterically hinder the presentation of MHC-peptides to the TCR, whereas the relatively small CD45R0 ectodomain would exert little or no hindrance (Davis and

Van der Merwe 1996). Indeed, as described above, it has been suggested that CD45 is completely excluded from the cSMAC, that part of the immune synapse where MHC-peptide engagement with the TCR is thought to occur (Johnson et al. 2000). However, as already mentioned, a more detailed kinetic analysis has revealed that CD45 is present in the cSMAC at the earliest time-points at which this structure can be detected (Freiberg et al. 2002). Furthermore, in the study in which the ability of CD45 chimaeras with ectodomains of varying size were assessed for their ability to promote TCR signalling, it was the chimaera with the largest and most heavily glycosylated ectodomain (CD43) that was most efficient (Irles et al. 2003). Therefore, a large ectodomain per se does not appear to hinder the actions of the CD45 cytoplasmic tail. One intriguing possibility is that the various CD45 isoforms may differentially localise within the immune synapse, thereby exerting different actions on the $p56^{lck}$ and $p59^{fyn}$ tyrosine kinases. Further work will be necessary to address this hypothesis. If confirmed, such findings might help to elucidate some of the disparities in the literature since, until now, CD45 in the context of the immune synapse has been analysed using pan-CD45 rather than CD45 isoform specific antibodies.

12.6.3 The cis-cis interaction model

There is an extensive literature describing the association of CD45 with other proteins in cis-cis interactions at the cell surface (the earlier literature is reviewed in Alexander 1997). Reconstitution of a CD45-deficient murine cell line with CD45 isoforms suggested that the CD45R0 isoform promoted greater IL-2 secretion upon engagement of the TCR with the cognate MHC-peptide as compared to other isoforms (Novak et al. 1994). Co-capping experiments in these cells revealed preferential CD4-CD45R0 association (Leitenberg et al. 1996). However, more detailed capping and co-immunoprecipitation studies indicated a basal association of CD45R0 with the TCR independent of CD4 expression and suggested that co-capping of CD4 with CD45R0 was mediated by this prior CD45R0-TCR association (Leitenberg et al. 1999). Nevertheless, CD4-CD45 association has also been described in primary $CD4^+$ T cells (Mittler et al. 1991; Bonnard et al. 1997). In a further study in which Fluorescence Resonance Energy Transfer (FRET) analysis was used to investigate cis-cis interactions, different CD45 isoforms were transfected into a CD45-deficient T cell line expressing the TCR, CD4, and CD8. The results suggested that CD45R0 preferentially associated with CD4 and CD8 relative to the CD45RBC and CD45RABC isoforms (Dornan et al. 2002). The CD45R0-CD4 association correlated with upregulated protein tyrosine phosphorylation events under both basal and TCR-stimulated conditions. These studies suggest a model in which the juxtaposition of CD45 PTPase activity in relation to its CD4 or CD8-associated $p56^{lck}$ kinase substrate promotes the action of the kinase, thereby upregulating TCR-stimulated intracellular phosphorylation events.

Attractive as such a model may be, one problem with such cell line studies is that the glycosylation status of CD45 isoforms and their associating glycoproteins may not be the same as in primary cells, resulting in associations that do not accurately reflect those found under more physiological conditions. To address this question, CD45$^{-/-}$ mice have been reconstituted with specific CD45 isoforms. In one study it was shown that either CD45R0 or CD45RABC transgenes, driven by a proximal ^{lck} promoter, restored T cell numbers to near normal in lymph nodes while no restoration of T cell populations in the spleen was found (Kozieradzki et al. 1997). In contrast, the Vav promoter has more recently been utilised to generate several lines of transgenic mice expressing either CD45R0 or CD45RB on the CD45$^{-/-}$ background in all haematopoietic lineages (Ogilvy, Alexander, and Holmes, unpublished). When expressed at levels comparable with the total CD45 expression found in hemizygous CD45$^{+/-}$ mice, thymic development was restored to apparently normal levels and reconstitution of mature T cells in both the lymph nodes and spleen was observed. In fact, expression of CD45R0 at a level only 6-8% of the total wild type CD45 expression level was sufficient to restore thymic development to levels equivalent to wild type. Mice expressing either the CD45R0 or CD45RB isoforms expressed at comparable levels were equally efficient in their ability to mount T-dependent antigenic responses. Interestingly, however, neither isoform was able to restore effective B cell maturation, nor BCR-stimulated proliferation, showing that CD45 isoforms play unique roles, which differ between the T and B lineages.

Overall, therefore, there is considerable support based largely on cell line studies for the cis-cis interaction model, but the model remains to be confirmed as being relevant to the function of CD45 isoforms in the whole animal context.

12.6.4 The homodimerisation model

The possibility that homodimerisation of transmembrane PTPases leads to the inhibition of phosphatase activity has received recent attention and is reviewed more extensively elsewhere in this volume (see Hertog, Section 3.3). An indication that the actions of CD45 might be regulated in this manner was suggested by studies using a chimaeric receptor with an EGFR ectodomain and a CD45 cytoplasmic tail. This chimaera successfully restored TCR signal transduction in a CD45-deficient cell line, but TCR signalling was largely abrogated upon subsequently dimerising the artificial receptor by the addition of EGF (Desai et al. 1993). Mutation of a putative wedge domain within the CD45 tail portion of the chimaera significantly blunted the ability of EGF to inhibit TCR-mediated signals, suggesting that the wedge domain might be critical in promoting homodimerisation (Majeti et al. 1998). Introduction of a point mutation at the same site in CD45 in a mouse model in situ resulted in a severe lymphoproliferative and lupus like syndrome with autoantibody production, consistent also with de-inhibition of CD45 PTPase activity (Majeti et al. 2000). However, since the point mutation was introduced into the germ-line, CD45 in all haematopoietic lineages is expected to bear the same mutation, so it will be important to determine which cell-type is responsible for the observed abnormalities.

CD45 has been shown to dimerise in several different studies. For example, CD45 was found to dimerise to a small extent after chemical cross-linking of a T cell line (Takeda et al. 1992) and recombinant fragments of the rat extracellular domain can exist as dimers as well as monomers (Symons et al. 1999). Is it possible that CD45 isoforms might differentially homodimerise, thereby inhibiting specific pools of CD45 PTPase activity? Interestingly, in a study using FRET it was found that CD45R0 preferentially homodimerises on the surface of a T cell line when compared with CD45RBC or CDRABC isoforms (Dornan et al. 2002). Similar findings were obtained using chemical cross-linking and a cysteine dimer-trapping method in which it was also shown that CD45R0 preferentially homodimerises in a manner hindered by sialylation and O-glycosylation (Xu and Weiss 2002).

As with the cis-cis interaction model, the homodimerisation model is also an attractive hypothesis for explaining the differential actions of CD45 isoforms. However, significant questions remain. For example, it remains to be shown that CD45 homodimerisation causes inhibition of its PTPase activity. Furthermore, in the FRET study cited above, both CD45R0 homodimers and CD4/CD8-CD45R0 heterodimers were detected on the surface of the same cell, yet the net effect on TCR signal transduction of CD45R0 expression appeared to be positive (Dornan et al. 2002). In addition, in a series of transgenic lines expressing increasing amounts of CD45R0 at the T cell surface, in which comparable numbers of peripheral T cells were observed in the spleen and lymph nodes, a quantitative effect was noted in which T-dependent antigenic responses correlated positively with the CD45R0 expression level (Ogilvy, Alexander, and Holmes, unpublished). This result is difficult to explain if the main role of CD45R0 is to exert a negative effect on T cell responses. Therefore, further work will be required to determine whether the homodimerisation model adequately explains the putative differential actions of CD45 isoforms in vivo.

12.7 Concluding remarks

It will be apparent from this Chapter that whereas much has been established with regard to the actions and functions of CD45, much also remains to be elucidated. What is clear is that CD45 exerts a dominantly positive regulatory role in both T and B cells, acting as a gatekeeper for signals mediated by the antigen receptors. When CD45 is absent, the threshold for signalling via the antigen receptors becomes significantly higher. In addition, it also seems very likely that CD45 can exert negative effects on T cell antigen receptor signalling. In T cells, at least, the key substrates of CD45 are the $p56^{lck}$ and $p59^{fyn}$ tyrosine kinases. Outside of such relative certainties, the hypothesising continues, but the data frequently remain contradictory or ambiguous. The subcellular localisation of CD45, the regulation of CD45 PTPase activity and the functions and molecular actions of CD45 isoforms, all remain topics for which much remains to be clarified. It was the biologist J.B.S. Haldane who once remarked that "the tragedy of science is a beautiful hypothesis slain by an ugly fact". Indeed, true Popperians should be delighted by such "tragedies" when their favourite hunches are finally rigorously tested and found wanting, as this is how science advances. No doubt, some of our current models are plain wrong, but as long as they keep being tested effectively, then verisimilitude will win the day in the long run.

Acknowledgements

The literature on CD45 is vast and I apologise to those whose work has not been cited due only to restrictions on space. This is a selective review and I have not attempted to cover every aspect of the field. I would like to thank the Biotechnology and Biological Sciences Research Council for financial support and my many colleagues, listed as co-authors on our publications, who have worked with me over the years on this fascinating topic.

References

Akbar AN, Terry L, Timms A, Beverley PCL, Janossy G (1988) Loss Of CD45R and Gain Of UCHL1 Reactivity Is a Feature Of Primed T-Cells. J Immunol 140:2171-2178

Alexander DR (1997) The role of the CD45 phosphotyrosine phosphatase in lymphocyte signalling. In: MM Harnett and KP Rigley (eds) Lymphocyte signalling: Mechanisms, subversion and manipulation. John Wiley & Sons Ltd., Sussex, England, pp 107-140.

Alexander DR (2000) The CD45 tyrosine phosphatase: a positive and negative regulator of immune cell function. Sem in Immunol 12:349-359

Ashwell JD, Doro U (1999) CD45 and Src-family kinases: and now for something completely different. Immunol Today 20:412-416

Baker M, Gamble J, Tooze R, Higgins D, Yang FT, O'Brien PCM, Coleman N, Pingel S, Turner M, Alexander DR (2000) Development of T-leukaemias in CD45 tyrosine phosphatase-deficient mutant [lck] mice. EMBO J 19:4644-4654

Baldwin TA, Ostergaard HL (2002) The protein tyrosine phosphatase CD45 reaches the cell surface via Golgi dependent and independent pathways. J Biol Chem 277:50333-50340

Batista FD, Iber D, Neuberger MS (2001) B cells acquire antigen from target cells after synapse formation. Nature 411:489-494

Benatar T, Carsetti R, Furlonger C, Kamalia N, Mak T, Paige CJ (1996) Immunoglobulin-Mediated Signal-Transduction In B-Cells From CD45-Deficient Mice. J Exp Med 183:329-334

Biffen M, McMichaelphillips D, Larson T, Venkitaraman A, Alexander D (1994) The CD45 Tyrosine Phosphatase Regulates Specific Pools Of Antigen Receptor-Associated P59(^{Fyn}) and CD4-Associated P56(^{Lck}) Tyrosine Kinases In Human T-Cells. EMBO J 13:1920-1929

Bonnard M, Maroun CR, Julius M (1997) Physical association of CD4 and CD45 in primary, resting CD4(+) T cells. Cell Immunol 175:1-11

Byth KF, Conroy LA, Howlett S, Smith AJH, May J, Alexander DR, Holmes N (1996) CD45-null transgenic mice reveal a positive regulatory role for CD45 in early thymocyte development, in the selection of CD4(+)CD8(+) thymocytes, and in B-cell maturation. J Exp Med 183:1707-1718

Chang HL, Lefrancois L, Zaroukian MH, Esselman WJ (1991) Developmental expression of CD45 alternate exons in murine T-cells - evidence of additional alternate exon use. J Immunol 147:1687-1693

Cheng PC, Dykstra ML, Mitchell RN, Pierce SK (1999) A role for lipid rafts in B cell antigen receptor signaling and antigen targeting. J Exp Med 190:1549-1560

Conroy LA, Byth KF, Howlett S, Holmes N, Alexander DR (1996) Defective depletion of CD45-null thymocytes by the Staphylococcus aureus enterotoxin B superantigen. Immunol Lett 54:119-122

Cyster JG, Healy JI, Kishihara K, Mak TW, Thomas ML, Goodnow CC (1996) Regulation of B-lymphocyte negative and positive selection by tyrosine phosphatase CD45. Nature 381:325-328

Davis SJ, Van der Merwe PA (1996) The structure and ligand interactions of CD2: Implications for T-cell function. Immunology Today 17:177-187

Desai DM, Sap J, Schlessinger J, Weiss A (1993) Ligand-mediated negative regulation of a chimeric transmembrane receptor tyrosine phosphatase. Cell 73:541-554

Desai DM, Sap J, Silvennoinen O, Schlessinger J, Weiss A (1994) The catalytic activity of the CD45 membrane-proximal domain is required for TCR signaling and regulation. EMBO J 13:4002-4010

Dornan S, Sebestyen Z, Gamble J, Nagy P, Bodnar A, Alldridge L, Doe S, Holmes N, Goff LK, Beverley P, Szollosi J, Alexander DR (2002) Differential association of CD45 isoforms with CD4 and CD8 regulates the actions of specific pools of p56^{lck} tyrosine kinase in T cell antigen receptor signal transduction. J Biol Chem 277:1912-1918

Doro U, Ashwell JD (1999) Cutting edge: The CD45 tyrosine phosphatase is an inhibitor of ^{Lck} activity in thymocytes. J Immunol 162:1879-1883

Doro U, Sakaguchi K, Appella E, Ashwell J D (1996) Mutational analysis of ^{Lck} in CD45-negative T-cells - dominant role of tyrosine-394 phosphorylation in kinase-activity. Mol Cell Biol 16:4996-5003

Edmonds SD, Ostergaard HL (2002) Dynamic Association of CD45 with Detergent-Insoluble Microdomains in T Lymphocytes. J Immunol 169:5036-5042

Ehrlich LI, Ebert PJ, Krummel MF, Weiss A, Davis MM (2002) Dynamics of p56^{lck} translocation to the T cell immunological synapse following agonist and antagonist stimulation. Immunity 17:809-822

Felberg J, Johnson P (1998) Characterization of recombinant CD45 cytoplasmic domain proteins - Evidence for intramolecular and intermolecular interactions. J Biol Chem 273:17839-17845

Felberg J, Johnson P (2000) Stable interdomain interaction within the cytoplasmic domain of CD45 increases enzyme stability. Biochem Biophys Res Commun 271:292-298

Fernandis AZ, Cherla RP, Ganju RK (2003) Differential Rrgulation of CXCR4-mediated T-cell chemotaxis and mitogen-activated protein kinase activation by the membrane tyrosine phosphatase, CD45. J Biol Chem 278:9536-9543

Freiberg BA, Kupfer H, Maslanik W, Delli J, Kappler J, Zaller DM, Kupfer A (2002) Staging and resetting T cell activation in SMACs. Nat Immunol 3:911-917

Fukuhara K, Okumura M, Shiono H, Inoue M, Kadota Y, Miyoshi S, Matsuda H (2002) A study on CD45 isoform expression during T-cell development and selection events in the human thymus. Hum Immunol 63:394-404

Greer SF, Wang Y, Raman C, Justement LB (2001) CD45 function is regulated by an acidic 19-amino acid insert in domain II that serves as a binding and phosphoacceptor site for casein kinase 2. J Immunol 166:7208-7218

Gupta N, DeFranco AL (2003) Visualizing lipid raft dynamics and early signaling events during antigen receptor-mediated B-lymphocyte activation. Mol Biol Cell 14:432-444

Hartley DA, Hurley TR, Hardwick JS, Lund TC, Medveczky PG, Sefton BM (1999) Activation of the [lck] tyrosine-protein kinase by the binding of the tip protein of herpesvirus saimiri in the absence of regulatory tyrosine phosphorylation. J Biol Chem 274:20056-20059

Hayami-Noumi K, Tsuchiya T, Moriyama Y, Noumi T (2000) Intra- and intermolecular interactions of the catalytic domains of human CD45 protein tyrosine phosphatase. FEBS Lett 468:68-72

Hayward AR, Lee J, Beverley PCL (1989) Ontogeny Of Expression Of Uchl1 Antigen On TCR-1+ (CD4/8) and TCR-Delta+ T-Cells. Eur J Immunol 19:771-773

He X, Woodford-Thomas TA, Johnson KG, Shah DD, Thomas ML (2002) Targeting of CD45 protein tyrosine phosphatase activity to lipid microdomains on the T cell surface inhibits TCR signaling. Eur J Immunol 32:2578-2587

Hermiston ML, Xu Z, Weiss A (2003) CD45: A critical regulator of signaling thresholds in immune cells. Annu Rev Immunol 21:107-137

Holdorf AD, Green JM, Levin SD, Denny MF, Straus DB, Link V, Changelian PS, Allen PM, Shaw AD (1999) Proline residues in CD28 and the Src homology (SH)3 domain of [lck] are required for T cell costimulation. J Exp Med 190:375-384

Holdorf AD, Lee KH, Burack WR, Allen PM, Shaw AS (2002) Regulation of [lck] activity by CD4 and CD28 in the immunological synapse. Nat Immunol 3:259-264

Hovis RR, Donovan JA, Musci MA, Motto DG, Goldman FD, Ross SE, Koretzky GA (1993) Rescue of signaling by a chimeric protein containing the cytoplasmic domain of CD45. Science 260:544-546

Ilangumaran S, Arni S, vanEchtenDeckert G, Borisch B, Hoessli DC (1999) Microdomain-dependent regulation of [lck] and [fyn] protein-tyrosine kinases in T lymphocyte plasma membranes. Mol Biol Cell 10:891-905

Irles C, Symons A, Michel F, Bakker TR, van der Merwe PA, Acuto O (2003) CD45 ecto-domain controls interaction with GEMs and [lck] activity for optimal TCR signaling. Nat Immunol 4:189-197

Janes PW, Ley SC, Magee AI (1999) Aggregation of lipid rafts accompanies signaling via the T cell antigen receptor. J Cell Biol 147:447-461

Johnson KG, Bromley SK, Dustin ML, Thomas ML (2000) A supramolecular basis for CD45 tyrosine phosphatase regulation in sustained T cell activation. Proc Natl Acad Sci USA. 97:10138-10143

Johnson P, Ostergaard HL, Wasden C, Trowbridge IS (1992) Mutational analysis of CD45 - a leukocyte-specific protein tyrosine phosphatase. J Biol Chem 267:8035-8041

Justement LB (2001) The role of the protein tyrosine phosphatase CD45 in regulation of B lymphocyte activation. Int Rev Immunol 20:713-738

Kashio N, Matsumoto W, Parker S, Rothstein DM (1998) The second domain of the CD45 protein tyrosine phosphatase is critical for interleukin-2 secretion and substrate recruitment of TCR-zeta in vivo. J Biol Chem 273:33856-33863

Katagiri T, Ogimoto M, Hasegawa K, Arimura Y, Mitomo K, Okada M, Clark MR, Mizuno K, Yakura H (1999) CD45 negatively regulates [lyn] activity by dephosphorylating both positive and negative regulatory tyrosine residues in immature B cells. J Immunol 163:1321-1326

Kishihara K, Penninger J, Wallace VA, Kundig TM, Kawai K, Wakeham A, Timms E, Pfeffer K, Ohashi PS, Thomas ML, Furlonger C, Paige CJ, Mak TW (1993) Normal B-lymphocyte development but impaired T-cell maturation in CD45-exon6 protein-tyrosine-phosphatase deficient mice. Cell 74:143-156

Kozieradzki I, Kundig T, Kishihara K, Ong CJ, Chiu D, Wallace VA, Kawai K, Timms E, Ionescu J, Ohashi P, Marth JD, Mak TW, Penninger JM (1997) T cell development in mice expressing splice variants of the protein tyrosine phosphatase CD45. J Immunol 158:3130-3139

Kung C, Pingel JT, Heikinheimo M, Klemola T, Varkila K, Yoo LI, Vuopala K, Poyhonen M, Uhari M, Rogers M, Speck SH, Chatila T, Thomas ML (2000) Mutations in the tyrosine phosphatase CD45 gene in a child with severe combined immunodeficiency disease. Nature Med 6:343-345

Leitenberg D, Boutin Y, Lu DD, Bottomly K (1999) Biochemical association of CD45 with the T cell receptor complex: regulation by CD45 isoform and during T cell activation. Immunity 10:701-711

Leitenberg D, Novak TJ, Farber D, Smith BR, Bottomly K (1996) The extracellular domain of CD45 controls association with the CD4-T cell-receptor complex and the response to antigen-specific stimulation. J Exp Med 183:249-259

Leupin O, Zaru R, Laroche T, Muller S, Valitutti S (2000) Exclusion of CD45 from the T-cell receptor signaling area in antigen-stimulated T lymphocytes. Curr Biol 10:277-280

Majeti R, Bilwes AM, Noel JP, Hunter T, Weiss A (1998) Dimerization-induced inhibition of receptor protein tyrosine phosphatase function through an inhibitory wedge. Science 279:88-91

Majeti R, Xu Z, Parslow TG, Olson JL, Daikh DI, Killeen N, Weiss A (2000) An inactivating point mutation in the inhibitory wedge of CD45 causes lymphoproliferation and autoimmunity. Cell 103:1059-1070

McFarland EDC, Hurley TR, Pingel JT, Sefton BM, Shaw A, Thomas ML (1993) Correlation between Src family member regulation by the protein-tyrosine-phosphatase CD45 and transmembrane signaling through the T-Cell receptor. Proc Natl Acad Sci USA 90:1402-1406

Mee PJ, Turner M, Basson MA, Costello PS, Zamoyska R, Tybulewicz VL (1999) Greatly reduced efficiency of both positive and negative selection of thymocytes in CD45 tyrosine phosphatase-deficient mice. Eur J Immunol 29:2923-2933

Minami Y, Stafford FJ, Lippincott-Schwartz J, Yuan LC, Klausner RD (1991) Novel redistribution of an intracellular pool of CD45 accompanies T cell activation. J Biol Chem 266:9222-9230

Mittler RS, Rankin BM, Kiener PA (1991) Physical associations between CD45 and CD4 or Cd8 occur as late activation events in antigen receptor-stimulated human T-cells. J Immunol 147:3434-3440

Monks CRF, Freiberg BA, Kupfer H, Sciaky N, Kupfer A (1998) Three-dimensional segregation of supramolecular activation clusters in T cells. Nature 395:82-86

Montixi C, Langlet C, Bernard AM, Thimonier J, Dubois C, Wurbel MA, Chauvin JP, Pierres M, He HT (1998) Engagement of T cell receptor triggers its recruitment to low-density detergent-insoluble membrane domains. EMBO J 17:5334-5348

Ng DHW, Maiti A, Johnson P (1995) Point mutation in the 2nd phosphatase domain of CD45 abrogates tyrosine phosphatase-activity. Biochem Biophys Res Commun 206:302-309

Novak TJ, Farber D, Leitenberg D, Hong SC, Johnson P, Bottomly K (1994) Isoforms of the transmembrane tyrosine phosphatase CD45 differentially affect T-cell recognition. Immunity 1:109-119

Ostergaard HL, Shackelford DA, Hurley TR, Johnson P, Hyman R, Sefton BM, Trowbridge IS (1989) Expression of CD45 alters phosphorylation of the ^{Lck}-encoded tyrosine protein-kinase in murine lymphoma T-cell lines. Proc Natl Acad Sci USA 86:8959-8963

Pao LI, Cambier JC (1997) Syk, but not ^{Lyn}, recruitment to B cell antigen receptor and activation following stimulation of CD45(-) B cells. J Immunol 158:2663-2669

Pelosi M, DiBartolo V, Mounier V, Mege D, Pascussi JM, Dufour E, Blondel A, Acuto O (1999) Tyrosine 319 in the interdomain B of ZAP-70 is a binding site for the Src homology 2 domain of ^{Lck}. J Biol Chem 274:14229-14237

Pingel S, Baker M, Turner M, Holmes N, Alexander DR (1999) The CD45 tyrosine phosphatase regulates CD3-induced signal transduction and T cell development in recombinase-deficient mice: restoration of pre-TCR function by active p56(^{lck}). Eur J Immunol 29:2376-2384

Pradhan D, Morrow J (2002) The spectrin-ankyrin skeleton controls CD45 surface display and interleukin-2 production. Immunity 17:303-315

Roach T, Slater S, Koval M, White L, McFarland EC, Okumura M, Thomas M, Brown E (1997) CD45 regulates Src family member kinase activity associated with macrophage integrin-mediated adhesion. Curr Biol 7:408-417

Rodgers W, Rose JK (1996) Exclusion of CD45 inhibits activity of p56(^{lck}) associated with glycolipid-enriched membrane domains. J Cell Biol 135:1515-1523

Rogers PR, Pilapil S, Hayakawa K, Romain PL, Parker DC (1992) CD45 alternative exon expression in murine and human CD4+ T-cell subsets. J Immunol 148:4054-4065

Seavitt JR, White LS, Murphy KM, Loh DY, Perlmutter RM, Thomas ML (1999) Expression of the p56(^{lck}) Y505F mutation in CD45-deficient mice rescues thymocyte development. Mol Cell Biol 19:4200-4208

Shenoi H, Seavitt J, Zheleznyak A, Thomas ML, Brown EJ (1999) Regulation of integrin-mediated T cell adhesion by the transmembrane protein tyrosine phosphatase CD45. J Immunol 162:7120-7127

Shiroo M, Goff L, Biffen M, Shivnan E, Alexander D (1992) CD45-tyrosine phosphatase-activated p59^{fyn} couples the T-cell antigen receptor to pathways of diacylglycerol production, protein-kinase-C activation and calcium influx. EMBO J 11:4887-4897

Shivnan E, Clayton L, Alldridge L, Keating KE, Gullberg M, Alexander DR (1996) CD45 monoclonal-antibodies inhibit TCR-mediated calcium signals, calmodulin-kinase-IV Gr-activation, and oncoprotein-18 phosphorylation. J Immunol 157:101-109

Sieh M, Bolen JB, Weiss A (1993) CD45 specifically modulates binding of ^{Lck} to a phosphopeptide encompassing the negative regulatory tyrosine of ^{Lck}. EMBO J 12:315-321

Stone JD, Conroy LA, Byth KF, Hederer RA, Howlett S, Takemoto Y, Holmes N, Alexander DR (1997) Aberrant TCR-mediated signalling in CD45-null thymocytes involves dysfunctional regulation of ^{Lck}, ^{Fyn}, TCR-z and ZAP-70. J Immunol 158:5773-5782

Stover DR, Charbonneau H, Tonks NK, Walsh KA (1991) Protein-tyrosine-phosphatase-CD45 is phosphorylated transiently on tyrosine upon activation of Jurkat T-cells. Proc Natl Acad Sci USA 88:7704-7707

Streuli M, Krueger NX, Thai T, Tang, M, Saito H (1990) Distinct functional roles of the two intracellular phosphatase like domains of the receptor-linked protein tyrosine phosphatases LCA and LAR. EMBO J 9:2399-2407

Symons A, Willis AC, Barclay AN (1999) Domain organization of the extracellular region of CD45. Prot Eng 12:885-892

Takeda A, Wu JJ, Maizel AL (1992) Evidence for monomeric and dimeric forms of CD45 associated with a 30-Kda phosphorylated protein. J Biol Chem 267:16651-16659

Tchilian EZ, Wallace DL, Wells RS, Flower DR, Morgan G, Beverley PC (2001) A deletion in the gene encoding the CD45 antigen in a patient with SCID. J Immunol 166:1308-1313

Thomas ML (1989) The leukocyte common antigen family. Ann Rev Immunol 7:339-369

Thomas ML (1999) The regulation of antigen-receptor signaling by protein tyrosine phosphatases: a hole in the story. Curr Opin Immunol 11:270-276

Thomas ML, Brown E J (1999) Positive and negative regulation of Src-family membrane kinases by CD45. Immunol Today 20:406-411

Tonks NK, Diltz CD, Fischer EH (1990) CD45, an integral membrane-protein tyrosine phosphatase - characterization of enzyme-activity. J Biol Chem 265:10674-10680

Trop S, Charron J, Arguin C, Hugo P (2000) Thymic selection generates T cells expressing self-reactive TCRs in the absence of CD45. J Immunol 165:3073-3079

Trowbridge IS, Thomas ML (1994) CD45 - an emerging role as a protein-tyrosine-phosphatase required for lymphocyte-activation and development. Annu Rev Immunol 12:85-116

Virts E, Barritt D, Raschke WC (1998) Expression of CD45 isoforms lacking exons 7, 8 and 10. Mol Immunol 35:167-176

Volarevic S, Niklinska BB, Burns CM, June CH, Weissman AM, Ashwell JD (1993) Regulation of TCR signaling by CD45 lacking transmembrane and extracellular domains. Science 260:541-544

Wallace VA, Penninger JM, Kishihara K, Timms E, Shahinian A, Pircher H, Kundig TM, Ohashi PS, Mak TW (1997) Alterations in the level of CD45 surface expression affect the outcome of thymic selection. J Immunol 158:3205-3214

Wang Y, Guo W, Liang LZ, Esselman WJ (1999) Phosphorylation of CD45 by casein kinase 2 - Modulation of activity and mutational analysis. J Biol Chem 274:7454-7461

Wu L, Fu J, Shen SH (2002) SKAP55 coupled with CD45 positively regulates T-cell receptor-mediated gene transcription. Mol Cell Biol 22:2673-2686

Xavier R, Brennan T, Li Q, McCormack C, Seed B (1998) Membrane compartmentation is required for efficient T cell activation. Immunity 8:723-732

Xu Z, Weiss A (2002) Negative regulation of CD45 by differential homodimerization of the alternatively spliced isoforms. Nat Immunol 3:764-771

Yanagi S, Sugawara H, Kurosaki M, Sabe H, Yamamura H, Kurosaki T (1996) CD45 modulates phosphorylation of both autophosphorylation and negative regulatory tyrosines of [Lyn] in B-cells. J Biol Chem 271:30487-30492

13 Receptor protein tyrosine phosphatases

Jeroen den Hertog

Abstract

The receptor protein-tyrosine phosphatases (RPTPs) form a subfamily of the classical protein-tyrosine phosphatases (PTPs). RPTPs are interesting because they have the ability to signal across the cell membrane due to their topology. Evidence is accumulating that RPTPs play important roles during embryonic development and in human disease. In addition, our insight into RPTP signalling is growing by identification of specific substrates and ligands. Finally, dimerization is emerging as an important regulatory mechanism of RPTPs and RPTPs may be regulated by redox signalling in an unexpected way.

13.1 Introduction

The receptor protein-tyrosine phosphatases (RPTPs) form a subfamily of the classical protein-tyrosine phosphatases that exclusively dephosphorylate phosphotyrosine (pTyr) in target proteins. The RPTPs have a single membrane-spanning domain and one or two conserved PTP domains that are highly homologous to the catalytic domains of the cytoplasmic PTPs (Tonks and Neel 2001). RPTPs are interesting because they have the potential to signal across the membrane. The human genome contains 22 distinct genes encoding transmembrane PTPs, many of which are differentially spliced, giving rise to a multitude of protein products. Phylogenetic trees indicate that in general the PTP domains of the RPTPs are more homologous to each other than to cytoplasmic PTPs (Andersen et al. 2001). Based on their PTP domains the RPTPs have been classified in eight subtypes (R1-R8). Interestingly, the different subtypes not only share homology in their cytoplasmic, catalytic domains, but also in the structure of their extracellular domains (Fig. 1). The extracellular domains of RPTPs vary from very short (e.g. PTPε, 27 amino acids) to very long, often with multiple functional domains, including fibronectin type III-like repeats, immunoglobulin domains, a MAM (Meprin/A5/μ) domain and a carbonic anhydrase-like domain. The ectodomains of many RPTPs are reminiscent of cell adhesion molecules (CAMs), suggesting a function for RPTPs in cell adhesion. Most RPTPs encode two catalytic PTP domains in the cytoplasmic region. The membrane-proximal domain, RPTP-D1, exhibits most of the catalytic activity, while the membrane-distal domain, RPTP-D2, has very little or no catalytic activity. The catalytic sites of RPTPs are highly conserved in sequence and architecture. The catalytic site cysteine is essential for

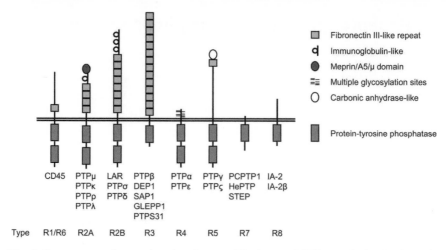

Fig. 1. Receptor protein-tyrosine phosphatases. The human RPTPs are depicted schematically with their extracellular domains above the double line, representing the cell membrane.

catalysis, because a cysteinyl-phosphate intermediate is formed during dephosphorylation. This cysteine is located in the phosphate binding loop (or PTP signature motif, [I/V]HCXAGXXR[S/T]G), flanked by the general acid (WPD) loop and the phosphotyrosine recognition loop (KNRY-motif) (Jackson and Denu 2001). The R7 and R8 subtypes are divergent from the other subtypes. The R7 RPTPs do not have a typical transmembrane domain and they do not target to the plasma membrane. The R8 RPTPs do not exhibit PTP activity because of a limited number of mutations in otherwise conserved residues, including the catalytic site cysteine. CD45, the R1/R6 subtype RPTP, will be reviewed elsewhere in this volume. Therefore, in this review, I will focus on the other five subtypes of bona fide transmembrane PTPs with catalytic activity, R2A - R5. I will review their in vivo function, and their role in signalling.

13.2 The function of RPTPs in vivo

RPTPs play an important role in vivo. Initially, the role of RPTPs in Drosophila axon pathfinding provided clear evidence that RPTPs play a crucial role in development (Van Vactor 1998; Van Vactor et al. 1998). Recently, the Ptprj gene, encoding the mouse homologue of DEP-1, PTPη (also known as CD148, F-36-12) was identified as the candidate for the mouse colon-cancer susceptibility locus. Moreover, in human colon, lung and breast cancers PTPRJ is often deleted, it shows allelic imbalance in loss of heterozygosity and it often contains missense mutations. Therefore, DEP-1 may be relevant to the development of several types of cancer in humans (Ruivenkamp et al. 2002). PTPζ (also known as RPTPβ, Ptprz) was recently identified as the receptor for VacA from Helicobacter pylori, which causes gastric ulcers. Another ligand of PTPζ, pleiotrophin, induces gastritis in mice – like VacA – while PTPζ knockout mice are not affected by either factor. These results demonstrate that erroneous PTPζ signalling causes gastruc ulcers (Fujikawa et al. 2003). Taken together, evidence is accumulating that RPTPs play an important role in development and disease.

13.2.1 DRPTPs in fly axonogenesis

Three type R2A and three type R3 RPTPs have been identified in Drosophila (Chien 1996; Schindelholz et al. 2001). One of the type R2A RPTPs, DLAR, is highly homologous in sequence and structure to mammalian LAR, and two type R3 RPTPs, DPTP10D and DPTP4E, are related to mammalian PTPβ. DPTP69D, DPTP99A, and DPTP52F do not have clear mammalian orthologues. All DRPTPs, except DPTP4E, are specifically expressed in CNS axons (Fitzpatrick et al. 1995; Tian et al. 1991; Yang et al. 1991). Mutants have been identified in all five DRPTPs that are expressed in the CNS and they all play a role in axon pathfinding (Chien 1996; Desai et al. 1996; Desai et al. 1997; Krueger et al. 1996; Sun et al. 2000; Sun et al. 2001; Schindelholz et al. 2001). Using double, triple, and quadruple mutants, complex genetic interactions were found among the five neuronal DRPTPs, in that they have redundant functions, as well as competitive and collaborative activities. For instance, most DRPTPs have redundant functions in outgrowth and bifurcation of the SNa nerve, because this is only affected in double mutants. Yet, DPTP10D antagonizes DLAR, DPTP99A, and DPTP69D in regulating extension of ISN motor axons past intermediate targets (Sun et al. 2001), demonstrating the complexity of the role of DRPTPs in axonogenesis. DRPTPs not only play a role in CNS and motor axon pathfinding, but also in retinal axon target selection. PTP69D functions in R1-R6 growth cones, and is required for proper projection to the lamina. Both the fibronectin type III-like repeats in the extracellular domain of PTP69D and intracellular PTP activity are required for correct targeting of R1-R6 cells, suggesting PTP69D functions by dephosphorylation of target proteins in the R1-R6 growth cones (Garrity et al. 1999). In DLAR mutants, R1-R6 photoreceptor neurons target to the lamina correctly, but fail to connect to the right target neurons (Clandinin et al. 2001). Mutant R7 axons initially target to the correct layer of the medulla, but later retract to the R8 target layer (Clandinin et al. 2001; Maurel-Zaffran et al. 2001). DLAR function in R1-R6 and R7 is cell autonomous, while its function in R8 is cell nonautonomous, as demonstrated using single cell mosaics. Interestingly, the cytoplasmic domain of DLAR is required for its function in R7, but not in R8, suggesting DLAR functions as a receptor in R7 and as a ligand in R8 (Clandinin et al. 2001; Maurel-Zaffran et al. 2001).

Several downstream effectors of the DRPTPs have been identified. The tyrosine kinase Abl antagonizes DLAR in the ISNb motor choice point pathway. Abl, together with its substrate Enabled (Ena), associate with DLAR and they are dephosphorylated by DLAR in vitro, suggesting a direct role for Abl and Ena in DLAR function (Wills et al. 1999). Trio and Liprin are two proteins that were identified in yeast two-hybrid screens to bind to mammalian LAR (Debant et al. 1996; Serra-Pages et al. 1995). The Drosophila orthologues of Trio and Liprin interact genetically with DLAR in axonogenesis and synapse morphogenesis, respectively (Bateman et al. 2000; Kaufmann et al. 2002). Trio contains guanine nucleotide exchange domains for Rho and Rac, and Trio interacts genetically with Rac, but not Rho, forming a link between DLAR and cytoskeletal events that underlie normal axonogenesis (Bateman et al. 2000). Fly genetics has provided strong evidence for the function of RPTPs in axon pathfinding. In leech, Hirudo medicinalis, HmLAR2 is highly expressed in the growth cones of comb cells and RNAi-mediated knock down of HmLAR2 expression leads to growth cone collapse, indicating that HmLAR2 is involved in similar processes in the leech as in the fly (Baker and Macagno 2000; Gershon et al. 1998). There are indications that vertebrate RPTPs have a similar role in growth cone morphology and axonogenesis. Chicken PTPσ (CRYPα) is closely related to DLAR. Disruption of the interaction of PTPσ with its ligand using antibodies or using soluble PTPσ ectodomain induces a dramatic change in the morphology of retinal growth cones and disrupts retinal axon growth (Ledig et al. 1999), suggesting that chicken PTPσ has a similar function as DLAR in growth cone morphology and axonogenesis.

13.2.2. RPTP function in the mouse

Roughly, half of the RPTPs have been reported to be inactivated by homologous recombination in the mouse (Table 1). Detailed analysis of these knockout mice is now starting to give insight into the function of RPTPs in mammalian development.

Inactivation of the type R2A RPTPs, PTPκ and PTPμ, did not induce an apparent phenotype; they develop normally, and are viable and fertile (Skarnes et al. 1995; Martijn Gebbink personal communication) Yet, these RPTPs may play important roles in development. For instance, PTPμ promotes neurite outgrowth of retinal ganglion cells when used as a substrate. Downregulation of PTPμ expression or overexpression of a catalytically inactive form of PTPμ significantly decreases neurite outgrowth on Ncadherin (Burden-Gulley and Brady-Kalnay 1999). PTPμ may play a similar role in vivo. However, the function of PTPμ may be taken over by (an)other RPTP(s), similar to the situation in the fly where often mutation of more than one DRPTP is required to detect axon pathfinding defects.

The type R2B RPTPs are highly homologous to Drosophila LAR and it was anticipated that the LAR subfamily knockouts would show axon pathfinding defects. However, LAR mutant mice identified in a gene trap screen do not show clear axon pathfinding defects, although a reduction in basal forebrain cholinergic neuron size and reduced hippocampal innervation were observed (Skarnes et al. 1995; Yeo et al. 1997). Proper gene targeting of LAR, resulting in deletion of almost the entire cytoplasmic domain, including both PTP domains (LAR-ΔP mice), does not induce obvious axon pathfinding defects either. LAR-ΔP mice display a defect in mammary gland development (Schaapveld et al. 1997). In addition, like the LAR mutant mice from the gene trap screen, LAR-ΔP mice have a decreased number of

Table 1. Phenotypes of RPTP knockout mice

type	name[a]	phenotype	reference
R1/R6	CD45	severe T-cell development defect	(Byth et al. 1996)
R2A	PTPμ	no apparent phenotype	Gebbink, pers. comm.
	PTPκ	no apparent phenotype	(Skarnes et al. 1995)
	PTPρ	not reported	
	PTPλ	not reported	
R2B	LAR	Reduced cholinergic neuron size, cholinergic innervation of the dentate gyrus decreased	(Skarnes et al. 1995) (Yeo et al. 1997) (Schaapveld et al. 1997)
	PTPσ	Mammary gland defects (impaired terminal differentiation of alveoli at late gestation)	(Elchebly et al. 1999)
	PTPδ	Pituitary dysplasia, olfactory lobe defects, reduction in the total central nervous system size and cell number	(Wallace et al. 1999)
		High neonatal mortality due to food intake defect and defect in hippocampal LTP formation	(Uetani et al. 2000)
R3	PTPβ	not reported	
	DEP1	vascularization defects, embryonic lethal (E11.5)	(Takahashi et al. 2003)
	GLEPP1	defects in regulation of the glomerular pressure filtration rate due to defects in podocyte function	(Wharram et al. 2000)
	PTPesp	not reported	
R4	PTPα	No apparent phenotype	(Su et al. 1999)
		Decrease in Src and Fyn activity in derived cell lines	(Ponniah et al. 1999)
	PTPε	Hypomyelination of sciatic nerve axon at early post-natal stage, decreased Kv channel activity and increased phosphorylation	(Peretz et al. 2000)
R5	PTPγ	not reported	
	PTPζ	reduced oligodendrocyte survival and recovery from demyelinating disease	(Harroch et al. 2002)
		resistant to gastric ulcer induction by VacA	(Fujikawa et al. 2003)
R7	PTPSL	not reported	
	STEP	not reported	
R8	IA-2	reduced glucose tolerance, depressed insulin release	(Saeki et al. 2002)
	IA-2β	not reported	

(a) For clarity, the names of the human orthologues are being used for these mouse RPTPs.

basal forebrain cholinergic neurons and reduced hippocampal cholinergic innervation (Van Lieshout et al. 2001), indicating that this is not a function of the LAR ectodomain and that PTP activity is required for this function of LAR.

The three mammalian type 2B RPTPs are highly homologous, suggesting they may have redundant functions. Nevertheless, the PTPσ knockout mice have a dramatic phenotype. They display retarded growth, increased neonatal mortality, and most remaining PTPσ knockout mice die from a wasting syndrome by approximately three weeks of age (Elchebly et al. 1999; Wallace et al. 1999). PTPσ knockout mice have a reduced brain size, olfactory lobes and architectural abnormalities in the brain and spinal cord (Elchebly et al. 1999; Wallace et al. 1999; Meathrel et al. 2002). Moreover, nerve regeneration is enhanced after sciatic nerve injury in PTPσ knockout mice, albeit errors in directional growth were reported (McLean et al. 2002). Taken together, PTPσ may be involved in neurogenesis, axon outgrowth and axon pathfinding in the CNS. PTPσ knockout mice also show neuroendocrine dysplasia with a severe depletion of luteinizing hormone-releasing hormone (LHRH)-positive cells (Elchebly et al. 1999). In addition, growth hormone (GH) and prolactin (PRL) expression are reduced in PTPσ knockout mice. Pancreatic islets are hypoplastic with reduced insulin immunoreactivity, resulting in hypoglycemia and death, which is caused by the reduced GH levels, since administration of GH rescues neonatal PTPσ knockout mice and normalizes the blood glucose (Batt et al. 2002). In conclusion, PTPσ plays a crucial role in establishing normal architecture of the CNS, and in differentiation and development of the neuroendocrine system.

Finally, the PTPδ knockout is semi-lethal due to reduced food intake of the mutant mice. Interestingly, learning of PTPδ knockout mice is impaired, while hippocampal long-term potentiation at CA1 and CA3 synapses is enhanced. Therefore, PTPδ may play a role in regulation of synaptic plasticity (Uetani et al. 2000). Despite their high homology, the three type R2B RPTPs have divergent functions in vivo. Still, similar signalling mechanisms may be underlying the different phenotypes. It will be interesting to see if signalling by these three RPTPs is similar at the cellular and molecular level.

DEP1 (CD148, PTPη) is a type R3 RPTP that is highly expressed in endothelial cells. Disruption of DEP-1 by replacement of the cytoplasmic domain by enhanced green fluorescent protein (EGFP) results in embryonic lethality at midgestation (E11.5) in homozygous mice. Dep-1 knockout embryos show vascularization defects with enlarged vessels, defective in vascular remodelling, branching, and impaired pericyte investment adjacent to endothelial structures. Therefore, DEP-1 may regulate endothelial proliferation and endothelium-pericyte interactions (Takahashi et al. 2003). GLEPP1, another type R3 RPTP is expressed on the apical cell surface of glomerular podocytes. Targeted disruption of GLEPP1 led to altered podocyte structure - amoeboid rather than octopoid - resulting in reduced glomerular filtration rate and hypertension (Wharram et al. 2000). The alterations in podocyte structure were accompanied by a change in distribution of vimentin, suggesting that GLEPP1 plays a role in regulating cytoskeletal architecture, similar to the type R3 RPTPs in the fly.

The two type R4 RPTPs are broadly expressed with elevated levels of expression in the nervous system. Phenotypically, there are no apparent defects in RPTPα knockout mice (Ponniah et al. 1999; Su et al. 1999). RPTPα knockout mice are born in accordance with Mendelian ratios, they survive to adulthood and are fertile, however, they display gender-specific changes in body size and nurturing defects (Su et al. 1999). The PTK Src is a substrate of RPTPα, and RPTPα-mediated dephosphorylation of the inhibitory C-terminal phosphorylation site, pTyr529 (mouse numbering) leads to activation of Src (den Hertog et al. 1993; Zheng et al. 1992). Src activity and the activity of the Src family member Fyn is reduced in brain lysates from RPTPα knockout mice and in mouse embryo fibroblasts, derived from RPTPα knockout mice (Ponniah et al. 1999; Su et al. 1999), clearly demonstrating that Src is a bona fide substrate of RPTPα. Nevertheless, reduced activity of Src family kinases apparently does not induce gross abnormalities in mice. PTPε knockout mice are also viable, they are born according to expected Mendelian ratios and they appear normal (Peretz et al. 2000). However, PTPε knockout mice display transient hypomyelination of sciatic nerve axons at an early postnatal age (5d). No difference in myelin sheath thickness is observed between wild type and PTPε knockout mice after 8 months. In PTPε knockout mice, the activity of two delayed rectifier voltage-gated potassium channels, Kv1.5 and Kv2.1, is increased and the Kv channel α-subunits are hyperphosphorylated. A substrate trapping mutant of PTPε binds tyrosine phosphorylated Kv2.1 channels, indicating that the Kv2.1 channel is a direct substrate of PTPε (Peretz et al. 2000). Despite the high sequence homology between PTPα and PTPε and despite the largely overlapping expression pattern of these two RPTPs, there is a difference in the phenotype of these two knockout mice. PTPε has a non-redundant function in Schwann cell growth and differentiation, while there are no gross developmental abnormalities in PTPα knockout mice.

Gene targeting of the type R5 RPTP, PTPζ (RPTPβ, Ptpz), in itself did not induce developmental defects (Harroch et al. 2000). PTPζ knockout mice are viable, fertile and do not show defects in the nervous system where it is most highly expressed. However, PTPζ knockout mice are more susceptible to experimental autoimmune encephalomyelitis (EAE), a model for multiple sclerosis, in that PTPζ knockout mice show impaired recovery from EAE, which is associated with increased apoptosis of mature oligodendrocytes in the spinal cord. Interestingly, human PTPζ is induced in remyelinating oligodendrocytes in multiple sclerosis lesions. Therefore, PTPζ may play a role in oligodendrocyte survival and in recovery from demyelinating disease (Harroch et al. 2002). In addition, as described above, PTPζ signalling is involved in VacA- or pleiotrophin-induced development of gastric ulcers (Fujikawa et al. 2003).

13.2.3 RPTPs and zebrafish

All mouse RPTPs have close homologues that are highly conserved and therefore may have redundant functions. The phenotype, induced by ablation of a given RPTP, may be masked because a close family member of the deleted RPTP takes over its function. Double, triple, or quadruple knockouts may be required to provide insight into the function of these conserved RPTPs. Several of these RPTP subfamily knockout mice are currently being generated and it will be interesting to see if these mice reveal redundant functions of the RPTPs. Yet, the function of a given RPTP may also be taken over by a non-related (R)PTP. There is ample evidence that tyrosine phosphorylated proteins are dephosphorylated by multiple (R)PTPs. Novel approaches will have to be developed to assess the function of

RPTPs in vivo. For instance, analysis of the phosphorylation state of specific RPTP substrates will provide insight into the function of the RPTPs. Another cause for the lack of apparent phenotypes in many of the described RPTP knock-outs may be the plasticity of the system. For instance, while PTPε knockout mice show a clear hypomyelination defect early after birth, this defect is rescued later in life, as there are no myelination defects in 8-month old PTPε knockout mice (Peretz et al. 2000). Perhaps many transient defects in the RPTP knockouts have not been noticed, because they were rescued by the time of analysis.

We have recently started to use the zebrafish as a model to study the function of RPTPs. The zebrafish allows us to circumvent redundancy of RPTP function. In zebrafish, one can use synthetic antisense modified oligonucleotides, morpholinos, to knock down expression of target genes. It is feasible to knock down multiple genes at the same time and therefore, entire subfamilies can be targeted. We have used antisense RPTPα morpholinos (RPTPα-mo), and found dramatic defects in eye development. Retinal lamination is disturbed, although differentiation still appears to occur (van der Sar et al. 2002). In addition, we found specific apoptosis in the mid-hindbrain boundary upon RPTPα knock down. Strikingly, antisense Src morpholinos (Src-mo) phenocopy the effect of RPTPα-mo, indicating that RPTPα and Src act in the same pathway in vivo (van der Sar, Rodriguez, Jopling, and JdH, unpublished observation). It is surprising that RPTPα knock down caused dramatic defects in zebrafish, while mouse PTPα knockouts do not show apparent phenotypes. Although it is not clear why there is such a difference in phenotypes between the mouse and zebrafish, the zebrafish apparently is more amenable for functional analysis of RPTPα signalling, and perhaps for RPTP signalling in general.

13.3 RPTP signalling

RPTPs have the potential to signal across the plasma membrane. However, clear examples of signalling intracellularly via RPTPs remain elusive. On the one hand, ligands have been identified to bind to the extracellular domain of RPTPs, but whether this leads to intracellular changes in PTP activity is often not clear. On the other hand, other mechanisms, including phosphorylation and oxidation of RPTPs are emerging as alternative regulatory mechanisms. However, proper analysis of the regulation of RPTPs is hampered by the lack of identified bona fide substrates of RPTPs.

13.3.1 RPTP substrates

One of the most important – if not the most important – question in elucidation of the molecular mechanisms underlying the function of RPTPs is identification of their substrates. Various approaches have led to identification of substrates and candidate substrates of RPTPs (Table 2). However, it is virtually impossible to-prove definitively that a protein is a bona fide in vivo RPTP substrate. A bona fide substrate must meet the following criteria: (1) the substrate is trapped by a mutant of the RPTP with a mutation in the catalytic site cysteine, or in the general acid/ general base aspartate (Flint et al. 1997), (2) phosphorylation of the substrate is enhanced in cells lacking the RPTP, e.g. RPTP knockout cells, and (3) re-expression of the RPTP in knockout cells results in reduced phosphorylation of the substrate. Taken together, substrate-trapping experiments in RPTP knockout cells/ tissues, similar to the strategy of Tremblay and co-workers with the non-receptor PTP-PEST (Cote et al. 1998), and subsequent analysis of the phosphorylation state of candidate substrates in these cells/ tissues before and after reexpression of the RPTP, should lead to identification of bona fide substrates of RPTPs. Next to the biochemical evidence, described above, biological evidence is highly valuable for confirmation of RPTP-substrate identity.

Several types of substrate have been identified (Table 2), based on their identity and/or function: adaptors, ion channels, PTKs, and RPTPs themselves. Adaptor proteins have been identified as substrates of multiple RPTPs. For instance, p130cas is a substrate of LAR, DEP1, and RPTPα (Buist et al. 2000; Noguchi et al. 2001; Wang et al. 2000; Weng et al. 1999). Since p130cas is phosphorylated on many (at least 13) sites, it is not surprising that it is bound by many RPTPs in sub-strate-trapping experiments, and thus identified as a substrate for many RPTPs. It will be interesting to see how p130cas phosphorylation is affected in RPTP knockout cells. In this respect, it is noteworthy that - surprisingly - p130cas phos-phorylation is reduced in RPTPα knockout cells (Su et al. 1999). Src and Fyn ac-tivity is reduced in RPTPα knockout cells as a result of enhanced phosphorylation of the C-terminal inhibitory phosphorylation site in Src and Fyn (see below). Src and Fyn phosphorylate p130cas, illustrating that unravelling RPTP-substrate inter-actions may be complex. Apparently, there is a dual involvement of RPTPα in p130cas phosphorylation. On the one hand, RPTPα induces phosphorylation by activation of Src and Fyn, and on the other hand, it reduces p130cas phosphoryla-tion by direct dephosphorylation (Buist et al. 2000).

Recently, several ion channels, both K+ channels as well as Na+ channels, have been identified as substrates of RPTPs (Peretz et al. 2000; Ratcliffe et al. 2000; Tsai et al. 1999), suggesting the interesting possibility that ion channels are regu-lated directly by RPTP-mediated dephosphorylation. In some cases, the evidence that the channels are direct substrates is limited (Table 2), but Kv1.5 and Kv2.1 ty-rosine phosphorylation is enhanced in PTPε knockout cells, and Kv2.1 binds to a substrate-trapping mutant of PTPε (Peretz et al. 2000), indicating that these chan-nels are bona fide substrates.

Src and Fyn have been identified definitively as substrates of RPTPα (den Her-tog et al. 1993; Peretz et al. 2000; Ponniah et al. 1999; Su et al. 1999; Zheng et al. 1992). Recently, Shalloway and co-workers demonstrated that Src binds to RPTPα pTyr789 through its SH2 domain, displacing GRB2, which leads to dephosphory-lation of Src pTyr529 and thus to activation of Src (Zheng et al. 2000). Moreover, displacement of GRB2 by Src apparently occurs specifically during mitosis through a mechanism that involves serine phosphorylation of RPTPα in the

Table 2. Substrates of type R2-R5 RPTPs

type	name	substrate	evidence[a]	B[b]	function	reference
R2A	PTPμ	cadherin	VT	Y	cell adhesion	Brady-Kalnay et al. 1995
		p120ctn	OE	Y	cell-cell contact	Zondag et al. 2000
R2B	LAR	β-catenin	IV, VT	Y	?	Kypta et al. 1996
		IR	AS, IV, OE	Y	IR signalling	Ahmad and Goldstein, 1997; Hashimoto et al. 1992; Mooney et al.1997
		IRS-1/2	OE, IV	N	IR signalling	Goldstein et al. 2000; Zabolotny et al. 2001
		p130cas	OE	?	apoptosis	Wang et al. 2000
		FRS-2, p180	OE	?	GF signalling	Weng et al. 1999
	PTPσ	EGFR	AS	N	EGFR signalling	Suarez et al. 1999
R3	GLEPP1	Paxillin	ST, OE	N	macrophage morphology	Pixley et al. 2001
	PTPβ	Tie-2	ST, OE	N	blood vessel morphogenesis?	Fachinger et al. 1999
	DEP1	PDGFR	OE, IV, PM	Y	PDGFR signalling	Kovalenko et al. 2000
		p130cas, FAK,	ST, IV, OE ST, OE	N	integrin signalling	Noguchi et al. 2001;
		Dok	OE, IV	N	TCR signalling	Baker et al. 2001
		PLCγ1, LAT		N		
R4	PTPα	Src/Fyn	OE, RE, KO, IV, PM, AS	Y	Src/Fyn activation cell adhesion	Zheng et al. 1992; den Hertog et al. 1993; Bhandari et al. 1998; Harder et al. 1998; Ponniah et al. 1999; Su et al. 1999; Arnott et al. 1999; den Hertog et al. 1994
		PTPα	IV, PM	Y	GRB2 binding	
		IR	OE	N	IR signalling	Moller et al. 1995
		Kv1.2	OE	Y	Kv1.2 activation	Tsai et al. 1999
		p130cas	ST, IV, OE	N	focal adhesion?	Buist et al. 2000
	PTPε	IR	OE	N		Moller et al. 1995
		Kv1.5, Kv2.1	KO, OE	N	IR signalling regulation of Kv channels	Peretz et al. 2000; Tiran et al. 2003
		Src	KO, ST	?	Src activation	Gil-Henn and Elson, 2003
R5	PTPζ	β-catenin	ST	Y	?	Meng et al. 2000
		sodium channel	ST	Y	sodium channel modulation	Ratcliffe et al. 2000
		GIT-1/ Cat-1	3H, IV, ST, OE	?	cytoskeletal remodelling?	Kawachi et al. 2001; Fujikawa et al. 2003

(Table 2 footnote) a) Evidence for these proteins being specific substrates: AS, immunoblotting of antisense expressing cells; IV, in vitro; KO, immunoblotting in knockout cells; OE, immunoblotting of overexpressing cells; PM, peptide mapping; PY, phosphotyrosine-specific antibodies; RE, immunoblotting of re-expressing cells; ST, substrate trapping; VT, vanadate treatment; 3H, yeast two-hybrid + kinase system.
b) This column indicates whether the RPTPs bind to their substrates: B, binding; Y, yes; N, no.

juxtamembrane domain, resulting in active mitotic Src (Zheng and Shalloway 2001; Zheng et al. 2002).

CD45 and RPTPα have been found to be phosphorylated on tyrosine themselves (Autero et al. 1994; den Hertog et al. 1994; Su et al. 1994), and it is highly likely that they have autodephosphorylation activity. Indeed, RPTPα dephosphorylates itself in vitro and tyrosine phosphorylation of catalytically inactive RPTPα is elevated, strongly suggesting that RPTPα has autodephosphorylation activity (den Hertog et al. 1994).

13.3.2 Ligand binding to RPTPs

The RPTPs have distinct extracellular domains (Fig. 1). It is appealing to hypothesize that the ectodomains act as ligand binding modules, transferring signals intracellularly, analogous to the receptor PTKs. Several types of interaction have been reported between RPTP ectodomains and protein ligands, including membrane bound proteins, extracellular matrix factors and soluble factors. Interestingly, pleiotrophin binding to PTPζ is the first example of ligand-induced modulation of RPTP activity, since pleiotrophin enhanced β-catenin phosphorylation in PTPζ expressing cells. Therefore, pleiotrophin binding may inhibit PTPζ activity (Meng et al. 2000). Similarly, VacA binds to PTPζ and induces enhanced Git1 phosphorylation (Fujikawa et al. 2003). PTPζ not only binds to soluble factors, but also to a membrane-bound protein, contactin. Binding of the PTPζ ectodomain to contactin is in trans and induces neurite outgrowth of primary tectal neurons (Peles et al. 1995). In fact, the PTPζ ectodomain binds to a protein complex, consisting of contactin and caspr (Peles et al. 1997). Contactin is a GPI-linked protein, while caspr is a transmembrane protein with a proline rich region in its cytoplasmic domain, capable of binding a subclass of SH3 domains. Therefore, PTPζ binding may trigger signalling in the contactin/caspr expressing cell. The cytoplasmic domain of PTPζ is not required for this function, suggesting that the PTPζ ectodomain functions as a ligand, rather than as a receptor. Such a function of RPTP ectodomains is consistent with the finding in the fly that the DLAR ectodomain is sufficient for its function in R8 photoreceptor cells (Clandinin et al. 2001; Maurel-Zaffran et al. 2001). The GPI-linked protein, contactin, not only interacts with PTPζ, but also with PTPα. However, the interaction with PTPα is in cis, when these proteins are expressed on the same cells (Zeng et al. 1999). The function of the interaction between PTPα and contactin remains to be determined.

PTPμ and PTPκ engage in homophilic interactions when expressed at high levels on apposing cells (Brady-Kalnay et al. 1993; Gebbink et al. 1993; Sap et al.

1994). The interaction is mediated by the MAM domain and the Ig-like domain. Despite high sequence homology between PTPμ and PTPκ, there is no heterodimerization between these two RPTPs (Zondag et al. 1995). PTPμ and PTPκ homophilic interactions may be involved in cell adhesion and/or cell recognition events.

Several RPTPs have been reported to interact with extracellular matrix proteins. Treatment of cells with Matrigel, a mixture of extracellular matrix proteins, enhanced DEP-1 specific activity, as assessed by immunoprecipitation/PTP assays. Immunoprecipitated DEP-1 is also activated in response to Matrigel, which is dependent on the ectodomain of DEP-1 (Sorby et al. 2001). These results suggest there is a direct effect of Matrigel on DEP-1 activity. The ectodomain of chicken PTPσ (CRYPα) binds to ligands in the retinal basal lamina and glial endfeet. PTPσ is a heparin-binding protein and the heparin sulphate proteoglycans (HSPGs) agrin and collagen XVIII are specific PTPσ ligands that bind to a specific site in the first Ig-like domain in PTPσ (Aricescu et al. 2002). Whether HSPG binding affects PTPσ catalytic activity remains to be determined. PTPσ also binds to ligands in developing muscle. However, this binding is independent of HSPGs, since heparinase treatment does not affect binding (Sajnani-Perez et al. 2003). These ligands remain to be identified.

Taken together, various classes of RPTP ligands have been identified that may affect RPTP catalytic activity positively or negatively. Furthermore, the RPTP ectodomains may act as ligands themselves, suggesting RPTPs may be involved in bidirectional signalling. Ligand binding to RPTPs may send a signal into the RPTP expressing cell and at the same time signal into the ligand expressing cell. It is noteworthy that many ectodomains of RPTPs are shed. They are cleaved off by specific proteases, close to the cell membrane and they remain bound to the rest of the RPTP after cleavage. However, the ectodomains may be mobilized in response to stimuli and act as ligands elsewhere.

13.3.3 Regulation of RPTPs by dimerization

Regulation of RPTP function and activity is not well understood. As described above, ligand binding may regulate RPTP activity, but to date only PTPζ has been shown to be regulated by ligand binding. Dimerization is emerging as an important regulatory mechanism of RPTP activity. In addition, phosphorylation may regulate RPTP activity in a dynamic fashion. Finally, there is evidence that redox signalling regulates RPTP activity.

Dimerization is an important regulatory mechanism for transmembrane proteins (Weiss and Schlessinger 1998). Evidence is accumulating that RPTPs are regulated by dimerization as well. The first indication for regulation of RPTPs by dimerization was that chimeric EGFR-CD45 with the extracellular domain of the epidermal growth factor (EGF) receptor fused to the cytoplasmic domain of CD45 is functionally inactivated by dimerization upon EGF binding (Fig. 2A) (Desai et al. 1993). The crystal structure of RPTPα-D1 provides structural evidence that

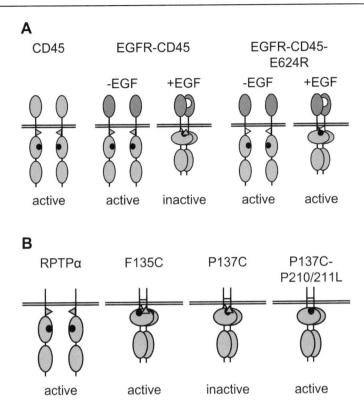

Fig. 2. Regulation of RPTPs by dimerization. (A) Chimeric EGFR-CD45 is functionally in-activated by EGF-induced dimerization, which is dependent on an intact wedge, since a single point mutation in the wedge (E624R) abolishes EGF-induced functional inactivation. (B) Introduction of a disulfide bridge in the extracellular domain leads to constitutive dimerization of RPTPα, and - depending on rotational coupling within the dimer - inactiva-tion, which is dependent on an intact wedge, since mutations that destabilize the wedge (P210/211L) abolish dimerization-induced inactivation. EGF is depicted as an open circle, the wedge as a triangle and the catalytic site as a black circle.

dimerization leads to inhibition of the catalytic activity of RPTPα since a helix-loop-helix wedge-like structure of one monomer inserts into the catalytic cleft of the other, thereby occluding the catalytic site (Bilwes et al. 1996). However, the crystal structures of two other RPTPs raise doubts over whether dimerization is a general regulatory mechanism for RPTPs. PTPμ-D1 did not form dimers like RPTPα-D1 in the crystal structure, despite the presence of a wedge-like structure (Hoffmann et al. 1997). In addition, the structure of the entire cytoplasmic domain of LAR did not show dimers (Nam et al. 1999) and the conformation sterically did not allow D1-D1 dimerization like the PTPα-D1 dimer. Nevertheless, experimen-tal evidence is accumulating that RPTPα and CD45 can be regulated by dimeriza-tion. In vivo activity studies using constitutively dimeric mutants of RPTPα with intermolecular disulfide bridges in the ectodomain confirm that dimerization can inhibit RPTPα activity (Jiang et al. 1999). Interestingly, dimerization per se does not inactivate RPTPα, since constitutively dimeric RPTPα-F135C is active, like wild type RPTPα, and RPTPα-P137C with a disulfide bridge two residues closer to the plasma membrane is inactive (Jiang et al. 1999). Apparently, dimerization of RPTPα only leads to inactivation of the catalytic activity when rotational coupling of the two monomers in the dimer is such that the wedge of one monomer inserts into the catalytic cleft of the other. Point mutations in the wedge of EGFR-CD45 chimeras and constitutive dimer mutants of RPTPα abolish dimerization-induced inhibition of PTP activity (Fig. 2) (Jiang et al. 1999; Majeti et al. 1998), demonstrating that the wedge plays a pivotal role in regulation of these RPTPs.

There is ample evidence that at least some RPTPs can be regulated by dimerization. However, do RPTPs dimerize in vivo? Cross-linking experiments and fluorescence resonance energy transfer (FRET) analysis indicate that RPTPα constitutively forms dimers in living cells (Jiang et al. 2000; Tertoolen et al. 2001). Similarly, CD45 forms dimers in living cells, as demonstrated by chemical cross-linking and FRET analysis (Dornan et al. 2002; Takeda et al. 1992; Xu and Weiss 2002). Interestingly, dimerization of distinct alternatively spliced isoforms of CD45 is different (Xu and Weiss 2002), suggesting that dimerization may be regulated at the level of alternative splicing. Another means of regulation of dimerization may be phosphorylation. RPTPα is phosphorylated on two serine residues in the juxtamembrane domain, very close to the wedge, which leads to activation of RPTPα (den Hertog et al. 1995; Tracy et al. 1995). It is likely that phosphorylation of these sites leads to disruption of the wedge-catalytic site interaction, opening up the catalytic site and thus activating RPTPα.

13.3.4 Regulation of RPTPs by redox signalling

Redox signalling may regulate RPTP activity directly. The pKa of the catalytic site cysteines is low due to the microenvironment in the catalytic pocket and therefore these cysteines are susceptible to oxidation. Since dephosphorylation proceeds through a cysteinyl-phosphate intermediate, oxidation of the catalytic site cysteine blocks catalysis completely. Several non-receptor PTPs are regulated directly by oxidation/ reduction. For instance, PTP1B is inactivated upon oxidation following EGF-treatment of cells (Lee et al. 1998), and Shp-2 is oxidized and inactivated upon PDGF-treatment (Meng et al. 2002). We found recently that dimerization of RPTPα is also regulated by redox signalling. H_2O_2 stabilizes RPTPα dimers as a result of a conformational change in RPTPα-D2. The catalytic site cysteine in RPTPα-D2 is required for this effect, suggesting that direct oxidation is involved (Blanchetot et al. 2002a). Interestingly, LAR-D2 also shows a conformational change in response to H_2O_2. Moreover, H_2O_2 induces heterodimerization between RPTPα and LAR, indicating that LAR can dimerize (Blanchetot et al. 2002b). Apparently, given the right experimental conditions, LAR is not sterically hindered to dimerize as suggested by the crystal structure (Nam et al. 1999). Interestingly, the H_2O_2-induced conformational change in RPTPα-D2 and stabilization of dimerization is reflected by a change in

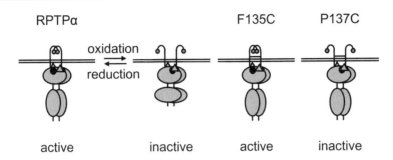

Fig. 3. Redox-regulated rotational coupling of RPTPα. H_2O_2 induces a reversible conformational change in RPTPα-D2, which leads to opening up of an epitope in the ectodomain (indicated by a circle), reflecting the situation in two constitutive dimers with different rotational coupling, F135C and P137C.

conformation in the ectodomain, as assessed by analysis of epitope accessibility in the ectodomain (Fig. 3). While the epitope is not accessible in wild type RPTPα or constitutively dimeric, active RPTPα-F135C, it is accessible in H_2O_2-treated RPTPα and in constitutively dimeric, inactive RPTPα-P137C. The change in accessibility in response to H_2O_2 is dependent on the catalytic site cysteine in RPTPα-D2, suggesting RPTPα transduces a signal from inside cells outwards in response to H_2O_2 treatment (van der Wijk et al. 2003). Since LAR also shows a conformational change in response to H_2O_2, it will be interesting to see if conformational changes in the cytoplasmic domain induce changes in the ectodomain of other RPTPs as well. Redox regulation of RPTPα activity is unusual. As expected, H_2O_2 treatment of cells leads to complete inactivation of RPTPα. However, RPTPα-C723S with a mutation of the catalytic site cysteine in RPTPα-D2 is only partially inhibited by H_2O_2 treatment, indicating that inactivation is not (only) caused by oxidation of the catalytic site cysteine (Cys433) in RPTPα-D1 (Blanchetot et al. 2002). We postulate that stabilization of dimerization as a result of the conformational change in RPTPα-D2 leads to complete inactivation of RPTPα. Whether dimerization is a general regulatory mechanism of RPTPs and whether oxidation/ reduction has similar effects on all RPTPs remains to be determined.

References

Ahmad F, Goldstein BJ (1997) Functional association between the insulin receptor and the transmembrane protein-tyrosine phosphatase LAR in intact cells. J Biol Chem 272:448-457

Andersen JN, Mortensen OH, Peters GH, Drake PG, Iversen LF, Olsen OH, Jansen PG, Andersen HS, Tonks NK, Moller NP (2001) Structural and evolutionary relationships among protein tyrosine phosphatase domains. Mol Cell Biol 21:7117-7136

Aricescu AR, McKinnell IW, Halfter W, Stoker AW (2002) Heparan sulfate proteoglycans are ligands for receptor protein tyrosine phosphatase sigma. Mol Cell Biol 22:1881-1892

Arnott CH, Sale EM, Miller J, Sale GJ (1999) Use of an antisense strategy to dissect the signaling role of protein-tyrosine phosphatase alpha. J Biol Chem 274:26105-26112

Autero M, Saharinen J, Pessa-Morikawa T, Soula-Rothhut M, Oetken C, Gassmann M Bergman M, Alitalo K, Burn P, Gahmberg CG (1994) Tyrosine phosphorylation of CD45 phosphotyrosine phosphatase by p50csk kinase creates a binding site for p56lck tyrosine kinase and activates the phosphatase. Mol Cell Biol 14:1308-1321

Baker JE, Majeti R, Tangye SG, Weiss A (2001) Protein tyrosine phosphatase CD148-mediated inhibition of T-cell receptor signal transduction is associated with reduced LAT and phospholipase Cgamma1 phosphorylation. Mol Cell Biol 21:2393-2403

Baker MW, Macagno ER (2000) RNAi of the receptor tyrosine phosphatase HmLAR2 in a single cell of an intact leech embryo leads to growth-cone collapse. Curr Biol 10:1071-1074

Bateman J, Shu H, Van Vactor D (2000) The guanine nucleotide exchange factor trio mediates axonal development in the *Drosophila* embryo. Neuron 26:93-106

Batt J, Asa S, Fladd C, Rotin D (2002) Pituitary, pancreatic and gut neuroendocrine defects in protein tyrosine phosphatase-sigma-deficient mice. Mol Endocrinol 16:155-169

Bhandari V, Lim KL, Pallen CJ (1998) Physical and functional interactions between receptor-like protein-tyrosine phosphatase alpha and p59fyn. J Biol Chem 273:8691-8698

Bilwes AM, den Hertog J, Hunter T, Noel JP (1996) Structural basis for inhibition of receptor protein-tyrosine phosphatase-alpha by dimerization. Nature 382:555-559

Blanchetot C, Tertoolen LG, den Hertog J (2002a) Regulation of receptor protein-tyrosine phosphatase alpha by oxidative stress. EMBO J 21:493-503

Blanchetot C, Tertoolen LG, Overvoorde J, den Hertog J (2002b) Intra- and intermolecular interactions between intracellular domains of receptor protein-tyrosine phosphatases. J Biol Chem 277:47263-47269

Brady-Kalnay SM, Flint AJ, Tonks NK (1993) Homophilic binding of PTP mu, a receptor-type protein tyrosine phosphatase, can mediate cell-cell aggregation. J Cell Biol 122:961-972

Brady-Kalnay SM, Rimm DL, Tonks NK (1995) Receptor protein tyrosine phosphatase PTPmu associates with cadherins and catenins in vivo. J Cell Biol 130:977-986

Buist A, Blanchetot C, Tertoolen LG, den Hertog J (2000) Identification of p130cas as an in vivo substrate of receptor protein-tyrosine phosphatase alpha. J Biol Chem 275:20754-20761

Burden-Gulley SM, Brady-Kalnay SM (1999) PTPmu regulates N-cadherin-dependent neurite outgrowth. J Cell Biol 144:1323-1336

Byth KF, Conroy LA, Howlett S, Smith AJ, May J, Alexander DR, Holmes N (1996) CD45-null transgenic mice reveal a positive regulatory role for CD45 in early thymocyte development, in the selection of CD4+CD8+ thymocytes, and B cell maturation. J Exp Med 183:1707-1718

Chien CB (1996) PY in the fly receptor-like tyrosine phosphatases in axonal pathfinding. Neuron 16:1065-1068

Clandinin TR, Lee CH, Herman T, Lee RC, Yang AY, Ovasapyan S, Zipursky SL (2001) Drosophila LAR regulates R1-R6 and R7 target specificity in the visual system. Neuron 32:237-248

Cote JF, Charest A, Wagner J, Tremblay ML (1998) Combination of gene targeting and substrate trapping to identify substrates of protein tyrosine phosphatases using PTP-PEST as a model. Biochemistry 37:13128-13137

Debant A, Serra-Pages C, Seipel K, O'Brien S, Tang M, Park SH, Streuli M (1996) The multidomain protein Trio binds the LAR transmembrane tyrosine phosphatase, contains a protein kinase domain, and has separate rac-specific and rho-specific guanine nucleotide exchange factor domains. Proc Natl Acad Sci USA 93:5466-5471

den Hertog J, Pals CE, Peppelenbosch MP, Tertoolen LG, de Laat SW, Kruijer W (1993) Receptor protein tyrosine phosphatase alpha activates pp60c-src and is involved in neuronal differentiation. EMBO J 12:3789-3798

den Hertog J, Sap J, Pals CE, Schlessinger J, Kruijer W (1995) Stimulation of receptor protein-tyrosine phosphatase alpha activity and phosphorylation by phorbol ester. Cell Growth Differ 6:303-307

den Hertog J, Tracy S, Hunter T (1994) Phosphorylation of receptor protein-tyrosine phosphatase alpha on Tyr789, a binding site for the SH3-SH2-SH3 adaptor protein GRB-2 in vivo. EMBO J 13:3020-3032

Desai CJ, Gindhart JGJ, Goldstein LS, Zinn K (1996) Receptor tyrosine phosphatases are required for motor axon guidance in the Drosophila embryo. Cell 84:599-609

Desai CJ, Krueger NX, Saito H, Zinn K (1997) Competition and cooperation among receptor tyrosine phosphatases control motoneuron growth cone guidance in Drosophila. Development 124:1941-1952

Desai DM, Sap J, Schlessinger J, Weiss A (1993) Ligand-mediated negative regulation of a chimeric transmembrane receptor tyrosine phosphatase. Cell 73:541-554

Dornan S, Sebestyen Z, Gamble J, Nagy P, Bodnar A, Alldridge L, Doe S, Holmes N, Goff LK, Beverley P, Szollosi J, Alexander DR (2002) Differential association of CD45 isoforms with CD4 and CD8 regulates the actions of specific pools of p56lck tyrosine kinase in T cell antigen receptor signal transduction. J Biol Chem 277:1912-1918

Elchebly M, Wagner J, Kennedy TE, Lanctot C, Michaliszyn E, Itie A, Drouin J, Tremblay ML (1999) Neuroendocrine dysplasia in mice lacking protein tyrosine phosphatase sigma. Nat Genet 21:330-333

Fachinger G, Deutsch U, Risau W (1999) Functional interaction of vascular endothelial-protein-tyrosine phosphatase with the angiopoietin receptor Tie-2. Oncogene 18:5948-5953

Fitzpatrick KA, Gorski SM, Ursuliak Z, Price JV (1995) Expression of protein tyrosine phosphatase genes during oogenesis in *Drosophila melanogaster*. Mech Dev 53:171-183

Flint AJ, Tiganis T, Barford D, Tonks NK (1997) Development of "substrate-trapping" mutants to identify physiological substrates of protein tyrosine phosphatases. Proc Natl Acad Sci USA 94:1680-1685

Fujikawa A, Shirasaka D, Yamamoto S, Ota H, Yahiro K, Fukada M, Shintani T, Wada A, Aoyama N, Hirayama T, Fukamachi H, Noda M (2003) Mice deficient in protein tyrosine phosphatase receptor type Z are resistant to gastric ulcer induction by VacA of Helicobacter pylori. Nat Genet 33:375-381

Garrity PA, Lee CH, Salecker I, Robertson HC, Desai CJ, Zinn K, Zipursky SL (1999) Retinal axon target selection in *Drosophila* is regulated by a receptor protein tyrosine phosphatase. Neuron 22:707-717

Gebbink MF, Verheijen MH, Zondag GC, van E I, Moolenaar WH (1993) Purification and characterization of the cytoplasmic domain of human receptor-like protein tyrosine phosphatase RPTP mu. Biochemistry 32:13516-13522

Gershon TR, Baker MW, Nitabach M, Macagno ER (1998) The leech receptor protein tyrosine phosphatase HmLAR2 is concentrated in growth cones and is involved in process outgrowth. Development 125:1183-1190

Gil-Henn H, Elson A (2003) Tyrosine phosphatase epsilon activates Src and supports the transformed phenotype of Neu-induced mammary tumor cells. J Biol Chem 278:15579-15586

Goldstein BJ, Bittner-Kowalczyk A, White MF, Harbeck M (2000) Tyrosine dephosphorylation and deactivation of insulin receptor substrate-1 by protein-tyrosine phosphatase 1B. Possible facilitation by the formation of a ternary complex with the Grb2 adaptor protein. J Biol Chem 275:4283-4289

Harder KW, Moller NP, Peacock JW, Jirik FR (1998) Protein-tyrosine phosphatase alpha regulates Src family kinases and alters cell-substratum adhesion. J Biol Chem 273:31890-31900

Harroch S, Furtado GC, Brueck W, Rosenbluth J, Lafaille J, Chao M, Buxbaum JD, Schlessinger J (2002) A critical role for the protein tyrosine phosphatase receptor type Z in functional recovery from demyelinating lesions. Nat Genet 32:411-414

Harroch S, Palmeri M, Rosenbluth J, Custer A, Okigaki M, Shrager P, Blum M, Buxbaum JD, Schlessinger J (2000) No obvious abnormality in mice deficient in receptor protein tyrosine phosphatase beta. Mol Cell Biol 20:7706-7715

Hashimoto N, Feener EP, Zhang WR, Goldstein BJ (1992) Insulin receptor protein-tyrosine phosphatases. Leukocyte common antigen-related phosphatase rapidly deactivates the insulin receptor kinase by preferential dephosphorylation of the receptor regulatory domain. J Biol Chem 267:13811-13814

Hoffmann KM, Tonks NK, Barford D (1997) The crystal structure of domain 1 of receptor protein-tyrosine phosphatase mu. J Biol Chem 272:27505-27508

Jackson MD, Denu JM (2001) Molecular reactions of protein phosphatases--insights from structure and chemistry. Chem Rev 101:2313-2340

Jiang G, den Hertog J, Hunter T (2000) Receptor-like protein tyrosine phosphatase alpha homodimerizes on the cell surface. Mol Cell Biol 20:5917-5929

Jiang G, den Hertog J, Su J, Noel J, Sap J, Hunter T (1999) Dimerization inhibits the activity of receptor-like protein-tyrosine phosphatase-alpha. Nature 401:606-610

Kaufmann N, DeProto J, Ranjan R, Wan H, Van Vactor D (2002) *Drosophila* liprin-alpha and the receptor phosphatase Dlar control synapse morphogenesis. Neuron 34:27-38

Kawachi H, Fujikawa A, Maeda N, Noda M (2001) Identification of GIT1/Cat-1 as a substrate molecule of protein tyrosine phosphatase zeta /beta by the yeast substrate-trapping system. Proc Natl Acad Sci USA 98:6593-6598

Kishihara K, Penninger J, Wallace VA, Kundig TM, Kawai K, Wakeham A, Timms E, Pfeffer K, Ohashi PS, Thomas ML (1993) Normal B lymphocyte development but impaired T cell maturation in CD45-exon6 protein tyrosine phosphatase-deficient mice. Cell 74:143-156

Kovalenko M, Denner K, Sandstrom J, Persson C, Gross S, Jandt E, Vilella R, Bohmer F, Ostman A (2000) Site-selective dephosphorylation of the platelet-derived growth factor beta-receptor by the receptor-like protein-tyrosine phosphatase DEP-1. J Biol Chem 275:16219-16226

Krueger NX, Van Vactor D, Wan HI, Gelbart WM, Goodman CS, Saito H (1996) The transmembrane tyrosine phosphatase DLAR controls motor axon guidance in *Drosophila*. Cell 84:611-622

Kypta RM, Su H, Reichardt LF (1996) Association between a transmembrane protein tyrosine phosphatase and the cadherin-catenin complex. J Cell Biol 134:1519-1529

Ledig MM, Haj F, Bixby JL, Stoker AW, Mueller BK (1999) The receptor tyrosine phosphatase CRYPalpha promotes intraretinal axon growth. J Cell Biol 147:375-388

Lee SR, Kwon KS, Kim SR, Rhee SG (1998) Reversible inactivation of protein-tyrosine phosphatase 1B in A431 cells stimulated with epidermal growth factor. J Biol Chem 273:15366-15372

Majeti R, Bilwes AM, Noel JP, Hunter T, Weiss A (1998) Dimerization-induced inhibition of receptor protein tyrosine phosphatase function through an inhibitory wedge. Science 279:88-91

Maurel-Zaffran C, Suzuki T, Gahmon G, Treisman JE, Dickson BJ (2001) Cell-autonomous and -nonautonomous functions of LAR in R7 photoreceptor axon targeting. Neuron 32:225-235

McLean J, Batt J, Doering LC, Rotin D, Bain JR (2002) Enhanced rate of nerve regeneration and directional errors after sciatic nerve injury in receptor protein tyrosine phosphatase sigma knock-out mice. J Neurosci 22:5481-5491

Meathrel K, Adamek T, Batt J, Rotin D, Doering LC (2002) Protein tyrosine phosphatase sigma-deficient mice show aberrant cytoarchitecture and structural abnormalities in the central nervous system. J Neurosci Res 70:24-35

Meng K, Rodriguez-Pena A, Dimitrov T, Chen W, Yamin M, Noda M, Deuel TF (2000) Pleiotrophin signals increased tyrosine phosphorylation of beta beta-catenin through inactivation of the intrinsic catalytic activity of the receptor-type protein tyrosine phosphatase beta/zeta. Proc Natl Acad Sci USA 97:2603-2608

Meng TC, Fukada T, Tonks NK (2002) Reversible oxidation and inactivation of protein tyrosine phosphatases in vivo. Mol Cell 9:387-399

Moller NP, Moller KB, Lammers R, Kharitonenkov A, Hoppe E, Wiberg FC, Sures I, Ullrich A (1995) Selective down-regulation of the insulin receptor signal by protein-tyrosine phosphatases alpha and epsilon. J Biol Chem 270:23126-23131

Mooney RA, Kulas DT, Bleyle LA, Novak JS (1997) The protein tyrosine phosphatase LAR has a major impact on insulin receptor dephosphorylation. Biochem Biophys Res Commun 235:709-712

Nam HJ, Poy F, Krueger NX, Saito H, Frederick CA (1999) Crystal structure of the tandem phosphatase domains of RPTP LAR. Cell 97:449-457

Noguchi T, Tsuda M, Takeda H, Takada T, Inagaki K, Yamao T, Fukunaga K, Matozaki T, Kasuga M (2001) Inhibition of cell growth and spreading by stomach cancer-associated protein-tyrosine phosphatase-1 (SAP-1) through dephosphorylation of p130cas. J Biol Chem 276:15216-15224

Peles E, Nativ M, Campbell PL, Sakurai T, Martinez R, Lev S, Clary DO, Schilling J, Barnea G, Plowman GD (1995) The carbonic anhydrase domain of receptor tyrosine phosphatase beta is a functional ligand for the axonal cell recognition molecule contactin. Cell 82:251-260

Peles E, Nativ M, Lustig M, Grumet M, Schilling J, Martinez R, Plowman GD, Schlessinger J (1997) Identification of a novel contactin-associated transmembrane receptor with multiple domains implicated in protein-protein interactions. EMBO J 16:978-988

Peretz A, Gil-Henn H, Sobko A, Shinder V, Attali B, Elson A (2000) Hypomyelination and increased activity of voltage-gated K(+) channels in mice lacking protein tyrosine phosphatase epsilon. EMBO J 19:4036-4045

Pixley FJ, Lee PS, Condeelis JS, Stanley ER (2001) Protein tyrosine phosphatase phi regulates paxillin tyrosine phosphorylation and mediates colony-stimulating factor 1-induced morphological changes in macrophages. Mol Cell Biol 21:1795-1809

Ponniah S, Wang DZ, Lim KL, Pallen CJ (1999) Targeted disruption of the tyrosine phosphatase PTPalpha leads to constitutive downregulation of the kinases Src and Fyn. Curr Biol 9:535-538

Ratcliffe CF, Qu Y, McCormick KA, Tibbs VC, Dixon JE, Scheuer T, Catterall WA (2000) A sodium channel signaling complex:modulation by associated receptor protein tyrosine phosphatase beta. Nat Neurosci 3:437-444

Ruivenkamp CA, van Wezel T, Zanon C, Stassen AP, Vlcek C, Csikos T, Klous AM, Tripodis N, Perrakis A, Boerrigter L, Groot PC, Lindeman J, Mooi WJ, Meijjer GA, Scholten G, Dauwerse H, Paces V, van Zandwijk N, van Ommen GJ, Demant P (2002) Ptprj is a candidate for the mouse colon-cancer susceptibility locus Scc1 and is frequently deleted in human cancers. Nat Genet 31:295-300

Saeki K, Zhu M, Kubosaki A, Xie J, Lan MS, Notkins AL (2002) Targeted disruption of the protein tyrosine phosphatase-like molecule IA-2 results in alterations in glucose tolerance tests and insulin secretion. Diabetes 51:1842-1850

Sajnani-Perez G, Chilton JK, Aricescu AR, Haj F, Stoker AW (2003) Isoform-specific binding of the tyrosine phosphatase ptpsigma to a ligand in developing muscle. Mol Cell Neurosci 22:37-48

Sap J, Jiang YP, Friedlander D, Grumet M, Schlessinger J (1994) Receptor tyrosine phosphatase R-PTP-kappa mediates homophilic binding. Mol Cell Biol 14:1-9

Schaapveld RQ, Schepens JT, Robinson GW, Attema J, Oerlemans FT, Fransen JA, Streuli M, Wieringa B, Hennighausen L, Hendriks WJ (1997) Impaired mammary gland development and function in mice lacking LAR receptor-like tyrosine phosphatase activity. Dev Biol 188:134-146

Schindelholz B, Knirr M, Warrior R, Zinn K (2001) Regulation of CNS and motor axon guidance in Drosophila by the receptor tyrosine phosphatase DPTP52F. Development 128:4371-4382

Serra-Pages C, Kedersha NL, Fazikas L, Medley Q, Debant A, Streuli M (1995) The LAR transmembrane protein tyrosine phosphatase and a coiled-coil LAR-interacting protein co-localize at focal adhesions. EMBO J 14:2827-2838

Skarnes WC, Moss JE, Hurtley SM, Beddington RS (1995) Capturing genes encoding membrane and secreted proteins important for mouse development. Proc Natl Acad Sci USA 92:6592-6596

Sorby M, Sandstrom J, Ostman A (2001) An extracellular ligand increases the specific activity of the receptor-like protein tyrosine phosphatase DEP-1. Oncogene 20:5219-5224

Su J, Batzer A, Sap J (1994) Receptor tyrosine phosphatase R-PTP-alpha is tyrosine-phosphorylated and associated with the adaptor protein Grb2. J Biol Chem 269:18731-18734

Su J, Muranjan M, Sap J (1999) Receptor protein tyrosine phosphatase alpha activates Src-family kinases and controls integrin-mediated responses in fibroblasts. Curr Biol 9:505-511

Suarez PE, Tenev T, Gross S, Stoyanov B, Ogata M, Bohmer FD (1999) The transmembrane protein tyrosine phosphatase RPTPsigma modulates signaling of the epidermal growth factor receptor in A431 cells. Oncogene 18:4069-4079

Sun Q, Bahri S, Schmid A, Chia W, Zinn K (2000) Receptor tyrosine phosphatases regulate axon guidance across the midline of the *Drosophila* embryo. Development 127:801-812

Sun Q, Schindelholz B, Knirr M, Schmid A, Zinn K (2001) Complex genetic interactions among four receptor tyrosine phosphatases regulate axon guidance in Drosophila. Mol Cell Neurosci 17:274-291

Takahashi T, Takahashi K, St John PL, Fleming PA, Tomemori T, Watanabe T, Abrahamson DR, Drake CJ, Shirasawa T, Daniel TO (2003) A mutant receptor tyrosine phosphatase, CD148, causes defects in vascular development. Mol Cell Biol 23:1817-1831

Takeda A, Wu JJ, Maizel AL (1992) Evidence for monomeric and dimeric forms of CD45 associated with a 30- kDa phosphorylated protein. J Biol Chem 267:16651-16659

Tertoolen LG, Blanchetot C, Jiang G, Overvoorde J, Gadella TWJ, Hunter T, den Hertog J (2001) Dimerization of receptor protein-tyrosine phosphatase alpha in living cells. BMC Cell Biol 2:8

Tian SS, Tsoulfas P, Zinn K (1991) Three receptor-linked protein-tyrosine phosphatases are selectively expressed on central nervous system axons in the Drosophila embryo. Cell 67:675-680

Tiran Z, Peretz A, Attali B, Elson A (2003) Phosphorylation-dependent regulation of Kv2.1 channel activity at tyrosine 124 by Src and by protein tyrosine phosphatase Epsilon. J Biol Chem: in press

Tonks NK, Neel BG (2001) Combinatorial control of the specificity of protein tyrosine phosphatases. Curr Opin Cell Biol 13:182-195

Tracy S, van der Geer P, Hunter T (1995) The receptor-like protein-tyrosine phosphatase, RPTP alpha, is phosphorylated by protein kinase C on two serines close to the inner face of the plasma membrane. J Biol Chem 270:10587-10594

Tsai W, Morielli AD, Cachero TG, Peralta EG (1999) Receptor protein tyrosine phosphatase alpha participates in the m1 muscarinic acetylcholine receptor-dependent regulation of Kv1.2 channel activity. EMBO J 18:109-118

Uetani N, Kato K, Ogura H, Mizuno K, Kawano K, Mikoshiba K, Yakura H, Asano M, Iwakura Y (2000) Impaired learning with enhanced hippocampal long-term potentiation in PTPdelta-deficient mice. EMBO J 19:2775-2785

van der Sar A, Zivkovic D, den Hertog J (2002) Eye defects in receptor protein-tyrosine phosphatase alpha knock-down zebrafish. Dev Dyn 223:292-297

van der Wijk T, Blanchetot C, Overvoorde J, den Hertog J (2003) Redox-regulated rotational coupling of receptor protein-tyrosine phosphatase alpha dimers. J Biol Chem 278:13968-13974

Van Lieshout EM, van der Heijden I, Hendriks WJ, Van der Zee CE (2001) A decrease in size and number of basal forebrain cholinergic neurons is paralleled by diminished hippocampal cholinergic innervation in mice lacking leukocyte common antigen-related protein tyrosine phosphatase activity. Neuroscience 102:833-841

Van Vactor D (1998) Protein tyrosine phosphatases in the developing nervous system. Curr Opin Cell Biol 10:174-181

Van Vactor D, O'Reilly AM, Neel BG (1998) Genetic analysis of protein tyrosine phosphatases. Curr Opin Genet Dev 8:112-126

Wallace MJ, Batt J, Fladd CA, Henderson JT, Skarnes W, Rotin D (1999) Neuronal defects and posterior pituitary hypoplasia in mice lacking the receptor tyrosine phosphatase PTPsigma. Nat Genet 21:334-338

Wang X, Weng LP, Yu Q (2000) Specific inhibition of FGF-induced MAPK activation by the receptor-like protein tyrosine phosphatase LAR. Oncogene 19:2346-2353

Weiss A, Schlessinger J (1998) Switching signals on or off by receptor dimerization. Cell 94:277-280

Weng LP, Wang X, Yu Q (1999) Transmembrane tyrosine phosphatase LAR induces apoptosis by dephosphorylating and destabilizing p130Cas. Genes Cells 4:185-196

Wharram BL, Goyal M, Gillespie PJ, Wiggins JE, Kershaw DB, Holzman LB, Dysko RC, Saunders TL, Samuelson LC, Wiggins RC (2000) Altered podocyte structure in GLEPP1 (Ptpro)-deficient mice associated with hypertension and low glomerular filtration rate. J Clin Invest 106:1281-1290

Wills Z, Bateman J, Korey CA, Comer A, Van Vactor D (1999) The tyrosine kinase Abl and its substrate enabled collaborate with the receptor phosphatase Dlar to control motor axon guidance. Neuron 22:301-312

Xu Z, Weiss A (2002) Negative regulation of CD45 by differential homodimerization of the alternatively spliced isoforms. Nat Immunol 3:764-771

Yang XH, Seow KT, Bahri SM, Oon SH, Chia W (1991) Two Drosophila receptor-like tyrosine phosphatase genes are expressed in a subset of developing axons and pioneer neurons in the embryonic CNS. Cell 67:661-673

Yeo TT, Yang T, Massa SM, Zhang JS, Honkaniemi J, Butcher LL, Longo FM (1997) Deficient LAR expression decreases basal forebrain cholinergic neuronal size and hippocampal cholinergic innervation. J Neurosci Res 47:348-360

Zabolotny JM, Kim YB, Peroni OD, Kim JK, Pani MA, Boss O, Klaman LD, Kamatkar S, Shulman GI, Kahn BB, Neel BG (2001) Overexpression of the LAR (leukocyte antigen-related) protein-tyrosine phosphatase in muscle causes insulin resistance. Proc Natl Acad Sci USA 98:5187-5192

Zeng L, D'Alessandri L, Kalousek MB, Vaughan L, Pallen CJ (1999) Protein tyrosine phosphatase alpha (PTPalpha) and contactin form a novel neuronal receptor complex linked to the intracellular tyrosine kinase fyn. J Cell Biol 147:707-714

Zheng XM, Resnick RJ, Shalloway D (2000) A phosphotyrosine displacement mechanism for activation of Src by PTPalpha. EMBO J 19:964-978

Zheng XM, Resnick RJ, Shalloway D (2002) Mitotic activation of protein-tyrosine phosphatase alpha and regulation of its Src-mediated transforming activity by its sites of protein kinase C phosphorylation. J Biol Chem 277:21922-21929

Zheng XM, Shalloway D (2001) Two mechanisms activate PTPalpha during mitosis. EMBO J 20:6037-6049

Zheng XM, Wang Y, Pallen CJ (1992) Cell transformation and activation of pp60c-src by overexpression of a protein tyrosine phosphatase. Nature 359:336-339

Zondag GC, Koningstein GM, Jiang YP, Sap J, Moolenaar WH, Gebbink MF (1995) Homophilic interactions mediated by receptor tyrosine phosphatases mu and kappa. A critical role for the novel extracellular MAM domain. J Biol Chem 270:14247-14250

Zondag GC, Reynolds AB, Moolenaar WH (2000) Receptor protein-tyrosine phosphatase RPTPmu binds to and dephosphorylates the catenin p120(ctn). J Biol Chem 275:11264-11269

14 The Shp-2 tyrosine phosphatase

Lisa A. Lai, Chunmei Zhao, Eric E. Zhang, and Gen-Sheng Feng

Abstract

Shp-2 functions as a cytoplasmic protein tyrosine phosphatase in multiple fundamental cellular signaling pathways, to regulate basic mechanisms such as proliferation, differentiation, apoptosis, and motility. Notably, one genetic disorder in
humans, Noonan Syndrome, has been linked to gain-of-function mutations in the
Shp-2/PTPN11 gene, and this phosphatase also appears to be a cellular target of
the Helicobacter pylori virulence factor CagA protein implicated in pathogenesis
of gastric carcinoma. The requirement for Shp-2 activity in development has been
demonstrated in vertebrate as well as in invertebrate models. While Shp-2 activity
has been implicated in numerous signaling pathways, including the Ras/MAPK
and JAK/STAT pathways, its physiological substrates remain elusive to investigators. This chapter provides a general background for Shp-2 signaling and discusses
its contribution to a number of diverse biological functions in humans and other
organisms.

14.1 Introduction

The Src homology 2 containing protein tyrosine phosphatase, Shp-2 (previously
known as SH-PTP2, Syp, PTPN11, PTP1D, and PTP2C), is a cytoplasmic enzyme. It belongs to a subfamily of protein tyrosine phosphatases, consisting of itself and Shp-1, both of which share a similar overall protein structure. Shp-1 and
Shp-2 phosphatases possess at their N-terminus two tandem SH2 domains--
conserved sequence motifs of approximately 100 amino acids which regulate interactions between signaling molecules through binding to phosphotyrosine (pTyr)
residues--and a catalytic phosphatase domain at the C-terminus. Despite their
similar architecture, their expression pattern is unique. Whereas Shp-1 expression
is limited to hematopoietic cells, Shp-2 is ubiquitously expressed. Also apparent
from the phenotypes in mutant mice is that these phosphatases are functionally
distinct, since neither Shp-1 nor Shp-2 can compensate for loss of the other.
Whereas Shp-1 functions primarily in negative regulation of signaling pathways,
Shp-2 can act either positively or negatively in regulation.

Corkscrew (CSW) is the *Drosophila* homologue of mammalian Shp-2. CSW
was identified as a positive regulator of the Torso signaling pathway, acting
downstream of Torso and in concert with D-raf (Perkins et al. 1992). The absence
of CSW suppressed a torso gain-of-function phenotype. CSW functions in other

Topics in Current Genetics, Vol. 5
J. Arino, D.R. Alexander (Eds.) Protein phosphatases
© Springer-Verlag Berlin Heidelberg 2004

RTK signaling as well, including *Drosophila* EGF receptor DER, FGFR Breathless and Sevenless (Herbst et al. 1996; Perkins et al. 1996; Raabe et al. 1996).

The crystal structure of Shp-2 (Hof et al. 1998) reveals that the N-terminal SH2 domain (SH2-N) folds in on itself to auto-inhibit the phosphatase domain, until which time it binds to a pTyr residue, when the inhibition is relieved. The most commonly used Shp-2 mutant is a dominant negative form in which active site residue Cysteine 459 has been substituted with Serine (C459S). This mutant functions as a dominant negative since the phosphatase domain can still bind phosphotyrosine residues, but is unable to remove the phosphate group. In addition, mutants within the SH2 domains, such as R32K or R138K, which disrupt SH2 interactions, have also been used to analyze Shp-2 activity (Pluskey et al. 1995). Mutations such as D61A or E76A result in constitutively active mutants—exhibiting 20-100 fold increases in phosphatase activity--due to loss of auto-inhibition (O'Reilly et al. 2000; Takai et al. 2002). Recently, a substrate trapping mutant that combined the D to A and C to S mutations was used in a screen to identify new Shp-2 interacting proteins (Agazie and Hayman 2003).

In this chapter, we will discuss how Shp-2 contributes to various cellular functions and how de-regulation of this phosphatase results in cellular apoptosis and disease states. We will then include an analysis of the various signaling pathways in which Shp-2 function has been implicated.

14.2 Shp-2 function in development

Shp-2 function has been shown to be required for gastrulation and in early vertebrate development. Homozygous deletion of the SH2-N domain of Shp-2 (Shp-$2^{\Delta 46-110}$) resulted in early embryonic lethality of mice at midgestation, at day E8.5-E10.5, with severe defects in posterior structures, and failure to complete gastrulation (Saxton et al. 1997). Analysis of chimeric mice revealed a requirement for Shp-2 to properly respond to fibroblast growth factor (FGF) signaling; defects in migration of mesodermal cells from the primitive streak were observed in these mice (Saxton and Pawson 1999). Studies of chimeric mice also demonstrated a functional requirement for Shp-2 in limb bud formation (Saxton et al. 2000). Previous experiments using microinjections also implicated Shp-2 function in the regulation of mesoderm formation in Xenopus (Tang et al. 1995; O'Reilly and Neel 1998), and more recent studies suggested a functional interaction between Shp-2 and the Src family kinase protein Laloo in this process (Weinstein and Hemmati-Brivanlou 2001).

Phenotypic analyses of chimeric animals and compound mutant animals have also provided clues about Shp-2 function in postnatal development and in adults. Although this phosphatase is widely expressed, Shp-2 appears to play a critical role in hematopoietic cell development as compared to other cell types (Qu et al. 1998). Genetic evidence for a signal-enhancing effect by Shp-2 on EGF signaling was obtained in compound mutant mice defective in both Shp-2 (Shp-$2^{\Delta 46-110}$) and EGFR (waved-2, wa-2) (Qu et al. 1999). A heterozygous Shp-2 mutation dramatically enhanced the defective phenotype of the wa-2 mice; mutant animals displayed growth retardation, and impaired epithelial, lung, and intestinal development (Qu et al. 1999). In another study using a targeted exon 2 deletion mutant, crossing of Shp-2 and Egfr$^{wa-2/wa-2}$ mutant mice demonstrated a specific developmental defect in semilunar, but not atrioventricular valves (Chen et al. 2000). Analysis of Shp-2$^{+/-}$:EGFR$^{wa-2/wa-2}$ mice revealed a relative increase in aortic valve diameter as a result of increased numbers of mesenchymal cells. Furthermore, mutant mice displayed characteristics of congestive heart failure, suggesting that these mice could be used as a model for cardiac heart disease. Consistently, Shp-2 mutation has been shown to negatively affect the differentiation potential of ES cells into cardiac muscle cells in vitro (Qu and Feng 1998).

14.3 Shp-2 and human diseases

With its ubiquitous expression and involvement in several key signaling pathways, defects in Shp-2 activity have long been thought to contribute to human diseases. Indeed, Shp-2 involvement has been implicated in a growing list of diseases, most notably the genetic disorder Noonan Syndrome, as well as in several contexts involving perturbed signaling, such as the immunodeficiency, X-linked lymphoproliferative disease (XLP), and neutropenia.

Noonan Syndrome (NS) is an autosomal dominant disorder; patients typically display some or all of these characteristics: dysmorphic facial features, short stature, heart disease, skeletal malformation, webbed neck and mental retardation. Several recent studies have shown that defects in Shp-2 function contribute to this disease state. Tartaglia et al. first mapped the candidate gene for NS to chromosome 12q24.1, where PTPN11/Shp-2 is located (Tartaglia et al. 2001). Subsequently, they identified two mis-sense mutations in exon 3 of that gene, specific to NS patients and not observed in un-afflicted family members or control individuals. An expanded screening of sporadic or familial NS patients revealed that, in fact, about 50% of patients tested had mis-sense mutation(s) located within the coding region of PTPN11/Shp-2 (Tartaglia et al. 2001; Maheshwari et al. 2002; Tartaglia et al. 2002). Furthermore, there was a correlation between Shp-2 mutation among NS patients and those suffering from pulmonary valve stenosis; no association was observed with hypertrophic cardiomyopathy. It appears that the majority of NS-related mutations likely result in Shp-2 gain-of-function since they clustered within interacting surfaces of the SH2-N and PTP domains (Hof et al. 1998). An association between NS patients with mutations in Shp-2 and the occurrence of LEOPARD syndrome or other skin pigmentation abnormalities has also been suggested (Legius et al. 2002). However, no correlation was observed with cardiofaciocutaneous (CFC) syndrome patients, a genetic disease long believed to be a more severe affliction of Noonan Syndrome (Ion et al. 2002).

The contribution of phosphatases, including Shp-2, to nephrogenesis and renal development were recently investigated in a model for renal cystic disease. Analysis of polycystic kidneys from bcl2$^{-/-}$ mice revealed a phenotype similar to other

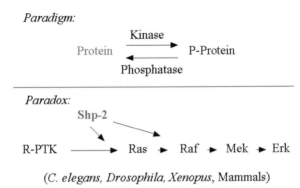

Paradigm:

Paradox:

(*C. elegans, Drosophila, Xenopus,* Mammals)

Fig. 1. It is well known that Shp-2 acts downstream of receptor tyrosine kinases (R-PTKs) in promoting the Ras/Erk pathway. This positive effect of a tyrosine phosphatase seems at odds with the paradigm for protein phosphorylation in regulation of cell activities, in which phosphatases act to reverse the activities of kinases. Although we do not have a clear answer yet to this "paradox", it should be emphasised that the positive effect of Shp-2 in the Ras/Erk pathway has been demonstrated in multiple systems, including *C. elegans, Drosophila, Xenopus* and mammals, by numerous research groups. Shp-2 exerts its positive effect in modulation of signals going through this pathway, apparently by removing a phosphate from a phospho-tyrosine residue.

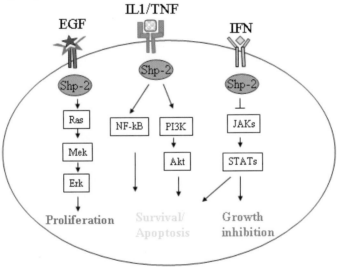

Fig. 2. The Shp-2 tyrosine phosphatase is apparently involved in multiple signaling pathways in the control of fundamental cellular activities such as cell proliferation and cell survival/apoptosis. In cellular response to growth factors, such as EGF, Shp-2 acts through the Ras/Erk pathway in promoting cell proliferation. The phosphatase also works via the NF-κB and PI3K/Akt pathways in the modulation of cell survival/apoptosis. Furthermore, Shp-2 acts as a negative regulator in the JAK/STAT pathway in inhibition of cell growth. In addition, Shp-2 may work in concert with Fak kinase in the control of cell motility, an aspect of its functions not shown in this figure.

renal cystic diseases, including an overall increase in apoptosis and proliferation, altered tubular morphology and incomplete differentiation of epithelial cells. Interestingly, Sorenson et al. observed reduced expression and activity of Shp-2, as well as altered cellular distribution in cystic kidneys isolated from postnatal day 20 bcl2$^{-/-}$ mice (Sorenson and Sheibani 2002). Similar to the observation made in Shp-2 mutant fibroblast cells (Yu et al. 1998), stable phosphorylation of the focal adhesion proteins FAK and paxillin was detected, suggesting that dephosphorylation of these proteins by Shp-2 positively affects both cell growth and differentiation in the kidney. Similar results were also observed during analysis of polycystic kidneys from *cpk* mice (Sorenson and Sheibani 2002), suggesting that increased apoptosis in these kidneys results from a loss of Shp-2 activity required for normal differentiation of epithelial cells, thus leading to renal cyst formation.

Defects in cellular signaling pathways result in immunodeficiency, one prominent example being X-linked lymphoproliferative disease (XLP), also known as Duncan syndrome, a progressive genetic disorder resulting from Epstein-Barr virus infection. This disease is typically characterized by uncontrolled lymphocyte proliferation resulting in fatal infectious mononucleosis, B cell lymphomas, and dys-gammaglobulinemia (Morra et al. 2001). The gene mutated in XLP patients was recently identified as the signaling lymphocytic activation molecule (SLAM)-associated protein (SAP), also known as DSHP or SH2D1A, and consists of a single SH2-domain and a C-terminal tail (Sayos et al. 1998). Engagement of SLAM and subsequent interactions with signaling proteins such as SAP and Shp-2 results in the activation of natural killer (NK) and T cells. *In vitro* experiments suggested that Shp-2 bound to phosphorylated SLAM in the absence of SAP and that the interaction was inhibited by addition of exogenous SAP (Sayos et al. 1998; Tangye et al. 1999). Therefore, defects in the signaling pathways resulting from SAP inhibition of Shp-2 affect helper- and cytotoxic-T cell responses, lead to accumulation of EBV-infected B cells in XLP patients and, ultimately, lymphoma development. However, the mechanism is unresolved—one question that remains to be answered is how the actions of Shp-2 upon SLAM and/or the Fyn kinase results in the specific phenotypes of XLP (Tangye et al. 1999; Chan et al. 2003; Latour et al. 2003).

Correlative evidence has also prompted investigation of how Shp-2 defects contribute to the development of severe congenital neutropenia, also known as Kostmann's Syndrome (Tidow et al. 1999). This disease is characterized by an incomplete differentiation of myeloid progenitor cells into mature neutrophils and an accumulation of myelocytic or promyelocytic cells (Palmblad et al. 2001). It was observed that Shp-2 protein levels are dramatically increased in neutrophils isolated from SCN patients, although by comparison there is a negligible difference in overall phosphatase activity (Tidow et al. 1999). These data support previous studies demonstrating increased phosphorylation and activity of the tyrosine kinase Jak2. However, it is still unclear whether this phenomenon results from prolonged treatment with granulocyte-colony stimulating factor (G-SCF) or is symptomatic of the disease.

It was recently reported that Shp-2 might be the cellular target of *Heliobacter pylori* virulence factor CagA protein (Higashi et al. 2002b). Amazingly, it has been estimated that *H. pylori* infects approximately half of the human population on the Earth, and this infection increases the risk of gastric adenocarcinoma and gastric mucosa-associated lymphoid tissue lymphoma. Association of Shp-2 with CagA leads to cellular acquisition of the "hummingbird phenotype" and CagA-dependent apoptosis (Higashi et al. 2002b; Tsutsumi et al. 2003). CagA binding to either Shp-2 or the C-terminal Src kinase, Csk, resulted in antagonism of the other's activity and supports chronic infection by *H. pylori* (Tsutsumi et al. 2003). Transfection of CagA along with wild type Shp-2 or a membrane-targeted Shp-2 mutant resulted in an increase in apoptosis of human gastric epithelial AGS cells, as evaluated by Annexin-V staining (Tsutsumi et al. 2003). Although this increase correlated with a decrease in colony formation, it suggested that Shp-2 may facilitate the growth inhibition and/or apoptotic response. The current view is that the genetic variations between *H. pylori* strains generate CagA products with differential Shp-2 binding activity, which may contribute to the broad range of diseases--from gastritis to cancer--resulting from *H. pylori* colonization of gastric epithelium (Higashi et al. 2002a).

Identification of Shp-2 involvement in the pathogenesis of these human diseases has provided further support that this phosphatase indeed acts to promote cell signaling. In the Noonan Syndrome, an autosomal dominant disorder, the *Shp-2* mutations identified are largely classified as "gain-of-function" mutations. Similarly, membrane targeting and aberrant activation of Shp-2 by CagA are implicated in CagA-induced gastric cancer. While the mechanisms remain to be elucidated for the most part, this is an exciting area of investigation and could potentially reveal the true physiological roles and/or targets of this enzyme.

14.4 Shp-2 in cell proliferation and differentiation

Transient activation of the Ras-Erk pathway by RTKs is believed to function in cell proliferation (Simon 2000). The first evidence indicating the involvement of Shp-2 came from the observation that insulin-induced cell proliferation was suppressed by ectopic expression of the mutant Shp-2^{cs} form (Milarski and Saltiel 1994; Noguchi et al. 1994). Similar results were obtained from other growth factors such as EGF and PDGF. As one might expect, due to an impaired mitogenic response, Shp-2 mutant cells proliferate at a slower rate than wild type cells (Shi et al. 1998). However, Shp-2 activity and interaction with a mitogenic G protein coupled receptor (GPCR), the B2 receptor, was enhanced following treatment with the anti-proliferative agent, bradykinin (Duchene et al. 2002).

Shp-2 mutants are defective in differentiation in a number of model systems. Shp-2$^{\Delta46-110}$ ES cell differentiation into erythroid and myeloid lineages was impaired, as was SCF-dependent proliferation and Erk activation (Qu et al. 1997). Stimulation of C2C12 myoblasts with FGF-2 resulted in a complex formation between Shp-2 and FRS2, and repression of myogenesis through an Erk-independent mechanism (Kontaridis et al. 2002). In addition, the phosphorylation of SHPS-1 and its association with Shp-2 correlated with MyoD expression in C2C12 myoblasts undergoing myogenesis (Kontaridis et al. 2001).

In addition to RTKs and integrin signaling, Shp-2 regulates the cytokine receptor-mediated pathways that control cell growth, survival and differentiation (Feng 1999). One shared cytokine receptor component, the glycoprotein gp130, physically interacts with Shp-2 through phosphorylation on Tyr759, located within a cytoplasmic tail region (Schmitz et al. 2000). Knock-in mice deficient on this tyrosine site (gp130$^{F759/F759}$) developed splenomegaly, lymphadenopathy and an enhanced acute phase reaction, suggesting that Shp-2 is a negative regulator of cytokine signaling pathways (Ohtani et al. 2000). This interaction is also dependent on SOCS3 recruitment, as shown with recent chimeric receptor studies (Fairlie et al. 2003). Combined with its role in RTK signaling, Shp-2 is apparently one of the key regulators in cell differentiation.

14.5 Shp-2 and cell motility

Shp-2 function in the regulation of cell motility has been demonstrated in a number of different cell line models through its activity on focal adhesion proteins, such as FAK, paxillin and p130^{Cas}. Analysis of Shp-2$^{\Delta46-110}$ mutant fibroblasts first revealed a defect in cell spreading and migration on fibronectin (Yu et al. 1998; Oh et al. 1999). Compared with wild type fibroblasts, Shp-2 mutant cells displayed delayed spreading, and cells appeared depolarized; their appearance more closely resembled epithelial cells. A modified chamber assay demonstrated that cell motility was dramatically reduced in mutant cells (Yu et al. 1998). Moreover, signaling through integrins and FAK was impaired in these cells. Aside from its involvement in FGF-stimulated migration, Shp-2 also functions positively in integrin-mediated signaling pathways (Manes et al. 1999; Oh et al. 1999).

Several studies have established a connection between Shp-2 and the small GTPase RhoA (Schoenwaelder et al. 2000; Kodama et al. 2001; Chang et al. 2002). Expression of Shp-2^{C459S} mutant in MDCK cells led to an increase in the formation of stress fibers and focal adhesions, which was blocked by *Clostridium botulinum* ADP-ribosyltransferase C3, suggesting the involvement of RhoA activity. Indeed, Y-27632, a selective inhibitor of Rho effector ROCK, was also able to block Shp-2^{C459S} induced stress fiber formation, as was the expression of a dominant negative form of Vav2, a guanine nucleotide exchange factor for RhoA (Kodama et al. 2001). However, the active Shp-2 DA mutant induced lamellipodia formation, which was inhibited by active RhoA or Vav2 expression (Kodama et al. 2001). Treatment of cells with calpeptin results in RhoA activation and actin reorganization, both of which were increased in Shp-2$^{\Delta46-110}$ mutant cells (Schoenwaelder et al. 2000). In addition, investigation of smooth muscle cells revealed that nitric oxide (NO) could both induce the expression of Shp-2 and increase cell motility, probably also via the suppression of RhoA activity (Brown et al. 2001; Chang et al. 2002).

Similar phenotypes, such as defects in polarized extension and migration of cells on fibronectin, were observed in SHPS-1 deficient fibroblast cells, suggesting that Shp-2 and SHPS-1 may work in the same signaling pathway controlling cell shape, adhesion and motility (Inagaki et al. 2000a, 2000b). A recent study of axon guidance molecules, ephrin/Eph, implied that EphA2 suppresses the Shp-2/SHPS-1-mediated integrin pathway by its inhibition of FAK (Miao et al. 2000). Shp-2 also potentially regulates axon guidance through modification of the UNC-5 receptor, RCM, following netrin stimulation (Tong et al. 2001).

14.6 Shp-2 and cell apoptosis

While the function of Shp-2 in positively affecting cell proliferation and differentiation has been extensively investigated, recent data suggest that Shp-2 may participate in caspase-dependent pro-apoptotic pathways as well as in cell survival pathways, such as those mediated by NF-κB.

Multiple independent studies have demonstrated that Shp-2 modulates cytokine-induced apoptotic responses through the Akt-dependent pathways and/or the phosphorylation of various focal adhesion proteins, suggestive of at least one general mechanism whereby Shp-2 contributes to apoptosis (Chauhan et al. 2000; Hakak et al. 2000; Hideshima et al. 2001). However, Shp-2 has also been shown to bind death receptors at the cell surface. Using peptides generated from the truncated immunoreceptor truncated inhibitory motif (ITIM), YXXL, identified in the TNF receptor, Daigle et al. showed association with Shp-1 and Shp-2 phosphatases in U937 cells (Daigle et al. 2002). These results were further verified by co-immunoprecipitation of Shp-2 with the TNF-R1 receptor following TNF-α stimulation.

The cytokine interleukin-6 (IL-6) has been shown to elicit both proliferative and apoptotic responses via the activation of the Ras/Erk dependent signaling pathway and Jak/Stat3 pathways, respectively. IL-6 treatment of multiple myeloma cells results in rapid phosphorylation of the phosphatase Shp-2, the protein kinase Akt, and the related adhesion focal tyrosine kinase (RAFTK) followed by apoptosis (Chauhan et al. 2000). Interestingly, the apoptotic response was abrogated by introduction of the Shp-2^{CS} mutant, suggestive of a positive role for Shp-2 (Hideshima et al. 2001). Immunoprecipitation assay confirmed that Shp-2 associates with activated RAFTK, although this interaction was not dependent on engagement of the Shp-2 SH2 domains (Chauhan et al. 2000; Hideshima et al. 2001). These results suggest that Shp-2 mediated downregulation of RAFTK as well as the coincident activation of Akt and PI3K results in an apoptotic response. Furthermore, studies using the inhibitor LY294002 further demonstrated that IL-6 treatment also affects the interaction of Shp-2 with PI3K, independent of signaling through the JAK1, JAK2, Tyk2, or PYK2 kinases (Hideshima et al. 2001).

Examination of IL-6 responses in premature myeloma and lymphoblast cell lines suggests that distinct pathways participate in responses to this cytokine in a cell type-specific manner. IL-6 induced proliferation in the ANBL6 myeloma cell line whereas, in contrast, apoptosis was triggered upon IL-6 treatment of the CESS lymphoblast cell line (Cheung and Van Ness 2002). Interestingly, these differences correlated with Shp-2 phosphorylation as well as association with the IL-6 receptor subunit, gp130. Altered Shp-2 activity and the subsequent change in the duration of Stat3 activation may contribute to the distinct signaling responses. It is unclear, however, how the terminal differentiation of cells influences the protective effect of Shp-2 or initiation of an apoptotic response.

A dramatic reduction in v-Src dependent transcriptional activity was observed in cells expressing a catalytically inactive Shp-2. Following transformation of wild type and Shp-2 mutant fibroblast cells with v-Src, differences in the tyrosine phosphorylation of focal adhesion proteins, such as paxillin, FAK and Cas, were observed by immunoblot analysis (Hakak et al. 2000). Furthermore, these phosphorylation differences were reflective of changes in cell morphology, caused by re-organization of the actin cytoskeleton. In addition, a coincident increase in the amount of floating/apoptotic cells was observed in cultures of Shp-2 mutant fibroblasts as compared to the wild type cells following v-Src transformation. Both the morphological changes and increase in apoptotic/floating cells were suppressed upon re-introduction of wild type Shp-2 into these mutant cells.

Shp-2 has also been demonstrated to positively regulate survival of mesencephalic dopaminergic neurons, which typically undergo degeneration in Parkinson's disease patients (Takai et al. 2002). Using recombinant adenoviruses to express wild type or Shp-2 mutants, it was demonstrated that loss of Shp-2 activity correlated with decreases in tyrosine hydroxylase (TH)-positive cells following treatment with brain-derived neurotrophic factor (BDNF) while no change in survival was observed in cells expressing constitutively active Shp-2 (Takai et al. 2002). Treatment with the PI3K inhibitor, LY294002, or the Erk inhibitor, U0126, abrogated the survival effect, suggesting that the survival pathways are dependent upon activation of PI3K and Erk, both of which are positively regulated by Shp-2. However, it is unclear whether Shp-2 activity affects both the differentiation and survival of cultured mesencephalic dopaminergic neurons since less dramatic effects were observed in a comparable analysis of cerebral cortical neurons.

14.7 Shp-2 and cytoplasmic signaling

The function of Shp-2 has been studied downstream of a variety of cell surface receptors. Receptor activation leads to several changes in Shp-2 protein. Initially, Shp-2 is recruited to activated receptor tyrosine kinases (RTKs) by binding to the pTyr residues within their cytoplasmic regions, at which time it becomes phosphorylated in many cases. This phosphorylation might modulate the Shp-2 activity as well as protein complex formation between Shp-2 and other SH2-containing proteins, such as Grb2. In certain pathways, Shp-2 catalytic activity is dispensable, suggestive of an additional role as an adaptor protein and the functional relevance of its subcellular localization (Bennett et al. 1994; O'Reilly and Neel 1998; Lacalle et al. 2002). Thus, the function of Shp-2 is likely dependent on both the catalytic activity of the phosphatase domain and its association with other cytoplasmic signaling proteins.

14.7.1 Shp-2 plays a positive role in the Ras/MAPK pathway

Shp-2 has been implicated in the Ras/MAPK pathway downstream of a variety of cell surface receptors, including RTKs, integrins, T cell receptor, angiotensin II induced G protein-coupled receptor, somatostatin receptor 1 (SSTR1), leptin receptors ObRa and ObRb, IL-2 receptor, leukemia inhibitory factor (LIF) receptor, and gp130 – the β chain of IL-6 receptor (Fuhrer et al. 1995; Frearson et al. 1996; Adachi et al. 1997; Reardon et al. 1997; Symes et al. 1997; Carpenter et al. 1998; Gadina et al. 1998; Marrero et al. 1998; Stofega et al. 1998; Florio et al. 1999; Kim and Baumann 1999; Li and Friedman 1999; Oh et al. 1999; Bjorbaek et al. 2001; Maile and Clemmons 2002a). Following a conserved mechanism of ligand-receptor engagement and subsequent receptor auto-activation, Shp-2 association and its activation occurs through interactions between the two SH2 domains of Shp-2 and pTyr-docking sites on the cytoplasmic domain of the respective receptors.

Shp-2 was proposed to act as an adapter protein through the two SH2 domains at its N-terminus and a potential phospho-tyrosine site (Y^{542}TNI of human Shp-2) for Grb2 SH2 domain binding. Indeed, Shp-2 was found to complex with Grb2 in many different cell types. The dominant negative mutant Shp-2^{C459S} was able to inhibit insulin-stimulated Erk activation and c-fos promoter-dependent gene transcription in CHO/IR cells. In a transient transfection study in COS7 cells, both the SH2 domains and the catalytic domain of Shp-2 were required for EGF stimulated Erk activation (Deb et al. 1998). By comparing signaling downstream of EGF in Shp-2$^{\Delta46-110}$ mutant cells, Shi et al. demonstrated that Shp-2 promotes Ras activation following protein complex formation with the adaptor protein, Gab1 (Shi et al. 2000). These and other studies established Shp-2 as a positive regulator of the Ras/Erk pathway, acting downstream of receptor tyrosine kinases, and also possibly upstream or in a parallel pathway to Ras.

Several studies also investigated Shp-2 regulation of Jnk, another module of the MAPK family, although with conflicting results. Insulin and EGF treatment induced Jnk activation in Rat-1 fibroblast cells expressing human insulin receptor, in a Shp-2 dependent manner, suggesting that Shp-2 functions upstream of Sos and Ras to modulate Jnk activation (Fukunaga et al. 2000). However, Shp-2^{C459S} was unable to inhibit EGF-induced Jnk activation in COS7 cells (Deb et al. 1998). In addition, Shp-2$^{\Delta46-100}$ mutant cells exhibited enhanced Jnk activity in response to stress conditions such as UV irradiation and heat shock treatment (Shi et al. 1998).

Overall, there are still many questions remaining open on how Shp-2 positively regulates Ras/MAPK pathways. One hypothesis is that dephosphorylation of pTyr residues by Shp-2 relieves some inhibition or attenuates a negative signal. In fact, studies of CSW function in Torso signaling suggested that CSW removal of a phosphate group from the Torso receptor tyrosine kinase negatively regulates Ras, since this activity results in loss of a binding site for the RasGAP, a negative regulator of Ras. Therefore, it follows that mutant Shp-2 would enhance, but not rescue, mutants with defective RTKs (Qu et al. 1999; Chen et al. 2000). On the other hand, the Shp-2 protein could act on additional phosphorylated proteins by removing negative regulators from those signaling complexes. The tissue-specific inhibitors of Ras/MAPK pathway, Sprouty proteins, display a direct uncoupling function by preventing the formation of FRS2/Grb2-SOS1/Shp-2 complex (Hanafusa et al. 2002). It will be interesting to determine whether the Sprouty proteins are Shp-2 substrates.

14.7.2 Shp-2 is involved in the regulation of the JAK/STAT pathway

Recent studies identified a role for Shp-2 in the regulation of multiple JAK/STAT pathways. Loss of Shp-2 has been shown to increase sensitivity to the cytotoxic effects of interferons α and γ through STAT activation (You et al. 1999). However, the mechanism of this regulation is unclear and Shp-2 may work on JAKs and/or STATs. Molecular association between Shp-2 and Jak2 was detected, although the SH2 domains appear to be not required for this interaction (Yin et al. 1997).

Ciliary neurotrophic factor (CNTF) induces STAT activation through gp130, a common receptor subunit for IL-6 family cytokines. The Shp-2^{C459S} mutant enhanced gp130-induced STAT-dependent gene transcription; similar effects were observed using a gp130 mutant impaired in Shp-2 binding. These results suggest that Shp-2 plays a negative role in gp130 mediated STAT activation (Symes et al. 1997; Servidei et al. 1998). These results were further corroborated by studies of chimeric receptors, combining the extracellular domain of G-CSF receptor with the transmembrane and cytoplasmic domain of gp130 (Kim et al. 1998). Increased intensity and duration of receptor phosphorylation, JAK activation, and STAT DNA binding activity were observed upon co-expression of either inactive Shp-2 with wild type receptor or a mutant G-gp130 chimeric receptor, with loss of the Shp-2 binding site (Kim et al. 1998). Using mutant receptors, Shp-2 was also indicated as a negative regulator of Stat3 activation, as induced by the leptin receptor, ObR, and growth hormone receptor (GHR)-mediated STAT5B activation (Carpenter et al. 1998; Stofega et al. 2000).

In contrast to the aforementioned studies, Shp-2 seems to play a positive role in STAT activation induced by the prolactin receptor (PRLR). Overexpression of Shp-2 truncation mutant containing only the two SH2 domains inhibited Jak2 kinase activity, phosphorylation of Stat5a and 5b, as well as the DNA binding activity of the STAT proteins (Berchtold et al. 1998). Shp-2 was found to form a complex with Stat5 through its C-terminal SH2 domain upon PRLR activation. The translocation of this complex into the nucleus is dependent on the catalytic activity of Shp-2 (Chughtai et al. 2002). Similarly, Shp-2 acts a positive regulator of IL-2 induced STAT-dependent gene transcription (Gadina et al. 1998).

The molecular mechanism of STAT regulation by Shp-2 phosphatase is not clear. It is suggested that STAT may be a direct substrate of the phosphatase activity. Shp-2, but not the closely related protein Shp-1, dephosphorylates Stat5 *in vi-*

tro. In addition, a Stat5-Shp-2 complex was isolated (Yu et al. 2000), and a phosphopeptide derived from Stat5A was able to fish out Shp-2 from cell lysates (Chen et al. 2003). A recent study suggested that Shp-2 might dephosphorylate STAT1 on both Tyr and Ser residues (Wu et al. 2002). It would be interesting to determine whether this proposed dual specificity is specific to Shp-2, or whether other protein tyrosine phosphatases, such as Shp-1 and PTP1B, can also target phosphorylated Ser/Thr residues.

14.7.3 Shp-2 functions in other signaling pathways

As described in section 1.6 of this chapter, Shp-2 may participate in the control of cell survival through modulation of the PI3K-Akt pathway. Furthermore, Shp-2 apparently has a positive role in the NF-κB pathway in mediating the induction of IL-6 by IL-1 and tumor necrosis factor (TNF), a cytokine circuit during inflammatory and immune responses (You et al. 2001). In addition, Shp-2 might be implicated in the DNA damage response (Yuan et al. 2003). Following treatment with cisplatin or gamma irradiation, Shp-2$^{\Delta 46-110}$ mutant cells displayed a reduced induction of the p53-related protein, p73, and the cyclin-dependent kinase inhibitor, p21. Shp-2 was also reported to play a role in PDGF-induced ROS reversible oxidation (Meng et al. 2002).

14.8 Interaction of Shp-2 with scaffold proteins

In addition to mediating interactions with activated receptors, the SH2 domains of Shp-2 also directly interact with a number of cellular proteins, including a unique family of adaptor proteins that are also known as docking or scaffold proteins, such as the GAB proteins, FRS2, and SHPS-1. Although these proteins lack enzymatic activity, they possess multiple protein-protein interaction motifs, such as SH2, PTB, and SH3 domains. In addition, many of these scaffold proteins contain a tandem Tyr motif at their C-terminal tails that are responsible for their interaction with Shp-2. Furthermore, by forming multi-protein complexes, these scaffold proteins may target Shp-2 to its physiological substrates. Dissection of these complexes will provide insight into how Shp-2 acts in various signaling pathways.

14.8.1 Interaction with GAB proteins

The Grb2 associated binder (GAB) proteins are characterized by possessing a pleckstrin homology (PH) domain at the N-terminus and multiple consensus Tyr residues that, when phosphorylated, serve as SH2 docking sites. In addition, these molecules contain one or more proline-rich motifs that interact with SH3 domains. Three GAB proteins (Gab1, 2, 3) have been identified in mammals, with homologues such as Daughter of Sevenless (DOS) in *Drosophila* and suppressor-of-Clr-

1 (SOC-1) in *C. elegans* also recently described (Holgado-Madruga et al. 1996; Raabe et al. 1996; Gu et al. 1998; Nishida et al. 1999; Zhao et al. 1999; Schutzman et al. 2001; Wolf et al. 2002). These proteins share a similar structure with the insulin receptor substrate (IRS) proteins, with the exception that GABs lack a PTB domain at their N-termini. By comparison, another docking protein consisting of multiple protein domains, FRS2, has a PTB domain but lacks an N-terminal PH domain. As a whole, these docking proteins are considered major Shp-2 partners in cell signaling.

Gab1 is involved in mediating signaling through a number of RTKs, such as c-Met, TrkA – the receptor for nerve growth factor (NGF), EGFR, IGF-1R, insulin receptor, B cell receptor (BCR), T cell receptor (TCR), and RET – the receptor for glial cell-derived neurotrophic factor (GDNF) (Holgado-Madruga et al. 1996; Weidner et al. 1996; Holgado-Madruga et al. 1997; Nishida et al. 1999; Wickrema et al. 1999; Shi et al. 2000; Winnay et al. 2000; Ingham et al. 2001). Gab1 is highly phosphorylated on Tyr residues in response to ligand stimulation at which time it forms protein complexes with Shp-2, PI3K, Grb2, and Shc, among others. Mutation of the two Gab1 residues, Y627 and Y695, abolished its interaction with Shp-2 (Cunnick et al. 2000, 2001). Consistent with previous observations, MDCK cells displayed sustained Erk activation after HGF stimulation, whereas cells expressing the Gab1 mutant deficient in Shp-2 binding showed a decrease in phospho-Erk levels (Maroun et al. 2000). In addition, expression of Gab1^{Y627F} inhibited LPA- and EGF-stimulated Erk2 activation (Cunnick et al. 2000, 2001). These results suggest that Gab1-Shp-2 interaction is critical for Erk1 activation in response to stimulation by growth factors.

Gab2 was identified by its association with Shp-2 in cells of hematopoietic lineages, and by its homology with Gab1 (Gu et al. 1998; Nishida et al. 1999; Zhao et al. 1999; Ali 2000; Gadina et al. 2000). In addition to TCR, many type I (or hematopoietic super family) cytokine receptors, including PRLR, EpoR, Mpl (receptor for Tpo), and receptors for IL-2, 3, 6 and 15 can induce the phosphorylation of Gab2 and its association with Shp-2 and other proteins (Gu et al. 1998; Nishida et al. 1999; Wickrema et al. 1999; Zhao et al. 1999; Ali 2000; Gadina et al. 2000). The phosphorylation of Gab2 can also be induced by the activation of RTKs, such as Fms (receptor for M-CSF or CSF1), c-Kit (receptor for stem cell factor), and ErbB family receptors (Nishida et al. 1999; Liu et al. 2001; Lynch and Daly 2002). There are two putative Shp-2 SH2 binding sites at the Gab2 C-terminus: YAL and YVQV (Gu et al. 1998). Mutant Gab2 fails to interact with both Grb2 and Shp-2 by immunoprecipitation (Sattler et al. 2002). Introduction of Gab2ΔShp-2 into Ba/F3 cells significantly downregulated Elk1 dependent gene transcription (Gu et al. 1998). Expression of this same Gab2 mutant in FD-Fms cells failed to enhance MAP kinase activation induced by M-CSF treatment (Liu et al. 2001).

Both Gab1 and Gab2 were originally viewed as potential substrates for Shp-2. Indeed, Shp-2 was shown to dephosphorylate both proteins *in vitro* (Nishida et al. 1999). However, no significant change in Gab1 tyrosine phosphorylation was observed in cells without functional Shp-2, suggesting it is not a major Shp-2 target (Shi et al. 2000). It remains possible that Shp-2 acts on one or more specific pTyr residues on the GAB proteins; yet, the increase of pTyr was not readily detectable in Shp-2 deficient cells. A recent study suggests that Shp-2 upon docking on Gab1 dephosphorylates the p85 binding sites on Gab1, thereby negatively regulating EGF-dependent PI3K activation (Zhang et al. 2002). The current view is that in response to stimuli, GAB proteins recruit Shp-2 to its specific targets, which addresses the mutual requirement of Gab1 and Shp-2 in the MAP kinase pathway. It is also consistent with the finding that a mutant form of DOS with the CSW (*Drosophila* Shp-2) binding sites retained was able to partially rescue a *dos* mutant phenotype (Herbst et al. 1999). Nevertheless, Shp-2 may be able to dephosphorylate its own binding sites on Gab1, leading to disassembly of the Gab1/Shp-2 complex.

14.8.2 Interaction with FRS proteins

In mammals, four FGFRs were identified and named as FGFR1-4. Genetic analysis of knockout mice revealed similarities between homozygous Shp-2 mutants and FGFR1 deficient mice, both of which are embryonic lethal and display multiple defects in mesodermal patterning (Saxton et al. 1997). Unlike EGFR and PDGFR, which have direct docking sites in the receptors, FGFRs recruit Shp-2 via adaptor proteins named FGF receptor substrate 2/3 (FRS2/3) (Ong et al. 2000). Of note, FRS2 is a ubiquitously expressed protein while FRS3 expression is restricted to embryonic day E7-E9.5. By comparison to FGFR1 and Shp-2 knockouts, FRS2 null mutant mice die on E7.0-E7.5 (Hadari et al. 1998). The structure of FRS2/3 consists of a PTB domain at the N-terminus and multiple pTyr-docking sites within its C-terminus, including four putative Grb2 binding sites and two sites for Shp-2. In response to FGF stimulation, the RTK-bound FRS2 undergoes phosphorylation on multiple tyrosine residues, followed by association with Grb2-SOS1 and Shp-2 (Ong et al. 2000).

In PC12 cells, the formation of this multi-protein complex is essential for FGF-induced differentiation. Ectopic expression of either the Shp-2[cs] form or a mutant FRS2 form deficient for all Grb2 and Shp-2 binding sites (6F) resulted in impairment in differentiation of PC12 cells, as well as shortened activation of Ras/Erk (Hadari et al. 2001). Therefore, depletion of Shp-2 from this complex resulted in attenuation of Ras activity, which is a key differentiation signal. Furthermore, reintroduction of the FRS2 6F or 2F (in which only the two putative Shp-2 binding sites were altered) mutants into FRS2-null fibroblasts did not rescue the mitogenic response, although the 4F (mutant for Grb2 binding sites) form could partially restore Ras activity. These experiments provided mechanistic insight into the role of Shp-2 in mediating FGF-induced myogenesis and limb development (Saxton et al. 2000; Kontaridis et al. 2002).

Interestingly, FRS2 adaptor protein can also bind to the activated neurotrophin receptor Trk (Dhalluin et al. 2000). In PC12 cells, NGF binds to TrkA and induces cell differentiation, and this signal is transduced via the FRS2/Grb2-SOS1/Shp-2 complex. Physical chemistry analysis demonstrated that association between FRS2 and FGFRs is primarily entropy-driven, while the binding with phosphorylated-Trk proteins depends on an enthalpy-contribution (Yan et al. 2002). This model suggests that FRS2 constitutively associates with FGFRs before growth factor induction. Following stimulation, FRS2 undergoes tyrosine-phosphorylation while either remaining associated with FGFR or switching to another receptor (i.e. NGFR), whereby it recruits the Grb2-SOS1/Shp-2 complex, and contributes to prolonged Ras/Erk activation and cell differentiation.

Further analysis on Shp-2 docking sites of FRS2 displayed that both molecules participate in FGF-initiated cell migration (Hadari et al. 2001). Reintroducing the 2F mutant of FRS2 into FRS2-deficient fibroblasts failed to restore the migration induced by FGF. Chimeric studies on Shp-2 mutant mice showed that mutant cells accumulated in the primitive streak of the epiblast and failed to commit gastrulation, demonstrating the interruption of FGF-initiated migration *in vivo* (Saxton and Pawson 1999).

14.8.3 Interaction with SHPS-1

The SH2 domain-containing tyrosine phosphatase substrate-1, SHPS-1, also known as SIRPα1 or BIT, belongs to a family of transmembrane glycoproteins, within the immunoglobulin superfamily, known as signal-regulatory proteins (Fujioka et al. 1996; Kharitonenkov et al. 1997). SHPS-1 was identified by virtue of its association with Shp-1 and Shp-2, and functions as a negative regulator of RTK-dependent cell proliferation and ITAM-mediated cell activation (Fujioka et al. 1996). SHPS-1 is phosphorylated on Tyr residues in response to various stimuli, including serum, insulin, LPA, cell adhesion to extracellular matrix, EGF, BDNF, growth hormone, and IGF-1 (Fujioka et al. 1996; Ochi et al. 1997; Takada et al. 1998; Takeda et al. 1998; Tsuda et al. 1998; Oh et al. 1999; Araki et al. 2000; Maile and Clemmons 2002b, 2002a). Once phosphorylated, SHPS-1 complexes with Shp-2; mutant SHPS-1 protein with four Tyr substituted with Phe was unable to recruit Shp-2. A tandem Tyr motif located within the cytoplasmic tail of SHPS-1 (Y449 and Y473) may be responsible for the interaction with Shp-2 SH2 domains (Takada et al. 1998). Overexpression of the catalytically inactive mutant of Shp-2^{C459S} resulted in enhanced phosphorylation of SHPS-1 (Inagaki et al. 2000a).

Dephosphorylation of SHPS-1 by Shp-2 may contribute to the positive role of Shp-2 in cell signaling. Expression of the SHPS-1 4F mutant enhanced BDNF-induced cell survival of cultured ventral mesencephalic dopaminergic neurons, as well as the co-expression of wild type SHPS-1 with Shp-2 (Takai et al. 2002). SHPS-1/Shp-2 interaction has also been implicated in regulation of cell adhesion and migration on extracellular matrices such as fibronectin and vitronectin (Inagaki et al. 2000a). In contrast to its negative role in RTK signaling, overexpression of SHPS-1 in 293 cells enhanced integrin-induced Erk activation; this activity required functional Shp-2 protein (Oh et al. 1999).

14.9 Concluding remarks

Even with the last decade of intensive study on Shp-2 phosphatase, many questions remain. For example, primary physiological Shp-2 substrates are yet to be identified. Both *in vitro* and *in vivo* data support the notion that Shp-2 is involved in fundamental cellular activities including proliferation, migration, differentiation, and survival. Apparently, the catalytic activity of Shp-2 is required for almost all of these functions, and the enzyme might act on different substrates in each of these cell activities. So far, a few phosphoproteins have been identified in complex with Shp-2 in cells and, furthermore, can be dephosphorylated by Shp-2 in an *in vitro* experimental system. However, a clear link between a site-specific dephosphorylation event and a Shp-2 function in modulation of a specific pathway has not been established. Scaffold proteins, such as members of IRS, FRS, and GAB families, presumably target Shp-2 to various substrates and therefore further investigation of Shp-2 interactions with these molecules may lead to specific clues in the hunt for physiologically relevant enzyme substrates. Targeted deletion of the *Shp-2* gene in various cell types will also contribute to elucidation of the Shp-2 functions in different pathways. Finally, genetic and molecular analyses of Shp-2 involvement in the pathogenesis of human diseases will help to elucidate further its basic biological actions.

Acknowledgements

The authors wish to apologize to the many colleagues whose contributions were not cited in this chapter due to space constraints. The work in the authors' laboratory has been supported by grants from National Institutes of Health (R01GM53660, R01CA78606, R01HL66208). L.A.L. is supported by a NIH training grant (T32AG00252) and G.S.F. was a recipient of a career development award from the American Diabetes Association.

References

Adachi M, Ishino M, Torigoe T, Minami Y, Matozaki T, Miyazaki T, Taniguchi T, Hinoda Y, Imai K (1997) Interleukin-2 induces tyrosine phosphorylation of SHP-2 through IL-2 receptor beta chain. Oncogene 14:1629-1633

Agazie YM, Hayman MJ (2003) Development of an efficient 'substrate trapping' mutant of SHP2, and identification of EGFR, Gab1, and three other proteins as target substrates. J Biol Chem 278:13952-13958

Ali S (2000) Recruitment of the protein-tyrosine phosphatase SHP-2 to the C-terminal tyrosine of the prolactin receptor and to the adaptor protein Gab2. J Biol Chem 275:39073-39080

Araki T, Yamada M, Ohnishi H, Sano SI, Hatanaka H (2000) BIT/SHPS-1 enhances brain-derived neurotrophic factor-promoted neuronal survival in cultured cerebral cortical neurons. J Neurochem 75:1502-1510

Bennett AM, Tang TL, Sugimoto S, Walsh CT, Neel BG (1994) Protein-tyrosine-phosphatase SHPTP2 couples platelet-derived growth factor receptor beta to Ras. Proc Natl Acad Sci USA USA 91:7335-7339

Berchtold S, Volarevic S, Moriggl R, Mercep M, Groner B (1998) Dominant negative variants of the SHP-2 tyrosine phosphatase inhibit prolactin activation of Jak2 (janus kinase 2) and induction of Stat5 (signal transducer and activator of transcription 5)-dependent transcription. Mol Endocrinol 12:556-567

Bjorbaek C, Buchholz, RM, Davis SM, Bates SH, Pierroz DD, Gu H, Neel BG, Myers MG, Flier JS (2001) Divergent roles of SHP-2 in ERK activation by leptin receptors. J Biol Chem 276:4747-4755

Brown C, Lin Y, Hassid A (2001) Requirement of protein tyrosine phosphatase SHP2 for NO-stimulated vascular smooth muscle cell motility. Amer J Physiol - Heart & Circ Phys 281:H1598-H1605

Carpenter LR, Farruggella TJ, Symes A, Karow ML, Yancopoulos GD, Stahl N (1998) Enhancing leptin response by preventing SH2-containing phosphatase 2 interaction with Ob receptor. Proc Natl Acad Sci USA USA 95:6061-6066

Chan B, Lanyi A, Song HK, Griesbach J, Simarro-Grande M, Poy F, Howie D, Sumegi J, Terhorst C, Eck MJ (2003) SAP couples Fyn to SLAM immune receptors. Nat Cell Biol 5:155-160

Chang Y, Ceacareanu B, Dixit M, Sreejayan N, Hassid A (2002) Nitric oxide-induced motility in aortic smooth muscle cells: role of protein tyrosine phosphatase SHP-2 and GTP-binding protein Rho. Circ Res 91:390-397

Chauhan D, Pandey P, Hideshima T, Treon S, Raje N, Davies FE, Shima Y, Tai YT, Rosen S, Avraham S, Kharbanda S, Anderson KC (2000) SHP2 mediates the protective effect of interleukin-6 against dexamethasone-induced apoptosis in multiple myeloma cells. J Biol Chem 275:27845-27850

Chen B, Bronson RT, Klaman LD, Hampton TG, Wang JF, Green PJ, Magnuson T, Douglas PS, Morgan JP, Neel BG (2000) Mice mutant for Egfr and Shp2 have defective cardiac semilunar valvulogenesis. Nat Genet 24:296-299

Chen Y, Wen R, Yang S, Schuman J, Zhang EE, Yi T, Feng GS, Wang D (2003) Identification of Shp-2 as a Stat5A phosphatase. J Biol Chem 278:16520-16527

Cheung WC, Van Ness B (2002) Distinct IL-6 signal transduction leads to growth arrest and death in B cells or growth promotion and cell survival in myeloma cells. Leuk 16:1182-1188

Chughtai N, Schimchowitsch S, Lebrun JJ, Ali S (2002) Prolactin induces SHP-2 association with Stat5, nuclear translocation, and binding to the beta-casein gene promoter in mammary cells. J Biol Chem 277:31107-31114

Cunnick JM, Dorsey JF, Munoz-Antonia T, Mei L, Wu J (2000) Requirement of SHP2 binding to Grb2-associated binder-1 for mitogen-activated protein kinase activation in response to lysophosphatidic acid and epidermal growth factor. J Biol Chem 275:13842-13848

Cunnick JM, Mei L, Doupnik CA, Wu J (2001) Phosphotyrosines 627 and 659 of Gab1 constitute a bisphosphoryl tyrosine-based activation motif (BTAM) conferring binding and activation of SHP2. J Biol Chem 276:24380-24387

Daigle I, Yousefi S, Colonna M, Green DR, Simon HU (2002) Death receptors bind SHP-1 and block cytokine-induced anti-apoptotic signaling in neutrophils. Nat Med 8:61-67

Deb TB, Wong L, Salomon DS, Zhou G, Dixon JE, Gutkind JS, Thompson SA, Johnson GR (1998) A common requirement for the catalytic activity and both SH2 domains of SHP-2 in mitogen-activated protein (MAP) kinase activation by the ErbB family of receptors. A specific role for SHP-2 in map, but not c-Jun amino-terminal kinase activation. J Biol Chem 273:16643-16646

Dhalluin C, Yan K, Plotnikova O, Lee KW, Zeng L, Kuti M, Mujtaba S, Goldfarb MP, Zhou MM (2000) Structural basis of SNT PTB domain interactions with distinct neurotrophic receptors. Mol Cell 6:921-929

Duchene J, Schanstra JP, Pecher C, Pizard A, Susini C, Esteve JP, Bascands JL, Girolami JP (2002) A novel protein-protein interaction between a G protein-coupled receptor and the phosphatase SHP-2 is involved in bradykinin-induced inhibition of cell proliferation. J Biol Chem 277:40375-40383

Fairlie WD, De Souza D, Nicola NA, Baca M (2003) Negative regulation of gp130 signaling mediated through Y757 is not dependent on the recruitment of Shp2. Biochem J 372:495-502

Feng GS (1999) Shp-2 tyrosine phosphatase: signaling one cell or many. Exp Cell Res 253:47-54

Florio T, Yao H, Carey KD, Dillon TJ, Stork PJ (1999) Somatostatin activation of mitogen-activated protein kinase via somatostatin receptor 1 (SSTR1). Mol Endocrinol 13:24-37

Frearson JA, Yi T, Alexander DR (1996) A tyrosine-phosphorylated 110-120-kDa protein associates with the C-terminal SH2 domain of phosphotyrosine phosphatase-1D in T cell receptor-stimulated T cells. Eur J Immunol 26:1539-1543

Fuhrer DK, Feng GS, Yang YC (1995) Syp associates with gp130 and Janus kinase 2 in response to interleukin-11 in 3T3-L1 mouse preadipocytes. J Biol Chem 270:24826-24830

Fujioka Y, Matozaki T, Noguchi T, Iwamatsu A, Yamao T, Takahashi N, Tsuda M, Takada T, Kasuga M (1996) A novel membrane glycoprotein, SHPS-1, that binds the SH2-domain-containing protein tyrosine phosphatase SHP-2 in response to mitogens and cell adhesion. Mol Cell Biol 16:6887-6899

Fukunaga K, Noguchi T, Takeda H, Matozaki T, Hayashi Y, Itoh H, Kasuga M (2000) Requirement for protein-tyrosine phosphatase SHP-2 in insulin-induced activation of c-Jun NH(2)-terminal kinase. J Biol Chem 275:5208-5213

Gadina M, Stancato LM, Bacon CM, Larner AC, O'Shea JJ (1998) Involvement of SHP-2 in multiple aspects of IL-2 signaling: evidence for a positive regulatory role. J Immunol 160:4657-4661

Gadina M, Sudashan C, Visconti R, Zhou YJ, Gu H, Neel BG, O'Shea JJ (2000) The docking molecule gab2 is induced by lymphocyte activation and is involved in signaling by interleukin-2 and interleukin-15 but not other common gamma chain-using cytokines. J Biol Chem 275:26959-26966

Gu H, Pratt JC, Burakoff SJ, Neel BG (1998) Cloning of p97/Gab2, the major SHP2-binding protein in hematopoietic cells, reveals a novel pathway for cytokine-induced gene activation. Mol Cell 2:729-740

Hadari YR, Gotoh N, Kouhara H, Lax I, Schlessinger J (2001) Critical role for the docking-protein FRS2 alpha in FGF receptor-mediated signal transduction pathways. Proc Natl Acad Sci USA USA 98:8578-8583

Hadari YR, Kouhara H, Lax I, Schlessinger J (1998) Binding of Shp2 tyrosine phosphatase to FRS2 is essential for fibroblast growth factor-induced PC12 cell differentiation. Mol Cell Biol 18:3966-3973

Hakak Y, Hsu YS, Martin GS (2000) Shp-2 mediates v-Src-induced morphological changes and activation of the anti-apoptotic protein kinase Akt. Oncogene 19:3164-3171

Hanafusa H, Torii S, Yasunaga T, Nishida E (2002) Sprouty1 and Sprouty2 provide a control mechanism for the Ras/MAPK signaling pathway. Nat Cell Biol 4:850-858

Herbst R, Carroll PM, Allard JD, Schilling J, Raabe T, Simon MA (1996) Daughter of sevenless is a substrate of the phosphotyrosine phosphatase Corkscrew and functions during sevenless signaling. Cell 85:899-909

Herbst R, Zhang X, Qin J, Simon MA (1999) Recruitment of the protein tyrosine phosphatase CSW by DOS is an essential step during signaling by the sevenless receptor tyrosine kinase. EMBO J 18:6950-6961

Hideshima T, Nakamura N, Chauhan D, Anderson KC (2001) Biologic sequelae of interleukin-6 induced PI3-K/Akt signaling in multiple myeloma. Oncogene 20:5991-6000

Higashi H, Tsutsumi R, Fujita A, Yamazaki S, Asaka M, Azuma T, Hatakeyama M (2002a) Biological activity of the Helicobacter pylori virulence factor CagA is determined by variation in the tyrosine phosphorylation sites. Proc Natl Acad Sci USA USA 99:14428-14433

Higashi H, Tsutsumi R, Muto S, Sugiyama T, Azuma T, Asaka M, Hatakeyama M (2002b) SHP-2 tyrosine phosphatase as an intracellular target of Helicobacter pylori CagA protein. Science 295:683-686

Hof P, Pluskey S, Dhe-Paganon S, Eck MJ, Shoelson SE (1998) Crystal structure of the tyrosine phosphatase SHP-2. Cell 92:441-450

Holgado-Madruga M, Emlet DR, Moscatello DK, Godwin AK, Wong AJ (1996) A Grb2-associated docking protein in EGF- and insulin-receptor signalling. Nature 379:560-564

Holgado-Madruga M, Moscatello DK, Emlet DR, Dieterich R, Wong AJ (1997) Grb2-associated binder-1 mediates phosphatidylinositol 3-kinase activation and the promotion of cell survival by nerve growth factor. Proc Natl Acad Sci USA USA 94:12419-12424

Inagaki K, Noguchi T, Matozaki T, Horikawa T, Fukunaga K, Tsuda M, Ichihashi M, Kasuga M (2000a) Roles for the protein tyrosine phosphatase SHP-2 in cytoskeletal organization, cell adhesion and cell migration revealed by overexpression of a dominant negative mutant. Oncogene 19:75-84

Inagaki K, Yamao T, Noguchi T, Matozaki T, Fukunaga K, Takada T, Hosooka T, Akira S, Kasuga M (2000b) SHPS-1 regulates integrin-mediated cytoskeletal reorganization and cell motility. EMBO J 19:6721-6731

Ingham RJ, Santos L, Dang-Lawson M, Holgado-Madruga M, Dudek, P, Maroun CR, Wong AJ, Matsuuchi L, Gold MR (2001) The Gab1 docking protein links the b cell antigen receptor to the phosphatidylinositol 3-kinase/Akt signaling pathway and to the SHP2 tyrosine phosphatase. J Biol Chem 276:12257-12265

Ion A, Tartaglia M, Song X, Kalidas K, van der Burgt I, Shaw AC, Ming JE, Zampino G, Zackai EH, Dean JC, Somer M, Parenti G, Crosby AH, Patton MA, Gelb BD, Jeffery S (2002) Absence of PTPN11 mutations in 28 cases of cardiofaciocutaneous (CFC) syndrome. Hum Genet 111:421-427

Kharitonenkov A, Chen Z, Sures I, Wang H, Schilling J, Ullrich A (1997) A family of proteins that inhibit signalling through tyrosine kinase receptors. Nature 386:181-186

Kim H, Baumann H (1999) Dual signaling role of the protein tyrosine phosphatase SHP-2 in regulating expression of acute-phase plasma proteins by interleukin-6 cytokine receptors in hepatic cells. Mol Cell Biol 19:5326-5338

Kim H, Hawley TS, Hawley RG, Baumann H (1998) Protein tyrosine phosphatase 2 (SHP-2) moderates signaling by gp130 but is not required for the induction of acute-phase plasma protein genes in hepatic cells. Mol Cell Biol 18:1525-1533

Kodama A, Matozaki T, Shinohara M, Fukuhara A, Tachibana K, Ichihashi M, Nakanishi H, Takai Y (2001) Regulation of Ras and Rho small G proteins by SHP-2. Genes to Cells 6:869-876

Kontaridis MI, Liu X, Zhang L, Bennett AM (2001) SHP-2 complex formation with the SHP-2 substrate-1 during C2C12 myogenesis. J Cell Sci 114:2187-2198

Kontaridis MI, Liu X, Zhang L, Bennett AM (2002) Role of SHP-2 in fibroblast growth factor receptor-mediated suppression of myogenesis in C2C12 myoblasts. Mol Cell Biol 22:3875-3891

Lacalle RA, Mira E, Gomez-Mouton C, Jimenez-Baranda S, Martinez AC, Manes S (2002) Specific SHP-2 partitioning in raft domains triggers integrin-mediated signaling via Rho activation. J Cell Biol 157:277-289

Latour S, Gish G, Helgason CD, Humphries RK, Pawson T, Veillette A (2003) Binding of SAP SH2 domain to Fyn SH3 domain reveals a novel mechanism of receptor singaling in immune regulation. Nat Cell Biol 5:149-154

Legius E, Schrander-Stumpel C, Schollen E, Pulles-Heintzberger C, Gewillig M, Fryns JP (2002) PTPN11 mutations in LEOPARD syndrome. J Med Genet 39:571-574

Li C, Friedman JM (1999) Leptin receptor activation of SH2 domain containing protein tyrosine phosphatase 2 modulates Ob receptor signal transduction. Proc Natl Acad Sci USA USA 96:9677-9682

Liu Y, Jenkins B, Shin JL, Rohrschneider LR (2001) Scaffolding protein Gab2 mediates differentiation signaling downstream of Fms receptor tyrosine kinase. Mol Cell Biol 21:3047-3056

Lynch DK, Daly RJ (2002) PKB-mediated negative feedback tightly regulates mitogenic signalling via Gab2. EMBO J 21:72-82

Maheshwari M, Belmont J, Fernbach S, Ho T, Molinari L, Yakub I, Yu F, Combes A, Towbin J, Craigen WJ, Gibbs R (2002) PTPN11 mutations in Noonan syndrome type I: detection of recurrent mutations in exons 3 and 13. Hum Mut 20:298-304

Maile LA, Clemmons DR (2002a) The alphaVbeta3 integrin regulates insulin-like growth factor I (IGF-I) receptor phosphorylation by altering the rate of recruitment of the Src-homology 2-containing phosphotyrosine phosphatase-2 to the activated IGF-I receptor. Endocrinol 143:4259-4264

Maile LA, Clemmons DR (2002b) Regulation of insulin-like growth factor I receptor dephosphorylation by SHPS-1 and the tyrosine phosphatase SHP-2. J Biol Chem 277:8955-8960

Manes S, Mira E, Gomez-Mouton C, Zhao ZJ, Lacalle RA, Martinez AC (1999) Concerted activity of tyrosine phosphatase SHP-2 and focal adhesion kinase in regulation of cell motility. Mol Cell Biol 19:3125-3135

Maroun CR, Naujokas MA, Holgado-Madruga M, Wong AJ, Park M (2000) The tyrosine phosphatase SHP-2 is required for sustained activation of extracellular signal-regulated kinase and epithelial morphogenesis downstream from the met receptor tyrosine kinase. Mol Cell Biol 20:8513-8525

Marrero MB, Venema VJ, Ju H, Eaton DC, Venema RC (1998) Regulation of angiotensin II-induced JAK2 tyrosine phosphorylation: roles of SHP-1 and SHP-2. Amer J Physiol 275:C1216-1223

Meng TC, Fukuda T, Tonks NK (2002) Reversible oxidation and inactivation of protein tyrosine phosphatases in vivo. Mol Cell 9:387-399

Miao H, Burnett E, Kinch M, Simon E, Wang B (2000) Activation of EphA2 kinase suppresses integrin function and causes focal-adhesion-kinase dephosphorylation. Nat Cell Biol 2:62-69

Milarski KL, Saltiel AR (1994) Expression of catalytically inactive Syp phosphatase in 3T3 cells blocks stimulation of mitogen-activated protein kinase by insulin. J Biol Chem 269:21239-21243

Morra M, Howie D, Grande MS, Sayos J, Wang N, Wu C, Engel P, Terhorst C (2001) X-linked lymphoproliferative disease: a progressive immunodeficiency. Ann Rev Immunol 19:657-682

Nishida K, Yoshida Y, Itoh M, Fukada T, Ohtani T, Shirogane T, Atsumi T, Takahashi-Tezuka M, Ishihara K, Hibi M, Hirano T (1999) Gab-family adapter proteins act downstream of cytokine and growth factor receptors and T- and B-cell antigen receptors. Blood 93:1809-1816

Noguchi T, Matozaki T, Horita K, Fujioka Y, Kasuga M (1994) Role of SH-PTP2, a protein-tyrosine phosphatase with Src homology 2 domains, in insulin-stimulated Ras activation. Mol Cell Biol 14:6674-6682

Ochi F, Matozaki T, Noguchi T, Fujioka Y, Yamao T, Takada T, Tsuda M, Takeda H, Fukunaga K, Okabayashi Y, Kasuga M (1997) Epidermal growth factor stimulates the tyrosine phosphorylation of SHPS-1 and association of SHPS-1 with SHP-2, a SH2 domain-containing protein tyrosine phosphatase. Biochem Biophys Res Commun 239:483-487

Oh ES, Gu H, Saxton TM, Timms JF, Hausdorff S, Frevert EU, Kahn BB, Pawson T, Neel BG, Thomas SM (1999) Regulation of early events in integrin signaling by protein tyrosine phosphatase SHP-2. Mol Cell Biol 19:3205-3215

Ohtani T, Ishihara K, Atsumi T, Nishida K, Kaneko Y, Miyata T, Itoh S, Narimatsu M, Maeda H, Fukada T, Itoh M, Okano H, Hibi M, Hirano T (2000) Dissection of signaling cascades through gp130 in vivo: reciprocal roles for STAT3- and SHP2-mediated signals in immune responses. Immun 12:95-105

Ong SH, Guy GR, Hadari YR, Laks S, Gotoh N, Schlessinger J, Lax I (2000) FRS2 proteins recruit intracellular signaling pathways by binding to diverse targets on fibroblast growth factor and nerve growth factor receptors. Mol Cell Biol 20:979-989

O'Reilly AM, Neel BG (1998) Structural determinants of SHP-2 function and specificity in Xenopus mesoderm induction. Mol Cell Biol 18:161-177

O'Reilly AM, Pluskey S, Shoelson SE, Neel BG (2000) Activated mutants of SHP-2 preferentially induce elongation of Xenopus animal caps. Mol Cell Biol 20:299-311

Palmblad J, Papadaki HA, Eliopoulos G (2001) Acute and chronic neutropenias. What is new? J Int Med 250:476-491

Perkins LA, Johnson MR, Melnick MB, Perrimon N (1996) The nonreceptor protein tyrosine phosphatase corkscrew functions in multiple receptor tyrosine kinase pathways in Drosophila. Dev Biol 180:63-81

Perkins LA, Larsen I, Perrimon N (1992) corkscrew encodes a putative protein tyrosine phosphatase that functions to transduce the terminal signal from the receptor tyrosine kinase torso. Cell 70:225-236

Pluskey S, Wandless TJ, Walsh CT, Shoelson SE (1995) Potent stimulation of SH-PTP2 phosphatase activity by simultaneous occupancy of both SH2 domains. J Biol Chem 270:2897-2900

Qu CK, Feng GS (1998) Shp-2 has a positive regulatory role in ES cell differentiation and proliferation. Oncogene 17:433-439

Qu CK, Shi ZQ, Shen R, Tsai FY, Orkin SH, Feng GS (1997) A deletion mutation in the SH2-N domain of Shp-2 severely suppresses hematopoietic cell development. Mol Cell Biol 17:5499-5507

Qu CK, Yu WM, Azzarelli B, Feng GS (1999) Genetic evidence that Shp-2 tyrosine phosphatase is a signal enhancer of the epidermal growth factor receptor in mammals. Proc Natl Acad Sci USA USA 96:8528-8533

Raabe T, Riesgo-Escovar J, Liu X, Bausenwein BS, Deak P, Maroy P, Hafen E (1996) DOS, a novel pleckstrin homology domain-containing protein required for signal transduction between sevenless and Ras1 in Drosophila. Cell 85:911-920

Reardon DB, Dent P, Wood SL, Kong T, Sturgill TW (1997) Activation in vitro of somatostatin receptor subtypes 2, 3, or 4 stimulates protein tyrosine phosphatase activity in membranes from transfected Ras-transformed NIH 3T3 cells: coexpression with catalytically inactive SHP-2 blocks responsiveness. Mol Endocrinol 11:1062-1069

Sattler M, Mohi MG, Pride YB, Quinnan LR, Malouf NA, Podar K, Gesbert F, Iwasaki H, Li S, Van Etten RA, Gu H, Griffin JD, Neel BG (2002) Critical role for Gab2 in transformation by BCR/ABL. Cancer Cell 1:479-492

Saxton TM, Ciruna BG, Holmyard D, Kulkarni S, Harpal K, Rossant J, Pawson T (2000) The SH2 tyrosine phosphatase shp2 is required for mammalian limb development. Nat Genet 24:420-423

Saxton TM, Henkemeyer M, Gasca S, Shen R, Rossi DJ, Shalaby F, Feng GS, Pawson T (1997) Abnormal mesoderm patterning in mouse embryos mutant for the SH2 tyrosine phosphatase Shp-2. EMBO J 16:2352-2364

Saxton TM, Pawson T (1999) Morphogenetic movements at gastrulation require the SH2 tyrosine phosphatase Shp2. Proc Natl Acad Sci USA USA 96:3790-3795

Sayos J, Wu C, Morra M, Wang N, Zhang X, Allen D, van Schaik S, Notarangelo L, Geha R, Roncarolo MG, Oettgen H, De Vries JE, Aversa G, Terhorst C (1998) The X-linked lymphoproliferative-disease gene product SAP regulates signals induced through the co-receptor SLAM. Nature 395:462-469

Schmitz J, Weissenbach M, Haan S, Heinrich PC, Schaper F (2000) SOCS3 exerts its inhibitory function on interleukin-6 signal transduction through the SHP2 recruitment site of gp130. J Biol Chem 275:12848-12856

Schoenwaelder SM, Petch LA, Williamson D, Shen R, Feng GS, Burridge K (2000) The protein tyrosine phosphatase Shp-2 regulates RhoA activity. Curr Biol 10:1523-1526

Schutzman JL, Borland CZ, Newman JC, Robinson MK, Kokel M, Stern MJ (2001) The Caenorhabditis elegans EGL-15 signaling pathway implicates a DOS-like multisubstrate adaptor protein in fibroblast growth factor signal transduction. Mol Cell Biol 21:8104-8116

Servidei T, Aoki Y, Lewis SE, Symes A, Fink JS, Reeves SA (1998) Coordinate regulation of STAT signaling and c-fos expression by the tyrosine phosphatase SHP-2. J Biol Chem 273:6233-6241

Shi ZQ, Lu W, Feng GS (1998) The Shp-2 tyrosine phosphatase has opposite effects in mediating the activation of extracellular signal-regulated and c-Jun NH2-terminal mitogen-activated protein kinases. J Biol Chem 273:4904-4908

Shi ZQ, Yu DH, Park M, Marshall M, Feng GS (2000) Molecular mechanism for the Shp-2 tyrosine phosphatase function in promoting growth factor stimulation of Erk activity. Mol Cell Biol 20:1526-1536

Simon MA (2000) Receptor tyrosine kinases: specific outcomes from general signals. Cell 103:13-15

Sorenson CM, Sheibani N (2002) Altered regulation of SHP-2 and PTP 1B tyrosine phosphatases in cystic kidneys from bcl-2 -/- mice. Amer J Physiol - Renal Fluid & Electro Physiol 282:F442-450

Stofega MR, Herrington J, Billestrup N, Carter-Su C (2000) Mutation of the SHP-2 binding site in growth hormone (GH) receptor prolongs GH-promoted tyrosyl phosphorylation of GH receptor, JAK2, and STAT5B. Mol Endocrinol 14:1338-1350

Stofega MR, Wang H, Ullrich A, Carter-Su C (1998) Growth hormone regulation of SIRP and SHP-2 tyrosyl phosphorylation and association. J Biol Chem 273:7112-7117

Symes A, Stahl N, Reeves SA, Farruggella T, Servidei T, Gearan T, Yancopoulos G, Fink JS (1997) The protein tyrosine phosphatase SHP-2 negatively regulates ciliary neurotrophic factor induction of gene expression. Curr Biol 7:697-700

Takada T, Matozaki T, Takeda H, Fukunaga K, Noguchi T, Fujioka Y, Okazaki I, Tsuda M, Yamao T, Ochi F, Kasuga M (1998) Roles of the complex formation of SHPS-1 with SHP-2 in insulin-stimulated mitogen-activated protein kinase activation. J Biol Chem 273:9234-9242

Takai S, Yamada M, Araki T, Koshimizu H, Nawa H, Hatanaka H (2002) Shp-2 positively regulates brain-derived neurotrophic factor-promoted survival of cultured ventral mesencephalic dopaminergic neurons through a brain immunoglobulin-like molecule with tyrosine-based activation motifs/Shp substrate-1. J Neurochem 82:353-364

Takeda H, Matozaki T, Fujioka Y, Takada T, Noguchi T, Yamao T, Tsuda M, Ochi F, Fukunaga K, Narumiya S, Yamamoto T, Kasuga M (1998) Lysophosphatidic acid-induced association of SHP-2 with SHPS-1: roles of RHO, FAK, and a SRC family kinase. Oncogene 16:3019-3027

Tang TL, Freeman RM, O'Reilly AM, Neel BG, Sokol SY (1995) The SH2-containing protein-tyrosine phosphatase SH-PTP2 is required upstream of MAP kinase for early Xenopus development. Cell 80:473-483

Tangye SG, Lazetic S, Woolatt E, Sutherland GR, Lanier LL, Phillips JH (1999) Cutting edge: human 2B4, an activating NK cell receptor, recruits the protein tyrosine phosphatase SHP-2 and the adaptor signaling protein SAP. J Immunol 162:6981-6985

Tartaglia M, Kalidas K, Shaw A, Song X, Musat DL, van der Burgt I, Brunner HG, Bertola DR, Crosby A, Ion A, Kucherlapati RS, Jeffery S, Patton MA, Gelb BD (2002) PTPN11 mutations in Noonan syndrome: molecular spectrum, genotype-phenotype correlation, and phenotypic heterogeneity. Amer J Hum Genet 70:1555-1563

Tartaglia M, Mehler EL, Goldberg R, Zampino G, Brunner HG, Kremer H, van der Burgt I, Crosby AH, Ion A, Jeffery S, Kalidas K, Patton MA, Kucherlapati RS, Gelb BD (2001) Mutations in PTPN11, encoding the protein tyrosine phosphatase SHP-2, cause Noonan syndrome. Nature Genet 29:465-468

Tidow N, Kasper B, Welte K (1999) SH2-containing protein tyrosine phosphatases SHP-1 and SHP-2 are dramatically increased at the protein level in neutrophils from patients with severe congenital neutropenia (Kostmann's syndrome). Exp Hema 27:1038-1045

Tong J, Killeen M, Steven R, Binns KL, Culotti J, Pawson T (2001) Netrin stimulates tyrosine phosphorylation of the UNC-5 family of netrin receptors and induces Shp2 binding to the RCM cytodomain. J Biol Chem 276:40917-40925

Tsuda M, Matozaki T, Fukunaga K, Fujioka K, Fujioka Y, Imamoto A, Noguchi T, Takada T, Yamao, T, Takeda H, Ochi F, Yamamoto T, Kasuga M (1998) Integrin-mediated tyrosine phosphorylation of SHPS-1 and its association with SHP-2. Roles of Fak and Src family kinases. J Biol Chem 273:13223-13229

Tsutsumi R, Higashi H, Higuchi M, Okada M, Hatakeyama M (2003) Attenuation of Helicobacter pylori CagA-SHP-2 signaling by interaction between CagA and C-terminal Src kinase. J Biol Chem 278:3664-3670

Weidner KM, Di Cesare S, Sachs M, Brinkmann V, Behrens J, Birchmeier W (1996) Interaction between Gab1 and the c-Met receptor tyrosine kinase is responsible for epithelial morphogenesis. Nature 384:173-176

Weinstein DC, Hemmati-Brivanlou AA (2001) Src family kinase function during early Xenopus development. Dev Dyn 220:163-168

Wickrema A, Uddin S, Sharma A, Chen F, Alsayed Y, Ahmad S, Sawyer ST, Krystal G, Yi T, Nishada K, Hibi M, Hirano T, Platanias LC (1999) Engagement of Gab1 and Gab2 in erythropoietin signaling. J Biol Chem 274:24469-24474

Winnay JN, Bruning JC, Burks DJ, Kahn CR (2000) Gab-1-mediated IGF-1 signaling in IRS-1-deficient 3T3 fibroblasts. J Biol Chem 275:10545-10550

Wolf I, Jenkins BJ, Liu Y, Seiffert M, Custodio JM, Young P, Rohrschneider LR (2002) Gab3, a new DOS/Gab family member, facilitates macrophage differentiation. Mol Cell Biol 22:231-244

Wu TR, Hong YK, Wang XD, Ling MY, Dragoi AM, Chung AS, Campbell AG, Han ZY, Feng GS, Chin YE (2002) SHP-2 is a dual-specificity phosphatase involved in Stat1 dephosphorylation at both tyrosine and serine residues in nuclei. J Biol Chem 277:47572-47580

Yan KS, Kuti M, Yan S, Mujtaba S, Farooq A, Goldfarb MP, Zhou MM (2002) FRS2 PTB domain conformation regulates interactions with divergent neurotrophic receptors. J Biol Chem 277:17088-17094

Yin T, Shen R, Feng GS, Yang YC (1997) Molecular characterization of specific interactions between SHP-2 phosphatase and JAK tyrosine kinases. J Biol Chem 272:1032-1037

You M, Flick LM, Yu D, Feng GS (2001) Modulation of the nuclear factor kappa B pathway by Shp-2 tyrosine phosphatase in mediating the induction of interleukin (IL)-6 by IL-1 or tumor necrosis factor. J Exp Med 193:101-110

You M, Yu DH, Feng GS (1999) Shp-2 tyrosine phosphatase functions as a negative regulator of the interferon-stimulated Jak/STAT pathway. Mol Cell Biol 19:2416-2424

Yu CL, Jin YJ, Burakoff SJ (2000) Cytosolic tyrosine dephosphorylation of STAT5. Potential role of SHP-2 in STAT5 regulation. J Biol Chem 275:599-604

Yu DH, Qu CK, Henegariu O, Lu X, Feng GS (1998) Protein-tyrosine phosphatase Shp-2 regulates cell spreading, migration, and focal adhesion. J Biol Chem 273:21125-21131

Yuan LP, Yu WM, Yuan Z, Haudenschild CC, Qu CK (2003) Role of Shp-2 tyrosine phosphatase in the DNA damage-induced cell death response. J Biol Chem 278:15208-15216

Zhang SQ, Tsiaras WG, Araki T, Wen G, Minichiello L, Klein R, Neel BG (2002) Receptor-specific regulation of phosphatidylinositol 3'-kinase activation by the protein tyrosine phosphatase Shp2. Mol Cell Biol 22:4062-4072

Zhao C, Yu DH, Shen R, Feng GS (1999) Gab2, a new pleckstrin homology domain-containing adapter protein, acts to uncouple signaling from ERK kinase to Elk-1. J Biol Chem 274:19649-19654

Zhao R, Zhao ZJ (1999) Tyrosine phosphatase SHP-2 dephosphorylates the platelet-derived growth factor receptor but enhances its downstream signalling. Biochem J 338:35-39

15 SHP-1 twelve years on: structure, ligands, substrates and biological roles

Jean Gerard Sathish, Reginald James Matthews

Abstract

SHP-1 is an SH2 domain-containing protein tyrosine phosphatase (PTP) implicated in the negative regulation of a diverse range of activatory signaling pathways in leukocytes. Structural studies have revealed that SHP-1 PTP activity is
subject to direct inhibition by its amino terminal SH2 domain whereby auto-
inhibition is predicted to be relieved by a series of allosteric changes involving the
sequential engagement of the carboxy and amino terminal SH2 domains of SHP-1
by phosphotyrosine-containing ligands. The major physiological ligands for SHP-
1 are a super-family of inhibitory receptors found in leukocytes that possess Immune receptor Tyrosine-based Inhibition Motifs (ITIMs) in their cytoplasmic domains which when tyrosine phosphorylated are capable of binding with high affinity to the SH2 domains of SHP-1. Hence, fully activated SHP-1 is localized to the
plasma membrane. SHP-1 appears to mediate inhibition of the early steps of activatory signaling pathways but the mechanisms of SHP-1 action still remain to be
elucidated.

15.1 Introduction

SHP-1, an intracellular member of the family of Protein Tyrosine Phosphatases
(PTPs), is an intriguing molecule from a number of perspectives but has primarily
attracted attention for its distinct mode of catalytic regulation and for the profound
manner with which it is able to regulate the behavior of leukocytes. SHP-1 is typical of the vast majority of PTPs that were identified in the 1990s by reverse genetics whereby the conservation of hallmark residues in the PTP domain facilitated
the isolation of cDNAs encoding these enzymes. The translated sequences of
cDNAs for human, mouse and rat SHP-1 (Matthews et al. 1992; Plutzky et al.
1992; Shen et al. 1991; Yi et al. 1992) each revealed a polypeptide predicted to
consist of a conserved domain structure encompassing tandem Src homology 2
(SH2) domains at the amino-terminus followed by a classical PTP domain and a
basic region at the carboxy-terminus (see Fig. 1). SHP-1 is an exemplar of a discrete sub-family of PTPs possessing tandem SH2 domains (Neel 1993) and has
approximately 60% overall sequence identity with a second vertebrate PTP, SHP-
2. Invertebrate orthologs of SHP-2 (Gutch et al. 1998; Perkins et al. 1992) but not
SHP-1 have been isolated and it therefore appears that SHP-1 origins are coinci-

Topics in Current Genetics, Vol. 5
J. Arino, D.R. Alexander (Eds.) Protein phosphatases
© Springer-Verlag Berlin Heidelberg 2004

dent with vertebrate evolution. It is worth recollecting that prior to the isolation of cDNAs encoding SHP-1 there was no indication of the potential existence of PTP domains fused to SH2 domains. However, since molecular biology approaches first deciphered the primary structure of SHP-1 some twelve years ago, remarkable advances have occurred in our understanding of the structure and regulation of this PTP in addition to a wide appreciation of its functional roles. An understanding of the physiological importance of SHP-1 has been greatly accelerated by the recognition that the spontaneous mouse mutants, motheaten (*me/me*) and viable motheaten (*me^v/me^v*), result from independent point-mutations in the SHP-1 gene leading to complete loss or significant reduction in SHP-1 PTP activity, respectively (Green and Shultz 1975; Shultz et al. 1984; Shultz et al. 1993; Tsui et al. 1993). Homozygous *me/me* and *me^v/me^v* mice exhibit multiple hematopoietic defects that give rise to a profound autoreactivity characterized primarily by tissue infiltration of myeloid lineage cells (Shultz et al. 1997). This review describes the significant advances that have been made regarding the structure and biology of SHP-1 and provides a limited critique of the extensive literature pertaining to SHP-1. The intention is also to highlight the major unanswered questions that still exist for SHP-1.

15.2 Expression

SHP-1 has a distinct pattern of expression that encompasses all hematopoietic cells, certain epithelial cells (Matthews et al. 1992; Plutzky et al. 1992; Yi et al. 1992) and astrocytes and oligodendrocytes of the CNS (Horvat et al. 2001; Massa et al. 2000). Two major forms of SHP-1 transcript have been identified in humans and mice and denoted as (I)SHP-1 and (II)SHP-1. The sequences of (I)SHP-1 and (II)SHP-1 are identical within each species with the exception of the 5'untranslated regions and the initial coding nucleotides (Banville et al. 1995; Martin et al. 1999; Tsui et al. 2002). MLSRG are the first five amino acids of (I)SHP-1 and MVR comprise the first 3 amino acids of (II)SHP-1 resulting in SHP-1 polypeptides of 597 and 595 amino acids, respectively. For both species, (I)SHP-1 and (II)SHP-1 transcripts are found primarily in epithelial and hematopoietic cells, respectively (Banville et al. 1995; Martin et al. 1999; Tsui et al. 2002), but SHP-1 transcript utilization in the CNS has not been defined. The hematopoietic pattern of expression of SHP-1 encompasses cells at different stages of differentiation (Matthews et al. 1992; Plutzky et al. 1992; Yi et al. 1992; Zhang et al. 1999a) and includes microglia within the CNS (Sorbel et al. 2002). Levels of SHP-1 transcript are considerably lower in cells of epithelial origin in comparison to hematopoietic cells (Plutzky et al. 1992).

15.3 Structure

The crystal structure of an experimentally generated isoform of human SHP-1 that includes a 61 amino acid truncation at the carboxy-terminus has recently been described and represents a significant advance in our understanding of this molecule (Yang et al. 2003). The three-dimensional structure at a resolution of 2.8Å confirms that SHP-1 consists of three distinct and compact domains as predicted from the primary sequence. Specifically, residues 1-108 and 116-208 fold as the amino (N) and carboxy (C) SH2 domains respectively and residues 270-532 comprise the PTP domain (Yang et al. 2003). Importantly, the three-dimensional structure of SHP-1 now provides a very well defined frame-work for interpreting the extensive biochemical observations that have been made for the enzyme especially regarding the SH2 and catalytic domain mediated interactions with phosphotyrosine-containing ligands and substrates respectively.

15.3.1 SH2 domains and their ligands

SH2 domains, as first defined in the Src-like kinases and subsequently in an extensive number of intracellular signaling adaptors and enzymes, have the capacity to bind with high affinity to phosphotyrosine embedded in the context of specific peptide sequences (Pawson and Scott 1997). It was correctly anticipated that the SH2 domains of SHP-1 would similarly bind to phosphotyrosine bearing polypeptide ligands thereby recruiting SHP-1 to its substrate(s). The molecular details of SHP-1 SH2 domain ligand binding are well characterized and are documented below. However, there were early clues as to an additional regulatory role for the SH2 domains of SHP-1 provided by the observation that bacterially expressed wild type SHP-1 is a very weak PTP in comparison to other isolated PTPs (Zhao et al. 1993a; Zhao et al. 1993b). This intrinsic inhibition of SHP-1 PTP activity can be released by truncation of both SH2 domains, or simply by deletion of the N SH2 domain (Pei et al. 1994; Townley et al. 1993). In contrast, removal of the C SH2 domain has minimal effects on basal PTP activity (Pei et al. 1996). Furthermore, a synthetic phosphopeptide derived from sequences in the erythropoietin receptor enhances the activity of SHP-1 by binding to the N but not the C SH2 domain (Pei et al. 1996). Finally, SHP-1 auto-inhibition has to be independent of phosphotyrosine because the SHP-1 used in the PTP activity assays was generated in prokaryotes in which it cannot undergo tyrosine phosphorylation. Based on these observations, a model of allosteric inhibition has been proposed whereby the N SH2 of SHP-1 is predicted to undergo a conformational change upon binding phosphopeptide ligand thereby releasing auto-inhibition of the PTP domain (Pei et al. 1996).

The crystal structure of SHP-1 provides an exquisite and satisfying rationalization of these biochemical data and mirrors a very similar structure obtained previously for SHP-2 (Hof et al. 1998; Yang et al. 2003). Each of the SH2 domains of SHP-1 has a typical SH2 domain fold whereby the phosphopeptide-binding site is exposed on the surface of the molecule and faces away from the PTP domain

(Yang et al. 2003). However, the orientation of each SH2 domain with respect to the PTP domain is distinct and with major implications for function (Yang et al. 2003). The N SH2 domain makes extensive contact with the PTP domain (Yang et al. 2003). In contrast, the C SH2 domain makes no contact with the PTP domain and is likely to be highly mobile and readily available for binding to phosphotyrosine-containing ligands (Yang et al. 2003). In its ligand-free state, the loop linking the β4 and β5 strands of the central β-sheet within the N SH2 domain of SHP-1 extends into the active site of the PTP domain of SHP-1 thereby sterically occluding any possible interaction with substrate and maintaining the enzyme in an inactive conformation (Yang et al. 2003). Further, in the inactive state of SHP-1, the phosphotyrosine-binding surface of the N SH2 domain is distorted thereby precluding any possible interaction with ligand (Yang et al. 2003). Ligand engagement of the C SH2 domain of SHP-1 is predicted to facilitate a conformational change in the molecule that restores the ligand-binding pocket of the N SH2 domain (Yang et al. 2003). The subsequent engagement of the phosphotyrosine-binding pocket of the N SH2 domain by ligand would be concomitant with an opening of the PTP domain thereby permitting substrate access (Yang et al. 2003). The N SH2 domain of SHP-1 is therefore predicted to exist in a dynamic equilibrium between two mutually exclusive conformations with the allosteric switch potentially regulated by the C SH2 domain.

The structural determination of SHP-1 leads to three important predictions. Firstly, the C SH2 domain is likely to initiate interaction with phosphorylated ligand. Secondly, optimal activation of the enzyme in vivo will most likely require simultaneous engagement of both SH2 domains of SHP-1. This is consistent with biochemical evidence that biphosphorylated peptides are the most potent phosphopeptide sequences for increasing SHP-1 PTP activity in vitro (Burshtyn et al. 1997). Consequently, the most physiologically relevant polypeptides for recruiting SHP-1 will possess two or more phosphopeptide sequences each capable of interacting with the SH2 domains of SHP-1. Alternatively, because of the large degree of flexibility of its C SH2, SHP-1 may have the capacity to bind with high affinity and be fully activated by molecules that have single phosphopeptide sequences but present more than one binding phosphopeptide sequence as a consequence of dimerization. A confirmation or otherwise of these predictions will await the solution of a complex of SHP-1 bound to a high affinity peptide ligand.

Knowledge of the three-dimensional structure of SHP-1 allows for a critical reassessment of the ligands that have been reported to interact with SHP-1. The identification of ligands for the SH2 domains of SHP-1 initially involved immunoprecipitation of SHP-1 from different hematopoietic cell lysates followed by analysis of the co-immunoprecipitating tyrosyl phosphoproteins. Such studies identified a number of cytokine and growth factor receptors including the IL-3β chain (Yi et al. 1993), erythropoietin receptor (Klingmuller et al. 1995; Yi et al. 1995), and c-Kit receptor (Yi and Ihle 1993) as ligands for SHP-1. However, an alignment of the intracellular domains of these individual receptors did not immediately reveal a consensus recognition sequence for the SH2 domains of SHP-1. Instead, the major breakthrough in our understanding of SHP-1 binding ligands came from two independent studies involving B cells. Firstly, it was demonstrated that following triggering of the B Cell Receptor (BCR), the α2,6-linked sialic acid binding co-receptor, CD22, can be readily immunoprecipitated with SHP-1 (Campbell and Klinman 1995; Doody et al. 1995; Lankester et al. 1995; Law et al. 1996). Furthermore, three phosphotyrosyl peptides corresponding to sequences within the intracellular tail of CD22 are capable of inhibiting interaction of SHP-1 and phosphorylated CD22 (Doody et al. 1995). The same CD22-derived phospho-tyrosyl peptides also increase the PTP activity of SHP-1 in a dose-dependent fashion when added to an in vitro PTP assay (Doody et al. 1995). Finally, mutagenesis of CD22 has confirmed that two of these three phosphotyrosine sequences within the intracellular domain of CD22 are required to bind to SHP-1 in vivo and conversely both SH2 domains of SHP-1 are required for efficient binding to CD22 (Blasioli et al. 1999; Otipoby et al. 2001). The second critical observation pertaining to SHP-1 ligands also originated in B cells whereby a 13-amino acid motif was identified in the cytoplasmic domain of the inhibitory receptor, FcγRIIB, as being responsible for mediating the inhibitory effects of this receptor (Muta et al. 1994). When synthesized as a phosphotyrosyl peptide, this immune receptor tyrosine-based inhibitor motif (ITIM) is capable of binding in vitro to SHP-1 (D'Ambrosio et al. 1995). Significantly, the SHP-1 binding sequences in CD22 resemble the ITIM identified in the FcγRIIB receptor and hence the ITIM was minimally defined as pYXXL whereby the phosphotyrosine residue (pY) within the ITIM motif is defined as position 0 and X indicates any amino acid (D'Ambrosio et al. 1995). Ironically, although the FcγRIIB receptor has subsequently been demonstrated not to interact with SHP-1 in vivo (Fong et al. 1996; Ono et al. 1996; Ono et al. 1997), the ITIM model has provided important insights into SHP-1 function (Blery and Vivier 1999; Thomas 1995).

Further significant advances in the recognition of the importance of the ITIM to SHP-1 biology came from a series of studies indicating that distinct families of inhibitory receptors (Ly49, KIR, CD94/NKG2A) specific for classical and non-classical MHC Class I exist on mouse and human Natural killer (NK) cells (Lanier 1998). Each of these receptors possesses one or more ITIMs on their intracellular domains (Burshtyn et al. 1996; Houchins et al. 1997; Olcese et al. 1996). Indeed, much of our current understanding of the molecular details of SHP-1 ligand recognition has been driven by studies on the ITIMs of NK cell inhibitory receptors. The KIR and Ly49/NKG2A families possess immunoglobulin-like or C-type lectin extracellular domains respectively (Lanier 1998). With one exception, the inhibitory isoforms of the monomeric KIRs possess tandem ITIMs separated by approximately thirty amino acids. The exception is the structurally distinct member of the KIR family, KIR2DL4, which possesses one ITIM and tellingly, does not bind to SHP-1 (Yusa et al. 2002). In contrast, the Ly49 sub-family of receptors each encodes single ITIMs but form dimers on the cell surface (Lanier 1998).

In all cases, the inhibitory receptor has to be tyrosine phosphorylated in order to associate with SHP-1 (Binstadt et al. 1996; Burshtyn et al. 1996; Campbell et al. 1996; Carretero et al. 1998; Le Drean et al. 1998; Nakamura et al. 1997). Furthermore, it has been formally demonstrated by mutagenesis or deletion studies that tandem phosphorylated ITIMs in the intracellular domains of KIR and NKG2A molecules are required for association with SHP-1 in vivo (Bruhns et al. 1999;

Burshtyn et al. 1999; Kabat et al. 2002). The implication of these findings is that both SH2 domains are required for association of SHP-1 with inhibitory receptors. This conflicts with observations that phosphoryl peptides corresponding to single ITIMs are capable of both adsorbing intact SHP-1 from cell lysates and activating recombinant wild type SHP-1 (Burshtyn et al. 1996; Campbell et al. 1996; Le Drean et al. 1998; Olcese et al. 1996). However, in these studies, high local concentrations of peptide were used, which may be pertinent, as in another study it was demonstrated that both ITIMs of a KIR derived peptide required tyrosine phosphorylation in order to associate with SHP-1 in vitro (Fry et al. 1996). In this latter study, it is envisaged that the ITIM peptide concentrations used more closely resemble physiological conditions, again suggesting the cooperativity of tandem ITIMs for optimal SHP-1 recruitment and activation. Likewise, experiments indicating that single isolated SH2 domains of SHP-1 can bind in vitro to phosphorylated receptors (Burshtyn et al. 1996; Vely et al. 1997) need to be interpreted with caution knowing that the N SH2 domain interaction with phosphotyrosine-containing ligands is highly regulated in intact SHP-1. Certainly, the activation of recombinant wild type SHP-1 is achieved most efficiently by long biphosphopeptides encompassing tandem ITIMs (Burshtyn et al. 1997). Nevertheless, it has been observed repeatedly in a cellular context that mutation of one or other ITIM within a given inhibitory receptor does not always correlate with a significant loss of inhibitory function as would be anticipated with a major decrease in SHP-1 association (Bruhns et al. 1999; Burshtyn et al. 1999; Fry et al. 1996; Kabat et al. 2002). One possible explanation for this discrepancy may be that SHP-2 is also contributing to the inhibitory function mediated by these receptors, especially as SHP-2 has a capacity for binding to single ITIMs (Bruhns et al. 1999). The involvement of SHP-2 in inhibitory receptor function may also complicate the interpretation of data suggesting that, within a juxtaposed pair of ITIM sequences, one ITIM may be more critical than another for inhibitory function (Burshtyn et al. 1999; Fry et al. 1996; Kabat et al. 2002). In summary, most experimental findings point to an obligatory engagement of both the N and C SH2 domains by tandem ITIMs on the inhibitory receptor for high affinity binding of SHP-1. The requirement for two ITIM sequences is predicted to greatly enhance the biological specificity of SHP-1 and tightly localize its PTP activity within the cell (see Fig. 1).

As a further refinement of the molecular definition of an ITIM for SHP-1, the specificity of residues upstream and downstream of the phosphotyrosine within an ITIM has been examined (Burshtyn et al. 1997; Vely et al. 1997). Such studies have highlighted the importance of aliphatic residues at positions -2 and $+3$ within an ITIM for SHP-1 recognition and have hence re-defined the ITIM for SHP-1 as (V/I/LXpYXXL/V) with a preference for leucine at the $+3$ position. Screens to determine the optimal binding specificity of the SH2 domains of SHP-1 are in excellent agreement with this consensus sequence (Beebe et al. 2000). As outlined above, the single ITIM of the FcγRIIB receptor is insufficient for recruiting SHP-1 in vivo (Fong et al. 1996; Ono et al. 1996; Ono et al. 1997) and subsequent analysis has identified the inositol 5-phosphatases (SHIPs) as mediating the inhibitory signaling of FcγRIIB (Nadler et al. 1997; Ono et al. 1997). Interestingly, the

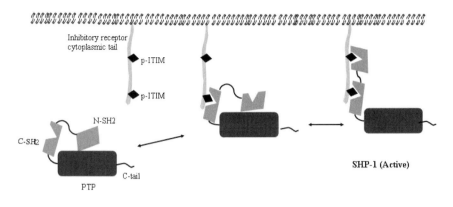

Fig. 1. Model depicting possible steps of SHP-1 activation: SHP-1 normally exists in an inactive configuration in the cytosol, in which the N SH2 domain occludes the substrate binding pocket of the PTP domain. A phosphorylated ITIM on an inhibitory receptor initially engages the C SH2 domain, an event that restores the phosphotyrosine binding pocket of the N SH2 domain thereby allowing the N SH2 domain to bind to a second ITIM on the inhibitory receptor. Binding of the N SH2 domain to a pITIM induces a conformational change that relieves the inhibition of the PTP domain and the enzyme is activated.

specificity of an ITIM for recruiting SHIP1 and SHIP2 includes a leucine at position +2 but with no absolute requirement for an aliphatic residue at position –2 (Bruhns et al. 2000). Although a structure of SHP-1 bound to a high affinity ligand is not yet available, the crystal structure of the isolated SH2 domains of SHP-2 bound to peptide ligands reveals an extended peptide binding groove accommodating residues –2 to +5 relative to the phosphotyrosine in each SH2 domain (Eck et al. 1996). Based on the overall sequence identities of SHP-1 and –2, it is envisaged that similar peptide binding grooves exist for the SH2 domains of SHP-1. Hence, the structural and mutagenesis data are consistent in implicating residues – 2 and +3 of the ITIM as critical for SHP-1 recognition of high-affinity ligands. The SH2 domains of SHP-1 are unusual in also recognizing residues upstream of the target phosphotyrosine. Based on mutagenesis and binding studies with SHP-2, the increased requirements for SHP-1 SH2 domain recognition are likely to be due, in part, to substitution of a glycine residue for a highly conserved arginine at position 2 in the αA helix of the SH2 domain (Huyer and Ramachandran 1998). The exact preferences and degrees of flexibility of the individual SH2 domains of SHP-1 for specific ITIMs are still not fully understood and hence the orientation of the N and C SH2 domains on tandem ITIMs remains to be resolved.

The identification of the prototypic inhibitory receptors on NK cells has led to the discovery (by a combination of reverse-genetic (Samaridis and Colonna 1997), low-stringency hybridization (Borges et al. 1997), bio-informatic (Wagtmann et al. 1997) and expression cloning (Meyaard et al. 1997) approaches) of a superfamily of inhibitory receptors expressed primarily on leukocytes, including PIR-B,

ILTs, MAFA, gp49, CEACAM1, LAIR-1, SIRPα, and IRTAs (Long 1999; Miller et al. 2002; Ravetch and Lanier 2000). These inhibitory receptors have been demonstrated to interact strongly with SHP-1 and, with minor exceptions, can be distinguished by the possession of two ITIM sequences in their intracellular domains that fit the prototypic sequence of a SHP-1 specific ITIM (V/I/LXpYXXL/V). Analysis of the individual ITIMs in the PIR-B and ILT2 receptors has indicated that serine may be occasionally tolerated in the –2 position of the membrane distal ITIM sequence (Bellon et al. 2002; Blery et al. 1998; Maeda et al. 1998). The CD33 sub-family of Siglec receptors are unusual in that the membrane distal ITIM does not conform to the consensus for a SHP-1 ITIM (Crocker and Varki 2001). Furthermore, limited data suggest that a mutant form of SHP-1 with a defective phosphotyrosine binding pocket in the C SH2 domain is capable of binding to CD33 (Ulyanova et al. 1999), which is inconsistent with the prevailing thinking on SHP-1 activation. Therefore, additional clarification of the mechanism of SHP-1 association with this set of receptors is needed. In addition, a limited number of receptors exist that bind preferentially to SHP-2 and are distinguished by the presence of a threonine at position –2 of the membrane distal ITIM (CD31/PECAM-1 (Newton-Nash and Newman 1999) and FD503/PILRα (Fournier et al. 2000; Mousseau et al. 2000). Hence, the number and sequence identity of specific ITIMs in a receptor may provide some indication of whether a given molecule preferentially interacts with SHP-1 or SHP-2. There is insufficient space to detail each of these receptors and excellent reviews on the biology of inhibitory receptors are available (Blery and Vivier 1999; Long et al. 1997; McVicar and Burshtyn 2001; Ravetch and Lanier 2000; Tomasello et al. 2000). However, a number of critical observations have been made with these additional receptors with regards to SHP-1 and are worthy of mention.

One of the co-inhibitory receptors expressed on B cells, PIR-B, associates with both SHP-1 and SHP-2 (Maeda et al. 1998). Furthermore, the introduction of chimeric receptors encoding the extracellular and the intracellular domains of FcγRIIB and PIR-B respectively, into SHP-1 or SHP-2 deficient isolates of the chicken B cell line, DT40, has indicated redundant roles for SHP-1 and SHP-2 in mediating inhibitory signaling from the PIR-B receptor (Maeda et al. 1998). The potential overlapping inhibitory roles of SHP-1 and SHP-2 in the context of PIR-B function further highlight the difficulty of specifically relating the structural importance of individual phosphotyrosyl sequences within inhibitory receptors to SHP-1 activation (Maeda et al. 1998).

The inhibitory receptor, gp49, is expressed on NK cells and mast cells, possesses tandem ITIMs and has been confirmed to associate with SHP-1 (Lu-Kuo et al. 1999; Rojo et al. 1997; Wang et al. 1997). In the absence of the PTP domain, the isolated SH2 domains of SHP-1 are able to tolerate mutations in the phosphotyrosine binding pockets of either the N or C SH2 domain in order to associate with gp49 (Wang et al. 1999). However, mutation of the phosphotyrosine binding pocket of the C SH2 domain in the context of recombinant full-length SHP-1 abolishes binding to gp49 but an equivalent mutation in the N SH2 domain still allows for interaction with gp49 (Wang et al. 1999). Together, these results support the hypothesis that in the context of wild type SHP-1, the C SH2 domain is critical for initiating interaction with receptor (Fig. 1).

In summary, current structural and biochemical evidence points to SHP-1 being preferentially recruited to molecules bearing tandem ITIMs with the hallmark sequence (S/V/I/LXpYXXL/V) whereby the spacing between individual ITIMs can range between 19 and 31 amino acids (Blery and Vivier 1999). In addition, SHP-1 may have the capacity to bind to ligands possessing single ITIMs provided two ITIMs are made available as a consequence of homodimerization of the target ligand. Both SH2 domains of SHP-1 are implicated in binding to target molecules with the C SH2 domain critically involved in promoting interactions. If indeed the requirements for SH2 domain engagement of SHP-1 to ligand in vivo are as stringent as indicated above, then direct SHP-1 association with a large number of other receptors that do not possess the required ITIMs must either be much weaker, SH2 domain-independent, be mediated by bridging molecules, or in some cases may require reinterpretation. These receptors include CD45 (Pani et al. 1996; Perez-Villar et al. 1999; Sen et al. 1999), EpoR (Klingmuller et al. 1995; Yi et al. 1995), IL-3 (Bone et al. 1997; Yi et al. 1993), CSF-R1 (Chen et al. 1996), c-Kit (Yi and Ihle 1993), TCR (Pani et al. 1996), BCR (Pani et al. 1995), EGF-R (Tomic et al. 1995), IFNα/β (David et al. 1995), IL-2R (Migone et al. 1998), PDGF-R (Yu et al. 1998), CD46 (Kurita-Taniguchi et al. 2000), CTLA-4 (Guntermann and Alexander 2002), Death receptors (Daigle et al. 2002), ERG K^+ channel receptor (Cayabyab et al. 2002), Somatostatin receptor (Bousquet et al. 1998) and intracellular signaling molecules including Vav (Kon-Kozlowski et al. 1996), Tyk-2 (Yetter et al. 1995), Jak-2 (Blanchette et al. 1999; Jiao et al. 1996), SLP-76 (Mizuno et al. 1996), p85 Yu (Yu et al. 1998), PYK2 (Kumar et al. 1999) PI3K (Cuevas et al. 1999) Lyn (Deehan et al. 2001; Somani et al. 2001). Indeed, in the context of one B cell line, SHP-1 inhibitory effects have been demonstrated to have an obligatory dependence on the availability of ITIM expressing inhibitory receptors (Adachi et al. 2001). Future research will focus on identifying ligands for many of the inhibitory receptors and achieving an appreciation of the physiological circumstances by which the inhibitory receptors become phosphorylated and capable of recruiting SHP-1/2.

15.3.2 PTP domain and substrate(s)

Crystals of the PTP domain of SHP-1 in isolation (Yang et al. 1998) and in conjunction with the adjacent tandem SH2 domains (Yang et al. 2003) have been resolved to reveal a structure with close similarities to the prototypic PTP domain as first defined for PTP1B (Barford et al. 1994) and possessing all the hallmark residues of PTPs (Andersen et al. 2001). The catalytic domain of SHP-1 is composed of twelve β strands and six α helices (Yang et al. 1998). Ten of the β strands form a highly twisted β sheet comprising the core of the domain and are flanked by two and four helices on either side. While sharing similar overall structures, SHP-1 and PTP1B differ in possessing a number of insertions and deletions in surface loops including one between the two remaining β strands, β5 and 6, that are each likely to contribute to substrate recognition. The binding site for phosphotyrosine lies in a deep pocket on the surface of the PTP domain as previously described for other PTPs. The active site of the enzyme, as defined by the PTP signature motif, HCXAGXGR(S/T), is found at the base of the phosphotyrosine binding pocket (Yang et al. 1998). The pocket is enclosed in part by residues from three variable surface loops ($\alpha 1/\beta 1$, $\beta 5/\beta 6$ and $\alpha 5/\alpha 6$) that together must account for the substrate specificity of SHP-1 (Yang et al. 1998). Residues from a fourth loop ($\beta 11/\alpha 3$) also line the binding pocket and encode for the invariant WPD motif that is involved in PTP catalysis (Jia et al. 1995; Stuckey et al. 1994). The active site residues are found on the end of β strand 12, the first turn of α helix 4 and the intervening loop between $\beta 12$ and $\alpha 4$. Incidentally, the mutation in viable motheaten mice results in a 5 amino acid insertion or a 23 amino acid deletion in the $\beta 5/\beta 6$ loop, thereby either shortening or extending the substrate binding pocket respectively (Yang et al. 1998). The WPD loop is a critical component of the PTP catalytic mechanism and in the context of PTP1B undergoes a dramatic movement from an open to a closed active conformation following binding of phosphotyrosyl peptide substrate (Jia et al. 1995). The aspartic acid residue located at the apex of the WPD loop acts as a general acid/base in the catalytic reaction and movement of the WPD loop upon substrate binding is critical for positioning the aspartic acid in close proximity to phosphotyrosine substrate (Barford 1999). Interestingly, constraints on the flexibility of the WPD loop appear to be different for SHP-1 in comparison to PTP1B and YopH from *Yersinia* (Yang et al. 1998). Crystals of the PTP domain of SHP-1 in its unbound state have revealed the WPD loop to be in a half-open, half-closed conformation (Yang et al. 1998). As the WPD loop needs to be in an open position to initially bind substrate, it is predicted that eventual movement of the WPD loop into an active closed conformation is likely to be energetically more costly for SHP-1 than for PTP1B. It has therefore been proposed that SHP-1 may have relatively high substrate specificity (Yang et al. 1998). Although structures of SHP-1 complexed with two phosphopeptides derived from SIRPα have been generated, they are unlikely to represent physiologically relevant complexes because the WPD loop remains incompletely closed in each case (Yang et al. 2000).

Biochemical approaches to defining substrate(s) for SHP-1 have been four-fold and a summary of the results obtained with these approaches is given below. Firstly, a reasonable assumption has been made that substrates of SHP-1 will be hyper-phosphorylated in cells in which the level of SHP-1 activity has been reduced. In this regard, the well documented role for SHP-1 in raising thresholds for signaling through the lymphocyte antigen-specific receptors has been scrutinized closely (see below for more details). Engagement of both the BCR and TCR results in the sequential activation of three families of intracellular PTKs (Src-, Syk- and Btk-like) and the inducible tyrosine phosphorylation of the PTKs themselves, the invariant polypeptides of the TCR and BCR complexes (CD3, zeta and Igα, and β chains respectively) and numerous downstream adaptors and effectors (Tomlinson et al. 2000). The challenge has been to distinguish direct SHP-1 substrate(s) from indirect ones because, depending on the particular study, many of the tyrosyl phosphoproteins listed here as implicated in antigen-receptor signaling have been invoked as SHP-1 substrates (Lorenz et al. 1996; Pani et al. 1996; Pani et al. 1995). In other studies, it has been extremely difficult to distinguish any differences in tyrosine phosphorylation between SHP-1 sufficient and deficient lymphocytes which may reflect a masking of the site of SHP-1 dephosphorylation on molecules with multiple sites of phosphorylation (Johnson et al. 1999; Sathish et al. 2001b). Alternatively, differences in function between SHP-1 sufficient and deficient cells, usually measured over several hours, may not always translate into biochemical differences that can be detected readily.

The second approach to defining SHP-1 substrates has been to examine the molecular consequences of triggering inhibitory receptors. This is premised on the basis that SHP-1 activation and subsequent dephosphorylation of substrate(s) critical to the activatory pathway is responsible for the negative sequelae of receptor engagement. Examination of inhibitory receptor engagement has implicated a wide range of tyrosyl phosphoproteins, including TCRζ, Zap-70, PLCγ (Binstadt et al. 1996; Dietrich et al. 2001), SLP-76 (Binstadt et al. 1998) and LAT (Dietrich et al. 2001; Valiante et al. 1996) as candidate SHP-1 substrates but with similarly conflicting results to those obtained with SHP-1 deficient lymphocytes. Both types of analyses are necessarily qualified by the indirect nature of the approach. However, a novel and intriguing perspective on the nature of SHP-1 substrates is the recent observation that KIR receptor engagement precludes the cytoskeletal-mediated polarization of lipid rafts upon NK cell activation (Lou et al. 2000) and consequently prevents the tyrosine phosphorylation of the activatory receptor, CD244 (Watzl and Long 2003). In addition, integrin mediated adhesion and conjugate formation between NK and target cells is reduced by inhibitory receptor engagement, further suggesting a potential link between SHP-1 and a component or regulator of the actin cytoskeleton (Burshtyn et al. 2000; Vyas et al. 2001). The third approach to substrate definition has been to demonstrate the SHP-1 mediated in vitro dephosphorylation of potential substrates such as the PTKs, ZAP-70, lyn, and lck but with the major caveat being the physiological relevance of such substrates (Chiang and Sefton 2001; Raab and Rudd 1996; Somani et al. 2001).

The fourth approach to substrate definition has been to generate substrate-trapping mutants of SHP-1 that are predicted to retain substrate binding but demonstrate impaired catalytic activity. Mutation of either the cysteine in the active site or aspartic acid on the WPD loop has been demonstrated to generate direct substrate trapping mutants in the context of other PTPs (Garton et al. 1996). Equivalent mutations have been introduced into the isolated PTP domain of SHP-1 but unfortunately have not allowed for the direct purification of stable enzyme-substrate complexes from cell lysates (Timms et al. 1998), and our own unpublished observations). Instead, the expression of putative trapping (C453S and D419A) and non-trapping (no substrate binding capacity) mutants of SHP-1 in cell lines has been used to make indirect inferences on potential SHP-1 substrate(s) (Berg et al. 1999; Dustin et al. 1999; Mizuno et al. 2000; Timms et al. 1998; Wu et al. 1998). The presumption is that in vivo substrates of SHP-1 should demonstrate enhanced tyrosine phosphorylation in the presence of trapping mutants in comparison to non-trapping equivalents. Such analyses have indicated that the inhibitory receptors, PIR-B, SIRPα (Timms et al. 1998) and CD72 (Wu et al. 1998) may be substrates of the SHP-1 PTP domain as well as being SH2 domain ligands. In addition, similar analyses have invoked the intracellular PTK, Syk (Dustin et al. 1999) and adaptors BLNK (Mizuno et al. 2000) and p62[DOK] (Berg et al. 1999). Finally, a bacterial fusion protein coding for the SHP-1 PTP domain encompassing a C453S mutation has been reported to bind directly to the adaptor SLP-76 in far Western analysis (Binstadt et al. 1998). Interestingly, SLP-76 is a centrally important molecule in the context of TCR signaling and as such would be a logical substrate of SHP-1. However, as NK killing occurs normally in the absence of SLP-76 it is difficult to advocate SLP-76 as being a critical substrate of SHP-1 in terms of defining the molecular basis of inhibition of NK cell killing (Peterson et al. 1999). As a potential SHP-1 substrate, similar criticisms can be applied to LAT (Zhang et al. 1999b).

15.3.3 Carboxy-terminus and potential adaptor role

Unfortunately, the recently determined structures for SHP-1 do not include the terminal sixty amino acids of the enzyme and therefore the structural relationship of the tail of SHP-1 to its other domains is currently not apparent. Nevertheless, the C-terminus of SHP-1 is distinguished by a number of interesting features that require further resolution. Firstly, two tyrosine residues at positions 536 and 564 can be inducibly phosphorylated in vitro and phosphotyrosine at position 564 has been detected in vivo (Lorenz et al. 1994). The functional significance of these sites of phosphorylation remains to be fully determined, but the 536 site matches the consensus sequence for Grb2 SH2 domain binding and recent evidence suggests that the 564 site may be an important determinant in the recruitment of lck by SHP-1 in T cells (Stefanova et al. 2003). In addition, there is limited evidence indicating that tyrosine phosphorylation of SHP-1 may influence the N SH2 mediated autoinhibition of PTP activity (Zhang et al. 2003). Anionic phospholipids such as phosphatidic acid (PA) and phosphatidylserine have the intriguing capacity of stimulating the PTP activity of SHP-1 in vitro (Frank et al. 1999; Zhao et al. 1993b) whereby direct high-affinity binding of PA has been mapped to the 41 C-terminal amino acids of SHP-1 (Frank et al. 1999). It will be important to demonstrate whether this phospholipid binding property of SHP-1 plays a physiological role in membrane recruitment and PTP domain activation. There is also evidence of a nuclear localization signal in a basic cluster ($K_{593}RK_{595}$) at the extreme C-terminus of SHP-1 that appears to be silenced in hematopoietic cells but becomes operational in non-hematopoietic cells (Craggs and Kellie 2001). The functional significance of SHP-1 localization in the nuclei of epithelial cells has not been fully explored but experimental results have implicated β-catenin as one possible substrate (Duchesne et al. 2003).

15.4 Function

Our current understanding of the biological roles of SHP-1 is primarily restricted to hematopoietic cells and has essentially been elucidated by studying three model systems:

a. cells that lack active SHP-1, notably primary cells from the me/me and the me^v/me^v mice or cell lines derived from such mice. Additionally, cell lines manipulated to lack endogenous SHP-1 such as the chicken B cell line, DT40, have been utilized.

b. cells in which SHP-1 activity has been blocked by introduction of a catalytically inactive, dominant negative form of SHP-1.

c. inhibitory receptors displaying inhibitory effects on cell function that have been demonstrated or are presumed to be mediated by SHP-1.

The application of each of these approaches to distinct hematopoietic cell types is summarized below.

15.4.1 T lymphocytes

Studies on SHP-1 deficient T cells from motheaten and viable motheaten mice have affirmed repeatedly that SHP-1 has negative regulatory roles on T cell function (Zhang et al. 2000). Primarily, SHP-1 deficient T cells are hyper-responsive to TCR stimulation. The cardinal responses of T cells to TCR engagement are upregulation of a limited number of cell-surface receptors, secretion of IL-2 and proliferation. Thymocytes from me/me or me^v/me^v mice, when stimulated with anti-TCR/CD3 antibodies, proliferate at a 3 to 5 fold higher level than control thymocytes and also secrete increased amounts of IL-2 (Lorenz et al. 1996; Pani et al. 1996; Sathish et al. 2001b; Zhang et al. 1999a). Consistent with lowered thresholds for TCR activation, SHP-1 deficiency leads to increased positive selection of thymocytes in mice expressing MHC Class I or II restricted transgenic TCRs (Carter et al. 1999; Johnson et al. 1999; Zhang et al. 1999a). In addition, analyses of peripheral T cells from mice expressing transgenic TCRs reveal an increased sensitivity and maximal proliferation of SHP-1 deficient T cells in response to stimulation with cognate peptide (Carter et al. 1999; Johnson et al. 1999; Zhang et al. 1999a). An intriguing finding from these studies is that SHP-1 deficient T cells proliferate more at all intensities of TCR stimulation suggesting that SHP-1 does not necessarily act on a pathway originating directly from the TCR. In this context, it is recognized that engagement of co-stimulatory receptors such as CD28 can lower the threshold of TCR responsiveness (Viola and Lanzavecchia 1996). Indeed, data indicate that increasing stimulus strength from CD28 eliminates the proliferative differences detected between the SHP-1 deficient and sufficient thymocytes (Sathish et al. 2001b). One interpretation of these findings is that the role of SHP-1 in the context of TCR responses is to negatively regulate signals from accessory pathways and thereby indirectly influence TCR signaling. The introduction of catalytically inactive, dominant negative (DN) SHP-1 into T cells of mice by transgenesis has recapitulated the findings obtained from me/me and me^v/me^v mice and confirms that SHP-1 effects on TCR are T cell autonomous (Plas et al. 1999; Zhang et al. 1999a). Furthermore, over-expression of DN SHP-1 in T cell hybridomas and the human Jurkat T cell line, also confers a TCR hyper-responsiveness (Carter et al. 1999; Plas et al. 1996). Incidentally, SHP-1 effects in T cells are not restricted to TCR signaling for it has been demonstrated that me^v/me^v T lymphocytes have a greater migratory response to the chemokine, SDF-1, possibly correlating with a higher degree of actin polymerization (Kim et al. 1999). Furthermore, the potential relevance of SHP-1 function to T cell mediated disease processes is supported by two recent observations in which lowered levels of SHP-1 activity in T cells from *me^v/+* and *me/+* heterozygous mice result in a more aggressive disease onset in experimental models for allergic airway inflammation and autoimmune encephalomyelitis (Deng et al. 2002; Kamata et al. 2003).

An indirect measure of the role of SHP-1 in T cell biology has been ascertained from studying the consequences of inhibitory receptor engagement. Subsets of human and mouse T lymphocytes have been shown to endogenously express a number of different inhibitory receptors (Ugolini and Vivier 2000) and the in vivo inhibitory effects of KIRs and CD94/NKG2A receptors on CD8$^+$ T cell mediated cytoxicity have been observed (Ikeda et al. 1997; Moser et al. 2002). Alternatively, inhibitory receptors have been introduced exogenously into T cells and the functional consequences observed. Transgenic T cells genetically engineered to express Ly49 cannot efficiently counter LCMV viral infection and also fail to clear tumor in vivo (Brawand et al. 2000; Zajac et al. 1999). Likewise, crosslinking the human inhibitory receptor, LAIR-1, on a subset of peripheral cytotoxic T cells inhibits the anti-CD3 triggered killing of target cells (Meyaard et al. 1999). It has been further demonstrated by co-precipitation experiments that LAIR-1 associates constitutively with SHP-1 at the membrane of the Jurkat T cell line and peripheral human T cells (Sathish et al. 2001a). The SHP-1 found at the membrane is active, indicating an ongoing engagement of its SH2 domains by LAIR-1. However, the association is not regulated by TCR triggering and has led to the proposal that a LAIR-1/SHP-1 complex exerts a basal tonic inhibition of TCR signaling, thereby raising the threshold for T cell activation. Furthermore, it is not apparent how SHP-1 would gain access to the TCR as the introduction of catalytically inactive SHP-1 into T cells results in an increased sensitivity to TCR/CD3 triggering without the required coengagement of any inhibitory receptors with the TCR (Carter et al. 1999; Plas et al. 1996; Zhang et al. 1999a). Likewise, expression of a MHC ClassI/SHP-1 chimera in T cells requires no ligation with the TCR in order to negatively influence TCR signaling outcomes (Musci et al. 1997). The conclusion from these observations is that recruitment of SHP-1 to the plasma membrane, without a direct physical juxtaposition to the TCR, is sufficient to affect TCR signaling. The inhibitory receptor, ILT2, is expressed in human T cells and when crosslinked on T cell clones is capable of inhibiting T cell proliferation and antigen specific target killing (Saverino et al. 2000). Interestingly, in addition to inhibition of TCR triggered early signaling events, ILT2 engagement suppresses the re-organization of the actin cytoskeleton, further suggesting that SHP-1 can target events that lead to actin polymerization (Dietrich et al. 2001).

15.4.2 B lymphocytes

As in T lymphocytes, SHP-1 has primarily been implicated in the negative regulation of antigen receptor signaling thresholds (Siminovitch and Neel 1998). B cells from motheaten mice hyperproliferate in response to stimuli through the BCR (Pani et al. 1995). Furthermore, examination of responses initiated by a transgenic BCR recognizing Hen Egg Lysozyme (HEL) reveals an exaggerated intracellular Ca^{2+} elevation in motheaten B cells resulting in cell death when encountering soluble HEL, an effect not seen in normal B cells (Cyster and Goodnow 1995). A chicken B cell line genetically manipulated to lack expression of SHP-1 further demonstrates that SHP-1 is required for inhibition of BCR induced Ca^{2+} elevation (Ono et al. 1997). Experiments involving the introduction of dominant negative SHP-1 in B cells similarly indicate an elevated BCR induced Ca^{2+} mobilization and MAPK activation (Dustin et al. 1999). Intriguingly, the presence of DN SHP-1 in these cells results in a marked increase in cell adhesiveness leading to the formation of cell aggregates upon BCR stimulation. Clearly, SHP-1 must also act on pathways that regulate the affinity/avidity of adhesion receptors in B cells.

SHP-1 has been demonstrated to bind to three ITIM-containing receptors in B cells, CD22, CD72, and PIR-B as mentioned earlier. The role of each of these receptors and SHP-1 in B cell function will be described briefly. The influence of CD22 on B cell activation is subtle whereby Ca^{2+} mobilization induced by BCR ligation is depressed when CD22 is simultaneously co-engaged (Smith et al. 1998). In addition, when CD22 along with its associated SHP-1 is segregated away from the site of BCR engagement, B cells are shown to hyper-proliferate (Doody et al. 1995). Thus, CD22 negatively regulates BCR activation thresholds by virtue of SHP-1 recruitment, activation and proximity to the BCR. This function is revealed more dramatically in B cells that are deficient in CD22. CD22$^{-/-}$ B cells are hyper-responsive to BCR triggering, a phenotype that is reminiscent of SHP-1 deficient B cells, albeit to a lesser degree (Nitschke et al. 1997; O'Keefe et al. 1996). Interestingly, stimulation through BCRs composed of IgG in contrast to IgM or IgD is not accompanied by CD22 phosphorylation and SHP-1 recruitment (Wakabayashi et al. 2002). This correlates with higher signal strength in B cells with IgG-BCR establishing that modulation of SHP-1 activity can lead to functional alterations in BCR signaling during B cell ontogeny.

CD72 is another ITIM containing receptor that when clustered has been shown to undergo tyrosine phosphorylation and SHP-1 recruitment (Adachi et al. 1998). Co-ligation of CD72 with the BCR results in a depression of Ca^{2+} mobilization and ERK activation reinforcing its primarily negative role in B cells (Adachi et al. 2000). CD72 deficient mice have been generated and B cells from these mice recapitulate, to a degree, the phenotype of motheaten B cells in that they demonstrate an elevated Ca^{2+} response and proliferation in response to BCR triggering (Pan et al. 1999). A ligand for CD72 has been identified as a member of the semaphorin family, CD100 (Kumanogoh et al. 2000). Contrary to the expectation that CD100 binding to CD72 might lead to an inhibition of B cell activation, CD100-CD72 interaction actually synergises with CD40 induced signals to enhance proliferative responses (Kumanogoh et al. 2000). This observation can be explained by the effect CD100 binding has on CD72, namely dephosphorylation of CD72 and uncoupling of SHP-1 from CD72 thereby resulting in a relief from inhibition. It indicates that B cells, by modulating CD72 phosphorylation through interaction with CD100, can further fine-tune thresholds for B cell activation.

PIR-B is an ITIM containing receptor expressed on the surface of mouse B cells and myeloid cells (Kubagawa et al. 1997). Phosphorylated PIR-B recruits SHP-1 and can inhibit Ca^{2+} mobilization triggered by BCR ligation (Blery et al. 1998). PIR-B is constitutively phosphorylated on splenic B cells thereby suggesting the possibility of a continual tonic inhibition of B cell responses (Ho et al. 1999). This is borne out from the phenotype of B cells from PIR-B deficient mice, which exhibit a constitutively active phenotype and also heightened responses to stimulation through the BCR (Ujike et al. 2002). As anticipated, the phenotype of motheaten derived B cells therefore appears to be a compound of the deficiency phenotypes for the individual inhibitory receptors, CD22, CD72, and PIR-B.

15.4.3 Natural killer cells

Natural killer cells are lymphoid cells that, unlike T and B lymphocytes, lack specific antigen receptors but are capable of killing tumor and virus infected cells by a mechanism that does not involve specific antigen recognition (Moretta et al. 2002). It has long been appreciated that NK cells preferentially kill targets lacking cell surface expression of MHC Class I receptors (Karre et al. 1986). This was subsequently established to result from an active inhibitory process mediated through the binding of MHC Class I receptors to a diverse range of Natural killer cell inhibitory receptors that includes KIRs, ILTs, gp49, CD94/NKG2A and Ly49 (Long 1999). As discussed earlier, these inhibitory receptors, when phosphorylated on their intracellular ITIMs, have the capacity to recruit and activate SHP-1. Inhibitory receptors exert a profound inhibition on the primary cytotoxic function of NK cells and the significant diminution of inhibition of NK cell cytotoxicity in motheaten mice is strong evidence that SHP-1 is a crucial component of NK inhibitory receptor function (Nakamura et al. 1997). In addition, NK inhibitory pathways also regulate lymphokine secretion (Ortaldo et al. 1997).

Although there is considerable evidence that inhibitory receptors utilize SHP-1 in the suppression of NK cell function (Binstadt et al. 1996; Burshtyn et al. 1996; Nakamura et al. 1997), the mechanisms that lead to SHP-1 mediated inhibition of NK cell function are more obscure. Interestingly, however, KIR engagement appears to disrupt an early adhesive step in the NK cell killing program, mediated primarily by integrins, that precludes the formation of stable NK cell-target cell conjugates (Burshtyn 2000). Substituting the intracellular domain of KIR with full-length SHP-1 also blocks conjugate formation thereby demonstrating a critical role for SHP-1 in the inhibition of adhesion (Burshtyn et al. 2000). Furthermore, detailed analysis of the topology of signal transduction molecules in the NK immune synapse within cytolytic interactions, reveals a rapid polarization of talin and lck to the centre of the synapse followed by the subsequent recruitment of protein kinase-C, SLP-76 and lysosomes (Vyas et al. 2002; Vyas et al. 2001). However, KIR engagement on the NK cell prevents the sustained polarization of lck and talin and precludes movement of signaling molecules into the centre of the synapse (Vyas et al. 2002; Vyas et al. 2001).

SHP-1 has also been implicated in inhibiting the activation-induced movement of membrane microdomains on the surface of NK cells. The plasma membrane of many cells, by virtue of its lipid composition, contains regions of lateral microheterogeneity. The presence of sphingolipids including glycosphingolipids and sphingomyelin along with cholesterol induces the assembly of liquid ordered phases distinct from the surrounding glycerolipids of the plasma membrane and these microdomains have been termed lipid rafts (Simons and Ikonen 1997). Lipid rafts are thought to be hotspots for signaling as they contain concentrations of signaling molecules, which are, either associated constitutively or are recruited upon activation (Janes et al. 2000). An investigation of the distribution of lipid rafts on NK cells has revealed that upon conjugation with a target tumor cell, an active PTK-dependent polarization of rafts towards the target cell occurs (Lou et al. 2000). However, simultaneous engagement of KIRs on NK cells produces a dramatic inhibition of raft polarization that is dependent upon the catalytic activity of SHP-1 (Lou et al. 2000). A related study has explored the effect of SHP-1 on the tyrosine phosphorylation and recruitment into lipid rafts of the NK cell activatory receptor, CD244. Whereas activation of NK cells by targets expressing the CD244 ligand, CD48, leads to CD244 phosphorylation and recruitment into rafts, KIR engagement on the NK cell prevents both events from occurring (Watzl and Long 2003). Interestingly, actin polymerization is required for CD244 phosphorylation and raft recruitment further implicating SHP-1 as an effector that acts proximal to actin polymerization in NK cells (Watzl and Long 2003).

15.4.4 Myeloid lineage cells

Myeloid lineage cells are, predominantly involved in the innate arm of the immune system, acting as mediators of inflammation and representing the first cellular response to infection. The critical role that SHP-1 plays in controlling the degree of activation of these cell types is dramatically revealed in the phenotype of motheaten and viable motheaten mice. These mice exhibit an over-expansion of cells derived from the myelomonocytic lineage and eventually succumb to an inflammatory pathology that includes hemorrhagic pneumonitis accompanied by the accumulation of expanded myeloid cell populations within the alveoli of the lung (Shultz et al. 1997). Therefore, the motheaten phenoptype is a vivid testimony to the central role played by SHP-1 in the biology of myeloid-lineage cells. The ex-vivo study of myeloid cells from SHP-1 deficient mice also confirms their hyper-

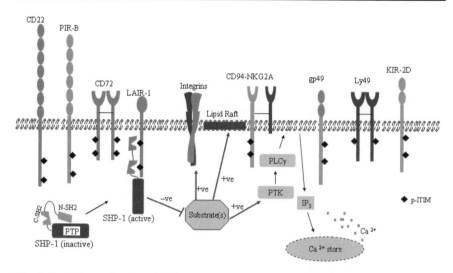

Fig. 2. Possible mechanism of SHP-1 action in hematopoeitic cells. SHP-1 is recruited and activated at the plasma membrane by phosphorylated ITIMs on inhibitory receptors. Active SHP-1 dephosphorylates a putative key substrate influencing integrin-mediated adhesion, lipid raft polarization, PTK activation, activatory receptor phosphorylation and calcium release.

reactivity. Bone marrow granulocytes from motheaten viable mice exhibit enhanced proliferative responses to the growth factor G-CSF (Tapley et al. 1997). In addition, the proliferation of macrophages from SHP-1 deficient mice is enhanced in response to GM-CSF (Jiao et al. 1997) and IL-3 (Yi et al. 1993). Nevertheless, the most profound effects of SHP-1 in myeloid cells appear to be manifest in the adhesive properties of these cells. SHP-1 deficient macrophages demonstrate an increased adhesion and spreading which is also accompanied by dysregulated detachment (Roach et al. 1998). Likewise, motheaten bone marrow neutrophils are hyperadherent to protein-coated plastic and demonstrate decreased chemotaxis that may be attributable to decreased de-adhesion (Kruger et al. 2000). These changes in macrophage and neutrophil function most certainly account for the majority of the lesions that form the pathology of the motheaten mouse. The ITIM-containing receptors, PIR-B/ILTs and SIRPα are thought to constitute the major activatory ligands for SHP-1 in macrophages and neutrophils (Timms et al. 1998).

SHP-1 also contributes to the delivery of inhibitory signals that impede mast cell activation. It is likely that SHP-1, by analogy with its role in the other cell types discussed above, might be involved in negative regulation of mast cells by association with mast cell ITIM-containing receptors. Indeed, an inhibitory receptor expressed on mast cells, gp49B, when co-ligated with the IgE receptor, FcεRI, can suppress intracellular Ca^{2+} release and FcεRI triggered exocytosis in an ITIM dependent manner (Lu-Kuo et al. 1999). This effect is at least partly dependent on SHP-1 as the inhibition of mast cell activation via gp49B is diminished in SHP-1 deficient mouse bone marrow derived mast cells. The central role gp49 plays in mast cell biology is evidenced by mast cells that have been made deficient in gp49B. These cells display exaggerated activation upon stimulation through FcεRI and also an exacerbation of Ig triggered anaphylaxis in vivo (Katz 2002).

15.5 Summary

Rapid and significant progress in the characterization of the ligands for the SH2 domains of SHP-1 has generated an important signaling paradigm whereby ITIM-containing receptor mediated recruitment and activation of SHP-1 results in the negative inhibition of a range of leukocyte functions. Any explanation for the mechanism of SHP-1 action must reconcile the pleiotropic nature of SHP-1 effects, including the apparent regulation of antigen, cytokine, growth factor, and chemokine receptor signaling in addition to influences on cytotoxicity and adhesive properties of leukocytes. Although there remains much uncertainty regarding the identity of physiological substrate(s) of SHP-1, an attractive explanation to account for the myriad effects (and proposed substrates) of SHP-1 may be the existence of a single central substrate(s), possibly a phosphotyrosine-containing molecule involved in regulating the actin cytoskeleton in leukocytes (see Fig. 2). Rigorous and comprehensive screens of cytoskeletal regulators will perhaps eventually yield the substrate(s) of SHP-1. However, conclusive proof of the identity of physiological SHP-1 substrate(s) will ultimately require the demonstration that a substrate can bind to the SHP-1 PTP domain and induce conformational changes that include closure of the WPD loop. This will necessarily require crystallography of SHP-1/substrate complexes.

Acknowledgements

RJM and JGS thank The Wellcome Trust (Project Grant 065556) and the Medical Research Council (Co-operative Component Grant G0100191) for their generous support.

References

Adachi T, Flaswinkel H, Yakura H, Reth M, Tsubata T (1998) The B cell surface protein CD72 recruits the tyrosine phosphatase SHP-1 upon tyrosine phosphorylation. J Immunol 160:4662-4665

Adachi T, Wakabayashi C, Nakayama T, Yakura H, Tsubata T (2000) CD72 negatively regulates signaling through the antigen receptor of B cells. J Immunol 164:1223-1229

Adachi T, Wienands J, Wakabayashi C, Yakura H, Reth M, Tsubata T (2001) SHP-1 requires inhibitory co-receptors to down-modulate B cell antigen receptor-mediated phosphorylation of cellular substrates. J Biol Chem 276:26648-26655

Andersen JN, Mortensen OH, Peters GH, Drake PG, Iversen LF, Olsen OH, Jansen PG, Andersen HS, Tonks NK, Moller NP (2001) Structural and evolutionary relationships among protein tyrosine phosphatase domains. Mol Cell Biol 21:7117-7136

Banville D, Stocco R, Shen SH (1995) Human protein tyrosine phosphatase 1C (PTPN6) gene structure: alternate promoter usage and exon skipping generate multiple transcripts. Genomics 27:165-173

Barford D (1999) Colworth Medal Lecture. Structural studies of reversible protein phosphorylation and protein phosphatases. Biochem Soc Trans 27:751-766

Barford D, Flint AJ, Tonks NK (1994) Crystal structure of human protein tyrosine phosphatase 1B. Science 263:1397-1404

Beebe KD, Wang P, Arabaci G, Pei D (2000) Determination of the binding specificity of the SH2 domains of protein tyrosine phosphatase SHP-1 through the screening of a combinatorial phosphotyrosyl peptide library. Biochemistry 39:13251-13260

Bellon T, Kitzig F, Sayos J, Lopez-Botet M (2002) Mutational analysis of immunoreceptor tyrosine-based inhibition motifs of the Ig-like transcript 2 (CD85j) leukocyte receptor. J Immunol 168:3351-3359

Berg KL, Siminovitch KA, Stanley ER (1999) SHP-1 regulation of p62(DOK) tyrosine phosphorylation in macrophages. J Biol Chem 274:35855-35865

Binstadt BA, Billadeau DD, Jevremovic D, Williams BL, Fang N, Yi T, Koretzky GA, Abraham RT, Leibson PJ (1998) SLP-76 is a direct substrate of SHP-1 recruited to killer cell inhibitory receptors. J Biol Chem 273:27518-27523

Binstadt BA, Brumbaugh KM, Dick CJ, Scharenberg AM, Williams BL, Colonna M, Lanier LL, Kinet JP, Abraham RT, Leibson PJ (1996) Sequential involvement of Lck and SHP-1 with MHC-recognizing receptors on NK cells inhibits FcR-initiated tyrosine kinase activation. Immunity 5:629-638

Blanchette J, Racette N, Faure R, Siminovitch KA, Olivier M (1999) Leishmania-induced increases in activation of macrophage SHP-1 tyrosine phosphatase are associated with impaired IFN-gamma-triggered JAK2 activation. Eur J Immunol 29:3737-3744

Blasioli J, Paust S, Thomas ML (1999) Definition of the sites of interaction between the protein tyrosine phosphatase SHP-1 and CD22. J Biol Chem 274:2303-2307

Blery M, Kubagawa H, Chen CC, Vely F, Cooper MD, Vivier E (1998) The paired Ig-like receptor PIR-B is an inhibitory receptor that recruits the protein-tyrosine phosphatase SHP-1. Proc Natl Acad Sci USA 95:2446-2451

Blery M, Vivier E (1999) How to extinguish lymphocyte activation, immunotyrosine-based inhibition motif (ITIM)-bearing molecules a solution? Clin Chem Lab Med 37:187-191

Bone H, Dechert U, Jirik F, Schrader JW, Welham MJ (1997) SHP1 and SHP2 protein-tyrosine phosphatases associate with betac after interleukin-3-induced receptor tyrosine phosphorylation. Identification of potential binding sites and substrates. J Biol Chem 272:14470-14476

Borges L, Hsu ML, Fanger N, Kubin M, Cosman D (1997) A family of human lymphoid and myeloid Ig-like receptors, some of which bind to MHC class I molecules. J Immunol 159:5192-5196

Bousquet C, Delesque N, Lopez F, Saint-Laurent N, Esteve JP, Bedecs K, Buscail L, Vaysse N, Susini C (1998) sst2 somatostatin receptor mediates negative regulation of insulin receptor signaling through the tyrosine phosphatase SHP-1. J Biol Chem 273:7099-7106

Brawand P, Lemonnier FA, MacDonald HR, Cerottini JC, Held W (2000) Transgenic expression of Ly49A on T cells impairs a specific antitumor response. J Immunol 165:1871-1876

Bruhns P, Marchetti P, Fridman WH, Vivier E, Daeron M (1999) Differential roles of N- and C-terminal immunoreceptor tyrosine-based inhibition motifs during inhibition of cell activation by killer cell inhibitory receptors. J Immunol 162:3168-3175

Bruhns P, Vely F, Malbec O, Fridman WH, Vivier E, Daeron M (2000) Molecular basis of the recruitment of the SH2 domain-containing inositol 5-phosphatases SHIP1 and SHIP2 by fcgamma RIIB. J Biol Chem 275:37357-37364

Burshtyn DN, Lam AS, Weston M, Gupta N, Warmerdam PA, Long EO (1999) Conserved residues amino-terminal of cytoplasmic tyrosines contribute to the SHP-1-mediated inhibitory function of killer cell Ig-like receptors. J Immunol 162:897-902

Burshtyn DN, Scharenberg AM, Wagtmann N, Rajagopalan S, Berrada K, Yi T, Kinet JP, Long EO (1996) Recruitment of tyrosine phosphatase HCP by the killer cell inhibitor receptor. Immunity 4:77-85

Burshtyn DN, Shin J, Stebbins C, Long EO (2000) Adhesion to target cells is disrupted by the killer cell inhibitory receptor. Curr Biol 10:777-780

Burshtyn DN, Yang W, Yi T, Long EO (1997) A novel phosphotyrosine motif with a critical amino acid at position -2 for the SH2 domain-mediated activation of the tyrosine phosphatase SHP-1. J Biol Chem 272:13066-13072

Campbell KS, Dessing M, Lopez-Botet M, Cella M, Colonna M (1996) Tyrosine phosphorylation of a human killer inhibitory receptor recruits protein tyrosine phosphatase 1C. J Exp Med 184:93-100

Campbell MA, Klinman NR (1995) Phosphotyrosine-dependent association between CD22 and protein tyrosine phosphatase 1C. Eur J Immunol 25:1573-1579

Carretero M, Palmieri G, Llano M, Tullio V, Santoni A, Geraghty DE, Lopez-Botet M (1998) Specific engagement of the CD94/NKG2-A killer inhibitory receptor by the HLA-E class Ib molecule induces SHP-1 phosphatase recruitment to tyrosine-phosphorylated NKG2-A: evidence for receptor function in heterologous transfectants. Eur J Immunol 28:1280-1291

Carter JD, Neel BG, Lorenz U (1999) The tyrosine phosphatase SHP-1 influences thymocyte selection by setting TCR signaling thresholds. Int Immunol 11:1999-2014

Cayabyab FS, Tsui FW, Schlichter LC (2002) Modulation of the ERG K+ current by the tyrosine phosphatase, SHP-1. J Biol Chem 277:48130-48138

Chen HE, Chang S, Trub T, Neel BG (1996) Regulation of colony-stimulating factor 1 receptor signaling by the SH2 domain-containing tyrosine phosphatase SHPTP1. Mol Cell Biol 16:3685-3697

Chiang GG, Sefton BM (2001) Specific dephosphorylation of the Lck tyrosine protein kinase at Tyr-394 by the SHP-1 protein-tyrosine phosphatase. J Biol Chem 276:23173-23178

Craggs G, Kellie S (2001) A functional nuclear localization sequence in the C-terminal domain of SHP-1. J Biol Chem 276:23719-23725

Crocker PR, Varki A (2001) Siglecs, sialic acids and innate immunity. Trends Immunol 22:337-342

Cuevas B, Lu Y, Watt S, Kumar R, Zhang J, Siminovitch KA, Mills GB (1999) SHP-1 regulates Lck-induced phosphatidylinositol 3-kinase phosphorylation and activity. J Biol Chem 274:27583-27589

Cyster JG, Goodnow CC (1995) Protein tyrosine phosphatase 1C negatively regulates antigen receptor signaling in B lymphocytes and determines thresholds for negative selection. Immunity 2:13-24

D'Ambrosio D, Hippen KL, Minskoff SA, Mellman I, Pani G, Siminovitch KA, Cambier JC (1995) Recruitment and activation of PTP1C in negative regulation of antigen receptor signaling by Fc gamma RIIB1. Science 268:293-297

Daigle I, Yousefi S, Colonna M, Green DR, Simon HU (2002) Death receptors bind SHP-1 and block cytokine-induced anti-apoptotic signaling in neutrophils. Nat Med 8:61-67

David M, Chen HE, Goelz S, Larner AC, Neel BG (1995) Differential regulation of the alpha/beta interferon-stimulated Jak/Stat pathway by the SH2 domain-containing tyrosine phosphatase SHPTP1. Mol Cell Biol 15:7050-7058

Deehan MR, Harnett W, Harnett MM (2001) A filarial nematode-secreted phosphorylcholine-containing glycoprotein uncouples the B cell antigen receptor from extracellular signal-regulated kinase-mitogen-activated protein kinase by promoting the surface Ig-mediated recruitment of Src homology 2 domain-containing tyrosine phosphatase-1 and Pac-1 mitogen-activated kinase-phosphatase. J Immunol 166:7462-7468

Deng C, Minguela A, Hussain RZ, Lovett-Racke, AE, Radu C, Ward ES, Racke MK (2002) Expression of the tyrosine phosphatase SRC homology 2 domain-containing protein tyrosine phosphatase 1 determines T cell activation threshold and severity of experimental autoimmune encephalomyelitis. J Immunol 168:4511-4518

Dietrich J, Cella M, Colonna M (2001) Ig-like transcript 2 (ILT2)/leukocyte Ig-like receptor 1 (LIR1) inhibits TCR signaling and actin cytoskeleton reorganization. J Immunol 166:2514-2521

Doody GM, Justement LB, Delibrias CC, Matthews RJ, Lin J, Thomas ML, Fearon DT (1995) A role in B cell activation for CD22 and the protein tyrosine phosphatase SHP. Science 269:242-244

Duchesne C, Charland S, Asselin C, Nahmias C, Rivard N (2003) Negative regulation of beta -catenin signaling by tyrosine phosphatase SHP-1 in intestinal epithelial cells. J Biol Chem

Dustin LB, Plas DR, Wong J, Hu YT, Soto C, Chan AC, Thomas ML (1999) Expression of dominant-negative src-homology domain 2-containing protein tyrosine phosphatase-1 results in increased Syk tyrosine kinase activity and B cell activation. J Immunol 162:2717-2724

Eck MJ, Pluskey S, Trub T, Harrison SC, Shoelson SE (1996) Spatial constraints on the recognition of phosphoproteins by the tandem SH2 domains of the phosphatase SH-PTP2. Nature 379:277-280

Fong DC, Malbec O, Arock M, Cambier JC, Fridman WH, Daeron M (1996) Selective in vivo recruitment of the phosphatidylinositol phosphatase SHIP by phosphorylated Fc gammaRIIB during negative regulation of IgE-dependent mouse mast cell activation. Immunol Lett 54:83-91

Fournier N, Chalus L, Durand I, Garcia E, Pin JJ, Churakova T, Patel S, Zlot C, Gorman D, Zurawski S, Abrams J, Bates EE, Garrone P (2000) FDF03, a novel inhibitory receptor of the immunoglobulin superfamily, is expressed by human dendritic and myeloid cells. J Immunol 165:1197-1209

Frank C, Keilhack H, Opitz F, Zschornig O, Bohmer FD (1999) Binding of phosphatidic acid to the protein-tyrosine phosphatase SHP-1 as a basis for activity modulation. Biochemistry 38:11993-12002

Fry AM, Lanier LL, Weiss A (1996) Phosphotyrosines in the killer cell inhibitory receptor motif of NKB1 are required for negative signaling and for association with protein tyrosine phosphatase 1C. J Exp Med 184:295-300

Garton AJ, Flint AJ, Tonks NK (1996) Identification of p130(cas) as a substrate for the cytosolic protein tyrosine phosphatase PTP-PEST. Mol Cell Biol 16:6408-6418

Green MC, Shultz LD (1975) Motheaten, an immunodeficient mutant of the mouse. I. Genetics and pathology. J Hered 66:250-258

Guntermann C, Alexander DR (2002) CTLA-4 suppresses proximal TCR signaling in resting human CD4(+) T cells by inhibiting ZAP-70 Tyr(319) phosphorylation: a potential role for tyrosine phosphatases. J Immunol 168:4420-4429

Gutch MJ, Flint AJ, Keller J, Tonks NK, Hengartner MO (1998) The Caenorhabditis elegans SH2 domain-containing protein tyrosine phosphatase PTP-2 participates in signal transduction during oogenesis and vulval development. Genes Dev 12:571-585

Ho LH, Uehara T, Chen CC, Kubagawa H, Cooper MD (1999) Constitutive tyrosine phosphorylation of the inhibitory paired Ig-like receptor PIR-B. Proc Natl Acad Sci USA 96:15086-15090

Hof P, Pluskey S, Dhe-Paganon S, Eck MJ, Shoelson SE (1998) Crystal structure of the tyrosine phosphatase SHP-2. Cell 92:441-450

Horvat A, Schwaiger F, Hager G, Brocker F, Streif R, Knyazev P, Ullrich A, Kreutzberg GW (2001) A novel role for protein tyrosine phosphatase shp1 in controlling glial activation in the normal and injured nervous system. J Neurosci 21:865-874

Houchins JP, Lanier LL, Niemi EC, Phillips JH, Ryan JC (1997) Natural killer cell cytolytic activity is inhibited by NKG2-A and activated by NKG2-C. J Immunol 158:3603-3609

Huyer G, Ramachandran C (1998) The specificity of the N-terminal SH2 domain of SHP-2 is modified by a single point mutation. Biochemistry 37:2741-2747

Ikeda H, Lethe B, Lehmann F, van Baren N, Baurain JF, de Smet C, Chambost H, Vitale M, Moretta A, Boon T, Coulie PG (1997) Characterization of an antigen that is recognized on a melanoma showing partial HLA loss by CTL expressing an NK inhibitory receptor. Immunity 6:199-208

Janes PW, Ley SC, Magee AI, Kabouridis PS (2000) The role of lipid rafts in T cell antigen receptor (TCR) signalling. Semin Immunol 12:23-34

Jia Z, Barford D, Flint AJ, Tonks NK (1995) Structural basis for phosphotyrosine peptide recognition by protein tyrosine phosphatase 1B. Science 268:1754-1758

Jiao H, Berrada K, Yang W, Tabrizi M, Platanias LC, Yi T (1996) Direct association with and dephosphorylation of Jak2 kinase by the SH2-domain-containing protein tyrosine phosphatase SHP-1. Mol Cell Biol 16:6985-6992

Jiao H, Yang W, Berrada K, Tabrizi M, Shultz L, Yi T (1997) Macrophages from motheaten and viable motheaten mutant mice show increased proliferative responses to GM-CSF: detection of potential HCP substrates in GM-CSF signal transduction. Exp Hematol 25:592-600

Johnson KG, LeRoy FG, Borysiewicz LK, Matthews RJ (1999) TCR signaling thresholds regulating T cell development and activation are dependent upon SHP-1. J Immunol 162:3802-3813

Kabat J, Borrego F, Brooks A, Coligan JE (2002) Role that each NKG2A immunoreceptor tyrosine-based inhibitory motif plays in mediating the human CD94/NKG2A inhibitory signal. J Immunol 169:1948-1958

Kamata T, Yamashita M, Kimura M, Murata K, Inami M, Shimizu C, Sugaya K, Wang CR, Taniguchi M, Nakayama T (2003) src homology 2 domain-containing tyrosine phosphatase SHP-1 controls the development of allergic airway inflammation. J Clin Invest 111:109-119

Karre K, Ljunggren HG, Piontek G, Kiessling R (1986) Selective rejection of H-2-deficient lymphoma variants suggests alternative immune defence strategy. Nature 319:675-678

Katz H (2002) Inhibition of anaphylactic inflammation by the gp49B1 receptor on mast cells. Mol Immunol 38:1301

Kim CH, Qu CK, Hangoc G, Cooper S, Anzai N, Feng GS, Broxmeyer HE (1999) Abnormal chemokine-induced responses of immature and mature hematopoietic cells from motheaten mice implicate the protein tyrosine phosphatase SHP-1 in chemokine responses. J Exp Med 190:681-690

Klingmuller U, Lorenz U, Cantley LC, Neel BG, Lodish HF (1995) Specific recruitment of SH-PTP1 to the erythropoietin receptor causes inactivation of JAK2 and termination of proliferative signals. Cell 80:729-738

Kon-Kozlowski M, Pani G, Pawson T, Siminovitch KA (1996) The tyrosine phosphatase PTP1C associates with Vav, Grb2, and mSos1 in hematopoietic cells. J Biol Chem 271:3856-3862

Kruger J, Butler JR, Cherapanov V, Dong Q, Ginzberg H, Govindarajan A, Grinstein S, Siminovitch KA, Downey GP (2000) Deficiency of Src homology 2-containing phosphatase 1 results in abnormalities in murine neutrophil function: studies in motheaten mice. J Immunol 165:5847-5859

Kubagawa H, Burrows PD, Cooper MD (1997) A novel pair of immunoglobulin-like receptors expressed by B cells and myeloid cells. Proc Natl Acad Sci USA 94:5261-5266

Kumanogoh A, Watanabe C, Lee I, Wang X, Shi W, Araki H, Hirata H, Iwahori K, Uchida J, Yasui T, Matsumoto M, Yoshida K, Yakura H, Pan C, Parnes JR, Kikutani H (2000) Identification of CD72 as a lymphocyte receptor for the class IV semaphorin CD100: a novel mechanism for regulating B cell signaling. Immunity 13:621-631

Kumar S, Avraham S, Bharti A, Goyal J, Pandey P, Kharbanda S (1999) Negative regulation of PYK2/related adhesion focal tyrosine kinase signal transduction by hematopoietic tyrosine phosphatase SHPTP1. J Biol Chem 274:30657-30663

Kurita-Taniguchi M, Fukui A, Hazeki K, Hirano A, Tsuji S, Matsumoto M, Watanabe M, Ueda S, Seya T (2000) Functional modulation of human macrophages through CD46 (measles virus receptor): production of IL-12 p40 and nitric oxide in association with recruitment of protein-tyrosine phosphatase SHP-1 to CD46. J Immunol 165:5143-5152

Lanier LL (1998) NK cell receptors. Annu Rev Immunol 16:359-393

Lankester AC, van Schijndel GM, van Lier RA (1995) Hematopoietic cell phosphatase is recruited to CD22 following B cell antigen receptor ligation. J Biol Chem 270:20305-20308

Law CL, Sidorenko SP, Chandran KA, Zhao Z, Shen SH, Fischer EH, Clark EA (1996) CD22 associates with protein tyrosine phosphatase 1C, Syk, and phospholipase C-gamma(1) upon B cell activation. J Exp Med 183:547-560

Le Drean E, Vely F, Olcese L, Cambiaggi A, Guia S, Krystal G, Gervois N, Moretta A, Jotereau F, Vivier E (1998) Inhibition of antigen-induced T cell response and antibody-induced NK cell cytotoxicity by NKG2A: association of NKG2A with SHP-1 and SHP-2 protein-tyrosine phosphatases. Eur J Immunol 28:264-276

Long EO (1999) Regulation of immune responses through inhibitory receptors. Annu Rev Immunol 17:875-904

Long EO, Burshtyn DN, Clark WP, Peruzzi M, Rajagopalan S, Rojo S, Wagtmann N, Winter, CC (1997) Killer cell inhibitory receptors: diversity, specificity, and function. Immunol Rev 155:135-144

Lorenz U, Ravichandran KS, Burakoff SJ, Neel BG (1996) Lack of SHPTP1 results in src-family kinase hyperactivation and thymocyte hyperresponsiveness. Proc Natl Acad Sci USA 93:9624-9629

Lorenz U, Ravichandran KS, Pei D, Walsh CT, Burakoff SJ, Neel BG (1994) Lck-dependent tyrosyl phosphorylation of the phosphotyrosine phosphatase SH-PTP1 in murine T cells. Mol Cell Biol 14:1824-1834

Lou Z, Jevremovic D, Billadeau DD, Leibson PJ (2000) A balance between positive and negative signals in cytotoxic lymphocytes regulates the polarization of lipid rafts during the development of cell-mediated killing. J Exp Med 191:347-354

Lu-Kuo JM, Joyal DM, Austen KF, Katz HR (1999) gp49B1 inhibits IgE-initiated mast cell activation through both immunoreceptor tyrosine-based inhibitory motifs, recruitment of src homology 2 domain-containing phosphatase-1, and suppression of early and late calcium mobilization. J Biol Chem 274:5791-5796

Maeda A, Kurosaki M, Ono M, Takai T, Kurosaki T (1998) Requirement of SH2-containing protein tyrosine phosphatases SHP-1 and SHP-2 for paired immunoglobulin-like receptor B (PIR-B)-mediated inhibitory signal. J Exp Med 187:1355-1360

Martin A, Tsui HW, Shulman MJ, Isenman D, Tsui FW (1999) Murine SHP-1 splice variants with altered Src homology 2 (SH2) domains. Implications for the SH2-mediated intramolecular regulation of SHP-1. J Biol Chem 274:21725-21734

Massa PT, Saha S, Wu C, Jarosinski KW (2000) Expression and function of the protein tyrosine phosphatase SHP-1 in oligodendrocytes. Glia 29:376-385

Matthews RJ, Bowne DB, Flores E, Thomas ML (1992) Characterization of hematopoietic intracellular protein tyrosine phosphatases: description of a phosphatase containing an SH2 domain and another enriched in proline-, glutamic acid-, serine-, and threonine-rich sequences. Mol Cell Biol 12:2396-2405

McVicar DW, Burshtyn DN (2001) Intracellular signaling by the killer immunoglobulin-like receptors and Ly49. Sci STKE 2001:RE1

Meyaard L, Adema GJ, Chang C, Woollatt E, Sutherland, GR, Lanier, LL, Phillips, JH (1997) LAIR-1, a novel inhibitory receptor expressed on human mononuclear leukocytes. Immunity 7:283-290

Meyaard L, Hurenkamp J, Clevers H, Lanier LL, Phillips JH (1999) Leukocyte-associated Ig-like receptor-1 functions as an inhibitory receptor on cytotoxic T cells. J Immunol 162:5800-5804

Migone TS, Cacalano NA, Taylor N, Yi T, Waldmann TA, Johnston JA (1998) Recruitment of SH2-containing protein tyrosine phosphatase SHP-1 to the interleukin 2 receptor; loss of SHP-1 expression in human T-lymphotropic virus type I-transformed T cells. Proc Natl Acad Sci USA 95:3845-3850

Miller I, Hatzivassiliou G, Cattoretti G, Mendelsohn C, Dalla-Favera R (2002) IRTAs: a new family of immunoglobulinlike receptors differentially expressed in B cells. Blood 99:2662-2669

Mizuno K, Katagiri T, Hasegawa K, Ogimoto M, Yakura H (1996) Hematopoietic cell phosphatase, SHP-1, is constitutively associated with the SH2 domain-containing leukocyte protein, SLP-76, in B cells. J Exp Med 184:457-463

Mizuno K, Tagawa Y, Mitomo K, Arimura Y, Hatano N, Katagiri T, Ogimoto M, Yakura H (2000) Src homology region 2 (SH2) domain-containing phosphatase-1 dephosphorylates B cell linker protein/SH2 domain leukocyte protein of 65 kDa and selectively regulates c-Jun NH2-terminal kinase activation in B cells. J Immunol 165:1344-1351

Moretta A, Bottino C, Mingari MC, Biassoni R, Moretta L (2002) What is a natural killer cell? Nat Immunol 3:6-8

Moser JM, Gibbs J, Jensen PE, Lukacher AE (2002) CD94-NKG2A receptors regulate antiviral CD8(+) T cell responses. Nat Immunol 3:189-195

Mousseau DD, Banville D, L'Abbe D, Bouchard P, Shen, SH (2000) PILRalpha, a novel immunoreceptor tyrosine-based inhibitory motif-bearing protein, recruits SHP-1 upon tyrosine phosphorylation and is paired with the truncated counterpart PILRbeta. J Biol Chem 275:4467-4474

Musci MA, Beaves SL, Ross SE, Yi T, Koretzky GA (1997) Surface expression of hemopoietic cell phosphatase fails to complement CD45 deficiency and inhibits TCR-mediated signal transduction in a Jurkat T cell clone. J Immunol 158:1565-1571

Muta T, Kurosaki T, Misulovin Z, Sanchez, M, Nussenzweig MC, Ravetch JV (1994) A 13-amino-acid motif in the cytoplasmic domain of Fc gamma RIIB modulates B-cell receptor signalling. Nature 369:340

Nadler MJ, Chen B, Anderson JS, Wortis HH, Neel BG (1997) Protein-tyrosine phosphatase SHP-1 is dispensable for FcgammaRIIB-mediated inhibition of B cell antigen receptor activation. J Biol Chem 272:20038-20043

Nakamura MC, Niemi EC, Fisher MJ, Shultz LD, Seaman WE, Ryan JC (1997) Mouse Ly-49A interrupts early signaling events in natural killer cell cytotoxicity and functionally associates with the SHP-1 tyrosine phosphatase. J Exp Med 185:673-684

Neel BG (1993) Structure and function of SH2-domain containing tyrosine phosphatases. Semin Cell Biol 4:419-432

Newton-Nash DK, Newman PJ (1999) A new role for platelet-endothelial cell adhesion molecule-1 (CD31): inhibition of TCR-mediated signal transduction. J Immunol 163:682-688

Nitschke L, Carsetti R, Ocker B, Kohler G, Lamers MC (1997) CD22 is a negative regulator of B-cell receptor signalling. Curr Biol 7:133-143

O'Keefe TL, Williams GT, Davies SL, Neuberger MS (1996) Hyperresponsive B cells in CD22-deficient mice. Science 274:798-801

Olcese L, Lang P, Vely F, Cambiaggi A, Marguet D, Blery M, Hippen KL, Biassoni R, Moretta A, Moretta L, Cambier JC, Vivier E (1996) Human and mouse killer-cell inhibitory receptors recruit PTP1C and PTP1D protein tyrosine phosphatases. J Immunol 156:4531-4534

Ono M, Bolland S, Tempst P, Ravetch JV (1996) Role of the inositol phosphatase SHIP in negative regulation of the immune system by the receptor Fc(gamma)RIIB. Nature 383:263-266

Ono M, Okada H, Bolland S, Yanagi S, Kurosaki T, Ravetch JV (1997) Deletion of SHIP or SHP-1 reveals two distinct pathways for inhibitory signaling. Cell 90:293-301

Ortaldo JR, Mason AT, Mason LH, Winkler-Pickett RT, Gosselin P, Anderson SK (1997) Selective inhibition of human and mouse natural killer tumor recognition using retroviral antisense in primary natural killer cells: involvement with MHC class I killer cell inhibitory receptors. J Immunol 158:1262-1267

Otipoby KL, Draves KE, Clark EA (2001) CD22 regulates B cell receptor-mediated signals via two domains that independently recruit Grb2 and SHP-1. J Biol Chem 276:44315-44322

Pan C, Baumgarth N, Parnes JR (1999) CD72-deficient mice reveal nonredundant roles of CD72 in B cell development and activation. Immunity 11:495-506

Pani G, Fischer KD, Mlinaric-Rascan I, Siminovitch KA (1996) Signaling capacity of the T cell antigen receptor is negatively regulated by the PTP1C tyrosine phosphatase. J Exp Med 184:839-852

Pani G, Kozlowski M, Cambier JC, Mills GB, Siminovitch KA (1995) Identification of the tyrosine phosphatase PTP1C as a B cell antigen receptor-associated protein involved in the regulation of B cell signaling. J Exp Med 181:2077-2084

Pawson T, Scott JD (1997) Signaling through scaffold, anchoring, and adaptor proteins. Science 278:2075-2080

Pei D, Lorenz U, Klingmuller U, Neel BG, Walsh CT (1994) Intramolecular regulation of protein tyrosine phosphatase SH-PTP1: a new function for Src homology 2 domains. Biochemistry 33:15483-15493

Pei D, Wang J, Walsh CT (1996) Differential functions of the two Src homology 2 domains in protein tyrosine phosphatase SH-PTP1. Proc Natl Acad Sci USA 93:1141-1145

Perez-Villar JJ, Whitney GS, Bowen MA, Hewgill DH, Aruffo AA, Kanner SB (1999) CD5 negatively regulates the T-cell antigen receptor signal transduction pathway: involvement of SH2-containing phosphotyrosine phosphatase SHP-1. Mol Cell Biol 19:2903-2912

Perkins LA, Larsen I, Perrimon N (1992) corkscrew encodes a putative protein tyrosine phosphatase that functions to transduce the terminal signal from the receptor tyrosine kinase torso. Cell 70:225-236

Peterson EJ, Clements JL, Ballas ZK, Koretzky GA (1999) NK cytokine secretion and cytotoxicity occur independently of the SLP-76 adaptor protein. Eur J Immunol 29:2223-2232

Plas DR, Johnson R, Pingel JT, Matthews RJ, Dalton M, Roy G, Chan AC, Thomas ML (1996) Direct regulation of ZAP-70 by SHP-1 in T cell antigen receptor signaling. Science 272:1173-1176

Plas DR, Williams CB, Kersh GJ, White LS, White JM, Paust S, Ulyanova T, Allen PM, Thomas ML (1999) Cutting edge: the tyrosine phosphatase SHP-1 regulates thymocyte positive selection. J Immunol 162:5680-5684

Plutzky J, Neel BG, Rosenberg RD (1992) Isolation of a src homology 2-containing tyrosine phosphatase. Proc Natl Acad Sci USA 89:1123-1127

Raab M, Rudd CE (1996) Hematopoietic cell phosphatase (HCP) regulates p56LCK phosphorylation and ZAP-70 binding to T cell receptor zeta chain. Biochem Biophys Res Commun 222:50-57

Ravetch JV, Lanier LL (2000) Immune inhibitory receptors. Science 290:84-89

Roach TI, Slater SE, White LS, Zhang X, Majerus PW, Brown EJ, Thomas ML (1998) The protein tyrosine phosphatase SHP-1 regulates integrin-mediated adhesion of macrophages. Curr Biol 8:1035-1038

Rojo S, Burshtyn DN, Long EO, Wagtmann N (1997) Type I transmembrane receptor with inhibitory function in mouse mast cells and NK cells. J Immunol 158:9-12

Samaridis J, Colonna M (1997) Cloning of novel immunoglobulin superfamily receptors expressed on human myeloid and lymphoid cells: structural evidence for new stimulatory and inhibitory pathways. Eur J Immunol 27:660-665

Sathish JG, Johnson KG, Fuller KJ, LeRoy FG, Meyaard L, Sims MJ, Matthews RJ (2001a) Constitutive association of SHP-1 with leukocyte-associated Ig-like receptor-1 in human T cells. J Immunol 166:1763-1770

Sathish JG, Johnson KG, LeRoy FG, Fuller KJ, Hallett MB, Brennan P, Borysiewicz LK, Sims MJ, Matthews RJ (2001b) Requirement for CD28 co-stimulation is lower in SHP-1-deficient T cells. Eur J Immunol 31:3649-3658

Saverino D, Fabbi M, Ghiotto F, Merlo A, Bruno S, Zarcone D, Tenca C, Tiso M, Santoro G, Anastasi G, Cosman D, Grossi CE, Ciccone E (2000) The CD85/LIR-1/ILT2 inhibitory receptor is expressed by all human T lymphocytes and down-regulates their functions. J Immunol 165:3742-3755

Sen G, Bikah G, Venkataraman C, Bondada S (1999) Negative regulation of antigen receptor-mediated signaling by constitutive association of CD5 with the SHP-1 protein tyrosine phosphatase in B-1 B cells. Eur J Immunol 29:3319-3328

Shen SH, Bastien L, Posner BI, Chretien P (1991) A protein-tyrosine phosphatase with sequence similarity to the SH2 domain of the protein-tyrosine kinases. Nature 352:736-739

Shultz LD, Coman DR, Bailey CL, Beamer WG, Sidman CL (1984) "Viable motheaten," a new allele at the motheaten locus. I. Pathology. Am J Pathol 116:179-192

Shultz LD, Rajan TV, Greiner DL (1997) Severe defects in immunity and hematopoiesis caused by SHP-1 protein-tyrosine-phosphatase deficiency. Trends Biotechnol 15:302-307

Shultz LD, Schweitzer PA, Rajan TV, Yi T, Ihle JN, Matthews RJ, Thomas ML, Beier DR (1993) Mutations at the murine motheaten locus are within the hematopoietic cell protein-tyrosine phosphatase (Hcph) gene. Cell 73:1445-1454

Siminovitch KA, Neel BG (1998) Regulation of B cell signal transduction by SH2-containing protein-tyrosine phosphatases. Semin Immunol 10:329-347

Simons K, Ikonen E (1997) Functional rafts in cell membranes. Nature 387:569-572

Smith KG, Tarlinton DM, Doody GM, Hibbs ML, Fearon DT (1998) Inhibition of the B cell by CD22: a requirement for Lyn. J Exp Med 187:807-811

Somani AK, Yuen K, Xu F, Zhang J, Branch DR, Siminovitch KA (2001) The SH2 domain containing tyrosine phosphatase-1 down-regulates activation of Lyn and Lyn-induced tyrosine phosphorylation of the CD19 receptor in B cells. J Biol Chem 276:1938-1944

Sorbel JD, Brooks DM, Lurie DI (2002) SHP-1 expression in avian mixed neural/glial cultures. J Neurosci Res 68:703-715

Stefanova I, Hemmer B, Vergelli M, Martin R, Biddison WE, Germain RN (2003) TCR ligand discrimination is enforced by competing ERK positive and SHP-1 negative feedback pathways. Nat Immunol 4:248-254

Stuckey JA, Schubert HL, Fauman EB, Zhang ZY, Dixon JE, Saper MA (1994) Crystal structure of Yersinia protein tyrosine phosphatase at 2.5 A and the complex with tungstate. Nature 370:571-575

Tapley P, Shevde NK, Schweitzer PA, Gallina M, Christianson SW, Lin IL, Stein RB, Shultz LD, Rosen J, Lamb P (1997) Increased G-CSF responsiveness of bone marrow cells from hematopoietic cell phosphatase deficient viable motheaten mice. Exp Hematol 25:122-131

Thomas ML (1995) Of ITAMs and ITIMs: turning on and off the B cell antigen receptor. J Exp Med 181:1953-1956

Timms JF, Carlberg K, Gu H, Chen H, Kamatkar S, Nadler MJ, Rohrschneider LR, Neel BG (1998) Identification of major binding proteins and substrates for the SH2-containing protein tyrosine phosphatase SHP-1 in macrophages. Mol Cell Biol 18:3838-3850

Tomasello E, Blery M, Vely F, Vivier E (2000) Signaling pathways engaged by NK cell receptors: double concerto for activating receptors, inhibitory receptors and NK cells. Semin Immunol 12:139-147

Tomic S, Greiser U, Lammers R, Kharitonenkov A, Imyanitov E, Ullrich A, Bohmer FD (1995) Association of SH2 domain protein tyrosine phosphatases with the epidermal growth factor receptor in human tumor cells. Phosphatidic acid activates receptor dephosphorylation by PTP1C. J Biol Chem 270:21277-21284

Tomlinson MG, Lin J, Weiss A (2000) Lymphocytes with a complex: adapter proteins in antigen receptor signaling. Immunol Today 21:584-591

Townley R, Shen SH, Banville D, Ramachandran C (1993) Inhibition of the activity of protein tyrosine phosphate 1C by its SH2 domains. Biochemistry 32:13414-13418

Tsui HW, Hasselblatt K, Martin A, Mok SC, Tsui FW (2002) Molecular mechanisms underlying SHP-1 gene expression. Eur J Biochem 269:3057-3064

Tsui HW, Siminovitch KA, de Souza L, Tsui FW (1993) Motheaten and viable motheaten mice have mutations in the haematopoietic cell phosphatase gene. Nat Genet 4:124-129

Ugolini S, Vivier E (2000) Regulation of T cell function by NK cell receptors for classical MHC class I molecules. Curr Opin Immunol 12:295-300

Ujike A, Takeda K, Nakamura A, Ebihara S, Akiyama K, Takai T (2002) Impaired dendritic cell maturation and increased T(H)2 responses in PIR-B(-/-) mice. Nat Immunol 3:542-548

Ulyanova T, Blasioli J, Woodford-Thomas TA, Thomas ML (1999) The sialoadhesin CD33 is a myeloid-specific inhibitory receptor. Eur J Immunol 29:3440-3449

Valiante NM, Phillips JH, Lanier LL, Parham P (1996) Killer cell inhibitory receptor recognition of human leukocyte antigen (HLA) class I blocks formation of a pp36/PLC-gamma signaling complex in human natural killer (NK) cells. J Exp Med 184:2243-2250

Vely F, Olivero S, Olcese L, Moretta A, Damen JE, Liu L, Krystal G, Cambier JC, Daeron M, Vivier E (1997) Differential association of phosphatases with hematopoietic co-receptors bearing immunoreceptor tyrosine-based inhibition motifs. Eur J Immunol 27:1994-2000

Viola A, Lanzavecchia A (1996) T cell activation determined by T cell receptor number and tunable thresholds. Science 273:104-106

Vyas YM, Maniar H, Dupont, B (2002) Cutting edge: differential segregation of the SRC homology 2-containing protein tyrosine phosphatase-1 within the early NK cell immune synapse distinguishes noncytolytic from cytolytic interactions. J Immunol 168:3150-3154

Vyas YM, Mehta KM, Morgan M, Maniar H, Butros L, Jung S, Burkhardt JK, Dupont B (2001) Spatial organization of signal transduction molecules in the NK cell immune synapses during MHC class I-regulated noncytolytic and cytolytic interactions. J Immunol 167:4358-4367

Wagtmann N, Rojo S, Eichler E, Mohrenweiser H, Long EO (1997) A new human gene complex encoding the killer cell inhibitory receptors and related mono-cyte/macrophage receptors. Curr Biol 7:615-618

Wakabayashi C, Adachi T, Wienands J, Tsubata T (2002) A distinct signaling pathway used by the IgG-containing B cell antigen receptor. Science 298:2392-2395

Wang LL, Blasioli J, Plas DR, Thomas ML, Yokoyama WM (1999) Specificity of the SH2 domains of SHP-1 in the interaction with the immunoreceptor tyrosine-based inhibitory motif-bearing receptor gp49B. J Immunol 162:1318-1323

Wang LL, Mehta IK, LeBlanc PA, Yokoyama WM (1997) Mouse natural killer cells express gp49B1, a structural homologue of human killer inhibitory receptors. J Immunol 158:13-17

Watzl C, Long EO (2003) Natural Killer Cell Inhibitory Receptors Block Actin Cytoskeleton-dependent Recruitment of 2B4 (CD244) to Lipid Rafts. J Exp Med 197:77-85

Wu Y, Nadler MJ, Brennan LA, Gish GD, Timms JF, Fusaki N, Jongstra-Bilen J, Tada N, Pawson T, Wither J, Neel BG, Hozumi N (1998) The B-cell transmembrane protein CD72 binds to and is an in vivo substrate of the protein tyrosine phosphatase SHP-1. Curr Biol 8:1009-1017

Yang J, Cheng Z, Niu T, Liang X, Zhao ZJ, Zhou GW (2000) Structural basis for substrate specificity of protein-tyrosine phosphatase SHP-1. J Biol Chem 275:4066-4071

Yang J, Liang X, Niu T, Meng W, Zhao Z, Zhou GW (1998) Crystal structure of the catalytic domain of protein-tyrosine phosphatase SHP-1. J Biol Chem 273:28199-28207

Yang J, Liu L, He D, Song X, Liang X, Zhao ZJ, Zhou GW (2003) Crystal Structure of Human Protein-tyrosine Phosphatase SHP-1. J Biol Chem 278:6516-6520

Yetter A, Uddin S, Krolewski JJ, Jiao H, Yi T, Platanias LC (1995) Association of the interferon-dependent tyrosine kinase Tyk-2 with the hematopoietic cell phosphatase. J Biol Chem 270:18179-18182

Yi T, Ihle JN (1993) Association of hematopoietic cell phosphatase with c-Kit after stimulation with c-Kit ligand. Mol Cell Biol 13:3350-3358

Yi T, Mui AL, Krystal G, Ihle JN (1993) Hematopoietic cell phosphatase associates with the interleukin-3 (IL-3) receptor beta chain and down-regulates IL-3-induced tyrosine phosphorylation and mitogenesis. Mol Cell Biol 13:7577-7586

Yi T, Zhang J, Miura O, Ihle JN (1995) Hematopoietic cell phosphatase associates with erythropoietin (Epo) receptor after Epo-induced receptor tyrosine phosphorylation: identification of potential binding sites. Blood 85:87-95

Yi TL, Cleveland JL, Ihle JN (1992) Protein tyrosine phosphatase containing SH2 domains: characterization, preferential expression in hematopoietic cells, and localization to human chromosome 12p12-p13. Mol Cell Biol 12:836-846

Yu Z, Su L, Hoglinger O, Jaramillo ML, Banville D, Shen SH (1998) SHP-1 associates with both platelet-derived growth factor receptor and the p85 subunit of phosphatidylinositol 3-kinase. J Biol Chem 273:3687-3694

Yusa S, Catina TL, Campbell KS (2002) SHP-1- and phosphotyrosine-independent inhibitory signaling by a killer cell Ig-like receptor cytoplasmic domain in human NK cells. J Immunol 168:5047-5057

Zajac AJ, Vance RE, Held W, Sourdive DJ, Altman JD, Raulet DH, Ahmed R (1999) Impaired anti-viral T cell responses due to expression of the Ly49A inhibitory receptor. J Immunol 163:5526-5534

Zhang J, Somani AK, Siminovitch KA (2000) Roles of the SHP-1 tyrosine phosphatase in the negative regulation of cell signalling. Semin Immunol 12:361-378

Zhang J, Somani AK, Yuen D, Yang Y, Love PE, Siminovitch KA (1999a) Involvement of the SHP-1 tyrosine phosphatase in regulation of T cell selection. J Immunol 163:3012-3021

Zhang W, Sommers CL, Burshtyn DN, Stebbins CC, DeJarnette JB, Trible RP, Grinberg A, Tsay HC, Jacobs HM, Kessler CM, Long EO, Love PE, Samelson LE (1999b) Essential role of LAT in T cell development. Immunity 10:323-332

Zhang Z, Shen K, Lu W, Cole PA (2003) The Role of C-terminal Tyrosine Phosphorylation in the Regulation of SHP-1 Explored via Expressed Protein Ligation. J Biol Chem 278:4668-4674

Zhao Z, Bouchard P, Diltz CD, Shen SH, Fischer EH (1993a) Purification and characterization of a protein tyrosine phosphatase containing SH2 domains. J Biol Chem 268:2816-2820

Zhao Z, Shen SH, Fischer EH (1993b) Stimulation by phospholipids of a protein-tyrosine-phosphatase containing two src homology 2 domains. Proc Natl Acad Sci USA 90:4251-4255

16 The dual-specific protein tyrosine phosphatase family

Andres Alonso, Ana Rojas, Adam Godzik, Tomas Mustelin

Abstract

Dual-specificity protein phosphatases (DSPs) belong to the protein tyrosine phosphatase (PTP) superfamily since they contain the conserved motif HCX2GX2R and share the same tertiary structure. The DSP family constitutes approximately half of all PTPs and includes a diverse group of proteins with a wide distribution among living organisms. They dephosphorylate proteins with phosphate on serine, threonine and/or tyrosine. The best-characterized substrates are the mitogen-activated protein kinases, which are dephosphorylated at tyrosine and threonine in a TXY motif. Additional substrates have been identified recently, ADF/cofilin for the slingshot DSP and glucokinase for DUSP12. DSPs can play key roles in multi-cellular organisms, as shown for the DSPs puckered and slingshot in the fruit fly and for LIP-1 in the worm. However, the physiological roles that DSPs play in higher vertebrates are largely unknown and there seems to be more redundancy. This chapter will discuss the evolution, structure, and function of the DSP family.

16.1 Introduction

The family of dual-specificity phosphatases (DSPs) is here defined as the members of the protein tyrosine phosphatase (PTP) superfamily that contain the extended consensus signature motif $DX_{26}VLVH\underline{C}X(A/M)\underline{G}(V/I)S\underline{R}SX_5AYLM$. DSPs belong to the group of VH1-like phosphatases according to the classification proposed by Fauman and Saper (Fauman and Saper 1996) and display, at least in vitro, activity against peptides with phosphate esters of Tyr, Ser, or Thr. Additional phosphatases, such as Cdc25 phosphatases, also show dual-specific activity (Honda et al. 1993), although they have a different topology (Fauman et al. 1998) and do not share sequence similarity with DSPs apart from the CX5R motif common to all Cys-based hydrolases. Thus, Cdc25 is not a member of the DSP family (as defined above), but became a dual-specific phosphatase as the result of convergent evolution. At least two additional VH1-like phosphatases, PTEN (phosphatase and tensin homology deleted on chromosome 10) (Myers et al. 1997) and PIR1 (phosphatase that interacts with RNA/RNP complex 1) (Deshpande et al. 1999), dephosphorylate PTyr, Pser, and PThr in vitro, but have non-protein substrates in vivo: the D3-phosphate of phosphatidylinositol(3,4,5)-trisphosphate

Topics in Current Genetics, Vol. 5
J. Arino, D.R. Alexander (Eds.) Protein phosphatases
© Springer-Verlag Berlin Heidelberg 2004

Table 1. DSPs published to date

Human protein	DSP	Species ortholog	Chromosomal localisation	Expression	Subcellular localization	Substrates	References
Classical DSPs							
CL100/ HVH1	DUSP1	MKP-1 (r), 3CH134 (m), PTPN10, erp	5q34	Wide	Nuclear	Erk, Jnk, p38	(Keyse and Emslie 1992; Sun et al. 1993; Kwak et al. 1994)
HVH2/TYP 1	DUSP4	MKP-2 (r)	8p12-p11	Wide	Nuclear	Erk, Jnk, p38	(Guan and Butch 1995; King et al. 1995; Misra-Press et al. 1995; Smith et al. 1997)
PAC-1	DUSP2		2q11	Hemapoietic cells	Nuclear	Erk and p38	(Rohan et al. 1993; Martell et al. 1994; Yi et al. 1995)
HVH3/B23	DUSP5		10q25	Wide	Nuclear	Erk	(Ishibashi et al. 1994; Kwak and Dixon 1995)
PYST1	DUSP6	MKP-3/rVH6 (r)	12q22-q23	Wide	Cytosolic	Erk	(Groom et al. 1996; Wiland et al. 1996; Furukawa et al. 1998)
PYST2/B59	DUSP7	MKP-X	3p21	Wide	Cytosolic	Erk	(Shin et al. 1997; Dowd et al. 1998)
MKP-4	DUSP9	Pyst3 (m)	Xq28	Placenta , Kidney and fetal liver	Nuclear/ Cytosolic	Erk and p38	(Muda et al. 1997; Dickinson et al. 2002b)
MKP-5	DUSP10	MKP-5	1q32	Wide	Nuclear/ Cytosolic	Jnk and p38 α & β	(Tanoue et al. 1999; Theodosiou et al. 1999; Masuda et al. 2000)
HVH5	DUSP8	M3/6 (m)	11p15.5	Brain	Nuclear/ Cytosolic	Jnk and p38	(Martell et al. 1995; Theodosiou et al. 1996; Nesbit et al. 1997)
MKP-7	DUSP16	MKP-M (m)	12p12	Wide	Cytosolic	Jnk and p38 α & β	(Masuda et al. 2001; Matsuguchi et al. 2001; Tanoue et al. 2001b; Montpetit et al. 2002)

Table 1: continued

Human protein	DSP	Species ortholog	Chromosomal localisation	Expression	Subcellular localization	Substrates	References
Atypical DSPs							
VHR	DUSP3	T-DSP11 (m)	17q21	Wide	Nuclear/ Cytosolic	Erk, Jnk	(Ishibashi et al. 1992)
MKPX/ JSP1/ VHX		LMW-DSP2 (m), TS-DSP2, JKAP	6p25.3	Wide	Cytosolic	Inhibits Erk, Jnk, p38 and activates Jnk	(Aoyama et al. 2001; Shen et al. 2001; Alonso et al. 2002; Chen et al. 2002)
DUSP13	DUSP13	TMDP (m)/ TS-DSP6 (m)	10q22.2	Testis and skeletal muscle		Unknown	(Nakamura et al. 1999)
HSSH-1		SSH1 (fly)	12q24.12	Wide		ADF/Cofilin	(Niwa et al. 2002)
HSSH2		SSH2 (fly)	17q11.2	Wide		ADF/Cofilin	(Niwa et al. 2002)
HSSH3		SSH3 (fly)	11q13.1	Wide		ADF/Cofilin	(Niwa et al. 2002)
HYVH1	DUSP12	GKAP (r), LMW_DSP4 (m)	1q21-q22	Wide	Cytosolic/ nuclear	Glucokinase	(Muda et al. 1999; Munoz-Alonso et al. 2000)
SKRP1	DUSP19	LDP-2 (m)/ TS-DSP1 (m)/ SKRP1 (m)	2q32.1	Wide	Cytosolic	Jnk	(Zama et al. 2002b; Zama et al. 2002a)
LMW-DSP20		BJ-HCC-26 tumor antigen	22q12.1	Testis	Nuclear/ Cytosolic	Unknown	(Hood et al. 2002)
LMW-DSP21			Xp11.4-p11.23	Testis	Nuclear/ Cytosolic	Unknown	(Hood et al. 2002)
MKP6/MK P-L	DUSP14	DSUP14 (m)	17q12	Wide	Nuclear/ Cytosolic	Erk and Jnk	(Marti et al. 2001)

(Maehama and Dixon 1998) and mRNA (Deshpande et al. 1999), respectively. Another distant relative, KAP (Kinase associated phosphatase) (Hannon et al. 1994) is highly specific for the positive regulatory site T160 in Cdk2 (Poon and Hunter 1995). Although PTEN, PIR1, and, KAP are VH1-like enzymes, they do not belong to the DSP family as defined here and they will not be discussed in this chapter.

Structurally, the catalytic domain of DSPs shows the same topology as other phosphatases of the PTP superfamily, such as PTP1B (Barford et al. 1994), PTEN (Lee et al. 1999), or KAP (Hanlon and Barford 1998). However, while the catalytic cleft of DSPs is only 6Å in depth to allow access of PSer and PThr to the catalytic Cys, the deeper pocket of conventional phosphatases (9Å) only permits access to the longer PTyr residues. On the other hand, the 9Å deep catalytic pocket of PTEN is wider to accommodate the inositol ring with phosphate at the D3 position.

The first DSP, the viral protein VH1 (Vaccinia virus H1 open reading frame), was identified by Dixon and colleges in 1991 (Guan et al. 1991). Since then, 21 human DSPs have been characterized and published. It has been predicted that the human genome contains the genes for between 29 (Lander et al. 2001) and 38 DSPs. Unfortunately, the nomenclature for these genes is varied and inconsistent: several phosphatases were cloned simultaneously by several research groups and given different names. The DUSP nomenclature used for human genes is perhaps most consistent, but is unlikely to replace the older and more familiar names of many enzymes. Table 1 contains all DSPs published to date, including their synonyms, species orthologs, chromosomal locations, expression profiles, and other characteristics.

DSPs have been detected in all main groups of organisms from bacteria to plants, yeast, and mammals. The best-characterized substrates, the mitogen-activated protein kinases (MAPKs), are also present from yeast to man. MAPKs are activated by dual-specificity MAPK kinases (MAPKKs), which phosphorylate the Thr and Tyr residues of a TXY motif present in the activation loop of all MAPKs (Chen et al. 2001b; Pearson et al. 2001). Dephosphorylation of either one or both of the residues in the TXY motif renders the MAPKs essentially inactive. Thus, conventional protein tyrosine phosphatases (PTPs), such as HePTP, can inactivate the MAPKs by dephosphorylating the Tyr in the TXY motif (Saxena et al., 1998,1999a), while Ser/Thr phosphatases like PP2A inactivate them by dephosphorylating the Thr residue (Saxena and Mustelin 2000). In contrast, MAPK-specific DSPs are able to dephosphorylate both residues. Judging from the large number of phosphatases that dephosphorylate MAPKs, these kinases are under much more elaborate control by phosphatases than by kinases.

While much of the early work on DSPs focused on the biochemical and structural aspects of these enzymes, more recent work has begun to unravel the physiological functions of DSPs, particularly in invertebrates. For example, the *puckered* (Martin-Blanco et al. 1998) and *slingshot* (Niwa et al. 2002) genes in *Drosophila melanogaster*, and LIP-1 (Berset et al. 2001) in *Caenorhabditis elegans*, have non-redundant functions revealing gene deletion. Our understanding of the physiological roles of DSPs in mammals is much more rudimentary: several reported

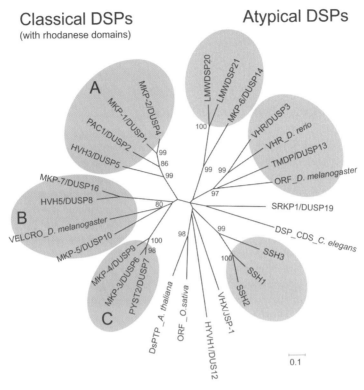

Fig. 1. Phylogenetic tree built using the catalytic domain sequences of 29 DSPs. Species origin is indicated, except for human sequences. Groups of closely related classical DSPs (A, B, and C) are in blue ovals, while groups of closely related atypical DSPs are joined in pink ovals. Bootstrap values (generated from 10,000 Bootstrap replicates) are shown when higher than 80%. Two DSPs with very close human orthologs were deleted from the final figure. The sequences used in this alignment were: 2499745 (MKP-2/DUSP4); 13124229 (chicken DUSP4; not shown); 1346900 (MKP-1/DUSP1); 14715265 (*Xenopus* MAPKXCL100; not shown); 4758206 (PAC1/DUSP2); 12707566 (HVH3/DUSP5); 20137933 (MKP-7/DUSP16); 4758212 (HVH5/DUSP8); 17737839 (VELCRO_*D. melanogaster*); 20138090 (MKP-5/DUSP10); 4503421 MKP-4/DUSP9); 2499747 (MKP-3/ DUSP6); 20141423 (PYST2/DUSP7); 4150963 (DsPTP_*A. thaliana*); 20805272 (ORF *O. sativa*); Q9UNI6 (HYVH1/DUSP12); 9910432 (VHX/JSP-1); 18376665 (SSH2); Q8WYL5 (SSH1); Q8WYL0 (SSH3); 17556208 (DSP_CDS_*C. elegans*); 18148909 (SKRP1/DUSP19); 19528569 (ORF_*D. melanogaster*); spQ9UII6 (TMDP/DUSP13); 22204201 (VHR_*D. rerio*); 423050 (VHR/DUSP3); 20137720 (MKP-6/DUSP14); 24415104 (LMWDSP21), and 24415102 (LMWDSP20).

knockout mice, such as MKP-1-/- (Dorfman et al. 1996) and JKAP-/- (Chen et al. 2002) animals, have no apparent phenotype. However, new functions in cell lines and new substrates have recently been found for DSPs. For example, the cytoskeletal protein ADF/cofilin was identified as the substrate for the slingshot

DSPs (Niwa et al. 2002) and glukokinase, an enzyme involved in glucose metabolism, has been shown to be a likely substrate for glukokinase associated protein (GKAP/DUSP12)(Munoz-Alonso et al. 2000). Thus, new breakthroughs and significant progress are to be expected within the next few years.

16.2 Phylogeny and classification

Figure 1 shows a phylogenetic tree that we have constructed from the catalytic domain of 29 selected DSP sequences (only 27 are shown in the figure). This tree shows that DSPs segregate into two main branches. One of these branches contains the MAPK phosphatases (MKPs), which we term the **classical DSPs**. They are characterized by the presence of a rhodanese domain located at the amino terminal of the catalytic domain. This group is subdivided in statistically well-defined clades that reflect recent gene duplications and species orthologs, for instance, chicken MKP-2/DUSP4 and frog MKP-1/DUSP1 (MKPXCL100), which fall very close to their human or mouse orthologs (and therefore not shown in Fig. 1). Conventional gene duplication appears to be the mechanism of evolution for this subgroup, as is also evident from a similar analysis of the rhodanese domain sequences (data not shown).

The classical DSPs range in size from 34 kDa (PAC-1) to 73 kDa (MKP-7) and share a highly similar rhodanese domain plus catalytic phosphatase domain architecture. They can be further classified into three groups based on phylogenetic analysis and gene structure. Group A contains the nuclear MKPs: DUSP1, DUSP2, DUSP4, and DUSP5, which target all three main MAPKs. Group B includes the nuclear and cytosolic MKPs that dephosphorylate the stress-activated kinases Jnk and p38, namely DUSP8, DUSP10, and DUSP16. Group C is formed by the cytosolic MKPs that mainly target Erk: DUSP6, DUSP7, and DUSP9. This classification was also proposed by Theodosiou and Asworth (Theodosiou and Ashworth 2002).

The other main branch of the DSP tree comprises a number of much more diverse enzymes devoid of rhodanese domains. We refer to this branch as the **atypical DSPs**. The divergence at the sequence level suggests that this group is phylogenetically considerably older than the classical DSPs and that the classical DSPs evolved from an ancestral atypical DSP that acquired a rhodanese domain. Evidence for recent gene duplications is also apparent within the atypical DSPs (Fig. 2). For example, LMW-DSP20 and LMW-DSP21 (Hood et al. 2002) are recent duplicons of MKP-6. Although many atypical DSPs are small enzymes (Mr around 20 kDa) and consist essentially only of a catalytic domain, there are a few larger proteins with additional domains, such as hYVH1/DUSP12 with a Zn-finger domain and the 1049-amino acid residue hSSH1L (human slingshot-1 long isoform).

It is interesting to note that DSPs become more numerous in higher eukaryotes, particularly in mammals. Thus, the DSP family has undergone a marked expansion during more recent evolution. This paralogy, often accompanied by domain accretion (Subramanian et al. 2001), is typical of higher eukarya (e.g. in immune response related proteins (Koonin et al. 2002)), and serves to increase functional diversity and the fine-tuning of cell physiology. In agreement with this notion, none of DSPs from plants group with the classical DSPs, indicating that a rhodanese domain was added to a DSP after the emergence of metazoans. Since the rhodanese domain adds specificity and affinity for MAPKs, the acquisition of this domain by DSPs was likely driven by evolutionary pressure to improve the regulation of MAPKs. A rhodanese domain is also present in the conventional protein tyrosine phosphatases PYP1 and PYP2 from *Schizosaccharomyces pombe* and PTP3 from *Saccharomyces cerevisiae* (Zhan and Guan 1999), which all target MAPKs. This represents an independent addition of a rhodanese domain to a phosphatase to improve specificity and affinity in the control over MAPKs. Interestingly, classical DSPs and MAPK-targeting yeast PTPs all have an inactive rhodanese domain, while Cdc25 consists of a catalytically active rhodanese domain that has evolved into a PTyr and PThr phosphatase (Fauman et al. 1998). Thus, the rhodanese domain has entered the phosphatase superfamily on three separate occasions during evolution.

16.3. Structural features of the DSP family

DSPs share high sequence similarity in the catalytic domain, but they have highly divergent noncatalytic N- and C-termini. They range in size from the 184 amino acids of VHX to the 1049 of hSSH1L. Whereas low molecular weight DSPs, such as VHR, VHX, MKP-6, SKRP-1, and TMDP, only contain a catalytic domain, DSPs of higher molecular weight have additional domains, such as rhodanese domains (classical DSPs), an extended C-terminus with PEST sequences (MKP-7 and hVH5), a Zn-finger domain (DUSP12), or long regions of unknown functions, such as the conserved domain A and domain B in the slingshot (SSH) phosphatases. We will first discuss the catalytic domain.

16.3.1 The catalytic domain

The DSPs catalytic domain (often called DSPc), consists of approximately 150 amino acid residues, which is considerably less than the 280 amino acids of the catalytic domain of classical PTPs (Andersen et al. 2001). Despite this difference in size and the relatively low degree of sequence similarity, the three-dimensional structures of the catalytic domains of the DSPs solved so far, VHR (Yuvaniyama et al. 1996), MKP-3 (Stewart et al. 1999), and PAC-1 (Farooq et al. 2003), are very similar to the structures obtained for other PTPs, such as PTP1B (Barford et al. 1994), YopH (Stuckey et al. 1994), PTEN (Lee et al. 1999) and KAP (Hanlon and Barford 9 1998). This conserved catalytic domain consists of an $\alpha+\beta$ domain with a central group of four twisted β-sheets sandwiched by α-helices. The catalytic site is located in the center of the molecule between the strand $\beta8$ and the

DSP catalytic domain alignment

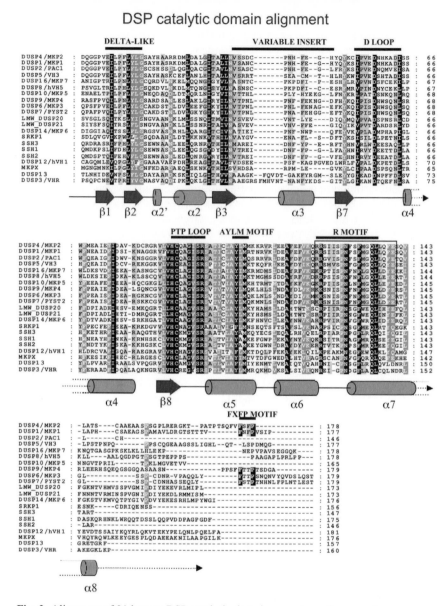

Fig. 2. Alignment of 21 human DSP catalytic domain sequences. Conserved sequence motifs are indicated, as are structural elements corresponding to the secondary structure of VHR as numbered by Yuvaniyama et al. The extended R motif in classical DSPs is boxed.

helix α5 (following the numbering used by Yuvaniyama et al. (Yuvaniyama et al. 1996)) and contains the consensus sequence HCX2G(V/I)SRS. The main structural differences between the catalytic domains of DSPs and PTPs reside in the loops that connect the conserved secondary elements. DSPs have a 6Å deep active site, compared to about 9Å in other PTPs. This allows access to both PSer/PThr and PTyr to the catalytic Cys.

Figure 2 shows an alignment of the catalytic domain of the 21 human DSPs published to date. The alignment reveals several regions with high sequence similarity, which we will discuss briefly. At the N-terminus of the DSP catalytic domain, we find the loop connecting α1-β1 that Saper and colleagues named recognition region (Yuvaniyama et al. 1996). This region contributes to the depth of the active site. In PTP1B, this loop is 41 amino acids long and forms the edge of the deep binding pocket, while in VHR it is only 10 amino acid long and forms a single loop that creates a shallower active site cleft. In addition to this effect on the depth of the catalytic pocket, this region is involved in substrate binding. In the case of VHR, the shorter loop exposes an Arg residue (at position 158), which is conserved in the PTP family, but not exposed in most PTPs. This Arg forms a positively charged pocket that binds the PThr of a diphosphorylated peptide corresponding to the activation loop of a MAPK (Schumacher et al. 2002). Moreover, it has been shown that some amino acids in this region in conventional PTPs are involved in binding to substrate (Jia et al. 1995).

Next, in the β1-β2 sheets, is the **Delta-like motif**, so named because of similarity with the δ domain in the transcription factor c-jun, where it functions as a docking site for Jnk kinase (Kallunki et al. 1996). A similar motif is present in the Ets transcription factor Elk-1 (Yang et al. 1998b), which functions as a docking site for Erk and Jnk, and in many additional proteins that interact with MAPKs. In the case of DSPs, the delta-like motif consists of the consensus sequence XI/VXPXLY/FLG. In MKP-7/DUSP16, HVH5/DUSP8, SSH3, and PYST2/DUSP7, this motif has a Lys or Arg in the first position, increasing the similarity with c-Jun. In SSH3, the Arg is preceded by additional positive residues further improving the similarities with the consensus for delta domains (Jacobs et al. 1999). The role of the delta-like motif has been tested experimentally in the mouse homolog of HVH5/DUSP8 (called M3/6). Deletion of the delta-like motif reduced its phosphorylation by JNK and its ability to dephosphorylate Jnk. Furthermore, mutation of the two Leu residues in the motif (L168 and L170) blocked the ability of HVH5/DUSP8 to dephosphorylate Jnk (Johnson et al. 2000). Similarly, in MKP-3/DUSP6, truncation of residues 207-214 abolished the interaction of the catalytic domain with Erk2 (Stewart et al. 1999).

The region between the sheets β3 and β7 is the **variable insert** (Fig. 3). In DSPs, this region is shorter than in other PTPs, e.g. in VHR it is one fourth of the corresponding region in PTP1B. This region helps orient the Arg present in the active site to interact with the phosphate of the substrate. In the inactive conformation of MKP-3, this contact is absent (Stewart et al. 1999).

The **D-loop**, or general acid loop, contains the Asp residue that functions as the general acid-base catalyst in the enzymatic reaction (Denu et al. 1996b). This loop is localized between the β7 sheet and the α4 helix, and is quite variable in sequence between two conserved hydrophobic 11 residues (Leu, Val, or Ile). In conventional PTPs, this loop contains the conserved WPD sequence, but in DSPs only the Asp residue is conserved. This Asp (D92 in VHR) serves as a general acid

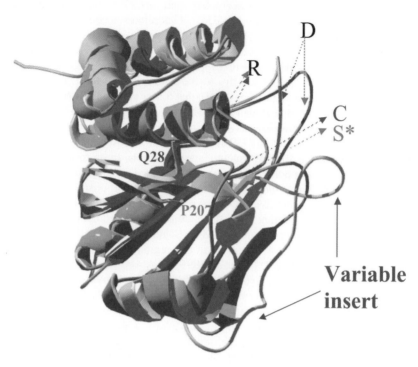

Fig. 3. Structural alignment of two DSPs. The polypeptide backbones of MKP-3/DUSP6 (green) and VHR/DUSP3 (blue) catalytic domains are shown. Alignment starts at P207 of MKP-3 and in Q28 of VHR. Residues involved in the catalytic mechanism are labelled in red with letters in green for MKP-3, in blue for VHR, or in black for both. Note that the catalytic cysteine in MKP-3 is mutated to Ser. The variable insert in VHR is also indicated.

catalyst for formation of the phosphoenzyme intermediate and as a general base in the hydrolysis of thiophosphate intermediate (Denu and Dixon 1998). Mutation of this Asp to Ala produces an inactive enzyme, often referred to as the substrate-trapping mutant (Flint et al. 1997), which can still bind substrate. For this reason, D-to-A mutant PTPs have been exploited to identify the substrates, so far with better luck for conventional PTPs than for DSPs.

The **PTP loop** sequence VHCX(A/M)G(V/I)SRS, located between the β8 sheet and the α5 helix, is highly conserved in DSPs and serves to characterize this group of phosphatases. Within this motif, only Cys, Gly, and Arg are conserved in the entire PTP superfamily. The Ser (S129 in VHR) that precedes the Arg is characteristic for DSPs. Other PTPs, such as PTEN, KAP, as well as conventional PTPs, have a Gly in this position, while Myotubularin contains an Asp. All DSPs also have a Ser after the Arg (S131 in VHR), while other PTPs have either Ser or Thr. The conserved active site cysteine (C124 in VHR) is located at the base of the catalytic pocket and is stabilized by an extensive network of hydrogen bonding in which the conserved His (H123 in VHR) plays an essential role (Yuvaniyama et al. 1996; Kim et al. 2001). The sulfhydryl group of C124 functions as a nucleophile forming a thiophosphate-enzyme intermediate with the substrate (Zhou et al. 1994). The arginine (residue 130 in VHR) helps in substrate binding by coordinating two of the oxygen atoms on the phosphoryl group. S131 in VHR has been proposed to donate a proton to stabilize the thiolate-leaving group; mutation of this residue to alanine decreases the rate of phosphoenzyme intermediate breakdown (Denu et al. 1996a).

The structure of VHR in complex with a diphosphorylated peptide corresponding to the activation loop of p38 MAPK was solved (Schumacher et al. 2002). This structure shows that the two phosphoresidues of this peptide bind to distinct pockets on the surface of VHR: while PTyr fits into the catalytic pocket, the PThr is bound to an adjacent basic pocket that contains R158, exposed in VHR because of a shorter recognition region (see above). This structure explains the unique selectivity of VHR for hydrolyzing PTyr in diphosphorylated peptides with the sequence pTXpY. The preference for PTyr over PThr is further increased by a narrow entrance to the active site created by the side chains of E126 and Y128, which are unique to VHR.

Following the PTP loop DSPs present another motif, the **AYLM motif**, which is highly conserved in DSPs and forms part of the extended signature sequence of this family. This motif is also present in other VH1-like phosphatases, like PTEN. In VHR, the Tyr residue of this motif (Y138) is exposed to the solvent and is phosphorylated in vitro and in vivo by the Zap-70 tyrosine kinase (Alonso et al. 2003). Mutation of this residue prevented VHR from dephosphorylating Erk and Jnk MAPKs. The exact mechanism by which this phosphorylation regulates the function of VHR remains unclear and it is not yet know if any other DSP is phosphorylated at the corresponding site. After the AYLM motif, there is another stretch of amino acids that are conserved in DSPs. We call this motif the **R motif** because starts with a highly conserved Arg in all family PTPs; indeed, this is one of the few amino acids conserved in the PTP superfamily, along with the residues involved in the catalytic mechanism of these enzymes. The consensus sequence is $RX_3PNX_2FX_2QL$ in all DSPs, and even more conserved in the classical DSPs: $(R/K)RX_2ISPNXFMGQLLX(F/Y)E$. In group C of classical DSPs (DUSP6/Pyst1, DUSP7/Pyst2, and DUSP9/MKP-4), this Arg has been substituted by a Lys. This Arg is on the surface in DSPs, since these phosphatases have a shorter $\alpha1 - \beta1$ region; however, conventional PTPs with a longer $\alpha1 - \beta1$ region have this residue hidden inside the molecule. The R motif is specific of DSPs and, thereby, serves to differentiate them from other non-conventional phosphatases such as PTEN, KAP or PIR1, that although close relatives of DSPs are classified in different families. Furthermore, additional residues in this motif allow dividing the MKPs in different groups that are in agreement with the clusters found in our phylogenetical study of this family.

In a C-terminal region of low sequence similarity, some classical DSPs contain the **FXFP motif** (Jacobs et al. 1999), also referred to as DEF (docking site for Erk, FXFP). This motif was first identified in the *C. elegans* protein LIN-1 (Jacobs et al. 1998), a member of the Ets family of transcription factors, where several gain-of-function mutations were found in the sequence FQFP in a C-terminal region that contains several Erk phosphorylation sites. The mutations impaired the negative regulation of LIN-1 by Erk phosphorylation, indicating that the FXFP motif was important for Erk interaction. The DEF docking motif is also present in other proteins that interact with MAP kinases, for example the GATA family of transcription factors, the KSR kinases (Jacobs et al. 1999) and other members of the Elk subfamily of Ets transcription factors (Yang et al. 1998b). Mutations in the FXFP sequence in MKP-3 (residues 364-367) lead to reduced interaction with Erk2 (Zhou et al. 2001). Conversely, addition of this motif to a peptide increases its phosphorylation efficiency by Erk.

16.3.2 The rhodanese domain

In 1991, Stephen Keyse pointed out the similarity between two stretches of amino acids in MKPs and the dual specificity phosphatase Cdc25 (Keyse and Ginsburg 1993). These motifs were called **Cdc25 homology motifs (CH2) A and B.** Whereas CH2 motifs are in the amino terminal regulatory domain of classical DSPs; they are flanking the catalytic pocket in Cdc25. These sequences represent rhodanese domains (reviewed by Bordo and Bork 2002), so named for the bacterial protein rhodanese, a sulfurtransferase involved in cyanide detoxification. Rhodanese domains are widespread among living organisms and are present in several kinds of enzymes, such as sulfurtransferases, phosphatases, ubiquitin ligases, or arsenate resistance proteins. Among phosphatases, rhodanese domains occur in classical DSPs, in Cdc25, and in the yeast tyrosine phosphatases that target MAPKs, PTP2 and PTP3 in *Sacharomyces cerevisiae*, and PYP1 and PYP2 in *Schizosacharomyces pombe* (Zhan and Guan 1999). Classical DSPs and yeast PTPs contain an inactive rhodanese domain N-terminal to the catalytic domain, while the rhodanese domain of Cdc25 has evolved into an active phosphatase domain, in which the phosphatase signature motif (CX5R) is placed in the region between the β4 and α4. The inactive rhodanese domain of classical DSPs is most similar to the inactive rhodanese domain found in ubiquitinating enzymes.

The three-dimensional structure of the DUSP6/MKP-3 rhodanese domain was recently solved. The core of this domain consists of a five-twisted β-sheet surrounded by five α-helices (Farooq et al. 2001). The degree of conservation of amino acid sequence among the rhodanese domain of classical DSPs (Fig. 4) implies this domain would fold in the same way in all the MKPs. M3/6 and MKP-3 are specific for Jnk and Erk, respectively (Muda et al. 1996). This specificity can be reversed by swapping their rhodanese domains (Muda et al. 1998), suggesting that the rhodanese domain is responsible for substrate selection. Furthermore, it was noticed that MKP-3 contained within its rhodanese domain a sequence resembling the **kinase interaction motif (KIM)** (Nichols et al. 2000) found in HePTP (Saxena et al. 1999a,b), STEP and PTP-SL (Pulido et al. 1998), tyrosine phosphatases that target MAPKs, and similar to a motif found in the C-terminus of Rsk kinase (Gavin and Nebreda 1999; Smith et al. 1999), a substrate of Erk kinases. A similar motif, characterized by the presence of two to four positively

Rhodanese domain alignment

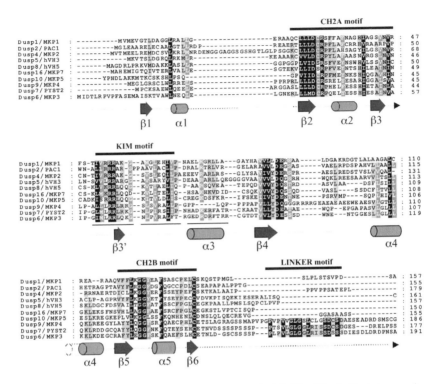

Fig. 4. Alignment of the rhodanese domain of classical DSPs. The indicated secondary structure elements correspond to the MKP-3/DUSP6 rhodanese domain. Conserved sequence motifs are indicated. The 3 modules of the extended KIM motif, as proposed by Nishida and colleagues, are underlined.

charged amino acids followed by a group of hydrophobic residues, is found in many substrates, activators and other regulators of MAP kinases (Tanoue et al. 2000). All of these motifs are involved in binding to Erk. For example, mutation of R65 in the KIM motif of MKP-3 abrogated its binding to Erk2 and reduced the dephosphorylation of Erk2 by MKP-3 (Nichols et al. 2000), Zhou et al. 2001). Similarly, R64 and L63 in DUSP6/Pyst1 also contribute to Erk2 binding (Farooq et al. 2001).

Further experimental evidence has expanded the region in the rhodanese domain that participates in the interaction with the MAPKs. NMR studies of the interaction between the rhodanese domain of MKP-3 and Erk2 revealed that the main region of interaction is located between the β3 sheet and the α3 helix of the rhodanese domain (Farooq et al. 2001), which represents an extended version of the KIM sequence. Mutagenesis of several residues in this extended motif reduced interaction (Zhou et al. 2001; Tanoue et al. 2002).

In addition to a KIM, the cytosolic phosphatases DUSP6/MKP-3, DUSP9/MKP-4 and DUSP7/PYST2 have a conserved stretch of amino acids described by Zhang and co-workers (Zhou et al. 2001), residues 161-177 in MKP-3, which we term the **linker motif**. This motif includes a proposed nuclear export signal, which may be responsible for the cytosolic localization of MKP-3. Replacement of Leu 167 and Leu 170 to Ala did not produce any change in Erk2 induced MKP-3 activation, but the binding affinity was 16-fold lower. Therefore, these amino acids play a role in Erk2 binding but not in Erk2-induced activation.

Although the CH2 motifs were described soon after the characterization of the first MKPs, their relevance still remains unclear in these phosphatases. The only experimental data available was obtained in the yeast tyrosine phosphatase PTP3, where mutations of conserved charged residues in the CH2A motif abrogated the binding of this phosphatase to the yeast MAPK Fus3 (Zhan and Guan 1999). Whether these residues are playing the same function in mammalian MKPs, still needs to be proved.

16.4 Interaction of DSPs with their substrates

16.4.1 Interaction of classical DSPs with MAPKs

As discussed above, it appears that the rhodanese domain is essential for interaction of classical DSPs with their MAPK substrates. Particularly the KIM region is important, as is the case for HePTP (Saxena et al., 1999a,b). However, it has also become clear that the interaction of classical DSPs with MAPKs is more complex and involves many additional contacts via motifs in both the rhodanese and catalytic domains, which also interact with each other (Farooq et al., 2001, 2003). For example, a peptide containing amino acids 1-222 of MKP-3 bound Erk2 with an affinity 15 times lower than the whole MKP-3 (Zhou et al. 2001). The interaction also results in conformational changes in the DSP and, as a consequence, catalytic activation. The KIM region is important, but not sufficient, for activation. MKP-3 mutated in the FXFP motif exhibited a 4-fold lower affinity toward Erk2 and was only activated only a third as much as the wild type MKP-3 (Zhou et al. 2001). The delta-like motif is also involved (Zhou et al. 2001). Taken together, the data suggest that multiple contacts are made between classical DSPs and their MAPK substrates, some of which serve to increase the affinity and/or specificity of binding, while an overlapping, but not identical, set of interactions serve to align catalytic amino acid residues of the DSP to optimize catalysis. This provides a structural basis for a high degree of substrate fidelity among classical DSPs.

While the interaction between MKP-3 and Erk2 is well understood and can be extended to a few other DSPs, many classical DSPs do not share all the involved motifs. For example, only group C DSPs (Fig. 1) have both FXFP and linker motifs, while a well-conserved FXFP motif is also found in MKP-1/DUSP1 and MKP-2/DUSP4, but not in other group A and B DSPs. Perhaps, these differences are indicative of different substrate selection, different substrate binding affinities, and different degrees of catalytic activation by substrate binding. MKP-5 (Tanoue et al. 1999) and MKP-7 (Tanoue et al. 2001b), for example, are not much activated by substrate binding. It also appears that the KIM region is not equally important in all DSPs: mutation of the highly conserved Arg in this motif in MKP-1 (Slack et al. 2001; Tanoue et al. 2002) and MKP-2 did not affect the binding of these phosphatases to Jnk.

The sites in MAPK that interact with DSPs include the common docking (CD) site (Tanoue et al. 2000), which is conserved in the whole MAPK family, and the ED docking site identified in p38 (Tanoue et al. 2001a). The CD motif is characterized by conserved acidic residues and is located in a grove ('docking grove') on the backside of the kinase domain, near the hinge between the upper and lower lobes, pointing away from the catalytic center. In Erk2, the CD motif contains the residues D316 and D319, the mutation of which results in the sevenmaker mutant of *Drosophila* ERK/Rolled, a gain-of-function mutant that is resistant to inactivation by phosphatases (Saxena and Mustelin 2000). The ED motif in p38 contains residues E160 and D161, and is important for interaction with downstream substrates, such as MAPKAPK-3, and with the MKP-5 and MKP-7 (Tanoue et al. 2001a). It seems that the acidic residues of the ED and CD motifs interact with the basic residues in the KIM motif of DSPs (Tanoue et al. 2002). However, not all DSP - MAPK interactions depend on these sequences: introduction of a sevenmaker mutation in Jnk did not block Jnk binding to MKP-2 (Chen et al. 2001a). Similarly, in the Jnk- and p38-specific MKP-5 and MKP-7, mutations in the KIM sequence Arg residues affected the interaction with p38 but had little influence on the interaction with Jnk (Tanoue et al. 2000; Tanoue et al. 2002). Perhaps, the KIM motif is more important for Erk and p38 interaction, while interaction with Jnk requires other motifs. The delta-like motif is a good candidate for a docking site between Jnk and MKPs, but there is not yet any experimental data to support this hypothesis.

An interesting new development in the topic of MAPK - DSP interaction, was the recent report that the scaffold protein JIP-1 (JNK interacting protein-1) binds MKP-7/DUSP16 and HVH5/DUSP8 (Willoughby et al. 2003), two classical MKPs that target Jnk and p38. PAC- 1/DUSP2, MKP-2/DUSP4, and MKP-4/DUSP9 did not bind. The interaction with JIP-1 involved the C-terminal tails of MKP-7/DUSP16 and HVH5/DUSP8, which is similar in these two DSPs, but not found in other DSPs. Moreover, this report showed that JIP-1 interaction with the Jnk-activating kinase MLK3 (mixed lineage kinase 3) facilitated Jnk activation and that JIP-1 interaction with MKP-7 promoted Jnk inactivation. It is not yet known whether MLK3 and MKP-7 are bound to JIP-1 at the same time or whether they bind in a sequential or regulated fashion. These data provides evidence for a role for scaffold proteins in organizing both activation and inactivation of MAPKs.

16.4.2 Interaction of atypical DSPs with their substrates

Much less is known about the interaction of atypical DSPs with their substrates. In addition, the physiological substrates are not known for most of these enzymes. Nevertheless, a few recent reports have provided some new insights. For example, the mouse ortholog of VHX (LMWDSP2) binds to p38 and Jnk (Aoyama et al. 2001). Another splice variant of the same enzyme was also shown to interact with Jnk in vivo, but not in vitro. In the same report, VHX was also shown to interact with MKK7, but not with Sek1 (Chen et al. 2002). Another atypical DSP, SKRP1 (SAPK pathway-regulating phosphatase 1), which is highly specific for JNK, has been shown to interact with MKK7, and through this kinase with Jnk. Co-precipitation analysis between SKRP1 and MKK7-activating MAPKK kinases (MAPKKKs) revealed that SKRP1 also interacted with ASK1 (apoptosis signal-regulating kinase 1), but not with MEKK1 (Zama et al. 2002b; Zama et al. 2002a). Consistent with these findings, SKRP1 expression increased the presence of ASK1-MKK7 complexes in a dose-dependent manner and specifically enhanced the activation of MKK7 by ASK1 (Zama et al. 2002b). The structural basis for any of these complexes remains unknown.

16.5 Regulation of DSPs

16.5.1 Transcriptional regulation of classical DSPs

The majority of (but not all) classical DSPs are regulated at the transcriptional level and are rapidly induced by a number of stimuli, similar to those that result in MAPK activation. Many of the genes that encode for classical DSPs were originally described as immediate early genes inducible by growth factors or cellular stresses. For instance, MKP-1 was identified as an immediate early gene rapidly induced by mitogens, heat shock or oxidative stress (Keyse and Emslie 1992; Charles et al. 1992). More recently, it has been reported that transcriptional induction of MKP-1 after stress is accompanied by chromatin remodeling (Li et al. 2001), phosphorylation and acetylation of histone H3-associated with the MKP-1 gene, and enhanced binding of RNA polymerase II to the promoter region, followed by transcription. Similarly, it has been reported that histone acetylation and chromatin remodeling are involved in transcription of MKP-M (Musikacharoen et al. 2003), a homolog of human MKP-7. In some cell types, induction of particular MKPs depends on the nature of the stimuli. For example, in PC12 cells, MKP-1, MKP-2, and hVH5 are induced by mitogens, whereas MKP-3 expression is induced by stimuli that promote neuronal differentiation, such as nerve growth factor (Charles et al. 1992; Camps et al. 1998a). Regulation at the transcriptional level has also been seen in intact animals. For instance in rats stimulated with kainic acid, MKP-1, hVH3, PAC-1, and MKP-3 are upregulated in discrete brain regions. Interestingly, different classical DSPs map to distinct brain areas that show neuronal plasticity, apoptotic cell death or survival (Misra-Press et al. 1995;

Boschert et al. 1998). MKP-1 also displays transient and highly regionalized induction within specific brain regions during normal embryonic development (Misra-Press et al. 1995).

The expression of DSPs is also regulated developmentally. During embryogenesis, MKP- 3/DUSP6 shows a pattern of expression similar to the fibroblast growth factor (FGF) family, particularly FGF8 (Dickinson et al. 2002a; Klock and Herrmann 2002). Interestingly, FGFs activate the Ras/MAPK signaling cascade and MKP-3 is a specific phosphatase for Erk2. MKP-3 is also induced by bFGF in PC12 cells (Camps et al. 1998a). MKP-4/DUSP9, on the other hand, is expressed in the mouse embryo in the areas involved in the development of placenta, liver, and muscle cells (Dickinson et al. 2002b). The expression pattern is similar to the pattern of the HGF (hepatocyte growth factor) receptor c-Met, perhaps indicating that MKP-4 may lie downstream of the c-Met.

The mechanisms that control gene induction are likely to be complex. Transcription of some classical DSP genes, including MKP-1, MKP-2, and PAC-1, is dependent at least in part on MAPK activation (Brondello et al. 1997; Boschert et al. 1998). This may constitute a negative feedback regulation loop and may allow for regulatory crosstalk between distinct MAPK modules pathways.

16.5.2 Catalytic activation of classical DSPs by substrate binding

An interesting feature of some classical DSPs is their catalytic activation induced by substrate binding (Camps et al. 1998b; Dowd et al. 1998). This mechanism was first described for MKP-3, where it was shown that direct binding to purified ERK2 activated MKP-3 by at least 30-fold (Camps et al. 1998b). In the absence of substrate, the catalytic domain adopts an open conformation in which the amino acids required for catalysis are not well positioned. For example, an Arg that should coordinate two oxygen atoms of the phosphate group is misplaced and the Asp that acts as a general acid in the catalysis is displaced by almost 5.5Å (Stewart et al. 1999) from the equivalent position in VHR (Fig. 3). These displacements impede catalysis when the enzyme is not bound to its substrate. Conversely, binding of Erk2 to MKP-3 induces a conformational change that converts the catalytic domain of MKP-3 into the optimal conformation for catalysis.

This model of substrate-induced activation has been demonstrated for a few other MKPs, such as MKP-1 (Hutter et al. 2000), MKP-4 (Camps et al. 1998b), and MKP-2 (Chen et al. 2001a). However, this is not a more general mechanism: MKP-5 and MKP-7 are not activated by MAPKs (Tanoue et al. 1999; Tanoue et al. 2001b). Moreover, the three-dimensional structure of the catalytic domain of PAC-1/DUSP2 (Farooq et al. 2003) shows that the residues involved in catalysis are much closer to the optimal active conformation and, thus, do not need to be rearranged by substrate binding.

16.5.3 Regulation of DSPs by phosphorylation

Although very few studies address the post-translational regulation of DSPs, it is clear that they can be phosphorylated on serine, threonine, or tyrosine. In the case of MKP-1, phosphorylation of two C-terminal Ser residues, S359, and S364, by Erk did not directly affect its ability to dephosphorylate Erk, but led to stabilization of the protein due to reduced degradation of MKP-1 by the proteasome (Brondello et al. 1999). S359 and S364 are C-terminal to the FXFP motif.

VHR has also been reported to be phosphorylated (Alonso et al. 2003). Phosphorylation of VHR at Y138 by the ZAP-70 tyrosine kinase was detected in both T cell lines and in normal T lymphocytes. It occurred in parallel with translocation of VHR to the immune synapse. Y138 is located in the AYLM motif, which is highly conserved in the DSP family (Fig. 2). Mutation of Y138 to Phe resulted in an increased phosphorylation of Erk and Jnk in T cells, and augmented downstream gene transactivation. Although the data indicate that VHR is functionally activated by Y138 phosphorylation, the molecular mechanism for this effect is not yet clear. However, there are many intriguing possibilities, including effects on subcellular localization or interaction with its substrates.

16.6 Physiological roles

Despite the many new insights into the structure and regulation of DSPs, there still are relatively little data to illuminate the physiological roles of DSPs in vertebrates. Only two DSP knockout mouse lines have been published: MKP-1/DUSP1 (Dorfman et al. 1996) and VHX/JKAP (Chen et al. 2002). Neither animal had any obvious phenotype, presumably due to redundancy within the DSP family. However, studies with some invertebrate DSPs, such as *puckered* and *slingshot* in *D. melanogaster,* and LIP-1 (lateral signal induced phosphatase 1) in the *C. elegans*, have demonstrated the unique and important role that some DSPs play in these organisms. One would predict that striking phenotypes may also be seen in mice if multiple DSPs were eliminated.

In *D. melanogaster*, it has been shown that the DSP *puckered* (puc) plays a critical role in dorsal closure during embryogenesis (Martin-Blanco et al. 1998). This process is regulated by the Jnk signaling pathway (Martin and Wood 2002). Several mutations in the Jnk pathway, like *hemipterous* (homolog of MKK7), *basket* (homolog of Jnk) and *Djun* (homolog of c-jun), result in problems with dorsal closure. *Puc* encodes a classical DSP with a catalytic domain most similar to MKP-5/DUSP10, but without rhodanese domain. It is highly specific for Jnk and is regulated transcriptionally in the *Drosophila* embryo. Mutation of *puc* led to cytoskeletal defects that resulted in a failure in dorsal closure (Martin-Blanco et al. 1998). The Jnk pathway and *puc* are also involved in other morphogenetic processes, like thorax closure (Zeitlinger and Bohmann 1999), optical disc morphogenesis (Agnes et al. 1999), and oogenesis (Dobens et al. 2001).

The other DSP studied in *Drosophila* is *slingshot* (*ssh*)(Niwa et al. 2002), so named because mutation of this gene produced a bifurcation of bristles and hairs. The loss of ssh increased the level of actin filaments and produced a disorganized epidermal cell morphogenesis. These effects are due to the physiological substrate for SSH, ADF (actin depolymerizing factor)/cofilin, a protein that accelerates depolymerization of actin filaments at pointed ends and separates long actin filaments. ADF/cofilin is inactivated by phosphorylation at S3 by LIM kinase and TESK (testicular protein kinases) (Yang et al. 1998a; Toshima et al. 2001). In mammalian cells, human SSH homologs also suppressed LIM kinase-induced actin reorganization. Furthermore, both *Drosophila* and human SSHs dephosphorylated phospho-cofilin in cultured cells (Niwa et al. 2002). Thus, SSHs appear to play an important role in actin dynamics by activating ADF/cofilin. There are three human genes homologous to *D. melanogaster ssh*, each producing at least two different proteins. Two of the human proteins hSSH-1L and hSSH2 were reported to be enzymatically active against p-nitro phenylphosphatate and ADF/cofilin in vitro, while hSSH-3 did not show any activity. Phospho-p38 or -JNK2 were not dephosphorylated, indicating that SSHs display distinct substrate specificities (Niwa et al. 2002).

A potential clue to the function of mammalian SSHs comes from a recent study, in which LIMK- 1 kinase, a neuronal specific LIM kinase family member, was deleted in mice by homologous recombination (Meng et al. 2002). Although the homozygous animals were superficially normal, they had deficits in spatial learning, alterations in long-term potentiation, and abnormalities in hippocampal dendritic spine structure. These finding are consistent with the proposed role for LIMK-1 in synapse formation and function. As SSHs counteract the action of LIM kinases, they may be involved in these CNS processes. A deletion of LIMK-1 has also been found in patients with Williams' syndrome (Morris and Mervis 2000), a neurodevelopmental disorder characterized by mental retardation, visual-spatial learning difficulties, and skeletal abnormalities.

Hajnal and colleagues reported recently that the classical DSP LIP-1 is involved in vulval development in *C. elegans* (Berset et al. 2001). LIP-1 contains a rhodanese domain and is most similar to the mammalian DSPs MKP-X/Pyst2/DUSP7 and Pyst-1/MKP-3/DUSP6. In *C. elegans*, the Notch signaling pathway up-regulates the expression of LIP-1, which then blocks the activation of the worm MAPK MPK-1. Total extracts from animals carrying a LIP-1 loss-of function mutation presented an increase level of MAPK activity. LIP-1 was also found to regulate meiotic cell cycle progression in the hermaphrodite germ line of *C. elegans* (Hajnal and Berset 2002). LIP-1 inactivates MPK-1 in germ cells that exit the pachytene stage of meiotic prophase I. Maintaining MPK-1 in an inactive state after pachytene exit is necessary to allow the developing oocytes to arrest the cell cycle in diakinesis until maturation is induced by the sperm signal. These findings provide an example of how DSPs that target MAPKs may play crucial roles in complex physiological processes.

16.7 Conclusions and future perspectives

The DSP family constitutes approximately half of all PTPs and is represented by some 40 genes in humans. Thus, it is very likely that they participate in numerous physiological processes, including all those that involve MAPKs. However, it is clear that the physiological functions even of the classical DSPs cannot simply be extrapolated from our current understanding of MAPKs. Rather, it seems that the classical DSPs control MAPK signaling in highly specific manners, each subject to a complex regulation of its own. Thus, classical DSPs are expressed in a cell- or tissue-specific manner, in temporally varying amounts, and in response to many different stimuli. They are also present in distinct compartments of the cells and they have different modes of post-translational regulation, substrate selection, and substrate affinity. It is likely that there are varying degrees of redundancy within the family, as also suggested by the lack of phenotype in two published knockout mice. It seems that double, triple, or higher order knockouts will be needed to sort out these issues. The picture is even more unpredictable for the atypical DSPs, most of which have unknown substrates.

It seems likely that the next few years will see a number of critical advances in our understanding of DSPs. New proteomics and mass spectrometry technologies for substrate identification and new approaches to manipulate gene expression (e.g. RNA interference) are likely to provide better insights into questions of substrate spectra and redundancy in function, at least in cell lines. We also predict that DSPs participate in protein complexes that include substrates, regulators, scaffolds, and other proteins that control spatial and temporal access to substrates. Most likely, some of the pathways and substrates will be unexpected and will be discovered by unbiased approaches, such as yeast two-hybrid screening and proteomics.

We predict that many DSPs will be found to contain phosphate on Ser, Thr, and/or Tyr, and that this phosphorylation, at least in some cases, serves important regulatory functions. It is also likely that this phosphate in DSPs is subject to autodephosphorylation and to dephosphorylation by other phosphatases, in some cases forming 'phosphatase cascades'. It is also possible that other post-translational modifications of DSPs will be discovered, such as lipid modifications, methylation, or ubiquitination.

Finally, the complete sequence of the human genome, as well as of several other genomes, have made it possible to study the entire set of DSPs in an organism. Investigations are now under way to determine if single nucleotide polymorphisms (SNPs) or mutations in PTPs and DSPs correlate with human disease. The pharmaceutical industry is also becoming increasingly interested in PTPs and DSPs as potential drug targets. This activity will intensify with the elucidation of the physiological function of these enzymes.

Acknowledgements

A.A. is supported by the Spanish Ministry of Education and Culture. Work in our laboratories is supported by grants GM60049 (A.G.), AI35603, AI48032, CA96949, AI55741 and AI53585 (T.M.) from the National Institutes of Health.

References

Agnes F, Suzanne M, Noselli S (1999) The *Drosophila* JNK pathway controls the morphogenesis of imaginal discs during metamorphosis. Development 126:5453-5462

Alonso A, Rahmouni S, Williams S, Van Stipdonk M, Jaroszewski L, Godzik A, Abraham RT, Schoenberger SP, Mustelin T (2003) Tyrosine phosphorylation of VHR phosphatase by ZAP-70. Nature Immunology 4:44-48

Andersen JN, Mortensen OH, Peters GH, Drake PG, Iversen LF, Olsen OH, Jansen PG, Andersen HS, Tonks NK, Moller NP (2001) Structural and evolutionary relationships among protein tyrosine phosphatase domains. Mol Cell Biol 21:7117-7136

Aoyama K, Nagata M, Oshima K, Matsuda T, Aoki N (2001) Molecular cloning and characterization of a novel dual specificity phosphatase, LMW-DSP2, that lacks the cdc25 homology domain. J Biol Chem 276:27575-27583

Barford D, Flint AJ, Tonks NK (1994) Crystal structure of human protein tyrosine phosphatase 1B. Science 263:1397-1404

Berset T, Hoier EF, Battu G, Canevascini S, Hajnal A (2001) Notch inhibition of RAS signaling through MAP kinase phosphatase LIP-1 during *C. elegans* vulval development. Science 291:1055-1058

Bordo D, Bork P (2002) The rhodanese/Cdc25 phosphatase superfamily. Sequence-structure-function relations. EMBO Rep 3:741-746

Boschert U, Dickinson R, Muda M, Camps M, Arkinstall S (1998) Regulated expression of dual specificity protein phosphatases in rat brain. Neuroreport 9:4081-4086

Brondello JM, Brunet A, Pouyssegur J, McKenzie FR (1997) The dual specificity mitogen-activated protein kinase phosphatase-1 and -2 are induced by the p42/p44MAPK cascade. J Biol Chem 272:1368-1376

Brondello JM, Pouyssegur J, McKenzie FR (1999) Reduced MAP kinase phosphatase-1 degradation after p42/p44MAPK-dependent phosphorylation. Science 286:2514-2517

Camps M, Chabert C, Muda M, Boschert U, Gillieron C, Arkinstall S (1998a) Induction of the mitogen-activated protein kinase phosphatase MKP3 by nerve growth factor in differentiating PC12. FEBS Lett 425:271-276

Camps M, Nichols A, Gillieron C, Antonsson B, Muda M, Chabert C, Boschert U, Arkinstall S (1998b) Catalytic activation of the phosphatase MKP-3 by ERK2 mitogen-activated protein kinase. Science 280:1262-1265

Charles CH, Abler AS, Lau LF (1992) cDNA sequence of a growth factor-inducible immediate early gene and characterization of its encoded protein. Oncogene 7:187-190

Chen AJ, Zhou G, Juan T, Colicos SM, Cannon JP, Cabriera-Hansen M, Meyer CF, Jurecic R, Copeland NG, Gilbert DJ, Jenkins NA, Fletcher F, Tan TH, Belmont JW (2002) The dual specificity JKAP specifically activates the c-Jun N-terminal kinase pathway. J Biol Chem 277:36592-36601

Chen P, Hutter D, Yang X, Gorospe M, Davis RJ, Liu Y (2001a) Discordance between the binding affinity of mitogen-activated protein kinase subfamily members for MAP kinase phosphatase-2 and their ability to activate the phosphatase catalytically. J Biol Chem 276:29440-29449

Chen Z, Gibson TB, Robinson F, Silvestro L, Pearson G, Xu B, Wright A, Vanderbilt C, Cobb MH (2001b) MAP kinases. Chem Rev 101:2449-2476

Denu JM, Dixon JE (1998) Protein tyrosine phosphatases: mechanisms of catalysis and regulation. Curr Opin Chem Biol 2:633-641

Denu JM, Lohse DL, Vijayalakshmi J, Saper MA, Dixon JE (1996a) Visualization of intermediate and transition-state structures in protein-tyrosine phosphatase catalysis. Proc Natl Acad Sci USA 93:2493-2498

Denu JM, Stuckey JA, Saper MA, Dixon JE (1996b) Form and function in protein dephosphorylation. Cell 87:361-364

Deshpande T, Takagi T, Hao L, Buratowski S, Charbonneau H (1999) Human PIR1 of the protein-tyrosine phosphatase superfamily has RNA 5'-triphosphatase and diphosphatase activities. J Biol Chem 274:16590-16594

Dickinson RJ, Eblaghie MC, Keyse SM, Morriss-Kay GM (2002a) Expression of the ERK-specific MAP kinase phosphatase PYST1/MKP3 in mouse embryos during morphogenesis and early organogenesis. Mech Dev 113:193-196

Dickinson RJ, Williams DJ, Slack DN, Williamson J, Seternes OM, Keyse SM (2002b) Characterization of a murine gene encoding a developmentally regulated cytoplasmic dual-specificity mitogen-activated protein kinase phosphatase. Biochem J 364:145-155

Dobens LL, Martin-Blanco E, Martinez-Arias A, Kafatos FC, Raftery LA (2001) *Drosophila puckered* regulates Fos/Jun levels during follicle cell morphogenesis. Development 128:1845-1856

Dorfman K, Carrasco D, Gruda M, Ryan C, Lira SA, Bravo R (1996) Disruption of the erp/mkp-1 gene does not affect mouse development: normal MAP kinase activity in ERP/MKP-1-deficient fibroblasts. Oncogene 13:925-931

Dowd S, Sneddon AA, Keyse SM (1998) Isolation of the human genes encoding the pyst1 and Pyst2 phosphatases: characterisation of Pyst2 as a cytosolic dual-specificity MAP kinase phosphatase and its catalytic activation by both MAP and SAP kinases. J Cell Sci 111 (Pt 22):3389-3399

Farooq A, Chaturvedi G, Mujtaba S, Plotnikova O, Zeng L, Dhalluin C, Ashton R, Zhou MM (2001) Solution structure of ERK2 binding domain of MAPK phosphatase MKP-3: structural insights into MKP-3 activation by ERK2. Mol Cell 7:387-399

Farooq A, Plotnikova O, Chaturvedi G, Yan S, Zeng L, Zhang Q, Zhou MM (2003) Solution structure of the MAPK phosphatase PAC-1 catalytic domain. insights into substrate-induced enzymatic activation of MKP. Structure 11:155-164

Fauman EB, Cogswell JP, Lovejoy B, Rocque WJ, Holmes W, Montana VG, Piwnica-Worms H, Rink MJ, Saper MA (1998) Crystal structure of the catalytic domain of the human cell cycle control phosphatase, Cdc25A. Cell 93:617-625

Fauman EB, Saper MA (1996) Structure and function of the protein tyrosine phosphatases. Trends Biochem Sci 21:413-417

Flint AJ, Tiganis T, Barford D, Tonks NK (1997) Development of "substrate-trapping" mutants to identify physiological substrates of protein tyrosine phosphatases. Proc Natl Acad Sci USA 94:1680-1685

Gavin AC, Nebreda AR (1999) A MAP kinase docking site is required for phosphorylation and activation of p90(rsk)/MAPKAP kinase-1. Curr Biol 9:281-284

Guan KL, Broyles SS, Dixon JE (1991) A Tyr/Ser protein phosphatase encoded by vaccinia virus. Nature 350:359-362

Hajnal A, Berset T (2002) The *C. elegans* MAPK phosphatase LIP-1 is required for the G(2)/M meiotic arrest of developing oocytes. EMBO J 21:4317-4326

Hanlon N, Barford D (1998) Purification and crystallization of the CDK-associated protein phosphatase KAP expressed in *Escherichia coli*. Protein Sci 7:508-511

Hannon GJ, Casso D, Beach D (1994) KAP: a dual specificity phosphatase that interacts with cyclin-dependent kinases. Proc Natl Acad Sci USA 91:1731-1735

Honda R, Ohba Y, Nagata A, Okayama H, Yasuda H (1993) Dephosphorylation of human p34cdc2 kinase on both Thr-14 and Tyr-15 by human cdc25B phosphatase. FEBS Lett 318:331-334

Hood KL, Tobin JF, Yoon C (2002) Identification and characterization of two novel low-molecular-weight dual specificity phosphatases. Biochem Biophys Res Commun 298:545-551

Hutter D, Chen P, Barnes J, Liu Y (2000) Catalytic activation of mitogen-activated protein (MAP) kinase phosphatase-1 by binding to p38 MAP kinase: critical role of the p38 C-terminal domain in its negative regulation. Biochem J 352 Pt 1:155-163

Jacobs D, Beitel GJ, Clark SG, Horvitz HR, Kornfeld K (1998) Gain-of-function mutations in the *Caenorhabditis elegans* lin-1 ETS gene identify a C-terminal regulatory domain phosphorylated by ERK MAP kinase. Genetics 149:1809-1822

Jacobs D, Glossip D, Xing H, Muslin AJ, Kornfeld K (1999) Multiple docking sites on substrate proteins form a modular system that mediates recognition by ERK MAP kinase. Genes Dev 13:163-175

Jia Z, Barford D, Flint AJ, Tonks NK (1995) Structural basis for phosphotyrosine peptide recognition by protein tyrosine phosphatase 1B. Science 268:1754-1758

Johnson TR, Biggs JR, Winbourn SE, Kraft AS (2000) Regulation of dual-specificity phosphatases M3/6 and hVH5 by phorbol esters. Analysis of a delta-like domain. J Biol Chem 275:31755-31762

Kallunki T, Deng T, Hibi M, Karin M (1996) c-Jun can recruit JNK to phosphorylate dimerization partners via specific docking interactions. Cell 87:929-939

Keyse SM, Emslie EA (1992) Oxidative stress and heat shock induce a human gene encoding a protein-tyrosine phosphatase. Nature 359:644-647

Keyse SM, Ginsburg M (1993) Amino acid sequence similarity between CL100, a dual-specificity MAP kinase phosphatase and cdc25. Trends Biochem Sci 18:377-378

Kim JH, Shin DY, Han MH, Choi MU (2001) Mutational and kinetic evaluation of conserved His-123 in dual specificity protein-tyrosine phosphatase vaccinia H1-related phosphatase: participation of Tyr-78 and Thr-73 residues in tuning the orientation of His-123. J Biol Chem 276:27568-27574

Klock A, Herrmann BG (2002) Cloning and expression of the mouse dual-specificity mitogen-activated protein (MAP) kinase phosphatase Mkp3 during mouse embryogenesis. Mech Develop 116:243-247

Koonin EV, Wolf YI, Karev GP (2002) The structure of the protein universe and genome evolution. Nature 420:218-223

Lander ES, et al. (2001) Initial sequencing and analysis of the human genome. Nature 409:860-921

Lee JO, Yang H, Georgescu MM, Di Cristofano A, Maehama T, Shi Y, Dixon JE, Pandolfi P, Pavletich NP (1999) Crystal structure of the PTEN tumor suppressor: implications for its phosphoinositide phosphatase activity and membrane association. Cell 99:323-334

Li J, Gorospe M, Hutter D, Barnes J, Keyse SM, Liu Y (2001) Transcriptional induction of MKP-1 in response to stress is associated with histone H3 phosphorylation-acetylation. Mol Cell Biol 21:8213-8224

Maehama T, Dixon JE (1998) The tumor suppressor, PTEN/MMAC1, dephosphorylates the lipid second messenger, phosphatidylinositol 3,4,5-trisphosphate. J Biol Chem 273:13375-13378

Marti F, Krause A, Post NH, Lyddane C, Dupont B, Sadelain M, King PD (2001) Negative-feedback regulation of CD28 costimulation by a novel mitogen-activated protein kinase phosphatase, MKP6. J Immunol 166:197-206

Martin P, Wood W (2002) Epithelial fusions in the embryo. Curr Opin Cell Biol 14:569-574

Martin-Blanco E, Gampel A, Ring J, Virdee K, Kirov N, Tolkovsky AM, Martinez-Arias A (1998) puckered encodes a phosphatase that mediates a feedback loop regulating JNK activity during dorsal closure in *Drosophila*. Genes Dev 12:557-570

Meng Y, Zhang Y, Tregoubov V, Janus C, Cruz L, Jackson M, Lu WY, MacDonald JF, Wang JY, Falls DL, Jia Z (2002) Abnormal spine morphology and enhanced LTP in LIMK-1 knockout mice. Neuron 35:121-133

Misra-Press A, Rim CS, Yao H, Roberson MS, Stork PJ (1995) A novel mitogen-activated protein kinase phosphatase. Structure, expression, and regulation. J Biol Chem 270:14587-14596

Morris CA, Mervis CB (2000) Williams syndrome and related disorders. Annu Rev Genomics Hum Genet 1:461-484

Muda M, Theodosiou A, Gillieron C, Smith A, Chabert C, Camps M, Boschert U, Rodrigues N, Davies K, Ashworth A, Arkinstall S (1998) The mitogen-activated protein kinase phosphatase-3 N-terminal noncatalytic region is responsible for tight substrate binding and enzymatic specificity. J Biol Chem 273:9323-9329

Muda M, Theodosiou A, Rodrigues N, Boschert U, Camps M, Gillieron C, Davies K, Ashworth A, Arkinstall S (1996) The dual specificity phosphatases M3/6 and MKP-3 are highly selective for inactivation of distinct mitogen-activated protein kinases. J Biol Chem 271:27205-27208

Munoz-Alonso MJ, Guillemain G, Kassis N, Girard J, Burnol AF, Leturque A (2000) A novel cytosolic dual specificity phosphatase, interacting with glucokinase, increases glucose phosphorylation rate. J Biol Chem 275:32406-32412

Musikacharoen T, Yoshikai Y, Matsuguchi T (2003) Histone acetylation and activation of CREB regulate transcriptional activation of MKP-M in LPS-stimulated macrophages. J Biol Chem

Myers MP, Stolarov JP, Eng C, Li J, Wang SI, Wigler MH, Parsons R, Tonks NK (1997) P-TEN, the tumor suppressor from human chromosome 10q23, is a dual-specificity phosphatase. Proc Natl Acad Sci USA 94:9052-9057

Nichols A, Camps M, Gillieron C, Chabert C, Brunet A, Wilsbacher J, Cobb M, Pouyssegur J, Shaw JP, Arkinstall S (2000) Substrate recognition domains within extracellular signal-regulated kinase mediate binding and catalytic activation of mitogen-activated protein kinase phosphatase-3. J Biol Chem 275:24613-24621

Niwa R, Nagata-Ohashi K, Takeichi M, Mizuno K, Uemura T (2002) Control of actin reorganization by Slingshot, a family of phosphatases that dephosphorylate ADF/cofilin. Cell 108:233-246

Pearson G, Robinson F, Beers Gibson T, Xu BE, Karandikar M, Berman K, Cobb MH (2001) Mitogen-activated protein (MAP) kinase pathways: regulation and physiological functions. Endocr Rev 22:153-183

Poon RY, Hunter T (1995) Dephosphorylation of Cdk2 Thr160 by the cyclin-dependent kinase-interacting phosphatase KAP in the absence of cyclin. Science 270:90-93

Pulido R, Zuniga A, Ullrich A (1998) PTP-SL and STEP protein tyrosine phosphatases regulate the activation of the extracellular signal-regulated kinases ERK1 and ERK2 by association through a kinase interaction motif. EMBO J 17:7337-7350

Saxena M, Mustelin T (2000) Extracellular signals and scores of phosphatases: all roads lead to MAP kinase. Semin Immunol 12:387-396

Saxena M, Williams S, Brockdorff J, Gilman J, Mustelin T (1999a) Inhibition of T cell signaling by MAP kinase-targeted hematopoietic tyrosine phosphatase (HePTP). J Biol Chem 274: 11693-11700

Saxena M, Williams S, Gilman J, Mustelin T (1998) Negative regulation of T cell antigen receptor signaling by hematopoietic tyrosine phosphatase (HePTP). J Biol Chem 273: 15340-15344.

Saxena M, Williams S, Tasken K, Mustelin T (1999b) Crosstalk between cAMP-dependent kinase and MAP kinase through a protein tyrosine phosphatase. Nat Cell Biol 1:305-311

Schumacher MA, Todd JL, Rice AE, Tanner KG, Denu JM (2002) Structural basis for the recognition of a bisphosphorylated MAP kinase peptide by human VHR protein Phosphatase. Biochemistry 41:3009-3017

Slack DN, Seternes OM, Gabrielsen M, Keyse SM (2001) Distinct binding determinants for ERK2/p38alpha and JNK map kinases mediate catalytic activation and substrate selectivity of map kinase phosphatase-1. J Biol Chem 276:16491-16500

Smith JA, Poteet-Smith CE, Malarkey K, Sturgill TW (1999) Identification of an extracellular signal-regulated kinase (ERK) docking site in ribosomal S6 kinase, a sequence critical for activation by ERK in vivo. J Biol Chem 274:2893-2898

Stewart AE, Dowd S, Keyse SM, McDonald NQ (1999) Crystal structure of the MAPK phosphatase Pyst1 catalytic domain and implications for regulated activation. Nat Struct Biol 6:174-181

Stuckey JA, Schubert HL, Fauman EB, Zhang ZY, Dixon JE, Saper MA (1994) Crystal structure of *Yersinia* protein tyrosine phosphatase at 2.5 A and the complex with tungstate. Nature 370:571-575

Subramanian G, Mural R, Hoffman SL, Venter JC, Broder S (2001) Microbial disease in humans: A genomic perspective. Mol Diagn 6:243-252

Tanoue T, Adachi M, Moriguchi T, Nishida E (2000) A conserved docking motif in MAP kinases common to substrates, activators and regulators. Nat Cell Biol 2:110-116

Tanoue T, Maeda R, Adachi M, Nishida E (2001a) Identification of a docking groove on ERK and p38 MAP kinases that regulates the specificity of docking interactions. EMBO J 20:466-479

Tanoue T, Moriguchi T, Nishida E (1999) Molecular cloning and characterization of a novel dual specificity phosphatase, MKP-5. J Biol Chem 274:19949-19956

Tanoue T, Yamamoto T, Maeda R, Nishida E (2001b) A Novel MAPK phosphatase MKP-7 acts preferentially on JNK/SAPK and p38 alpha and beta MAPKs. J Biol Chem 276:26629-26639

Tanoue T, Yamamoto T, Nishida E (2002) Modular structure of a docking surface on MAPK phosphatases. J Biol Chem 277:22942-22949

Theodosiou A, Ashworth A (2002) MAP kinase phosphatases. Genome Biol 3:REVIEWS3009

Toshima J, Toshima JY, Takeuchi K, Mori R, Mizuno K (2001) Cofilin phosphorylation and actin reorganization activities of testicular protein kinase 2 and its predominant expression in testicular Sertoli cells. J Biol Chem 276:31449-31458

Willoughby EA, Perkins GR, Collins MK, Whitmarsh AJ (2003) The JIP-1 scaffold protein targets MKP-7 to dephosphorylate JNK. J Biol Chem

Yang N, Higuchi O, Ohashi K, Nagata K, Wada A, Kangawa K, Nishida E, Mizuno K (1998a) Cofilin phosphorylation by LIM-kinase 1 and its role in Rac-mediated actin reorganization. Nature 393:809-812

Yang SH, Yates PR, Whitmarsh AJ, Davis RJ, Sharrocks AD (1998b) The Elk-1 ETS-domain transcription factor contains a mitogen-activated protein kinase targeting motif. Mol Cell Biol 18:710-720

Yuvaniyama J, Denu JM, Dixon JE, Saper MA (1996) Crystal structure of the dual specificity protein phosphatase VHR. Science 272:1328-1331

Zama T, Aoki R, Kamimoto T, Inoue K, Ikeda Y, Hagiwara M (2002a) A novel dual specificity phosphatase SKRP1 interacts with the MAPK kinase MKK7 and inactivates the JNK MAPK pathway. Implication for the precise regulation of the particular MAPK pathway. J Biol Chem 277:23909-23918

Zama T, Aoki R, Kamimoto T, Inoue K, Ikeda Y, Hagiwara M (2002b) Scaffold role of a mitogen-activated protein kinase phosphatase, SKRP1, for the JNK signaling pathway. J Biol Chem 277:23919-23926

Zeitlinger J, Bohmann D (1999) Thorax closure in *Drosophila*: involvement of Fos and the JNK pathway. Development 126:3947-3956

Zhan XL, Guan KL (1999) A specific protein-protein interaction accounts for the in vivo substrate selectivity of Ptp3 towards the Fus3 MAP kinase. Genes Dev 13:2811-2827

Zhou B, Wu L, Shen K, Zhang J, Lawrence DS, Zhang ZY (2001) Multiple regions of MAP kinase phosphatase 3 are involved in its recognition and activation by ERK2. J Biol Chem 276:6506-6515

Zhou G, Denu JM, Wu L, Dixon JE (1994) The catalytic role of Cys124 in the dual specificity phosphatase VHR. J Biol Chem 269:28084-28090

17 Tyrosine phosphatases in cancer: Targets for therapeutic intervention

Daniel F. McCain and Zhong-Yin Zhang

Abstract

Protein-tyrosine phosphatases (PTPases) function to remove the phosphoryl group from phosphotyrosine, phosphoserine, and phosphothreonine residues and are important regulators of cellular signal transduction. A number of enzymes from the PTPase superfamily are potential drug targets for cancer chemotherapy. Those that are discussed include: the receptor-like PTPases, PTPα, and PTPε, which dephosphorylate and activate Src; the cytoplasmic PTPase SHP-2, which is essential for growth factor-mediated signaling; JSP-1 (a.k.a. VHX, MKPX, or JKAP), which is required for the activation of growth factor and/or cytokine signaling; the PTPase that suppresses apoptosis, FAP-1; and several dual specificity phosphatases involved in the cell cycle, including Cdc25, KAP, Cdc14, and PRL.

17.1 Introduction

Protein-tyrosine phosphatases (PTPases) serve the important regulatory function of removing the phosphoryl group from phosphorylated tyrosine residues, thus opposing the action of protein-tyrosine kinases. Rather than just "house-keeping enzymes" that "turn off" tyrosine-kinase cascades as was originally assumed, it is now clear that PTPases can perform very specific regulatory functions in certain pathways and sometimes "turn on" signaling pathways. In fact, the human genome encodes over 100 proteins that belong to the PTPases superfamily, which is on the order of the number of protein tyrosine kinases, reflecting this specificity (Lander et al. 2001). Tyrosine phosphorylation has been shown to play a role in many aspects of cellular function (Hunter 2000). Given their now recognized importance in signal transduction, it is not surprising that a number of PTPases have been associated with specific diseases such as diabetes and cancer (Zhang 2001). This review will focus on some protein-tyrosine phosphatases that are potential drug targets for cancer.

As cancer is marked by aberrant tyrosine phosphorylation states, it is logical that some PTPases have been shown to play a role in oncogenesis and tumor progression. Several of these PTPases have demonstrated roles in cell cycle regulation, and have been validated as drug targets, while others are just potential drug targets based upon the role they play in cellular function.

Topics in Current Genetics, Vol. 5
J. Arino, D.R. Alexander (Eds.) Protein phosphatases
© Springer-Verlag Berlin Heidelberg 2004

The protein-tyrosine phosphatases all contain the PTPase signature motif, CX_5R (one-letter amino acid abbreviations), which is an indispensable part of the PTPase active site. All PTPases examined thus far employ the same general mechanism of catalysis, namely, direct nucleophilic attack of the phosphate of the phosphotyrosyl substrate by the active site Cys, which exists as a thiolate, facilitated by the active site Arg, which stabilizes the transition state by coordination of the phosphotyrosyl oxygens (for a review see: Zhang 1998). The PTPases can be split into two groups: the tyrosine-specific enzymes, which only work on phosphotyrosine, and the dual specific, which can hydrolyze phosphoserine and phosphothreonine as well as phosphotyrosine. The tyrosine-specific enzymes, of which there are about 40 members, can be further divided into the receptor-like, cytosolic, and low-molecular weight PTPases. The dual-specific enzymes, of which there are about 50, generally share less homology than the tyrosine-specific enzymes and therefore, are less easily subdivided, although they appear to employ the same general mechanism of catalysis (Wang et al. 2003). Interestingly, one member of the PTPase superfamily shown to be involved with cancer is the phosphoinositide phosphatase PTEN. PTEN, however, is a tumor suppressor rather than a therapeutic target and will not be covered in detail in this review (for a review see: Mayo and Donner 2002).

17.2 PTPases that activate SRC

Several receptor-like phosphatases that have been shown to play a role in cell proliferation are PTPα, PTPε, and CD45. All appear to dephosphorylate and activate Src-family tyrosine kinases, which initiates downstream events that can lead to oncogenic transformation. The receptor-like PTPases are all composed of extracellular domains, which contain various ligand binding sites, followed by a transmembrane region, and finally a cytosolic region, which contains two tyrosine-phosphatase domains. The membrane proximal tyrosine phosphatase domain (referred to as D1) contains most or all of the phosphatase activity for the enzyme, whereas the membrane distal domain (D2) usually shares a high degree of homology with D1, but has several mutations, which render it much less active or inactive (Buist et al. 1999). The function of the D2 domain of receptor-like PTPases is currently unknown. Several studies have shown that the D2 domain can mediate hetero- or homodimerization between various receptor-like PTPases which points to a mechanism of regulation similar to that of the protein-tyrosine kinases (Blanchetot and den Hertog 2000). However, in the case of receptor-like PTPases, the ultimate consequences of these interactions are sometimes unclear and may differ from enzyme to enzyme, whereas for the receptor tyrosine kinases, dimerization generally leads to activation of the kinase (Weiss and Schlessinger 1998).

It has been shown that overexpression of PTPα leads to the dephosphorylation and activation of Src and causes transformation (Zheng et al. 1992; den Hertog et al. 1993). Conversely, PTPα disruption in mice leads to diminished kinase activities of Src and Fyn (another Src family kinase), which strongly suggests that

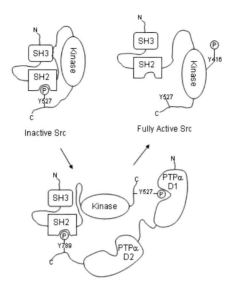

Fig. 1. Regulation of Src activity by PTPα. PTPα displaces tyrosine 527 from the SH2 domain of Src and then dephosphorylates it, allowing Src to autophosphorylate and attain its fully activated state. The cell membrane has been omitted for clarity.

PTPα is a physiological positive regulator of Src kinase activity in vivo (Brown and Cooper 1996; Ponniah et al. 1999; Su et al. 1999). Activation of Src kinase activity leads to tyrosine phosphorylation of various targets and can lead to transformation when unregulated (Brown and Cooper 1996). Src tyrosine kinases can be held in an inactive state by phosphorylation at tyrosine 527. The SH2 domain of Src intramolecularly interacts with phosphotyrosine 527 and renders the kinase inactive (Brown and Cooper 1996; Sicheri et al. 1997; Songyang et al. 1997; Williams et al. 1997; Young et al. 2001). It has been shown that PTPα can dephosphorylate phosphotyrosine 527 of Src, thus relieving this autoinhibitory mechanism (Zheng et al. 1992; den Hertog et al. 1993). Interestingly, this happens by a phosphotyrosine displacement mechanism, as shown in Figure 1. PTPα is phosphorylated at tyrosine 789 (den Hertog et al. 1994; Su et al. 1994), and this phosphotyrosine can bind to the SH2 domain of Src, which displaces the Src phosphotyrosine 527 and makes it accessible to dephosphorylation by PTPα (Zheng et al. 2000). Mutation of tyrosine 789 of PTPα into phenylalanine not only abrogates the ability of PTPα to bind to Src but also abrogates the ability of PTPα to cause transformation when overexpressed (Zheng et al. 2000). The fact that PTPα can cause the activation of Src, which can lead to transformation and that overexpression of PTPα alone is sufficient to cause transformation makes PTPα a potential cancer therapeutic target.

Another receptor-like PTPase that activates Src is PTPε. It has recently been reported that PTPε dephosphorylates and activates Src. In cells that were transformed with the neu oncogene, PTPε supports the transformed phenotype (Gil-Henn and Elson 2003). These results suggest that PTPε may also be a therapeutic target for cancer.

17.3 PTPases involved in growth factor or cytokine signaling

As their name implies, growth factors and cytokines activate signaling pathways, which can lead to the growth and proliferation of cells. Inappropriate activation of these pathways can lead to cell transformation and cancer. Two members of the PTPase family, SHP-2 and JSP-1 (also known as VHX, MKPX, or JKAP), have been shown to activate these pathways and may serve as potential therapeutic targets. SHP–2 has been reviewed more extensively elsewhere (Stein-Gerlach et al. 1998; Feng 1999; Qu 2002). In summary, SHP-2 is a PTPase that has two tandem N-terminal SH2 phosphotyrosine binding domains. SHP-2 interacts with a variety of activated growth factor receptor tyrosine kinases such as EGFR, which activates the phosphatase activity of SHP-2. The interactions of SHP-2 with the tyrosine phosphorylated receptors are necessary for the activation of a variety of downstream pathways, such as the Erk pathway (a MAP kinase pathway) or Akt/PI3 kinase pathway (Wu et al. 2001), which often promote cell proliferation (for reviews of MAP kinases see: Widmann et al. 1999; Johnson and Lapadat 2002). Interestingly, SHP-2 downregulates the JNK pathway (another MAP kinase pathway) and the Jak/STAT pathway (Wu et al. 2002), which are often associated with promoting apoptosis or growth suppression. Since, in general, it appears that inhibiting SHP-2 would tend to inhibit cell proliferation and make cells more responsive to apoptosis, SHP-2 may be a valid drug target. It should be kept in mind, however, that SHP-2 has many potential substrates and regulates several different pathways. Further, the MAP kinase pathways that SHP-2 regulates often result in different cellular responses depending upon the cell type or the initial state of the cell (Widmann et al. 1999).

Another phosphatase recently implicated in the regulation of a MAP kinase cascade is JSP-1 (Shen et al. 2001b) also known as VHX/MKPX (Alonso et al. 2002) or JKAP (Chen et al. 2002). Unlike other dual specificity phosphatases, which directly inactivate MAP kinase phosphatases, JSP-1 causes the activation of the JNK pathway. In fact, JSP stands for Jnk-stimulating phosphatase. JSP-1 appears to work upstream of JNK itself, possibly by activating MKK7 (Chen et al. 2002) or MKK4 (Shen et al. 2001b) kinases, which phosphorylate and activate JNK. Given the broad roles of the JNK pathway in cellular response to various forms of stresses, growth stimulation, and apoptosis, the effect of JSP-1 on JNK signal pathway makes it worth investigating as a potential drug target. In contrast to JSP-1, the MAP kinase phosphatases MKP1 and MKP2 are direct negative regulators of the JNK activity. Both MKP1 and MKP2 are over-expressed in human beast cancer, which may explain the reduced pro-apoptotic activity of JNK1 in the tumor (Wang et al. 2003). Consequently, suppression of MKP activity therapeutically could restore the pro-apoptotic activity of JNK in malignant cells.

17.4 FAP-1: A PTPase that suppresses apoptosis

A particularly compelling case for a cancer chemotherapeutic target can be made for FAP-1. FAP-1 is a cytosolic PTPase (Sato et al. 1995), which negatively regulates apoptosis signaling through the Fas receptor (Lee et al. 1999). Fas-mediated apoptosis is initiated when the Fas receptor trimerizes on the cell surface, either through binding of its ligand, FasL, or through high Fas concentrations on the cell surface (Micheau et al. 1999). This signal is transduced through a number of proteins and eventually results in the activation of caspases, which begin degrading cellular proteins resulting in apoptosis (Miller 1997; Muzio et al. 1997). The mechanism of action of several cancer drugs, such as etoposide and cisplatin, seems to lie with their ability to induce apoptosis (Micheau et al. 1999). It has been hypothesized that resistance to these drugs correlates with the resistance of the tumor cells to Fas-mediated apoptosis. Further, many studies have shown a correlation between FAP-1 expression and Fas-mediated apoptosis resistance in several types of cancer (Li et al. 2000b; Lee et al. 2001; Meinhold-Heerlein et al. 2001; Ungefroren et al. 2001). Taken together, targeting FAP-1 for inhibitor design may provide a way to induce apoptosis in cancer cells that are apoptosis resistant and thus augment drugs such as etoposide and cisplatin.

17.5. The cell cycle

The cell division cycle (Cdc), the process by which a cell replicates its DNA and splits into two daughter cells, which then ready themselves for another round of replication if necessary, is a complex and tightly regulated process. Progression through the cell cycle is largely driven by the activity of Cyclin-dependent kinases (Cdks) (Morgan 1997). In order to progress correctly through the cell cycle, Cdk activity must be rapidly activated at a specific time and then rapidly inactivated at another. Several mechanisms regulate Cdk activity. First, as the name implies, Cdks by themselves have very low activity and must be bound to a Cyclin subunit to be activated. The Cyclins are synthesized and degraded at specific times in the cell cycle, which makes their concentrations fluctuate in specific wave-like patterns, which give them their name. The Cyclin/Cdk must then be further activated by phosphorylation at Thr-160 (Thr-161 for Cdk1, also known as Cdc2) in order to effect progression through the cell cycle. Cyclin/Cdks can also be inhibited by phosphorylation at Thr-14 and/or Tyr-15 or by binding of endogenous inhibitors such as p21. As summarized in Figure 2, several PTPases have been found to be important regulators of cell cycle progression by controlling the activity of Cdks. Cdc25 dual specificity phosphatases activate Cdks by removing the inhibitory

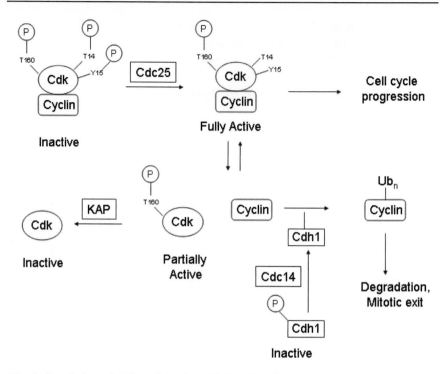

Fig. 2. Regulation of Cdk activity by PTPases. Cdc25 dephosphorylates and activates Cdk2. KAP dephosphorylates and inactivates Cdk2. Cdc14 causes the degradation of Cyclin by dephosphorylating and activating Cdh1, a cofactor of the anaphase promoting complex (APC). The APC has been omitted for clarity.

phosphorylations at Thr-14 and Tyr-15, thereby driving cell cycle progression (Dunphy and Kumagai 1991; Gautier et al. 1991; Millar et al. 1991; Strausfeld et al. 1991). KAP is a dual specificity phosphatase, which removes the activating phosphorylation at Thr-160, and thereby may contribute to the rapid inactivation of Cdk activity, which may be important in exiting one cell cycle phase and entering another. Cdc14, another dual specificity phosphatase, activates the degradation of certain Cyclins, and may have additional functions, which also serve to inactivate Cdks and allow exit from mitosis. These three PTPases will be discussed in detail below.

17.5.1 CDC25

Cdc25 was first characterized as a positive regulator of cell division in the fission yeast, *S. pombe* (Russell and Nurse 1987). Cdc25 phosphatases play a critical role in cell cycle progression (Coleman and Dunphy 1994; Hoffmann and Karsenti 1994; Nilsson and Hoffmann 2000). Humans have three isoforms of Cdc25 termed Cdc25A, Cdc25B, and Cdc25C, and each plays a distinct role in the cell cycle

(Nilsson and Hoffmann 2000). Cdc25 enzymes have been shown to be potential oncogenes under certain circumstances (Galaktionov et al. 1995). Cdc25A and Cdc25B have also been shown to be overexpressed in several types of cancer cells (Broggini et al. 2000; Cangi et al. 2000; Hernandez et al. 2000; Hernandez et al. 2001; Hu et al. 2001; Nishioka et al. 2001; Sasaki et al. 2001; Ito et al. 2002). Importantly, it has also been shown that treatment of tumor cell lines with Cdc25 inhibitors inhibits tumor growth or causes cell cycle arrest (Tamura et al. 1999; Wu and Sun 1999; Takahashi et al. 2000; Lazo et al. 2001; Peng et al. 2001), which validates Cdc25 as a drug target and makes inhibitors of Cdc25 potentially valuable.

17.5.2 KAP

KAP was discovered through yeast 2-hybrid studies as a Cdk-interacting protein, and shown to be a dual specificity phosphatase (Gyuris et al. 1993; Hannon et al. 1994). Further KAP was shown to dephosphorylate phospho-Thr-160 of Cdk2, but only in the absence of Cyclin. Interestingly, it was also shown that KAP can bind to Cyclin/Cdk complexes and this binding does not inhibit Cyclin/Cdk activity. It is only when the Cyclin subunit dissociates that KAP finally dephosphorylates Thr-160 and totally inactivates the kinase (Poon and Hunter 1995). One hallmark of the cell cycle is the rapid activation and inactivation of Cyclin-dependent kinases at appropriate times. In this way, KAP contributes to this "all or nothing" kinase activity. When the Cdk is in its highest activity state, with Cyclin bound and Thr-160 phosphorylated, KAP does not interfere. However, when the Cyclin dissociates leaving the phospho-Thr160-Cdk in an intermediate activity state, KAP returns the kinase to its basal, low activity state. This may be important for transiting from one stage of the cell cycle to the next. The transition from one stage of the cell cycle to the next is mediated by the degradation of one type of Cyclin, and the accumulation of another. It has been speculated that the dephosphorylation of phospho-Thr160-Cdk2 by KAP would ensure that Cdk2 is not immediately reactivated inappropriately when the latter Cyclin begins to accumulate (Poon and Hunter 1995). KAP overexpression has been shown to delay progression through G1 of the cell cycle (Gyuris et al. 1993; Hannon et al. 1994) and decrease phosphorylation levels of Thr160 on Cdk2 (Poon and Hunter 1995). This suggests that there is a sufficiently dynamic equilibrium in the binding between the G1 Cyclins and Cdk2 such that overexpression of KAP leads to the net decrease in phospho-Thr160 even when the G1 Cyclins are present.

Since KAP overexpression delays cell cycle progression and reduces Cdk activity, it may seem counterintuitive to consider KAP as a drug target for cancer chemotherapy. In general, the strategy is to inhibit proteins that play a positive role in activating cell cycle progression not delaying it. And in fact, one study has shown a correlation of expression of aberrant, nonfunctional KAP transcripts with hepatocellular carcinoma (Yeh et al. 2000). However, the authors noted that some of the KAP mutations also occurred in noncancerous cells, which indicates that factors other than KAP must be contributing to the transformation process. One study found high KAP expression associated with breast and prostate malignancies (Lee et al. 2000). Further, the authors showed that reducing KAP levels by antisense-KAP expression in HeLa and a prostate cancer cell line inhibited the transformed phenotype, retarded cell cycle progression, reduced Cdk activity, and reduced the tumorigenic potential of these cell lines. This study, therefore, suggests that KAP may in fact be a valid drug target, at least for some types of cancer. How exogenous KAP overexpression in one case and decreased KAP expression in another could both cause cell cycle retardation remains to be resolved. Lee and coworkers suggest that KAP may have other substrates or undiscovered roles in governing cell cycle progression such that reducing KAP levels ultimately leads to a reduction in Cdk activity. That KAP can dephosphorylate substrates other than phospho-Thr160-Cdk2 is certainly possible. As there are many more phosphorylated residues in the cell than there are phosphatases (viz. there are only about 100 protein-tyrosine phosphatases in the genome, and an even smaller number of phosphatases that work on phosphothreonine and phosphoserine), it stands to reason that most phosphatases will have more than one substrate. In any case, small molecule inhibitors for KAP are highly desirable. For one thing, the work by Lee et al. suggests that they may be useful for breast or prostate cancer treatment, and if nothing else may help to tease out the precise role of KAP in the cell cycle.

17.5.3 CDC14

Cdc14 is another inhibitor of Cdk activity that is a potential drug target in cancer. It was originally discovered as a temperature sensitive mutant in the budding yeast S. cerevisiae. At the restrictive temperature, yeast harboring the Cdc14 temperature sensitive mutant arrest in late mitosis, which suggests Cdc14 is essential in completing mitosis (Schild and Byers 1980; Wood and Hartwell 1982; Shirayama et al. 1996). CDC14 was shown to encode a dual specific phosphatase in yeast (Taylor et al. 1997). Two human CDC14 genes were discovered and shown to encode dual specificity phosphatases (Cdc14A and Cdc14B) (Li et al. 1997). In yeast, Cdc14 was shown to inhibit Cdk activity by three mechanisms, thereby effecting exit from mitosis. First, it dephosphorylates Sic1, an endogenous Cdk inhibitor, which prevents the degradation of Sic1 leading to its accumulation and the inhibition of the Cdk (called Cdc28 in yeast). Second, it dephosphorylates Swi5, the transcription factor for Sic1, which allows Swi5 to enter the nucleus and transcribe this Cdk inhibitor (Visintin et al. 1998). Third, Cdc14 dephosphorylates Cdh1/Hct1, a cofactor of the anaphase promoting complex (APC), which activates the mitotic Cyclin (Clb in yeast) ubiquitin ligase activity of APC (Visintin et al. 1998; Jaspersen et al. 1999). The APC is an ubiquitin ligase, which ubiquitinates cell cycle proteins, thereby targeting them for degradation by the ubiquitin-dependent proteosome. The action of Cdc14 dephosphorylating Cdh1/Hct1 causes the degradation of Cyclin and therefore the inhibition of Cdk activity. Interestingly, it appears that Cdc14 may generally function to reverse Cdk phosphorylation, as all three of the phosphorylation sites that Cdc14 dephosphorylates are known sites of Cdc28 phosphorylation (Jaspersen et al. 1999).

Although it now appears that Cdc14 enzymes may have different functions depending upon the organism (Trautmann and McCollum 2002), some parallels can be drawn between its functions in humans and in yeast. For one thing, it is now known that human Cdh1 is phosphorylated by Cdc2/CyclinB and that this phosphorylation prevents Cdh1 from activating the human APC. Further, Cdc14A dephosphorylates Cdh1 and allows it to activate the APC and to ubiquinate (Bembenek and Yu 2001), so causing the degradation of substrates such as Cyclin B and exit from mitosis. The fact that Cdc14 reverses Cdk phosphorylation in yeast has prompted researchers to look at known human Cdk substrates as potential Cdc14 substrates in humans. Cdc14 has been shown to be specific for Cdk-phosphorylated proteins in vitro (Kaiser et al. 2002). In fact, it has been found that Cdc14A and Cdc14B were able to dephosphorylate the tumor suppressor p53 at Ser-315 (Li et al. 2000a), a known Cdc2 phosphorylation site (Wang and Prives 1995). The protein p53 has been shown to induce the transcription of p21WAF1, an endogenous Cdk inhibitor (Dulic et al. 1994). At the outset, this would seem to mirror the dephosphorylation of the yeast Swi5; however, the role of Ser-315 phosphorylation in p53 function appears to be a lot more complex. Some studies suggest that phoshorylation at Ser-315 activates the DNA binding and transcriptional activity of p53 (Wang and Prives 1995; Blaydes et al. 2001), which suggests that dephosphorylation of this residue by Cdc14 may actually inhibit transcription of p21WAF1, meaning that this dephosphorylation is not analogous to the dephosphorylation of Swi5 in yeast.

Regardless of its precise role in promoting exit from mitosis, some studies suggest targeting Cdc14 for inhibitor development as a potential strategy in cancer treatment. Downregulation of Cdc14A by small interfering RNA (siRNA) caused mitotic defects including impaired centrosome separation and failure to undergo cytokinesis, and often cell death and detachment from the plate (Mailand et al. 2002). Clearly, this is another case where small molecule inhibitors of Cdc14 would be useful in probing its function, and perhaps even serve as a cancer treatment.

17.5.4 PRL

The precise function of PRL is not known but it has been shown to be associated with metastasis of colorectal cancer (Saha et al. 2001). The PRL-1 gene was discovered as a gene that was induced in the regenerating mouse liver and shown to encode a protein-tyrosine phosphatase (Diamond et al. 1994). The fact that it is induced in regenerating cells suggests that it may function in promoting the cell cycle. Diamond and coworkers further showed that PRL-1 can be phosphorylated by Src, and that it can dephosphorylate itself and phospho-Src, but the physiological relevance of this is not known. They also showed that cells overexpressing PRL-1 exhibit the transformed phenotype (Diamond et al. 1994). Humans have three PRL phosphatases called PRL-1, 2, and 3. Subsequently, it was shown that these proteins contain the CaaX motif that targets them for farnesylation (Cates et al. 1996; Zeng et al. 2000). Farnesylation is a posttranslational conjugation of a hydrophobic, isoprenoid group to the C-terminal end of the proteins (Gelb et al. 1998; Sinensky 2000), which is required for the proper localization and functions of PRL. A groundbreaking study showed that PRL-3 was overexpressed specifically in liver cancer cells that had metastasized from colorectal cancer cells (Saha et al. 2001). It has also been shown to be overexpressed in prostate tumor cells (Wang et al. 2002) and a number of other human tumor cells including HeLa cells (Wang et al. 2002). Although, the precise function and physiological substrates for PRL are not known, the fact that it is overexpressed in several cancers and associated with metastasis makes it a potential drug target, as it is metastasis that makes cancer so deadly (Saha et al. 2001).

17.6 Conclusions

Although much progress has been made in understanding the role that PTPases play in signal transduction and in diseases such as cancer, much remains to be known. For example, the identification of the physiological substrates of phosphatases remains a major challenge in the field. Two approaches can be used to identify physiological substrates of PTPases: the use of substrate-trapping mutant PTPases, and the use of specific PTPase inhibitors. The former approach has already been used widely in the study of PTPase function. While the latter approach has been used successfully to identify EGFR as a potential substrate for Cdc25A (Wang et al. 2002b), it is currently limited by the availability of specific inhibitors. As PTPases have emerged as drug targets, many groups have undertaken the design, synthesis and discovery of PTPase inhibitors. These efforts have been particularly focused on Cdc25 (see Section 5.1. Cdc25) and PTP1B (Shen et al. 2001a). Potent and selective inhibitors have been found for both enzymes, suggesting that it is feasible to expand the search for potent and selective PTPase inhibitors to include the enzymes mentioned in this review. Targeting specific protein tyrosine phosphatases for small molecule and drug development is an important undertaking for two reasons, the potential treatment of diseases such as cancer, and the development of tools to study signal transduction by PTPases.

Acknowledgements

This work was supported by NIH Grant CA69202 and the G. Harold and Leila Y. Mathers Charitable Foundation.

References

Alonso A, Merlo JJ, Na S, Kholod N, Jaroszewski L, Kharitonenkov A, Williams S, Godzik A, Posada JD, Mustelin T (2002) Inhibition of T cell antigen receptor signaling by

VHR-related MKPX (VHX), a new dual specificity phosphatase related to VH1 related (VHR). J Biol Chem 277:5524-5528

Bembenek J, Yu H (2001) Regulation of the anaphase-promoting complex by the dual specificity phosphatase human Cdc14a. J Biol Chem 276:48237-48242

Blanchetot C, den Hertog J (2000) Multiple interactions between receptor protein-tyrosine phosphatase (RPTP) alpha and membrane-distal protein-tyrosine phosphatase domains of various RPTPs. J Biol Chem 275:12446-12452

Blaydes JP, Luciani MG, Pospisilova S, Ball HM, Vojtesek B, Hupp TR (2001) Stoichiometric phosphorylation of human p53 at Ser315 stimulates p53-dependent transcription. J Biol Chem 276:4699-4708

Broggini M, Buraggi G, Brenna A, Riva L, Codegoni AM, Torri V, Lissoni AA, Mangioni C, D'Incalci M (2000) Cell cycle-related phosphatases CDC25A and B expression correlates with survival in ovarian cancer patients. Anticancer Res 20:4835-4840

Brown MT, Cooper JA (1996) Regulation, substrates and functions of src. Biochim Biophys Acta 1287:121-149

Buist A, Zhang YL, Keng YF, Wu L, Zhang ZY, den Hertog J (1999) Restoration of potent protein-tyrosine phosphatase activity into the membrane-distal domain of receptor protein-tyrosine phosphatase alpha. Biochemistry 38:914-922

Cangi MG, Cukor B, Soung P, Signoretti S, Moreira G Jr, Ranashinge M, Cady B, Pagano M, Loda M (2000) Role of the Cdc25A phosphatase in human breast cancer. J Clin Invest 106:753-761

Cates CA, Michael RL, Stayrook KR, Harvey KA, Burke YD, Randall SK, Crowell PL, Crowell DN (1996) Prenylation of oncogenic human PTP(CAAX) protein tyrosine phosphatases. Cancer Lett 110:49-55

Chen AJ, Zhou G, Juan T, Colicos SM, Cannon JP, Cabriera-Hansen M, Meyer CF, Jurecic R, Copeland NG, Gilbert DJ, Jenkins NA, Fletcher F, Tan TH, Belmont JW (2002) The dual specificity JKAP specifically activates the c-Jun N-terminal kinase pathway. J Biol Chem 277:36592-36601

Coleman TR, Dunphy WG (1994) Cdc2 regulatory factors. Curr Opin Cell Biol 6:877-882.

den Hertog J, Pals CE, Peppelenbosch MP, Tertoolen LG, de Laat SW, Kruijer W (1993) Receptor protein tyrosine phosphatase alpha activates pp60c-src and is involved in neuronal differentiation. EMBO J 12:3789-3798

den Hertog J, Tracy S, Hunter T (1994) Phosphorylation of receptor protein-tyrosine phosphatase alpha on Tyr789, a binding site for the SH3-SH2-SH3 adaptor protein GRB-2 in vivo. EMBO J 13:3020-3032

Diamond RH, Cressman DE, Laz TM, Abrams CS, Taub R (1994) PRL-1, a unique nuclear protein tyrosine phosphatase, affects cell growth. Mol Cell Biol 14:3752-3762

Dulic V, Kaufmann WK, Wilson SJ, Tlsty TD, Lees E, Harper JW, Elledge SJ, Reed SI (1994) p53-dependent inhibition of cyclin-dependent kinase activities in human fibroblasts during radiation-induced G1 arrest. Cell 76:1013-1023

Dunphy WG, Kumagai A (1991) The cdc25 protein contains an intrinsic phosphatase activity. Cell 67:189-196

Feng GS (1999) Shp-2 tyrosine phosphatase: signaling one cell or many. Exp Cell Res 253:47-54

Galaktionov K, Lee AK, Eckstein J, Draetta G, Meckler J, Loda M, Beach D (1995) CDC25 phosphatases as potential human oncogenes. Science 269:1575-1577

Gautier J, Solomon MJ, Booher RN, Bazan JF, Kirschner MW (1991) cdc25 is a specific tyrosine phosphatase that directly activates p34cdc2. Cell 67:197-211

Gelb MH, Scholten JD, Sebolt-Leopold JS (1998) Protein prenylation: from discovery to prospects for cancer treatment. Curr Opin Chem Biol 2:40-48

Gil-Henn H, Elson A (2003) Tyrosine phosphatase epsilon activates Src and supports the transformed phenotype of Neu-induced mammary tumor cells. J Biol Chem 21:21

Gyuris J, Golemis E, Chertkov H, Brent R (1993) Cdi1, a human G1 and S phase protein phosphatase that associates with Cdk2. Cell 75:791-803

Hannon GJ, Casso D, Beach D (1994) KAP: a dual specificity phosphatase that interacts with cyclin-dependent kinases. Proc Natl Acad Sci USA 91:1731-1735

Hernandez S, Bessa X, Bea S, Hernandez L, Nadal A, Mallofre C, Muntane J, Castells A, Fernandez PL, Cardesa A, Campo E (2001) Differential expression of cdc25 cell-cycle-activating phosphatases in human colorectal carcinoma. Lab Invest 81:465-473

Hernandez S, Hernandez L, Bea S, Pinyol M, Nayach I, Bellosillo B, Nadal A, Ferrer A, Fernandez PL, Montserrat E, Cardesa A, Campo E (2000) cdc25a and the splicing variant cdc25b2, but not cdc25B1, -B3 or -C, are over-expressed in aggressive human non-Hodgkin's lymphomas. Int J Cancer 89:148-152

Hoffmann I, Karsenti E (1994) The role of cdc25 in checkpoints and feedback controls in the eukaryotic cell cycle. J Cell Sci Suppl 18:75-79

Hu YC, Lam KY, Law S, Wong J, Srivastava G (2001) Identification of differentially expressed genes in esophageal squamous cell carcinoma (ESCC) by cDNA expression array: overexpression of Fra-1, Neogenin, Id-1, and CDC25B genes in ESCC. Clin Cancer Res 7:2213-2221

Hunter T (2000) Signaling--2000 and beyond. Cell 100:113-127

Ito Y, Yoshida H, Nakano K, Kobayashi K, Yokozawa T, Hirai K, Matsuzuka F, Matsuura N, Kakudo K, Kuma K, Miyauchi A (2002) Expression of cdc25A and cdc25B proteins in thyroid neoplasms. Br J Cancer 86:1909-1913

Jaspersen SL, Charles JF, Morgan DO (1999) Inhibitory phosphorylation of the APC regulator Hct1 is controlled by the kinase Cdc28 and the phosphatase Cdc14. Curr Biol 9:227-236

Johnson GL, Lapadat R (2002) Mitogen-activated protein kinase pathways mediated by ERK, JNK, and p38 protein kinases. Science 298:1911-1912

Kaiser BK, Zimmerman ZA, Charbonneau H, Jackson PK (2002) Disruption of centrosome structure, chromosome segregation, and cytokinesis by misexpression of human Cdc14A phosphatase. Mol Biol Cell 13:2289-2300

Lander ES et al. (2001) Initial sequencing and analysis of the human genome. Nature 409:860-921

Lazo JS, Aslan DC, Southwick EC, Cooley KA, Ducruet AP, Joo B, Vogt A, Wipf P (2001) Discovery and biological evaluation of a new family of potent inhibitors of the dual specificity protein phosphatase Cdc25. J Med Chem 44:4042-4049

Lee SH, Shin MS, Lee HS, Bae JH, Lee HK, Kim HS, Kim SY, Jang JJ, Joo M, Kang YK, Park WS, Park JY, Oh RR, Han SY, Lee JH, Kim SH, Lee JY, Yoo NJ (2001) Expression of Fas and Fas-related molecules in human hepatocellular carcinoma. Hum Pathol 32:250-256

Lee SH, Shin MS, Lee JY, Park WS, Kim SY, Jang JJ, Dong SM, Na EY, Kim CS, Kim SH, Yoo NJ (1999) In vivo expression of soluble Fas and FAP-1: possible mechanisms of Fas resistance in human hepatoblastomas. J Pathol 188:207-212

Lee SW, Reimer CL, Fang L, Iruela-Arispe ML, Aaronson SA (2000) Overexpression of kinase-associated phosphatase (KAP) in breast and prostate cancer and inhibition of the transformed phenotype by antisense KAP expression. Mol Cell Biol 20:1723-1732

Li L, Ernsting BR, Wishart MJ, Lohse DL, Dixon JE (1997) A family of putative tumor suppressors is structurally and functionally conserved in humans and yeast. J Biol Chem 272:29403-29406

Li L, Ljungman M, Dixon JE (2000a) The human Cdc14 phosphatases interact with and dephosphorylate the tumor suppressor protein p53. J Biol Chem 275:2410-2414

Li Y, Kanki H, Hachiya T, Ohyama T, Irie S, Tang G, Mukai J, Sato T (2000b) Negative regulation of Fas-mediated apoptosis by FAP-1 in human cancer cells. Int J Cancer 87:473-479

Mailand N, Lukas C, Kaiser BK, Jackson PK, Bartek J, Lukas J (2002) Deregulated human Cdc14A phosphatase disrupts centrosome separation and chromosome segregation. Nat Cell Biol 4:317-322

Mayo LD, Donner DB (2002) The PTEN, Mdm2, p53 tumor suppressor-oncoprotein network. Trends Biochem Sci 27:462-467

Meinhold-Heerlein I, Stenner-Liewen F, Liewen H, Kitada S, Krajewska M, Krajewski S, Zapata JM, Monks A, Scudiero DA, Bauknecht T, Reed JC (2001) Expression and potential role of Fas-associated phosphatase-1 in ovarian cancer. Am J Pathol 158:1335-1344

Micheau O, Solary E, Hammann A, Dimanche-Boitrel MT (1999) Fas ligand-independent, FADD-mediated activation of the Fas death pathway by anticancer drugs. J Biol Chem 274:7987-7992

Millar JB, McGowan CH, Lenaers G, Jones R, Russell P (1991) p80cdc25 mitotic inducer is the tyrosine phosphatase that activates p34cdc2 kinase in fission yeast. EMBO J 10:4301-4309

Miller DK (1997) The role of the Caspase family of cysteine proteases in apoptosis. Semin Immunol 9:35-49

Morgan DO (1997) Cyclin-dependent kinases: engines, clocks, and microprocessors. Annu Rev Cell Dev Biol 13:261-291

Muzio M, Salvesen GS, Dixit VM (1997) FLICE induced apoptosis in a cell-free system. Cleavage of caspase zymogens. J Biol Chem 272:2952-2956

Nilsson I, Hoffmann I (2000) Cell cycle regulation by the Cdc25 phosphatase family. Prog Cell Cycle Res 4:107-114

Nishioka K, Doki Y, Shiozaki H, Yamamoto H, Tamura S, Yasuda T, Fujiwara Y, Yano M, Miyata H, Kishi K, Nakagawa H, Shamma A, Monden M (2001) Clinical significance of CDC25A and CDC25B expression in squamous cell carcinomas of the oesophagus. Br J Cancer 85:412-421

Peng H, Xie W, Otterness DM, Cogswell JP, McConnell RT, Carter HL, Powis G, Abraham RT, Zalkow LH (2001) Syntheses and biological activities of a novel group of steroidal derived inhibitors for human Cdc25A protein phosphatase. J Med Chem 44:834-848

Ponniah S, Wang DZ, Lim KL, Pallen CJ (1999) Targeted disruption of the tyrosine phosphatase PTPalpha leads to constitutive downregulation of the kinases Src and Fyn. Curr Biol 9:535-538

Poon RY, Hunter T (1995) Dephosphorylation of Cdk2 Thr160 by the cyclin-dependent kinase-interacting phosphatase KAP in the absence of cyclin. Science 270:90-93

Qu CK (2002) Role of the SHP-2 tyrosine phosphatase in cytokine-induced signaling and cellular response. Biochim Biophys Acta 1592:297-301

Russell P, Nurse P (1987) Negative regulation of mitosis by wee1+, a gene encoding a protein kinase homolog. Cell 49:559-567

Saha S, Bardelli A, Buckhaults P, Velculescu VE, Rago C, St Croix B, Romans KE, Choti MA, Lengauer C, Kinzler KW, Vogelstein B (2001) A phosphatase associated with metastasis of colorectal cancer. Science 294:1343-1346

Sasaki H, Yukiue H, Kobayashi Y, Tanahashi M, Moriyama S, Nakashima Y, Fukai I, Kiriyama M, Yamakawa Y, Fujii Y (2001) Expression of the cdc25B gene as a prognosis marker in non-small cell lung cancer. Cancer Lett 173:187-192

Sato T, Irie S, Kitada S, Reed JC (1995) FAP-1: a protein tyrosine phosphatase that associates with Fas. Science 268:411-415

Schild D, Byers B (1980) Diploid spore formation and other meiotic effects of two cell-division-cycle mutations of *Saccharomyces cerevisiae*. Genetics 96:859-876

Shen K, Keng YF, Wu L, Guo XL, Lawrence DS, Zhang ZY (2001a) Acquisition of a specific and potent PTP1B inhibitor from a novel combinatorial library and screening procedure. J Biol Chem 276:47311-47319

Shen Y, Luche R, Wei B, Gordon ML, Diltz CD, Tonks NK (2001b) Activation of the Jnk signaling pathway by a dual-specificity phosphatase, JSP-1. Proc Natl Acad Sci USA 98:13613-13618

Shirayama M, Matsui Y, Toh-e A (1996) Dominant mutant alleles of yeast protein kinase gene CDC15 suppress the lte1 defect in termination of M phase and genetically interact with CDC14. Mol Gen Genet 251:176-185

Sicheri F, Moarefi I, Kuriyan J (1997) Crystal structure of the Src family tyrosine kinase Hck. Nature 385:602-609

Sinensky M (2000) Functional aspects of polyisoprenoid protein substituents: roles in protein-protein interaction and trafficking. Biochim Biophys Acta 1529:203-209

Songyang Z, Fanning AS, Fu C, Xu J, Marfatia SM, Chishti AH, Crompton A, Chan AC, Anderson JM, Cantley LC (1997) Recognition of unique carboxyl-terminal motifs by distinct PDZ domains. Science 275:73-77

Stein-Gerlach M, Wallasch C, Ullrich A (1998) SHP-2, SH2-containing protein tyrosine phosphatase-2. Int J Biochem Cell Biol 30:559-566

Strausfeld U, Labbe JC, Fesquet D, Cavadore JC, Picard A, Sadhu K, Russell P, Doree M (1991) Dephosphorylation and activation of a p34cdc2/cyclin B complex in vitro by human CDC25 protein. Nature 351:242-245

Su J, Batzer A, Sap J (1994) Receptor tyrosine phosphatase R-PTP-alpha is tyrosine-phosphorylated and associated with the adaptor protein Grb2. J Biol Chem 269:18731-18734

Su J, Muranjan M, Sap J (1999) Receptor protein tyrosine phosphatase alpha activates Src-family kinases and controls integrin-mediated responses in fibroblasts. Curr Biol 9:505-511

Takahashi M, Dodo K, Sugimoto Y, Aoyagi Y, Yamada Y, Hashimoto Y, Shirai R (2000) Synthesis of the novel analogues of dysidiolide and their structure-activity relationship. Bioorg Med Chem Lett 10:2571-2574

Tamura K, Rice RL, Wipf P, Lazo JS (1999) Dual G1 and G2/M phase inhibition by SC-alpha alpha delta 9, a combinatorially derived Cdc25 phosphatase inhibitor. Oncogene 18:6989-6996

Taylor GS, Liu Y, Baskerville C, Charbonneau H (1997) The activity of Cdc14p, an oligomeric dual specificity protein phosphatase from Saccharomyces cerevisiae, is required for cell cycle progression. J Biol Chem 272:24054-24063

Trautmann S, McCollum D (2002) Cell cycle: new functions for Cdc14 family phosphatases. Curr Biol 12:R733-R735

Ungefroren H, Kruse ML, Trauzold A, Roeschmann S, Roeder C, Arlt A, Henne-Bruns D, Kalthoff H (2001) FAP-1 in pancreatic cancer cells: functional and mechanistic studies on its inhibitory role in CD95-mediated apoptosis. J Cell Sci 114:2735-2746

Visintin R, Craig K, Hwang ES, Prinz S, Tyers M, Amon A (1998) The phosphatase Cdc14 triggers mitotic exit by reversal of Cdk-dependent phosphorylation. Mol Cell 2:709-718

Wang Hy, Cheng Z, Malbon CC (2003) Overexpression of mitogen-activated protein kinase phosphatases MKP1, MKP2 in human breast cancer. Cancer Lett 191:229-237

Wang J, Kirby CE, Herbst R (2002) The tyrosine phosphatase PRL-1 localizes to the endoplasmic reticulum and the mitotic spindle and is required for normal mitosis. J Biol Chem 277:46659-46668

Wang WQ, Sun JP, Zhang ZY (2003) An overview of the protein tyrosine phosphatase superfamily. Curr Top Med Chem 3:739-748

Wang Y, Prives C (1995) Increased and altered DNA binding of human p53 by S and G2/M but not G1 cyclin-dependent kinases. Nature 376:88-91

Wang Z, Wang M, Lazo JS, Carr BI (2002b) Identification of epidermal growth factor receptor as a target of Cdc25A protein phosphatase. J Biol Chem 277:19470-19475

Weiss A, Schlessinger J (1998) Switching signals on or off by receptor dimerization. Cell 94:277-280

Widmann C, Gibson S, Jarpe MB, Johnson GL (1999) Mitogen-activated protein kinase: conservation of a three-kinase module from yeast to human. Physiol Rev 79:143-180

Williams JC, Weijland A, Gonfloni S, Thompson A, Courtneidge SA, Superti-Furga G, Wierenga RK (1997) The 2.35 A crystal structure of the inactivated form of chicken Src: a dynamic molecule with multiple regulatory interactions. J Mol Biol 274:757-775

Wood JS, Hartwell LH (1982) A dependent pathway of gene functions leading to chromosome segregation in Saccharomyces cerevisiae. J Cell Biol 94:718-726

Wu CJ, O'Rourke DM, Feng GS, Johnson GR, Wang Q, Greene MI (2001) The tyrosine phosphatase SHP-2 is required for mediating phosphatidylinositol 3-kinase/Akt activation by growth factors. Oncogene 20:6018-6025

Wu FY, Sun TP (1999) Vitamin K3 induces cell cycle arrest and cell death by inhibiting Cdc25 phosphatase. Eur J Cancer 35:1388-1393

Wu TR, Hong YK, Wang XD, Ling MY, Dragoi AM, Chung AS, Campbell AG, Han ZY, Feng GS, Chin YE (2002) SHP-2 is a dual-specificity phosphatase involved in Stat1 dephosphorylation at both tyrosine and serine residues in nuclei. J Biol Chem 277:47572-47580

Yeh CT, Lu SC, Chen TC, Peng CY, Liaw YF (2000) Aberrant transcripts of the cyclin-dependent kinase-associated protein phosphatase in hepatocellular carcinoma. Cancer Res 60:4697-4700

Young MA, Gonfloni S, Superti-Furga G, Roux B, Kuriyan J (2001) Dynamic coupling between the SH2 and SH3 domains of c-Src and Hck underlies their inactivation by C-terminal tyrosine phosphorylation. Cell 105:115-126

Zeng Q, Si X, Horstmann H, Xu Y, Hong W, Pallen CJ (2000) Prenylation-dependent association of protein-tyrosine phosphatases PRL-1, -2, and -3 with the plasma membrane and the early endosome. J Biol Chem 275:21444-21452

Zhang ZY (1998) Protein-tyrosine phosphatases: biological function, structural characteristics, and mechanism of catalysis. Crit Rev Biochem Mol Biol 33:1-52

Zhang ZY (2001) Protein tyrosine phosphatases: prospects for therapeutics. Curr Opin Chem Biol 5:416-423

Zheng XM, Resnick RJ, Shalloway D (2000) A phosphotyrosine displacement mechanism for activation of Src by PTPalpha. EMBO J 19:964-978

Zheng XM, Wang Y, Pallen CJ (1992) Cell transformation and activation of pp60c-src by overexpression of a protein tyrosine phosphatase. Nature 359:336-339

Index

Printing: Druckhaus Berlin-Mitte GmbH
Binding: Buchbinderei Stein & Lehmann, Berlin